Paramagnetic Organometallic Species in Activation/Selectivity, Catalysis

NATO ASI Series

Advanced Science Institutes Series

A Series presenting the results of activities sponsored by the NATO Science Committee, which aims at the dissemination of advanced scientific and technological knowledge, with a view to strengthening links between scientific communities.

The Series is published by an international board of publishers in conjunction with the NATO Scientific Affairs Division

A Life Sciences **B Physics**	Plenum Publishing Corporation London and New York
C Mathematical and Physical Sciences **D Behavioural and Social Sciences** **E Applied Sciences**	Kluwer Academic Publishers Dordrecht, Boston and London
F Computer and Systems Sciences **G Ecological Sciences** **H Cell Biology**	Springer-Verlag Berlin, Heidelberg, New York, London, Paris and Tokyo

Series C: Mathematical and Physical Sciences - Vol. 257

Paramagnetic Organometallic Species in Activation/Selectivity, Catalysis

edited by

Michel Chanon

Michel Julliard

and

Jean Claude Poite

Faculty of Sciences Saint-Jérôme,
Marseille, France

Kluwer Academic Publishers

Dordrecht / Boston / London

Published in cooperation with NATO Scientific Affairs Division

Proceedings of the NATO Advanced Research Workshop on
Paramagnetic Organometallic Species in Activation/Selectivity, Catalysis
St. Maximin, France
4–9 October 1987

Library of Congress Cataloging in Publication Data

Paramagnetic organometallic species in activation/selectivity, catalysis / edited by Michel Chanon, Michel Julliard, and Jean Claude Poite.
p. cm. -- (NATO ASI series. Series C, Mathematical and physical sciences ; vol. 257)
"Proceedings of the NATO advanced research workshop held in St. Maximin, France, 4-9 October 1987"--P. iv.
Includes index.

1. Catalysis--Congresses. 2. Organometallic compounds--Congresses. 3. Paramagnetism--Congresses. I. Chanon, Michel, 1940- . II. Julliard, Michel. III. Poite, Jean Claude. IV. North Atlantic Treaty Organization. V. Series: NATO ASI series. Series C, Mathematical and physical sciences ; no. 257.
QD505.P36 1988
541.3'95--dc19 88-30782
CIP

ISBN-13: 978-94-010-6882-6 e-ISBN-13: 978-94-009-0877-2
DOI: 10.1007/978-94-009-0877-2

Published by Kluwer Academic Publishers,
P.O. Box 17, 3300 AA Dordrecht, The Netherlands.

Kluwer Academic Publishers incorporates the publishing programmes of
D. Reidel, Martinus Nijhoff, Dr W. Junk, and MTP Press.

Sold and distributed in the U.S.A. and Canada
by Kluwer Academic Publishers,
101 Philip Drive, Norwell, MA 02061, U.S.A.

In all other countries, sold and distributed
by Kluwer Academic Publishers Group,
P.O. Box 322, 3300 AH Dordrecht, The Netherlands.

Softcover reprint of the hardcover 1st edition 1989

This book contains the proceedings of a NATO Advanced Research Workshop held within the programme of activities of the NATO Special Programme on Selective Activation of Molecules running from 1983 to 1988 as part of the activities of the NATO Science Committee.

Other books previously published as a result of the activities of the Special Programme are

BOSNICH, B. (Ed.) - *Asymmetric Catalysis* (E103), 1986

PELIZZETTI, E. and SERPONE, N. (Eds.) - *Homogeneous and Heterogeneous Photocatalysis* (C174) 1986

SCHNEIDER, M. P. (Ed.) - *Enzymes as Catalysts in Organic Synthesis* (C178) 1986

SETTON, R. (Ed.) - *Chemical Reactions in Organic and Inorganic Constrained Systems* (C165) 1986

VIEHE, H. G., JANOUSEK, Z. and MERÉNYI, R. (Eds.) - *Substituent Effects in Radical Chemistry* (C189) 1986

BALZANI, V. (Ed.) - *Supramolecular Photochemistry* (C214) 1987

FONTANILLE, M. and GUYOT, A. (Eds.) - *Recent Advances in Mechanistic and Synthetic Aspects of Polymerization* (C215) 1987

LAINE, R. M. (Ed.) - *Transformation of Organometallics into Common and Exotic Materials: Design and Activation* (E141) 1988

BASSET, J.-M., et al. (Eds.) - *Surface Organometallic Chemistry: Molecular Approaches to Surface Catalysis* (C231) 1988

WHITEHEAD, J. C. (Ed.) - *Selectivity in Chemical Reactions* (C245) 1988

CONTENTS

LIST OF CONTRIBUTORS

C. AMATORE, ENS, Laboratoire de Chimie, 24 rue Lhomond, 75231 PARIS Cédex, FRANCE.

D. ASTRUC, Laboratoire de Chimie Moléculaire des Métaux de Transition, LA 35, Université de BORDEAUX I, 351 Cours de la Libération, 33405 TALENCE Cédex, FRANCE.

D. BALLIVET-TKATCHENKO, CNRS, 205 Route de Narbonne, 31077 TOULOUSE Cédex, FRANCE.

G. BARTOLI, Department of Chemical Sciences, Via San Agostino, 1, CAMERINO (MACERATA) , 40 136 BOLOGNA, ITALIE.

D.H.R. BARTON, Texas A&M University, Department of Chemistry, COLLEGE STATION, TX 77843-3255, USA.

T.L. BROWN, Beckman Institute, University of Illinois, 1304 West Clark St., URBANA, IL 61801, USA.

M.I. BRUCE, Department of Physical and Inorganic Chemistry, University of ADELAIDE, ADELAIDE, SOUTH AUSTRALIA 5001.

M. CHANON, Universitè AIX-MARSEILLE III, Centre de Saint-Jérome, 13397 Marseille Cédex 13, FRANCE.

C. CHATGILIALOGLU, Istituto dei Composti del Carbonio, Contenenti Eteroatomi - CNR, 89 Via Tolara di Sotto, 40064 OZZANO EMILIA , BOLOGNA, ITALIE.

H. CHERMETTE, Institut de Pysique Nucléaire de Lyon, 43 bd du 11 Novembre, 69 622 VILLEURBANNE, FRANCE.

N.G. CONNELLY, Department Inorganic Chemistry, University of BRISTOL, School of Chemistry, BS 81TS BRISTOL, UK.

R.H. CRABTREE, Department of Chemistry, YALE University, PO Box 6666, NEW-HAVEN, CN 06511, USA.

P. DIXNEUF, Université de RENNES 1, Laboratoire de Chimie de Coordination Organique, Campus de BEAULIEU, 35042 RENNES Cédex, FRANCE.

M.H. GARCIA, Istituto Superior Technico, Av. Rovisco Pais, 1096 LISBOA Cédex, PORTUGAL .

C. GIANNOTTI, ICSN CNRS, 99190 GIF-SUR-YVETTE, FRANCE.

J. HALPERN, Department of Chemistry, University of CHICAGO, 5735 South Ellis Ave., CHICAGO, Il 60637, USA.

H. KAGAN, Université de PARIS-SUD, Centre d'ORSAY, 91405 ORSAY Cédex, FRANCE.

W. KAIM, Institut für Anorganische Chemie, Universität, 6000 FRANKFURT AM MAIN 50, NIEDERUSELER HANG 1949, RFA.

J.K. KOCHI, Department of Chemistry, University of HOUSTON, HOUSTON, TX 77 004, USA.

J. KOTZ, Department of Chemistry, State University College, ONEONTA, N.Y. 13820, USA.

B. KRÄUTLER *, Laboratory of Organic Chemistry, ETH-Zürich, Universiätstr. 16, CH-8092 ZÜRICH, SWITZERLAND.

H. KUROSAWA *, Department of Applied Chemistry, Faculty of Engineering, OSAKA University, SUITA, OSAKA 565, JAPAN.

P. LEMOINE, Institut de Chimie, Université Louis Pasteur, 67008 STRASBOURG, FRANCE.

F. MINISCI, Politecnico di MILANO, Dipartimento di Chimica, 20133 MILANO, ITALIE.

J.-C. NEGREL *, Université d'AIX-MARSEILLE III, Centre de Saint-Jérome, 13397, MARSEILLE Cédex, FRANCE.

M. NOJIMA *, Department of Applied Chemistry, Faculty of Engineering, OSAKA University, SUITA, OSAKA 565, JAPAN.

G.F. PEDULLI, Università degli Studi di BOLOGNA, Dipartimento di Chimica Organica, 40127 BOLOGNA, ITALIE.

A.J. POE, J. TUZO WILSON Laboratories, Frindale College, University of TORONTO, MISSIAUGA ONTARIO, CANADA L5L IC6.

P.H. RIEGER, Department of Chemistry, BROWN University, PROVIDENCE, Rhode Island 02912, USA.

B. H. ROBINSON *, Department of Chemistry, University of OTAGO, P.O.Box 56, DUNEDIN, NEW ZEALAND.

E. ROMAN, Facultad de Quimica, Pontificia Universidad Catolica de Chile, Casilia 6177, SANTIAGO DE CHILE, CHILI.

G.A. RUSSELL, Department Chemistry, IOWA State University, AMES, IOWA 50011, USA.

J. SAN FILIPPO Jr, Wright Rienam Chemistry Laboratories, Rutgers State University of New Jersey, NEW BRUNSWICK, NJ 08903, USA.

M.C.R. SYMONS, Department of Chemistry, University Road, LEICESTER, LE1 7RH, UK.

D. TYLER, Department of Chemistry, University of OREGON, EUGENE, OREGON 97403, USA.

G. VAN KOTEN, Vakgroep Organische Chemie, University of UTRECHT, PADUALAAN 8 Trans III, UTRECHT, THE NETHERLANDS.

* Contributors invited to a written symposium to complete the contributions presented by participants to the workshop.

PREFACE

When one considers the overall representation of frontier orbital filling of hexacoordinate (Oh) and tetracoordinate (Td) inorganic and organometallic complexes, it clearly appears that out of 26 cases covering both high spin and low spin situations, 21 represent paramagnetic species (K. Purcell, J. Kotz, "Inorganic Chemistry", Saunders, 1977, p561). This would suggest that, if there is a part in chemistry to illustrate the reactivity of radical species, this part certainly is inorganic organometallic chemistry. In contrast with these expectations, and whereas the standard Organic Chemistry textbook (J. March ,"Advanced Organic Chemistry", J. Wiley, N.Y., 1985) has a specific chapter devoted to free radical reactivity, neither the inorganic standard (F.A. Cotton, G. Wilkinson, "Advanced Inorganic Chemistry", Wiley, 1988), nor the Organometallic one (J.P. Collman, L.S. Hegedus, J.R. Norton, R.G. Finke, "Principles and Applications of Organotransition Metal Chemistry", University Science Books Mill Valley C.A., 1987) possess such a specific chapter. The balance is partly restored because the two last cited books have a more comprehensive treatment of electron transfer phenomena. These comparisons show unambiguously that the importance of paramagnetic species in chemical reactivity still lacks a consistent treatment transcending the artificial barriers between branches of Chemistry. This book, which brings together experimental facts and concepts originating from organometallic and organic reactivities, is a step in the direction of bridging this gap.

The unifying thread which connects the 35 chapters throughout this book is Activation/Selectivity and Catalysis by means of radical chemistry. Indeed, the long lasting contempt of synthetic chemists for radical reactivity has certainly to do with the poor selectivities and yields most often obtained for reactions involving such species. The quantitative studies pursued by several groups very active in kinetics, now available in Landolt-Börnstein tables of reactivity (Radical Reaction Rates in Liquids, Ed. H. Fischer Springer-Verlag, Berlin, Vol. 13, 1984), have opened the way to a better mastering of radical reactivity and now the bad synthetic reputation is rapidly changing into a golden one.

The first chapter in this book illustrates how one may now tame radical species to transform the too often poor yield chain reactions into highly efficient synthetic tools (see also D.P. Curran, Synthesis, 1988, p 417 and 489). This new trend is also well illustrated by Giese's recent book(B. Giese, "Radicals in Organic Synthesis; Formation of Carbon-Carbon Bonds", Organic Chemistry Series, J.F. Baldwin, Pergamon Press, Oxford, 1986), by Russell's chapter and references cited therein and by other recent NATO series (H.G. Viehe, Z. Janousek and R. Merenyi, "Substituent Effects in Radical Chemistry", NATO ASI Series, Serie C, D. Reidel Publishing Company, Dordrecht, Vol. 189, 1986; F. Minisci and A. Citterio, "Free Radicals in Synthesis and Biology", NATO ASI Series, Serie C, Kluwer Academic Publishers, Dordrecht, 1988). Meanwhile, organometallic paramagnetic species have been completely independently developed as synthetic tools. The early works by J. Kwiatek (Catal. Rev. 1967, p. 37), J. Halpern (JACS 1969, p. 582), S. Feldberg (J. Phys. Chem., 1971, p. 2381), J. Osborn (JACS 1972, p. 4043), Hegedus (JACS 1974, p. 3250) and T.L. Brown (JACS 1975, p. 947) paved the way toward recognition of the importance of radical paramagnetic chain reactions in the activation of organometallic substrates. In 1978, J.K. Kochi's pioneering book enhanced this recognition. (Organometallic Mechanisms and Catalysis Academic Press, N.Y. 1978) and for 10 years a generous harvest of new reactions has confirmed this trend.

Living systems have developed highly selective reactions and it is essential that a good concept of activation find its ultimate validation in biosynthetic schemes. The two chapters on B12 reactivity show the possible role of paramagnetic organometallic species in bioactivation. On the organic side the most spectacular example is provided by the exploration of possible biosynthetic pathways of prostaglandins by Porter (Tetrahedron Lett. 1982, 2289).

We had the honour to organize this workshop because some years ago (Bull. Soc. Chim. Fr. II, 1982, p. 197) we pointed out the intrinsic unity brought to organic and inorganic/organometallic reactivities by the electron transfer catalysis (E.T.C.) concept. In this formulation, the electron is viewed as a catalyst (non classical in the sense that it is not a molecular or atomic substance) and this leads to a natural classification of catalytic reactions into those involving even- and odd-electron catalysts (the second including most of the radical chain reactions).

The difficulty in organizing a meeting of this sort appeared when we began to compile lists of experts in organometallic and in organic radical chemistry. The number clearly exceeded that possible for a NATO workshop. The chemists who finally accepted to participate wrote contributions which, we believe, give a fair account of this rapidly moving field, and they actively participated in lively discussions aimed at cross-fertilization. The term "fair account" must be tempered however : from the start, we decided to eliminate exhaustive treatment of O_2 activation (International NATO workshop in Galzignano, G. Modena Chairman, June 1984).

Thanks to the generous grant from NATO Scientific Affairs Division and complementary sponsorship by the Centre National de la Recherche Scientifique, the Unité Enseignement-Recherche de l'Université de Droit, Economie et Sciences d'Aix-Marseille, and various french industrial firms (Rhône-Poulenc, Roussel-Uclaf, Casanis, Maison des Agriculteurs) a workshop to gather 40 specialists became possible at St-Maximin (Var, France). It is a pleasure to acknowledge here the untiring patience that Dr. Sinclair offered us on the administrative side of the conference organization. Dr. I. Tkatchenko and the Scientific Committee (Sir D.H.R. Barton, G.A. Russell, A. Kaim, G. Balavoine and J.M. Surzur) wisely helped in the exacting job of selection of experts.

The lectures and posters were focused on the following topics :

- Activation of Organic Substrates using paramagnetic organometallic species (Chap. 1 to 10)
- Activation of Organometallic substrates using paramagnetic organometallic species (Chap. 11 to 31)
- Mechanistic border between odd- and even-electron processes in organometallic chemistry (Chap.32 to 35) .

The structures of paramagnetic organometallic species, their theoretical treatment, the photochemistry of organometallics and their intervention in heterogeneous catalysis were marginally represented in this conference.

A latent unity emerges from these contributions written by experts in different fields and we hope that this unity will raise up further experimental and theoretical explorations as well as other interdisciplinary meetings on the reactivity of Paramagnetic Species.

M. Chanon
M. Julliard
J.C. Poite

SELECTIVITY AND STEREOSPECIFICITY IN SYNTHETIC APPLICATIONS OF RADICAL REACTIONS

Derek H. R. Barton and Nubar Ozbalik
Texas A&M University
Chemistry Department
College Station, Texas 77843-3255

ABSTRACT The advantages of radical reactions for Organic Synthesis are summarised. Based on the idea of the disciplined radical, it is now possible to design radical reactions which afford a good yield of a single desired product. The system needs to contain a disciplinary group. In the case of radical reactions involving tin hydrides, it is the weak tin-hydrogen bond that is the disciplinary group. For the esters of thiohydroxamic acids, the disciplinary group is the thione function. Examples are given. Recent work has involved the design of stereospecific radical reactions. The hindrance around one chiral center is used to control the formation of the second. The ketal of (+)-(2R,3R) tartaric acid gives excellent stereospecificity (≃24:1) with retention of configuration.

Organic Synthesis, as it is practised at present, is largely based on ionic reactions. For some years, now, we have advocated the use of radical reactions, especially for the manipulation of sensitive natural products.[1,2] In contrast to ionic reactions, their radical counterparts take place under neutral conditions, often at room temperature or lower, and are less labile to the elimination, rearrangement, and neighbouring group participation phenomena seen in ionic chemistry.

However, until recently, radical reactions have been considered to be unselective and to give poor yields. The change that has taken place in the last few years is the recognition that free radicals can be ruled by a disciplinary group.[3] When this requirement is satisfied, good yields of single products can often be obtained and there is no reason why such reactions should not be used routinely in Organic Synthesis.

The first reaction of this kind was the reduction of halides by tin hydride derivatives, a process discovered accidentally by Van der Kerk in 1957.[4] In 1975, in response to a challenge from chemists interested in the deoxygenation of aminoglycoside antibiotics, we invented a process in which a suitable thiocarbonyl group derivative of a secondary alcohol was reduced with tributyltin hydride. There

M. Chanon et al. (eds.), Paramagnetic Organometallic Species in Activation / Selectivity, Catalysis, 1–12.

have been some spectacular examples of the application of this reaction.[5]

This was followed by an efficient method for the same type of compound using reduction by tributyl tin hydride of the isonitrile derivatives of the amine function.[6]

Another efficient process, introduced by Clive, is the reduction of phenylseleno-derivatives also by tin hydride reagents.[7]

All these synthetically useful radical chain reactions have in common the use of the tin-hydrogen bond as the functional group which disciplines the otherwise unruly free radicals.

Although all these reactions replace a functional group by hydrogen, the intermediate carbon radical can be captured intramolecularly by a suitably placed double bond.[8]

However, a more general approach to the generation and application of carbon and other radicals has recently been invented. The basic concept was that esters (mixed anhydrides) of thiohydroxamic acids would act as generators of carbon radicals. To take a specific example, the esters 1 (R=alkyl or alicyclyl) of N-hydroxy-2-thiopyridone rearrange with loss of carbon dioxide by a radical chain mechanism when heated to 80°, or when irradiated at any temperature with tungsten light to give thiopyridines 2 (Scheme 1, path A). The reaction occurs with other thio- hydroxamic esters having an appropriate structure such as 3.

Scheme I

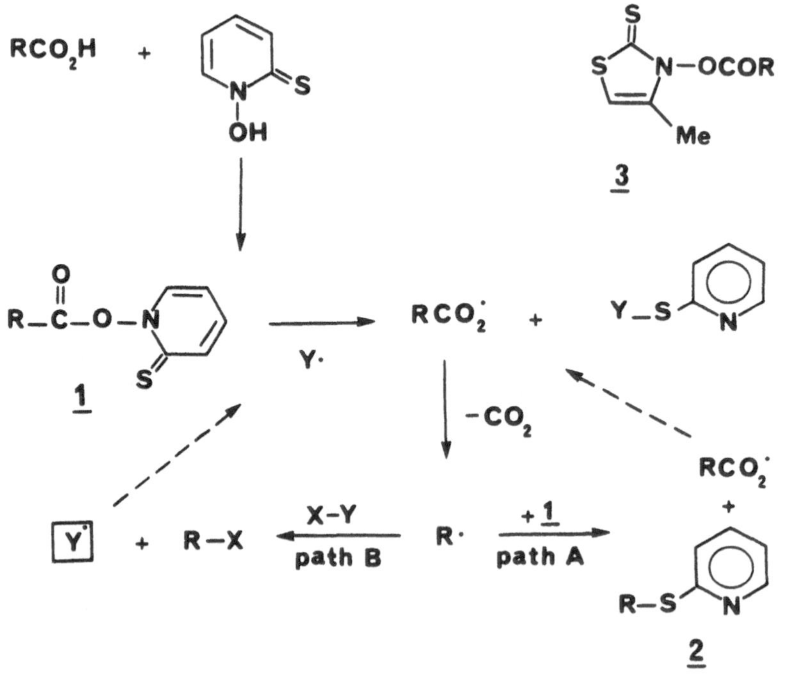

With the availability of a convenient source of carbon radicals, we considered inclusion of external radical traps into the system (X-Y in scheme 1, path B) thus diverting the reaction from its normal course of rearrangement to other useful transformations. Indeed, application of this idea provided a simple high yielding conversion of carboxylic acids into various synthetically valuable intermediates under mild conditions[3] (Table 1).

Table 1. Types of Transformations with Various Radical Traps

Radical Trap (X-Y)	Product (R-X)
None	R-SPY
nBu_3SnH *t*BuSH	R-H
Br-CCl_3 Cl-CCl_3 I-CHI_2	R-Br R-Cl R-I
O_2 $(PhS)_3Sb$-O_2	R-OOH R-OH
CH_2=CH-Z	RCH_2-CH(Z)(S-Py)
CH_2=C(COOEt)(CH_2-S*t*Bu)	R-CH_2-C(CO_2Et)=CH_2
$(Ar\text{-}X)_2$ X=S, Se, Te	Ar-X-R

This type of chemistry is well suited for the manipulation of amino acids and peptides which tend to undergo racemization at the α carbon under ionic conditions. We first examined the decarboxylation of N-protected amino acids by using a thiol as hydrogen atom transfer reagent (Scheme 2).[9] Alcoholic phenolic and even indolic groups do not need protection in this reaction. Manipulation of the side chain carboxyl groups are also possible when the α-carboxyl is appropriately protected. This is exemplified in the synthesis of optically pure vinylglycine 4, an important natural amino acid, from readily available glutamic acid (Scheme 3).[10]

Scheme II

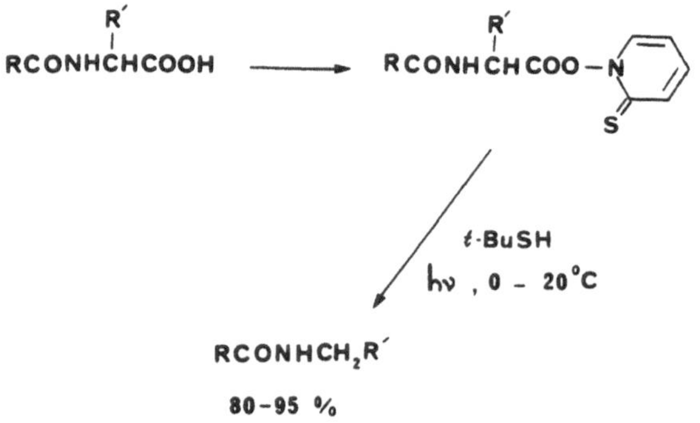

Scheme III

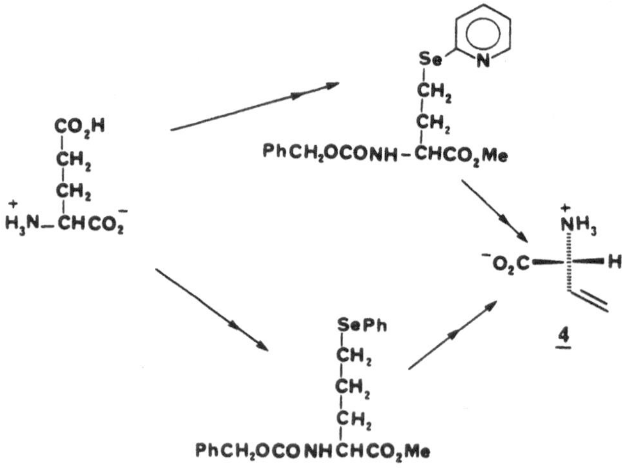

We had shown that interception of carbon radicals by electron poor olefines was an efficient process provided the background reactions of rearrangement and polymerization of the olefine can be avoided (Table 1). Extension of this reaction to amino acids would provide convenient means of increasing the carbon chain without loss of optical purity. Such a transformation is highly desirable in the synthesis of optically pure α-aminoadipic and a-aminopimelic acids. The mixed anhydrides (Scheme 4) derived from the protected amino acids 5 and 6 are photolyzed in presence of methyl acrylate. The resulting addition products are subjected to successive steps of saponification, reduction and deprotection to give the optically pure amino acids 7 and 8[11].

We next carried out the synthesis of two most important seleno-amino acids, L-selenomethionine 12 and L-selenocystin 15, starting from the readily available glutamic and aspartic acid derivatives. (Scheme 5). Irradiation of the mixed anhydride 9 in the presence of excess dimethylselenide afforded the selenomethionine derivative 10, which was deprotected after the steps of saponification and exposure to trifluoroacetic acid. An alternate route to compound 10 involved preparation of bromo derivative 11 and displacement of the bromine with sodium methylselenide.[12]

Scheme IV

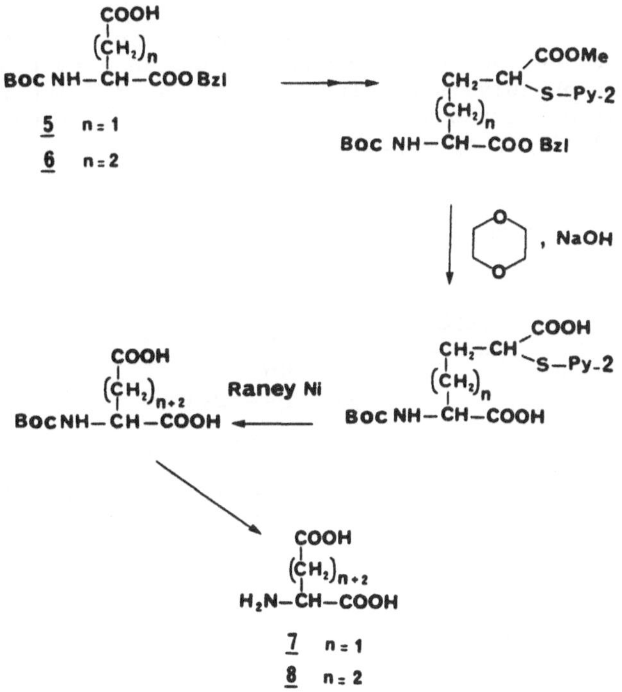

Scheme V

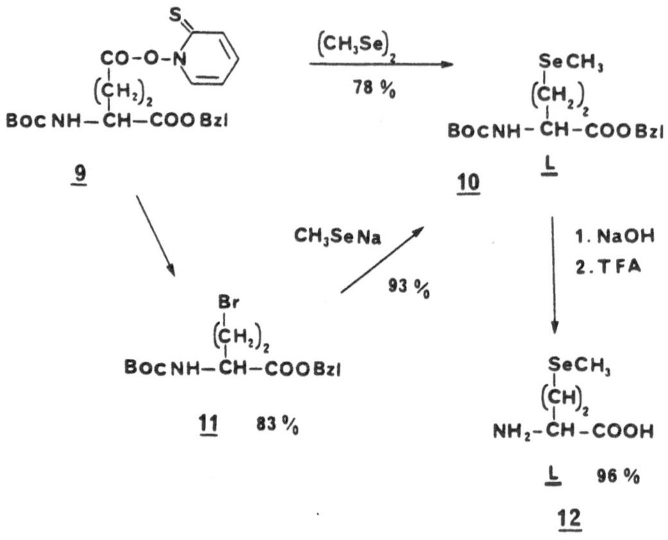

In the synthesis of selenocystine (Scheme 6), dicyanogen triselenide was found to be the most convenient substance to trap the radical generated by photolysis of mixed anhydride 13. The resulting selenocyanate 14 was converted into selenocystine 15 by treatment with sodium borohydride followed by deprotection.[12]

Scheme VI

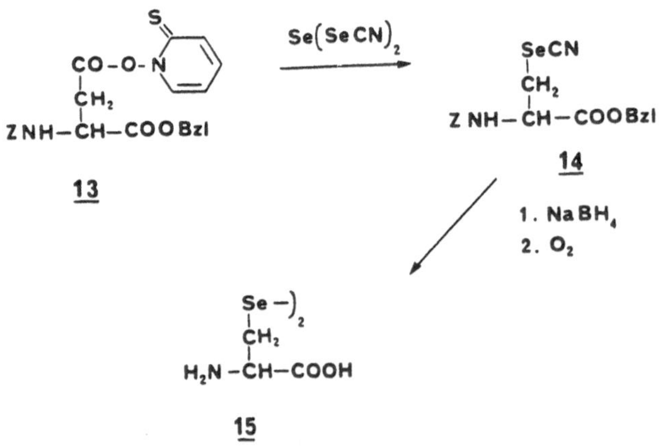

Another interesting application of the radical decarboxylation method in modification of the aspartic acid side chain permitted an easy synthesis of perhydroindole-2-carboxylic acid derivatives. The sequence indicated (Scheme 7) involves the incorporation of the chiral centre of L-aspartic acid into the 2-position of the title compounds with complete conservation of stereochemistry.[13] Photolysis of the N-hydroxy-2-thiopyridone derivative from 16 gave the cyclized product as a mixture of two stereoisomers. The stereochemistry assigned to 17a is based on the X-ray study of an analog. The stereochemistry at ring junction of 17b is deduced from its NMR spectrum. The configuration at C-4 is an assumption. Reductive removal of the thiopyridyl function and deprotection in the usual way gave the two desired perhydroindole derivatives.

In conclusion, the application of the radical decarboxylation reaction has provided simple efficient synthesis of many important amino acids starting from readily available glutamic and aspartic acid derivatives. Given the mildness of the reaction conditions coupled with good yields and selectivity, this method will find wide use in manipulation of complex and often fragile natural products.

Scheme VII

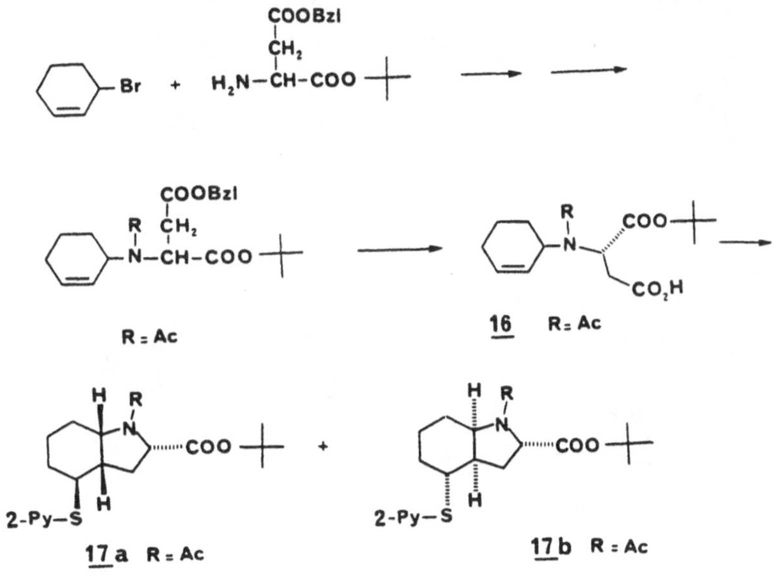

As has been illustrated above, we now have a new system of radical chemistry where yields can be excellent and where the reactions are selective. Why is this so? It is because the system has a new disciplinary group, the thiocarbonyl function. The carbon radicals that are generated either react with an added trap that then reacts with the thiocarbonyl function or they are disciplined by the thiocarbonyl group and give the decarboxylated rearranged thiopyridine derivatives.

The system has been modified by Newcomb and his colleagues[14] to an efficient synthesis of nitrogen radicals from photolysis of the easily prepared carbamate esters of type 18. Even more important was the fact that addition of trifluoroacetic acid to protonate the nitrogen radicals as they were formed produced aminium radical cations. The latter have a rich chemistry, but are normally prepared under strongly acidic conditions. As an example, the aminium radical cation 19, generated by photolysis of the appropriate 18, cyclised two times to give, in the presence of *t*-butyl thiol, the pyrrolizidine 20 in satisfactory (60%) yield. This work opens a new chapter in nitrogen radical chemistry.

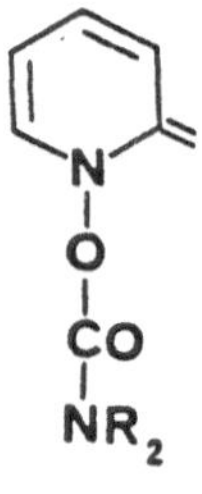

18

19 20

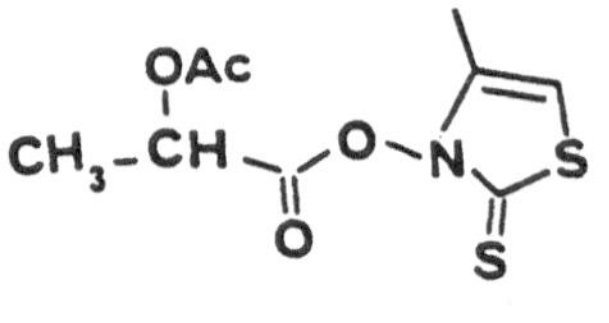

21

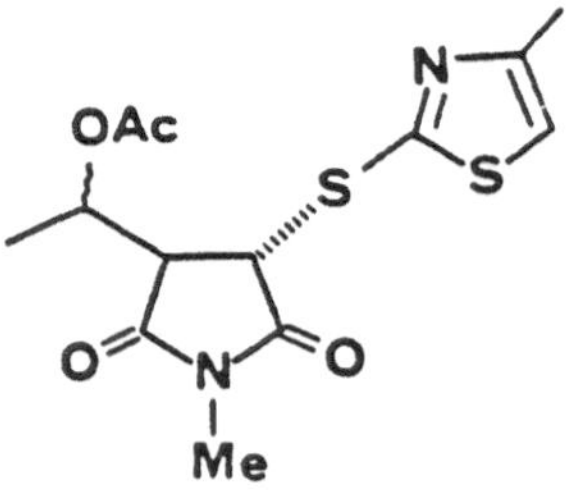

22

Now that the generation of carbon and nitrogen radicals can be accomplished smoothly and that these radicals can be disciplined to give single products, the second criticism of radical reactions, namely that they are not stereospecific, can be examined again. It is true that the addition of a radical from the α-carboxyl of a protected amino-acid to any kind of activated olefin gives products which are completely inactive[15]. Also, the L(+)-lactic acid derivative 21 affords on thermolysis a radical which adds very cleanly to N-methylmaleimide to give an adduct 22 (87%) with complete racemisation. It is clear, then, that stereospecificity can only be secured by using one chiral centre to direct the formation of another. We turned, therefore, to optically active tartaric acid, a compound with a venerable history.

We noted that [16] more stereoselective reactions are seen in the neighbouring positions in a five-membered ring rather than a six-membered ring. We decided, therefore, to use the known[17] R,R-monoester ketal 23. The ester 24 was prepared by the mixed anhydride method[18].

Irradiation of 24 with a tungsten lamp gave sulphide 25 (78%) as the only isomer detectable by N.M.R. This technique also indicated the trans configuration. A more rigorous proof of the retention of configuration was secured by studying the addition of the tartaric acid derived radical to methyl acrylate. This gave an adduct (26) (70%). Oxidation of the sulphide to sulphoxide followed by thermolysis in boiling toluene gave cleanly the trans olefin 27 ($[\alpha]_D$-35.4 in $CDCl_3$). Cleavage of the double bond with RuO_2-$NaIO_4$ in acetone-water and methylation with diazomethane gave the dimethyl tartrate derivative 28, $[\alpha]_D$-58°(c,0·86) in MeOH, identical to on an authentic sample, $[\alpha]_D$-58·7° (c,0·81) in MeOH in all respects (I.R. and N.M.R.). The striking retention of configuration in this radical reaction was thus confirmed. Although none of the meso-isomer 29 could be detected by N.M.R., a careful H.P.L.C. analysis[19] of the crude degradation product showed the presence of 4% of the meso-compound 29.

An identical sequence of reactions was carried on the meso-derivative 29. Half hydrolysis of 29 gave monomethyl ester 30. Addition of the derived radical to methyl acrylate followed by degradation gave racemic dimethyl tartarate (I.R. and N.M.R. comparison). Again H.P.L.C. analysis slowed the presence of 4% of the meso-diester 29.

The retention of configuration in this reaction is remarkable. The stereoselectivity is high enough (25:1) for most practical purposes, but could undoubtedly be improved, if necessary, by replacing either the methyl ester with a more bulky ester, or by making the ketal function more bulky.

We have briefly examined two other olefins, phenyl vinyl sulphone and N-methylmaleimide. Both of these olefins are non-polymerisable and thus easier to work with than methyl acrylate, which always gives a small percentage of the two-fold adduct. The former gave 70% of 31, whilst the latter gave 93% of 32. Both were mixtures of stereoisomers since the asymmetric centres created beyond the tartaric acid moiety cannot be controlled. Heating 32 with copper powder caused a smooth elimination of the pyridyl-sulphide group to give the olefin 33. From the N.M.R. spectrum this was again a single isomer.

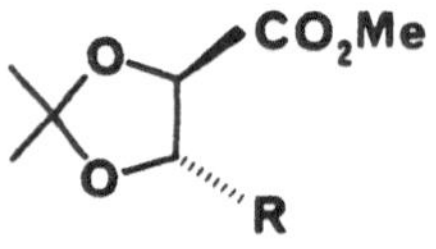

23 R = COOH

24 R = $-CO_2-N$ 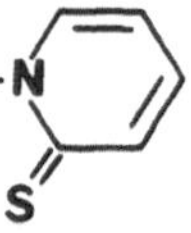

25 R = S-Py

26 R = $CH_2CH(SPy)\ CO_2Me$

27 R =  CO_2Me

28 R = CO_2Me

31 R = $CH_2CH(SPy)-SO_2Ph$

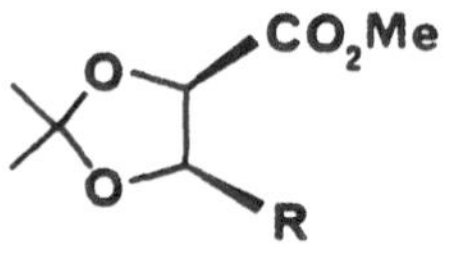

29 R = CO_2Me

30 R = COOH

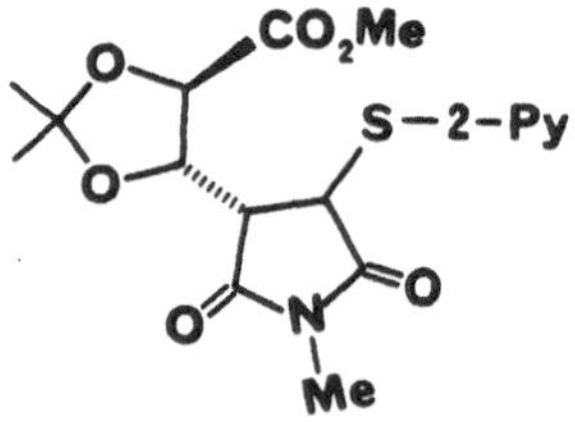

32

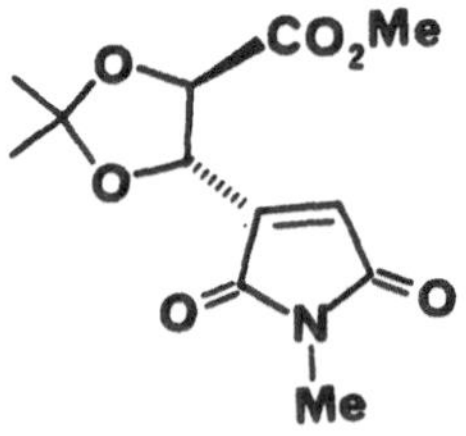

33

References

1. D.H.R. Barton and S.W. McCombie, J. Chem. Soc., Perkin Trans 1, 1574 (1975).

2. D.H.R. Barton, W.B. Motherwell and A. Stange, Synthesis, 743 (1981).

3. D.H.R. Barton and S.Z. Zard, Pure and Appl. Chem., **58**, 675, (1986); idem, Phil. Trans. R. Soc. London., B311, 505 (1985).

4. G.J.M. Van der Kerk, J.G. Noltes and J.G.A. Luitjen, J. Appl. Chem., **7**, 366 (1957).

5. a) T. Hayashi, T. Iwaoka, N. Takeda and E. Ohki, Chem. Pharm. Bull, **26**, 1786 (1978); b) R.E. Carney, J.B. McAlpine, M. Jackson, R.S. Stanaszek, W.H. Washburn, M. Cirovic and S.L. Mueller, J. Antibiotics, **31**, 441 (1978); c) J. Defaye, H. Driguez, B. Henrissat and E. Bar-Guilloux, Nouveau J. de Chimie, 1980, **4**, 59; d) V. Pozsgay and A. Nesmelyi, Carbohydr. Res., **85**, 143 (1980); e) D.J. Hart, J. Org. Chem. **46**, 367 (1981).

6. a) T. Saegusa, S. Kobayashi, T. Ito and N. Yasuda, J. Am. Chem. Soc., **90**, 4182 (1968); b) D. Ivor John, E.J. Thomas and N.D. Tyrrell, J. Chem. Soc. Chem. Commun., 345 (1979); c) D.H.R. Barton, G. Bringmann and W.B. Motherwell, J. Chem. Soc., Perkin Trans 1., 2665 (1980).

7. D.L.J. Clive, G.J. Chiattattu, V. Farina, W.A. Kiel, S.M. Menchen, C.G. Russell, A. Singh, C.K. Wong and N.J. Curtiss, J. Am. Chem. Soc., **102**, 4438 (1980).

8. a) G. Stork, R. Jr. Mook, S.A. Biller, and S. D. Rychnovsky, J. Am. Chem. Soc., **105**, 3741 (1983); b) D.L.J. Clive and P.L. Beaulieu, Chem. Soc. Chem. Commun. 307 (1983); c) T.V. Rajan Babu, J. Am. Chem. Soc., **109**, 609 (1987); d) R. Tsang and B. Fraser-Reid, J. Am. Chem. Soc. **108**, 2116 (1986).

9. D.H.R. Barton, Y. Hervé, P. Potier and J. Thierry, J. Chem. Soc. Chem. Commun., 1294 (1984).

10. D.H.R. Barton, D. Crich, Y. Hervé, P. Potier and J. Thierry, Tetrahedron, **41**, 4347 (1985).

11. Y. Hervé, Thèse de Docteur de L'Université Paris XI Orsay en Science, 1986.

12. D.H.R. Barton, D. Bridon, Y. Hervé, P. Potier, J. Thierry and S.Z. Zard, Tetrahedron, **42**, 4983 (1986).

13. D.H.R. Barton, J. Guilhem, Y. Hervé, P. Potier and J. Thierry, Tetrahedron Lett., **28**, 1413 (1987).

14. M. Newcomb, S.-V. Park, J. Kaplan and D.J. Marquardt, Tetrahedron Letts., 1985, **26**, 5651; M. Newcomb and T.M. Deeb, J. Am. Chem. Soc., **109**, 3163 (1987).

15. D.H.R. Barton, Y. Hervé, P. Potier and J. Thierry, Tetrahedron, in press.

16. B. Giese in "Radicals in Organic Synthesis: Formation of Carbon-Carbon Bonds", p.28, Pergamon Press, Oxford, 1986.

17. J.A. Musich and H. Rapoport, J. Am. Chem. Soc., **100**, 48 (1978).

18. D.H.R. Barton, D. Bridon, Y. Hervé, P. Potier and J. Thierry, ibid., **41**, 4347 (1985).

19. A.M. Krstuvlovic, J.-P. Porziensky and A. Wick (Synthelabo, Paris) and D.H.R. Barton, K. Tachdjian, A. Gateau-Olesker, S.D. Gero and S.Z.Zard, in preparation.

CHAIN REACTIONS OF ORGANOMERCURY HALIDES INVOLVING ELECTRON TRANSFER

Glen A. Russell
Department of Chemistry
Iowa State University
Ames, Iowa 50011
U.S.A.

ABSTRACT. Radical anions or neutral radicals which are easily oxidized will undergo a dissociative electron transfer reaction with RHgX. The rate of this process increases significantly from R = 1°-alkyl to 3°-alkyl (k(t-Bu):k(n-Bu ≈ 1000:1). Chain reactions ensue when the alkyl radical thus formed can regenerate a donor radical (D•) or radical anion ($D•^-$) by (a) S_H2 reaction with Y-D, (b) addition to an electron-rich alkene or to an anion, or (c) by addition-elimination in the system ZCH=CHD.. 1-Alkenylmercury halides will react with radicals (R• or Q•) by the addition-elimination sequence to form monomeric ClHg• which will undergo electron transfer with easily oxidized anions (Q^-) or $Hg(Q)_2$ to generate Q•. Carbon-centered acceptor radicals (A•), formed by addition of R• to an electron-poor alkene, will react with RHgCl by S_H2 substitution at Hg. For some carbon-centered radicals, such as enolate-type radicals, this process is inefficient. In these cases, addition of an easily oxidized anion, such as I^-, can result in an efficient chain process in which A• oxidizes I^- to I• which readily reacts with RHgCl to cycle R• in a chain process.

1. INTRODUCTION

Organomercurials (R_2Hg, RHgX) are readily available organometallic reagents with moderate reactivity in electrophilic substitution at carbon but with extremely low reactivity in nucleophilic substitution.[1] Electron-accepting heteroatom-centered radicals (Q•) and atoms readily attack the mercury atoms of R_2Hg or RHgX with the displacement of R•,[2] in a process whose rate is controlled by the stability of R•.[3] Reactions with halogen atoms are involved in the well known homolytic cleavage of alkyl-mercury bonds by molecular

M. Chanon et al. (eds.), Paramagnetic Organometallic Species in Activation / Selectivity, Catalysis, 13–28.

halogens (Scheme Ia)[4,5] or in the dehalogenation of vicinal dihalides by the alkylmercurials (Scheme Ib).[6]

SCHEME I

(a) $I\bullet + RHgX \longrightarrow IHgX + R\bullet$

$R\bullet + I_2 \longrightarrow RI + I\bullet$

(b) $R\bullet + XCR_2CR_2X \longrightarrow RX + \bullet CR_2CR_2X$

$\bullet CR_2CR_2X \longrightarrow R_2C{=}CR_2 + X\bullet$

$X\bullet + R_2Hg \longrightarrow XHgX + R\bullet$

Attack upon R_2Hg or RHgX by Q•, or by a carbon-centered acceptor radical (A•), is now recognized to occur in many chain reactions of R_2Hg or RHgX in which the attacking radical (A• or Q•) can be halogen atoms, RY• (Y = S, Se, Te), ArSO•, $ArSO_2\bullet$, $CCl_3\bullet$, $RCH_2\dot{C}HA'$ (A' = $(EtO)_2PO$, $ArSO_2$, p-$O_2NC_6H_4$, EtO_2C, MeCO) or a variety of substituted vinyl radicals.[2,3] Reaction 1 and the analogous reaction with Q•,

$$A\bullet + RHgR\ (\text{or } X) \longrightarrow AHgR\ (\text{or } X) + R\bullet \quad (1)$$

occurs when A• or Q• has a high electron affinity but at a rate which increases dramatically with the stability of the incipient alkyl radical (Table I). The reaction appears to occur in a concerted fashion involving a transition state with considerable polar character, e.g., I.

$$A\bullet + R_2Hg \longrightarrow [A^- \ R\bullet \ {}^+HgR] \longrightarrow AHgR + R\bullet$$

I

Acceptor radicals which will participate in Reaction 1 can be formed in chain reactions by attack of R• upon substrates such as Y-Q (e.g., PhSSPh or I_2, see Scheme Ia),[2] by abstraction followed by a β-elimination of Q• (e.g., 1,2-dihalides in Scheme Ib), by the homolytic addition-elimination process of Scheme IIa,b,[7,8] (and the analogous reactions of PhC≡CQ or $HC{\equiv}CCH_2Q$),[9] or by the addition of R• to a radicaphile (π_A) which yields an acceptor adduct radical (e.g., $RCH_2\dot{C}HP(O)(OEt)_2$, $Ph\dot{C}{=}CHR$),[3] Scheme IIc.

TABLE I. Relative Reactivities of Alkylmercury Chlorides Towards Free Radicals at 35-40 °C

Attacking Radical[a]	Solvent	Reaction No.	Relative Reactivity *t*-BuHgCl:*i*-PrHgCl:*n*-BuHgCl
$RCMe_2NO_2{\bullet}^-$	Me_2SO	2	1.0[b]:0.07:<0.005
$Bu_3Sn\bullet$	PhH	2	1.0:0.025[c]:0.005
ClHg•	PhH	3	1.0:0.011[c]:0.0001
PhS•	Me_2SO	1	1.0:0.08:<0.003
PhSe•	PhH	1	1.0 : <0.004
$RCH_2\dot{C}HP(O)(OEt)_2$	Me_2SO	1	1.0:0.07:0.014
I•	Me_2SO	1	1.0:0.006[c]:<0.0001

[a] Radical precursors were $Me_2C{=}NO_2^-$, (*E*)-PhCH=CHSnBu_3, (*E*)-PhCH=C(H)HgCl, PhSSPh, PhSeSePh, $CH_2{=}CHP(O)(OEt)_2$ and (*E*)-PhCH=CHI or $Ph_2C{=}CHI$. [b] $PhCH_2HgCl$ is 4.7-times as reactive. [c] *c*-$C_6H_{11}\bullet$.

SCHEME II

(a) R• + PhCH=CHQ ⟶ Ph$\dot{C}$H-C(H)(R)Q

Ph$\dot{C}$H-C(H)(R)Q ⟶ PhCH=CHR + Q•

(b) R• + $CH_2{=}CHCH_2Q$ ⟶ $RCH_2\dot{C}HCH_2Q$

$RCH_2\dot{C}HCH_2Q$ ⟶ $RCH_2CH{=}CH_2$ + Q•

(c) R• + π_A ⟶ $R\pi_A\bullet$

$R\pi_A\bullet$ + RHgCl ⟶ $R\pi_A$-HgCl + R•

Reaction of alkylmercury salts with alkaline $NaBH_4$ generates unstable RHgH intermediates which undergo free radical decomposition via RHg• (Scheme IIIa).[10-12] In the presence of reactive radicaphiles, new carbon-carbon bonds can be formed (Scheme IIIb).[13] In reactions involving RHg• as an intermediate, the structure of R has little effect upon the reactivity of its precursor, i.e., RHgH. Thus, in the competitive reactions of an excess of a 1:1 mixture of *t*-BuHgCl and *n*-BuHgCl with $CH_2{=}C(Cl)CN$ and $NaBH_4$ (via Scheme IIIb), *t*-$BuCH_2CH(Cl)CN$ and *n*-$BuCH_2CH(Cl)CN$ are formed in approximately a 1:1 ratio.[3,14]

SCHEME III

(a) $R\bullet + RHgH \longrightarrow RH + RHg\bullet$

$RHg\bullet \longrightarrow R\bullet + Hg^\circ$

(b) $R\bullet + CH_2{=}CH(EWG) \longrightarrow RCH_2\dot{C}H(EWG)$

$RCH_2\dot{C}H(EWG) + RHgH \longrightarrow RHg\bullet + RCH_2CH_2(EWG)$

Mercury(I) intermediates are also involved in the free radical chain decomposition of 1-alkenyl, 1-alkynyl or allyl mercurials containing an alkyl or thioalkyl ligand on mercury (Scheme IV).[7,15]

SCHEME IV (R = alkyl or thioalkyl)

$R\bullet + ArCH{=}C(H)HgR \longrightarrow Ar\dot{C}H{-}C(H)(R)HgR$

$Ar\dot{C}H{-}C(H)(R)HgR \longrightarrow ArCH{=}CHR + RHg\bullet$

2. ELECTRON TRANSFER OF DONOR RADICALS TO ALKYLMERCURY SALTS

Donor radicals or radical anions react with alkylmercury salts by electron transfer to form Hg° and the alkyl radical (Reaction 2).[16-18] Towards $D\bullet = Bu_3Sn\bullet$ or $D\bullet^- = RCMe_2NO_2\bullet^-$, the relative

$$D\bullet\ (\text{or } D\bullet^-) + RHgX \longrightarrow D^+\ (\text{or } D) + R\bullet + Hg^\circ + X^- \qquad (2)$$

reactivities of alkylmercury chlorides measured in competition experiments (Table I) increase from R = 1°-alkyl to 2°-alkyl to 3°-alkyl ($k(t\text{-Bu}):k(n\text{-Bu}) \approx 1000:1$).[3] This indicates that Reaction 2 occurs by dissociative electron transfer with a rate controlled by the stability of the incipient R•.

Donor radicals capable of reacting with RHgX can be formed in chain reactions by the attack of R• upon Y-D (e.g., $HSnBu_3$, Scheme Va),[14] by addition of R• to nucleophiles (Scheme Vb)[16-21] or neutral π_D systems capable of forming an adduct which is a donor radical (e.g., $RCH_2\dot{C}(C_6H_4OMe\text{-}p)_2$, Scheme Vc),[18,22] or by the addition-elimination process of Scheme Vd.[7,8] A variety of substrates which react with alkyl radicals to form donor radicals capable of electron transfer to RHgX (X = halogen, carboxylate) are listed in Table II.

TABLE II. Reactions of Various Substrates with *t*-BuHgCl Involving Reaction 2 and Scheme V

Donor Radical	Precursor/Substrate	Isolated Products	Scheme	Ref.
$Bu_3Sn\bullet$	$HSnBu_3$ and RHgCl(OAc)	Bu_3SnX and RH	Va	23, 24
$Bu_3Sn\bullet$	$HSnBu_3$ and CH_2=CH(EWG)	Bu_3SnX and RCH_2CH_2(EWG)	Va	13
N-Bn-3-(NH_2CO)Py•	N-benzyl-1,4-dihydro-nicotinamide	N-Bn-3-(NH_2CO)Py^+ and RH	Va	25
$RC(R^1)(R^2)NO_2\bullet^-$	$(R^1)(R^2)C{=}NO_2^-$	$RC(R^1)(R^2)NO_2$	Vb	16, 17
$RNO_2\bullet^-$; RN_3^-	NO_2^-; N_3^-	RNO_2; RN_3	Vb	19, 20
$RC(R^1)(R^2)\dot{C}OPh\bullet^-$	$(R^1)(R^2)C{=}C(O^-)Ph$	$RC(R^1)(R^2)COPh$	Vb	19, 20
$RCPh_3\bullet^-$; $RPPh_2\bullet^-$	Ph_3C^-; Ph_2P^-	$RCPh_3$; $RPPh_2$	Vb	19, 21
$Ph_2C(R)CN\bullet^-$	Ph_2CCN^-	$Ph_2C(R)CN$; $Ph_2C{=}C{=}NR$	Vb	19, 20
$RN(CO)_2C_6H_4\bullet^-$	phthalimide$^-$	N-R-phthalimide	Vb	19, 20
$PhCOCH_2\dot{P}(OMe)_3$	$P(OMe)_3$	$PhCOP(O)(OMe)_2$ and MeX	Vc	26
$(p\text{-}MeOC_6H_4)_2\dot{C}CH_2R$	$(p\text{-}MeOC_6H_4)_2C{=}CH_2$	$(p\text{-}MeOC_6H_4)_2C{=}CHR$	Vc	22
$PhCH_2N{=}\dot{C}R$	$PhCH_2NC$	$PhCH_2NHCOR$	Vc	18
$PhCH_2C_6H_4(NMe_2)_2\bullet$	$p\text{-}Me_2NC_6H_4NMe_2$	$(PhCH_2)(Me_2N)_2C_6H_3$	Vc	18
o- and *p*-$RC_5H_5N\bullet$	pyridine; PyH^+	*o*- and *p*-RC_5H_4N	Vc	18
$Bu_3Sn\bullet$	$ZCH{=}CHSnBu_3$, Z = Ph, $PhSO_2$, MeO_2C	ZCH=CHR and Bu_3SnCl	Vd	7, 8,
$Bu_3Sn\bullet$	$PhC{\equiv}CSnBu_3$	$PhC{\equiv}CR$ and Bu_3SnCl	Vd	9, 27
$Bu_3Sn\bullet$	$CH_2{=}CHCH_2SnBu_3$	$RCH_2CH{=}CH_2$ and Bu_3SnCl	Vd	27

SCHEME V

(a) $R\bullet + Bu_3SnH \longrightarrow RH + Bu_3Sn\bullet$

$Bu_3Sn\bullet + RHgX \longrightarrow Bu_3Sn^+ + R\bullet + Hg^\circ + X^-$

(b) $R\bullet + Me_2C{=}NO_2^- \longrightarrow RCMe_2NO_2\bullet^-$

$RCMe_2NO_2\bullet^- + RHgX \longrightarrow RCMe_2NO_2 + R\bullet + Hg^\circ + X^-$

(c) $R\bullet + (p\text{-}MeOC_6H_4)_2C{=}CH_2 \longrightarrow (p\text{-}MeOC_6H_4)_2\dot{C}CH_2R$

$(p\text{-}MeOC_6H_4)_2\dot{C}CH_2R + RHgX \longrightarrow (p\text{-}MeOC_6H_4)_2\overset{+}{C}CH_2R +$

$+ R\bullet + Hg^\circ + X^-$

(d) $R\bullet + PhCH{=}CHD \longrightarrow Ph\dot{C}H\text{-}C(H)(R)D$

$Ph\dot{C}H\text{-}C(H)(R)D \longrightarrow PhCH{=}CHR + D\bullet$

$D\bullet + RHgX \longrightarrow D^+ + R\bullet + Hg^\circ + X^-$

For synthetic purposes, the RHgX reagents, used for the alkylation of π_D or PhCH=CHD substrates, do not have to be isolated. Thus, solvomercuration products of alkenes can be prepared in situ, treated with the substrate (e.g., pyridine) and photostimulated to achieve alkylation by a chain mechanism (e.g., Scheme VI).

SCHEME VI

$Me_2C{=}CH_2 + Hg(O_2CCF_3)_2 \xrightarrow{MeOH} Me_2C(OMe)CH_2HgO_2CCF_3\ (RHgX)$

$RHgX \xrightarrow{h\nu} R\bullet + XHg\bullet \xrightarrow{RHgX} R\bullet + HgX_2 + Hg^\circ$

$R\bullet + PyH^+ \longrightarrow o\text{- and } p\text{-}RC_5H_5NH\bullet^+ \rightleftharpoons RC_5H_4NH\bullet$

$RC_5H_4NH\bullet + RHgX \longrightarrow RPyH^+\ Cl^- + Hg^\circ + R\bullet$

Kinetic chain lengths have been measured by the $(t\text{-Bu})_2NO\bullet$ inhibition method for the reactions of 1-5 equivalents of t-BuHgCl with 0.05 M substrate for $PhCH{=}CHSnBu_3$, $Me_2C{=}NO_2^-$, $(p\text{-}MeOC_6H_4)_2C{=}CH_2$ as well as Y-Q, PhCH=CHQ, and π_A substrates.[3] Table III summarizes data from ^1H NMR measurements in d_6-Me_2SO at 35-40 °C with photostimulation from a 275 W fluorescent sunlamp through Pyrex.

TABLE III. Kinetic Chain Lengths in Photostimulated Reaction of t-BuHgCl at 35-40°C in Me$_2$SO[a]

Substrate (Equiv)	Product	Initial Kinetic Chain Length
PhSSPh (0.2)	t-BuSPh	400[b]
Me$_2$C=NO$_2^-$ (1)	t-BuCMe$_2$NO$_2$	50[c]
CH$_2$=CHP(O)(OEt)$_2$ (0.25)	t-BuCH$_2$C(H)(HgCl)P(O)(OEt)$_2$	105 (95)[d]
CH$_2$=CHSO$_2$Ph (0.4)	t-BuCH$_2$CH(HgCl)SO$_2$Ph	103
HC≡CCOCH$_3$ (0.3)	t-BuCH=C(HgCl)COCH$_3$	32 (26)[d]
HC≡CPh (0.3)	t-BuCH=C(HgCl)Ph	15 (14)[d]
Ph$_2$C=CHI (0.3)	t-BuCH=CPh$_2$	100
(E)-PhCH=CHSnBu$_3$ (0.3)	(E)-t-BuCH=CHPh	71
(E)-PhCH=CHHgCl (0.3)	(E)-t-BuCH=CHPh	56
CH$_2$=C(C$_6$H$_4$OMe-p)$_2$ (0.3)	t-BuCH=C(C$_6$H$_4$OMe-p)$_2$	18

[a] $[t$-BuHgCl$]_0$ = 0.5 M; irradiation by a 275 W fluorescent sunlamp in 6 mm NMR tubes. Irradiation conditions were only approximately the same for different substrates. [b] 420 in the presence of 1 equiv of n-BuHgCl. [c] In presence of 18-crown-6; longer kinetics chains were observed in its absence. [d] PhH solvent.

The initial kinetic chain lengths (initial reaction rate/rate of initiation) were found to be >20 in nearly all cases. In view of the kinetic chain lengths involved, the competitive reactions of t-BuHgCl with two of the donor radical precursors from Table II (or with the acceptor radical precursors from Schemes I and II) should be a realistic measure of the reactivities of the precursors towards t-Bu•. In the competitive reactions of Scheme VII, the significant kinetic chain lengths require that essentially every (>95%) RA• or RB• formed continues the chain reaction to yield the product (P$_A$ or

SCHEME VII

$$R\bullet + \begin{matrix} A & \xrightarrow{k_A} & RA\bullet \\ B & \xrightarrow{k_B} & RB\bullet \end{matrix} \xrightarrow{RHgX} \begin{matrix} P_A \\ P_B \end{matrix} + R\bullet$$

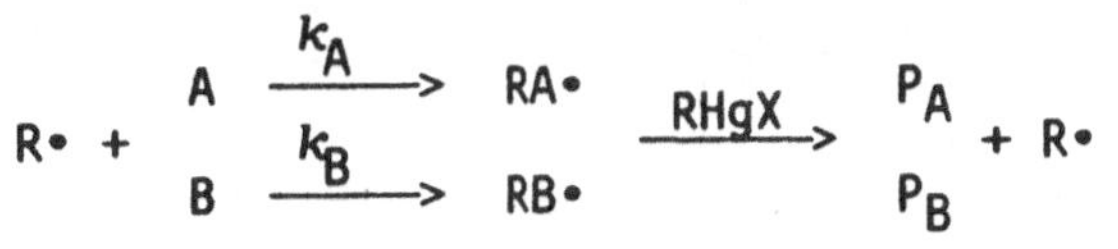

TABLE IV. Relative Reactivities Towards *tert*-Butyl Radical in Substitution or Addition

Substrate A (mmol)	Substrate B (mmol)	Conditions[a]	k_A/k_B[b]	Rel. React. of A[c]
CH_2=C(Cl)CN (2)	CH_2=$CHCO_2Et$ (2)	CH_2Cl_2, $NaBH_4$	53.0	1.6×10^3
$PhSO_2Cl$ (1)	*n*-BuSSBu-*n* (1)	PhH, S, 30 min	134.0	1.1×10^2
CH_2=$CHSO_2Ph$ (1)	CH_2=$CHCO_2Et$ (1)	CH_2Cl_2, $NaBH_4$	2.5	74
PhSSPh (1)	Ph_2C=CHI (1)	PhH, S, 10 min	43.0	43
CH_2=$CHCO_2Et$ (2)	CH_2=CHP(O)$(OEt)_2$ (2)	CH_2Cl_2, $NaBH_4$	3.0	30
$EtO_2CC{\equiv}CCO_2Et$ (1)	CH_2=$CHCO_2Et$ (1)	CH_2Cl_2, $NaBH_4$	0.90	27
$PhCH_2Br$ (0.5)	*n*-BuSSBu-*n* (3)	PhH, S, 50 min	18.0	14
CH_2=CHCN (1)	CH_2=$CHCO_2Et$ (1)	CH_2Cl_2, $NaBH_4$	0.38	11
CH_2=CHP(O)$(OEt)_2$ (0.5)	Ph_2C=CHI (0.5)	Me_2SO, R, 21 h	8.0	8
CH_2=CHP(O)$(OEt)_2$ (0.5)	PhSSPh (0.5)	Me_2SO, R, 21 h	0.25	11
$HC{\equiv}CCO_2Et$ (1)	CH_2=$CHCO_2Et$ (1)	CH_2Cl_2, $NaBH_4$	0.21	6
CH_2=CHPh (1)	CH_2=$CHCO_2Et$ (1)	CH_2Cl_2, $NaBH_4$	0.15	4.5
CH_2=C(C_6H_4OMe-*p*)$_2$ (1)	CH_2=CPh_2 (1)	CH_2Cl_2, $NaBH_4$	1.20	5
CH_2=CPh_2 (1)	CH_2=CHP(O)$(OEt)_2$ (1)	CH_2Cl_2, $NaBH_4$	0.40	4
CH_2=CPh_2 (0.5)	*n*-BuSSBu-*n* (0.5)	CH_2Cl_2, $NaBH_4$	4.2	3
MeSSMe (1)	Ph_2C=CHI (1)	Me_2SO, S, 30 min	3.0	3
CH_2=CHSOPh (1)	CH_2=CPh_2 (1)	CH_2Cl_2, $NaBH_4$	0.51	2
CH_2=CCl_2 (2)	CH_2=$CHCO_2Et$ (2)	CH_2Cl_2, $NaBH_4$	0.062	2
CH_2=CHSPh (1)	CH_2=CPh_2 (1)	CH_2Cl_2, $NaBH_4$	0.30	1
$(MeO)_3P$ (0.5)	CH_2=CHP(O)$(OEt)_2$ (0.5)	Me_2SO, S, 3 h	0.084	0.8
n-BuSSBu-*n* (1)	Ph_2C=CHI (1)	Me_2SO, S, 30 min	0.80	0.8
$PhCH_2SSCH_2Ph$ (1)	Ph_2=CHI (1)	Me_2SO, S, 30 min	0.60	0.6
$PhC{\equiv}CCO_2Et$ (1)	CH_2=CPh_2 (1)	CH_2Cl_2, $NaBH_4$	0.013	0.4

TABLE IV. (Continued)

Substrate A (mmol)	Substrate B (mmol)	Conditions[a]	k_A/k_B[b]	Rel. React. of A[c]
HC≡CPh (1)	CH_2=$CHCO_2Et$ (1)	CH_2Cl_2, $NaBH_4$	0.20	0.6
HC≡CPh (1)	CH_2=CPh_2 (1)	CH_2Cl_2, $NaBH_4$	0.077	0.3
Me_2C=NO_2K, 18-c-6 (0.5)	CH_2=CPh_2 (0.5)	Me_2SO, S, 2 h	0.066	0.3
Me_2C=NO_2Li (0.5)	*n*-BuSSBu-*n* (0.5)	Me_2SO, S, 2 h	0.30	0.2
i-PrSSPr-*i* (10)	Ph_2C=CHI (1)	Me_2SO, S, 30 min	0.035	4×10^{-2}
t-BuSSBu-*t* (21)	Ph_2C=CHI (0.5)	Me_2SO, S, 2 h	0.0045	5×10^{-3}
Pyridine (12 M)	Me_2C=NO_2K (0.1 M)	Py, S, 6 h	6.6×10^{-5}	2×10^{-6}
Norbornene (1)	CH_2=$CHCO_2Et$ (1)	CH_2Cl_2, $NaBH_4$	<0.02	<0.6
CH_2=CHZCl (1) ; Z = OEt, NH_2, $SiMe_3$	CH_2=$CHCO_2Et$ (1)	CH_2Cl_2, $NaBH_4$	<0.01	<0.3
CH_2=CHZ (1); Z = *t*-Bu, Me_3Si, Me_3Sn	CH_2=CPh_2 (1)	CH_2Cl_2, $NaBH_4$	<0.02	<0.1

[a] The mercurial (0.1 mmol) was reacted with an excess of two substitutes in 3-10 mL of deoxygenated solvent; S, sunlamp irradiation; R, irradiation at 350 mm in a Rayonet Photoreactor. Photostimulated reactions were performed at 35-40 °C. Reaction with $CH_2Cl_2/NaBH_4$ were not irradiated and were carried out at 25 °C following the procedure of Giese, B.; Kretzschmar, G.; Meixner, J. *Chem. Ber.* 1980, 113, 2787. [b] From the ratio of substitution or addition products which accounted for more than 80% of the RHgCl. With disulfides, values of k_A/k_B were calculated by extrapolating the rates of product formation to t = 0. [c] Reactivity of Ph_2C=CHI = 1.0. Towards *t*-Bu•, the absolute rate constant for addition to CH_2=$CHP(O)(OEt)_2$ is 5.9×10^4 $M^{-1}s^{-1}$ at 233 K; Baban, J. A.; Roberts, B. P. *J. Chem. Soc., Perkin Trans. 2* 1981, 161; using E_a = 4 kcal/mol yields 4.8×10^5 $M^{-1}s^{-1}$ and a rate constant for attack of *t*-Bu• upon PhCH=CHI of 4.8×10^4 $M^{-1}s^{-1}$ at 35 °C.

P_B). Under such conditions, the ratio of P_A/P_B is a fair measure of the relativity reactivity of A and B (k_A/k_B). Table IV summarizes competitive reactivity studies towards t-Bu• of Y-Q (Scheme I), PhCH=CHQ (Scheme IIa), π_A (Scheme IIc), and anions (Scheme Vb), as well as reactivities of CH_2=CH(EWG) determined in Scheme IIIb using t-BuHgCl and alkaline $NaBH_4$ in CH_2Cl_2 solution. Extensive relative reactivity data in the latter system has been previously reported using a variety of alkyl radicals[28] as well as the relative reactivities of a variety of anions towards both t-Bu• and $PhCOCH_2$• via the Scheme Vb approach.[20,29]

3. REACTIONS OF 1-ALKENYLMERCURY HALIDES INVOLVING ELECTRON TRANSFER

Monomeric ClHg• formed in a β-elimination process, e.g., from PhĊH-C(H)(R)HgCl, reacts readily with RHgX and Y-A reagents and with some easily oxidized anions, Q^- (e.g., Q^- = $PhSO_2^-$, PhS^-, $(EtO)_2PO^-$, $RP(OR)O^-$), Reactions 3-5.[7,15]

$$\text{ClHg}\bullet + \text{RHgCl} \longrightarrow \text{R}\bullet + \text{HgCl}_2 + \text{Hg}^\circ \qquad (3)$$

$$\text{ClHg}\bullet + \text{Y-Q} \longrightarrow \text{ClHgY} + \text{Q}\bullet \qquad (4)$$

$$\text{ClHg}\bullet + \text{Q}^- \longrightarrow \text{Cl}^- + \text{Hg}^\circ + \text{Q}\bullet \qquad (5)$$

Use of Reactions 3-5 as propagation steps in chain processes are illustrated in Scheme VIII.

SCHEME VIII

(a) $\text{ClHg}\bullet + \text{RHgCl} \longrightarrow \text{R}\bullet + \text{Hg}^\circ + \text{HgCl}_2$

$\text{R}\bullet + \text{PhCH=C(H)HgCl} \longrightarrow \text{Ph}\dot{\text{C}}\text{H-C(H)(R)HgCl}$

$\text{Ph}\dot{\text{C}}\text{H-C(H)(R)HgCl} \longrightarrow \text{PhCH=CHR} + \text{ClHg}\bullet$

(b) $\text{ClHg}\bullet + \text{PhSSPh} \longrightarrow \text{ClHgSPh} + \text{PhS}\bullet$

$\text{PhS}\bullet + \text{PhCH=C(H)HgCl} \longrightarrow \text{Ph}\dot{\text{C}}\text{H-C(H)(SPh)HgCl}$

$\text{Ph}\dot{\text{C}}\text{H-C(H)(SPh)HgCl} \longrightarrow \text{PhCH=CHSPh} + \text{ClHg}\bullet$

(c) $\text{ClHg}\bullet + \text{PhSO}_2^- \longrightarrow \text{PhSO}_2\bullet + \text{Hg}^\circ + \text{Cl}^-$

$\text{PhSO}_2\bullet + \text{PhCH=C(H)HgCl} \longrightarrow \text{Ph}\dot{\text{C}}\text{H-C(H)(SO}_2\text{Ph)HgCl}$

$\text{Ph}\dot{\text{C}}\text{H-C(H)(SO}_2\text{Ph)HgCl} \longrightarrow \text{PhCH=CHSO}_2\text{Ph} + \text{ClHg}\bullet$

Scheme VIIIc involves electron transfer from the anion Q^- = $PhSO_2^-$ to the easily reduced ClHg•. The reaction may involve the formation of QHgCl•⁻ and QHg• although such species are excluded in electron transfer from D• or D•⁻ to RHgCl. This electron transfer step thus plays a key role in the nucleophilic substitution of Reaction 6 (R =

$$RCH{=}C(H)HgCl + Q^- \xrightarrow{h\nu} RCH{=}CHQ + Cl^- + Hg^\circ \qquad (6)$$

alkyl or aryl) which occurs only with easily oxidized anions.[15] The reaction appears similar to the nucleophilic substitution reaction of saturated alkylmercury halides (Scheme Vb), Reaction 7,[16] but occurs

$$RHgCl + Nu^- \longrightarrow RNu + Hg^\circ + Cl^- \qquad (7)$$

by a completely different mechanism although both processes are radical chain reactions involving electron transfer, i.e., ClHg• + Q^- ⟶ Q• + Hg° + Cl^- in Reaction 6; RNu•⁻ + RHgX ⟶ RNu + R• + Hg° + X^- in Reaction 7. With RS^-, ligand exchange occurs readily with RCH=C(H)HgCl to give RCH=C(H)HgSPh which decomposes by the chain process of Scheme IV.[15] This process, in a way, involves an electron transfer reaction in the decomposition of PhSHg• to PhS• and Hg°, products which are formally the result of electron transfer between PhS^- and Hg^+.

The reaction of ClHg• with RHgCl seems to occur by a concerted process not involving RHg•. Thus, the relative reactivity of *t*-BuHgCl:*c*-C_6H_{11}HgCl:*n*-BuHgCl towards ClHg• are 1.0:0.011:0.0001 (Table I).[3] Furthermore, by the nitroxide inhibition method, the initial k.c.l. for the reaction of PhCH=C(H)HgCl with 3 equivalents of *t*-BuHgCl is ~60 (Table III). The reaction between ClHg• and RHgCl appears to involve the reduction of ClHg• to Cl^- and Hg° with oxidation of RHgCl to R• and $HgCl^+$.

Photolysis of mixtures of PhCH=C(H)HgCl or Ph_2C=C(H)HgCl with *t*-BuMgCl forms PhCH=CHBu-*t* or Ph_2CH=CHBu-*t* under conditions where PhCH=C(H)HgBu-*t* or Ph_2C=C(H)HgBu-*t* cannot be detected (PhCH=CHMgCl and *t*-BuHgCl readily react to form the isolable PhCH=C(H)HgBu-*t*).[26] Apparently, Reaction 8 occurs. Reaction of an excess of a 1:1 mixture of *t*-BuMgCl and *n*-BuHgCl with (*E*)-PhCH=C(H)HgCl yields only PhCH=CHBu-*t* (*E*/*Z* >50) upon photolysis.

$$ClHg\bullet + t\text{-}BuMgCl \longrightarrow t\text{-}Bu\bullet + MgCl_2 + Hg^\circ \qquad (8)$$

Mercury(II) salts (HgQ_2) will also react with D• (e.g., Bu_3Sn•), electron accepting heteroatom-centered radicals or atoms (e.g., I•)

or ClHg• to produce Q• = PhS•, $PhSO_2$•, $(EtO)_2PO$•, Reactions 9-12.[7-9,27] The resulting Q• will regenerate I•, Bu_3Sn• or ClHg• by

$$Bu_3Sn\bullet + HgQ_2 \longrightarrow Bu_3SnQ + Hg^\circ + Q\bullet \quad (9)$$

$$I\bullet + HgQ_2 \longrightarrow IHgQ + Q\bullet \quad (10)$$

$$ClHg\bullet + HgQ_2\bullet \longrightarrow ClHgQ + Hg^\circ + Q\bullet \quad (11)$$

the homolytic addition-elimination sequence involving $PhCH{=}CHSnBu_3$, PhCH=CHI or PhCH=C(H)HgCl. In the case of $(RCO_2)_2Hg$, attack of I• forms RCO_2• which decarboxylates to yield R•, Reaction 12.[8]

$$PhCH{=}CHI + (RCO_2)_2Hg \xrightarrow{h\nu} PhCH{=}CHR + RCO_2HgI + CO_2 \quad (12)$$

4. CONVERSION OF ADDUCT RADICALS TO DONOR OR ACCEPTOR RADICALS

Powerful electron-accepting radicals or atoms react readily with R_2Hg or RHgCl by S_H2 attack at Hg (Reaction 1) while powerful electron-donating species react by electron transfer to RHgX (Reaction 2). The alkylmercurials are thus excellent sources of alkyl radicals for reactivity studies or for synthesis. Unfortunately, in many cases, trapping of an alkyl radical by a good radicaphile leads to an adduct radical which fails to react readily with RHgX because of the lack of polar character or perhaps steric effects. Loss or gain of a proton will increase the electron-donating or electron-accepting ability of the adduct radical, e.g., Scheme IX.[19] However, alkylmercurials are

SCHEME IX

$$R\bullet + C_5H_5NH^+ \longrightarrow o\text{- and } p\text{-}RC_5H_5NH\bullet^+$$

$$RC_5H_5NH\bullet^+ \rightleftharpoons RC_5H_4NH\bullet \text{ (pyridinyl)} + H^+$$

$$RC_5H_4NH\bullet + RHgX \longrightarrow o\text{- and } p\text{-}RC_5H_4NH^+ X^- + R\bullet + Hg^\circ$$

often unstable in the presence of the strong acids or bases required for protonation or deprotonation of many adduct radicals (e.g., $RCH_2\dot{C}Ph_2$).

The polar requirements for Reactions 1 and 2 are nicely illustrated using 1,1-diarylethylenes as the radicaphiles for *t*-Bu•.[22] With the electron-rich 1,1-di-*p*-anisylethylene, the adduct radical is a donor (Table II), and electron transfer to *t*-BuHgCl occurs via Reaction 2. With 1,1-di-*p*-nitrophenylethylene, the adduct radical is an acceptor, and a chain reaction with *t*-BuHgCl occurs to yield after hydrolysis t-BuCH$_2$CH(C$_6$H$_4$NO$_2$-p)$_2$. Here, reaction of the adduct radical with t-BuHgCl may well involve electron transfer from the mercurial (Reaction 13) rather than S_H2 substitution at mercury

$$t\text{-BuCH}_2\dot{C}(C_6H_4NO_2\text{-}p)_2 + t\text{-BuHgCl} \longrightarrow$$
$$t\text{-BuCH}_2C(C_6H_4NO_2\text{-}p)_2^- + t\text{-Bu}\bullet + \text{ClHg}^+ \qquad (13)$$

(Reaction 1). However, with CH$_2$=CPh$_2$ the adduct radical t-BuCH$_2\dot{C}$Ph$_2$ has little driving force for electron donation or acceptance. Photochemical reaction with t-BuHgCl forms t-BuCH$_2$CHPh$_2$ and t-BuCH=CPh$_2$ in approximately a 1:1 ratio in the nonchain sequence of Scheme X.

SCHEME X

$$t\text{-BuHgCl} \xrightarrow{h\nu} t\text{-Bu}\bullet + \text{ClHg}\bullet$$
$$\text{ClHg}\bullet + t\text{-BuHgCl} \longrightarrow t\text{-Bu}\bullet + \text{HgCl}_2 + \text{Hg}^\circ$$
$$t\text{-Bu}\bullet + \text{CH}_2\text{=CPh}_2 \longrightarrow t\text{-BuCH}_2\dot{C}\text{Ph}_2$$
$$2\ t\text{-BuCH}_2\dot{C}\text{Ph}_2 \longrightarrow t\text{-BuCH}_2\text{CHPh}_2 + t\text{-BuCH=CPh}_2$$

A chain reaction reaction between CH$_2$=CPh$_2$ and t-BuHgCl can be achieved by the introduction of another propagation step which leads to either a donor (e.g., Bu$_3$Sn•) or acceptor (e.g., PhS•) species. Reaction of t-BuCH$_2\dot{C}$Ph$_2$ with Bu$_3$SnH forms t-BuCH$_2$CHPh$_2$ and recycles the reaction via Reaction 2 with D• = Bu$_3$Sn•. Reaction with PhSH will recycle the chain sequence via Reaction 1 (A• = PhS•), but now the intermediate PhS• will also add to CH$_2$=CPh$_2$ in competition with attack upon t-BuHgCl.

Another approach to finesse an unreactive adduct radical to participate in a chain reaction with RHgCl involves the reaction of the adduct radical with an electron transfer agent which upon loss or gain of an electron forms a donor or acceptor species now capable of undergoing Reaction 1 or 2. Iodide ion appears to react in this manner in reactions where the adduct radical is formed from α,β-unsaturated ketones or esters. Thus, although the chain reactions of CH$_2$=CHZ with excess RHgCl occurs readily via Scheme IIc with Z = (EtO)$_2$PO, PhSO$_2$ or p-O$_2$NC$_6$H$_4$ (to yield RCH$_2$CH(HgCl)Z), the reactions are poor with Z = MeCO or EtO$_2$C (the acrylates yield nearly exclusively telomers).[3] However, in the presence of NaI in Me$_2$SO, the photostimulated reactions occur readily to form t-BuCH$_2$CH$_2$Z with Z = C(O)Me or CO$_2$Et (Table V), particularly when MeOH is added as a proton donor.[26] Since telomers are not observed in the presence of I$^-$, the adduct radical (t-BuCH$_2\dot{C}$HZ) must be trapped before it can add

TABLE V. Photostimulated Reactions of *t*-BuHgX with CH_2=CH(EWG) to give *t*-BuCH_2CH_2(EWG) upon Workup

Substrate (mmol)	% Yield (equiv *t*-BuHgX; equiv NaI, hν, time)[a]		
	t-BuHgCl/Me_2SO[b]	*t*-BuHgCl/NaI/Me_2SO[c]	*t*-BuHgI/Me_2SO-MeOH[c,d]
CH_2=CHSO_2Ph (0.2)	87 (4; 0; S; 24 h)	95 (2; 4; R; 24 h)[d]	98 (2; 0; R; 2 h)
CH_2=CHSO_2Ph (0.2)	39 (1; 0; R; 4 h)	85 (1; 2; R; 4 h)[d]	---
CH_2=CHP(O)$(OEt)_2$ (0.4)	98 (4; 0; S; 2 h)	---	98 (2; 0; R; 3 h)
CH_2=CHP(O)$(OEt)_2$ (0.4)	30 (1; 0; S; 2 h)	86 (1; 2; R; 2 h)[d]	98 (2; 4; R; 3 h)
CH_2=CHCO_2Et (0.4)	5.0 (2; 0; R; 10 h)[e]	80 (1; 2; R; 6 h)[d,f]	88 (2; 0; R; 2 h)[g]
CH_2=CHC(O)Me (0.4)	6.5 (2; 0; R; 10 h)	85 (2; 2; R; 6 h)[d]	70 (2; 0; R; 2 h)
2-cyclohexenone (0.2)	35 (2; 0; R; 10 h)	85 (2; 4; R; 6 h)[d]	82 (2; 0; R; 2 h)
2,3-dichloromaleic anhydride (0.2)	27 (5; 0; S; 8 h)[h]	98 (4; 8; S; 5 h)[h]	---
(*E*)-PhCH=CHSO_2Ph (0.1)	43 (5; 0; S; 24 h)[i]	73 (5; 10; S; 24 h)[d,i]	---
	16 (5; 0; S; 24 h)[j]	27 (5; 10; S; 24 h)[d,j]	---
(*E*)-PhCOCH=CHCl (0.1)	68 (5; 0; S; 2 h)[k]	100 (5; 10; S; 1 h)[k]	---
(*E*)-PhCOCH=CHCl (0.1)	<10 (5; 0; S; 18 h)[l]	62 (5; 2.5; S; 2 h)[l]	---

[a] Substrate in 1-10 mL of deoxygenated solvent at 35-40 °C: S, 275 W sunlamp ca. 20 cm from Pyrex reaction vessel; R, Rayonet Photoreactor, 350 nm. [b] Workup by $NaBH_4$ or H_3O^+. [c] Workup with 1% hydrochloric acid or $Na_2S_2O_3$. [d] Me_2SO (60%) - MeOH (40%). [e] Major product was *t*-BuCH_2CH(CO_2Et)$CH_2CH_2CO_2$Et. [f] Product was *t*-Bu$CH_2CH_2CO_2$Me and no telomers were observed. [g] Me_2SO (60%) - EtOH (40%). [h] Product was 2,3-di-*t*-butylmaleic anhydride. [i] Substitution product, PhCH=CHBu-*t* (*E/Z* >50). [j] Addition product, PhCH(Bu-*t*)CH_2SO_2Ph. [k] Product was (*E*)-PhCOCH=CHBu-*t*. [l] *i*-PrHgCl yielding PhCOCH=CHPr-*i*, *E/Z* = 32.

to another molecular of CH_2=CHZ. The mechanism of Scheme XI is suggested.

SCHEME XI (Z = $COCH_3$, CO_2Et, $PhSO_2$, $(EtO)_2PO$

$$t\text{-Bu}\bullet + CH_2{=}CHZ \longrightarrow t\text{-BuCH}_2\dot{C}HZ$$

$$t\text{-BuCH}_2\dot{C}HZ + I^- \rightleftharpoons t\text{-BuCH}_2CHZ^- + I\bullet$$

$$t\text{-BuCH}_2CHZ^- + H^+ \longrightarrow t\text{-BuCH}_2CH_2Z$$

$$I\bullet + t\text{-BuHgI} \longrightarrow HgI_2 + t\text{-Bu}\bullet$$

Photostimulated reaction of CH_2=CPh_2 with *t*-BuHgCl in Me_2SO in the presence of $MeC(CO_2Et)_2^-$ gives a high yield of products derived from *t*-$BuCH_2CPh_2^-$ (e.g., *t*-$BuCH_2C(Ph)_2Bu$-*t* via Scheme Vb). Here, also, electron transfer is suggested (Reaction 14) to yield a new acceptor radical capable of reacting readily with *t*-BuHgCl.[22]

$$t\text{-BuCH}_2\dot{C}Ph_2 + MeC(CO_2Et)_2^- \rightleftharpoons t\text{-BuCH}_2CPh_2^- + Me\dot{C}(CO_2Et)_2 \quad (14)$$

5. ACKNOWLEDGMENT

It is a pleasure to acknowledge the contributions, many of which are unpublished, of my collaborators, J. Hershberger, H. Tashtoush, P. Ngoviwatchai, W. Jiang, S. Hu, R. K. Khanna, S. Herron, D. Guo, Y.-W. Wu and A. Pla-Dalmau. Financial support has been provided by the National Science Foundation and The Petroleum Research Fund.

6. REFERENCES

(1) Larock, R. C. *Organomercury Compounds in Organic Syntheses*, Springer-Verlag: Heidelberg and New York, 1985.
(2) Russell, G. A.; Tashtoush, H. *J. Am. Chem. Soc.* 1983, **105**, 1398.
(3) Russell, G. A.; Jiang, W.; Hu, S. S.; Khanna, R. K. *J. Org. Chem.* 1986, **51**, 5498.
(4) Winstein, S.; Traylor, T. G. *J. Am. Chem. Soc.* 1956, **78**, 2597.
(5) Jensen, F. R.; Gale, L. H. *J. Am. Chem. Soc.* 1960, **82**, 148. Jensen, F. R.; Gale, L. H. Rogers, J. E. *J. Am. Chem. Soc.* 1968, **90**, 5793.
(6) Nugent, W. A.; Kochi, J. K. *J. Organomet. Chem.* 1977, **124**, 349.

(7) Russell, G. A.; Tashtoush, H.; Ngoviwatchai, P. *J. Am. Chem. Soc.* 1984, **106**, 4622.
(8) Russell, G. A.; Ngoviwatchai, P. *Tetrahedron Lett.* 1985, **26**, 4975.
(9) Russell, G. A.; Ngoviwatchai, P. *Tetrahedron Lett.* 1986, **27**, 3479.
(10) Bordwell, F. G.; Douglas, B. *J. Am. Chem. Soc.* 1966, **88**, 993.
(11) Brown, H. C.; Geoghegan, P. J. *J. Am. Chem. Soc.* 1967, **89**, 1522. Brown, H. C.; Rei, M.-H. *J. Am. Chem. Soc.* 1969, **91**, 5646.
(12) Whitesides, G. M.; San Filippo, J. *J. Am. Chem. Soc.* 1970, **92**, 6611.
(13) Giese, B.; Meister, J. *Angew. Chem.* 1977, **89**, 178; 1979, **91**, 167.
(14) Russell, G. A. *Preprints of Papers, ACS Div. of Petroleum Chem.* 1986, **31**, 891.
(15) Russell, G. A.; Hershberger, J. *J. Am. Chem. Soc.* 1980, **102**, 7603.
(16) Russell, G. A.; Hershberger, J.; Owens, K. *J. Am. Chem. Soc.* 1979, **101**, 1312.
(17) Russell, G. A. Hershberger, J. Owens, K. *J. Organomet. Chem.* 1982, **225**, 43.
(18) Russell, G. A.; Guo, D.; Khanna, R. K. *J. Org. Chem.* 1985, **50**, 3423.
(19) Russell, G. A.; Khanna, R. K. *J. Am. Chem. Soc.* 1985, **107**, 1450.
(20) Russell, G. A.; Khanna, R. K. *Tetrahedron* 1985, **41**, 4133.
(21) Russell, G. A.; Khanna, R. K. *Phosphorus and Sulfur* 1987, **29**, 271.
(22) Russell, G. A.; Khanna, R. K.; Guo, D. *Chem. Commun.* 1986, 632.
(23) Quirk, R. P. *J. Org. Chem.* 1972, **37**, 3554.
(24) Bloodworth, A. J.; Courtneidge, J. *J. Chem. Soc., Perkin 1* 1982, 1797.
(25) Kurusawa, H.; Okada, H.; Yusuda, M. *Tetrahedron Lett.* 1980, **21**, 959. Kurusawa, H.; Okada, H.; Hattori, T. *Tetrahedron Lett.* 1981, **22**, 4495.
(26) Unpublished observations.
(27) Russell, G. A. Ngoviwatchai, P.; Tashtoush, H.; Hershberger, J. *Organometallics* 1987, **6**, 1414.
(28) Giese, B. *Angew. Chem. Int. Ed. Engl.* 1983, **22**, 753.
(29) Russell, G. A.; Khanna, R. K in *Nucleophilicity, Adv. in Chem. Ser., Am. Chem. Soc.* 1987, **215**, 355.

REDOX CATALYSIS AND ELECTRON-TRANSFER PROCESSES IN SELECTIVE ORGANIC SYNTHESES

Francesco Minisci*, Elena Vismara and
Francesca Fontana
Dipartimento di Chimica, Politecnico di Milano
Piazza Leonardo da Vinci, 32
20133 Milano
Italy

ABSTRACT - Redox catalysis and electron-transfer processes play an important role in selective organic syntheses by free-radicals.

Four new general, simple and cheap sources of alkyl radicals from alkyl iodides have been developped. They are based on the iodine abstraction from alkyl iodides by the methyl radical, which is generated from acetone-H_2O_2, dimethyl sulphoxide-H_2O_2, t-BuOOH or MeCOOH and $S_2O_8^{2-}$. These new radical sources have allowed to develop a large variety of selective reactions by nucleophilic and electrophilic carbon-centered radicals with aromatics, olefins, diazonium salts, quinones etc.

Moreover, hydroxy and alkoxy radicals generated in redox processes have been utilized for simple and cheap syntheses of industrial interest, such as hydroxylation of phenol, hydroxymethylation, formylation, carbamoylation of heterocyclic aromatics respectively by methanol, formaldehyde, formamide.

An overlap area in substitutions by ionic nucleophiles and nucleophilic radicals is related to the HOMO-LUMO and SOMO-LUMO interactions.

The meaning of the base catalysis in the selective oxidation of p-methyl phenols to the corresponding aldehydes by molecular oxygen is discussed.

The redox catalysis by metal salts has large synthetic involvements also from industrial point of view. On the other hand electron-transfer processes can give high selectivity when the redox characteristics of the starting compounds and of the reaction products are different.

M. Chanon et al. (eds.), Paramagnetic Organometallic Species in Activation / Selectivity, Catalysis, 29–60.

The catalysis is particularly interesting when effective redox chains can be established[1].

New synthetic developments, based on these concepts, form the subject of this paper.

Reactions involving oxy radicals

Hydrogen peroxide is one of the most unexpensive, simple and non-polluting reagents available in the chemical industry. Its use for selective organic syntheses is therefore of undoubted interest. It is known from long time that the hydroxyl radical can be easily generated from H_2O_2 by redox decomposition[2] (eq 1)

$$H_2O_2 + M^+ \longrightarrow HO\cdot + HO^- + M^{2+} \qquad (1)$$

However the hydroxyl radical is a very reactive and unselective species in its reactions in agreement with the Reactivity-Selectivity Principle: the rate constants of most of the reactions with organic and inorganic compounds are included in the range of 10^7-10^{10} $M^{-1}s^{-1}$. The enthalpic factor is one of the main reasons of these high rates. Thus the hydrogen abstraction from C-H bonds (eq 2) is always an exothermic process, due to the high energy of the bond HO-H, and, unless strong polar effects are working, the process is quite unselective[2].

$$R\text{-}H + \cdot OH \longrightarrow R\cdot + H_2O \qquad (2)$$

Similarly the addition of the hydroxyl radical to unsaturated systems are often exothermic, fast and unselective[2].

Moreover the high electron affinity of the hydroxyl radical makes fast the electron-transfer oxidation of a variety of inorganic cations[3] and anions[4] (i.e. eqs 3 and 4).

$$HO\cdot + M^{n+} \xrightarrow{k} HO^- + M^{(n+1)+} \qquad (3)$$

M^{n+}	log k
Ce^{3+}	8.3
Mn^{2+}	8.2

$$HO\cdot + X^- \xrightarrow{k} HO^- + X\cdot \qquad (4)$$

X^-	log k
NO_2^-	9.5
N_3^-	9.8
CNS^-	9.1
Cl^-	8.6
Br^-	10.5

The selective utilization of the hydroxyl radical in organic synthesis is therefore related to two conditions:

(a) The use of the hydroxyl radical as selective precursor of a much more versatile and selective radical.

(b) Reactions with simple molecules, which have no problem of regioselectivity, and can be in practical way used in large excess in order to control the chemoselectivity.

We have utilized both these conditions to develop selective syntheses.

The recent years have brought a rapid development in the use of alkyl radicals for the formation of C-C bonds and in the synthesis of target molecules[5]. The availability of general, simple, cheap and selective sources of alkyl radicals is therefore of great synthetic interest.

We[6] have recently developped selective homolytic alkylation of heteroaromatic compounds and diazonium salts by using as radical source the iodine abstraction from alkyl iodides by aryl radicals, a very fast and selective reaction (eq 5)

$$R{-}I + Ar^{\bullet} \xrightarrow{k} R^{\bullet} + I{-}Ar \qquad k \sim 10^{9}\ M^{-1}s^{-1} \qquad (5)$$

The Scheme 1 shows the mechanism of the alkylation of heteroaromatic bases by using benzoyl peroxide and alkyl iodides

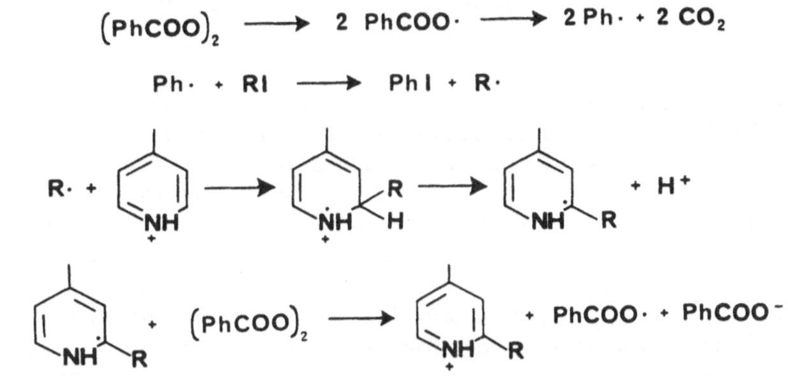

Scheme 1

The overall stoichiometry of the reaction is shown by eq 6

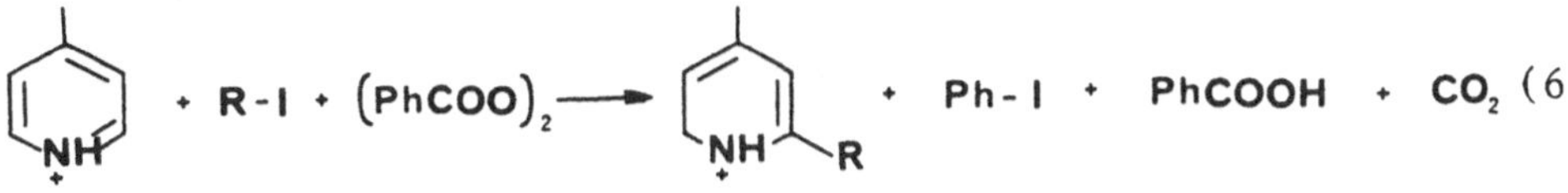 (6)

In the Scheme 2 a similar reaction is shown with diazonium salts as radical sources

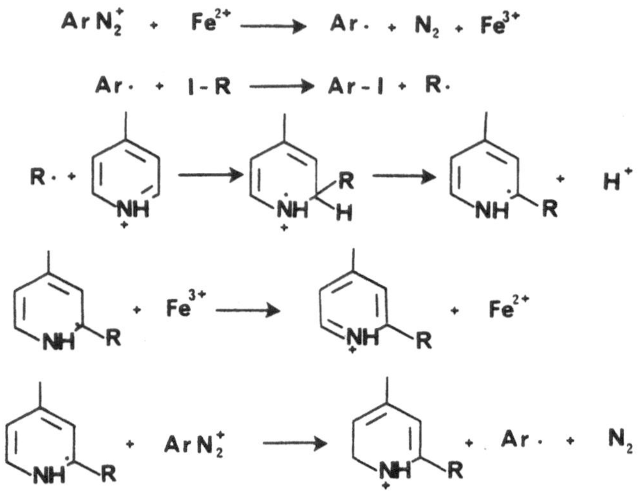

Scheme 2

Eq 7 shows the overall stoichiometry.

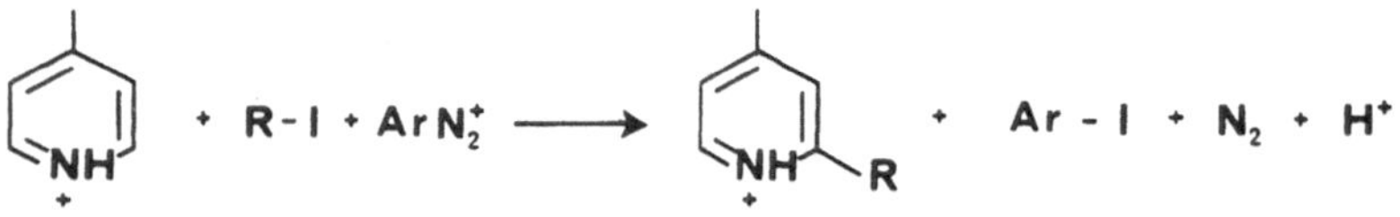

(7)

Even if both reactions (eqs 6 and 7) can have synthetic applications, there are several structural limitations. Thus aroyl peroxides are very useful as initiators of free-radical chains, but their use in stoichiometric amounts is less practical, above all for the preparation of large amounts of products. Moreover, the reaction does not work with t-alkyl iodides because it does not lead to t-alkyl radicals. A further limitation occurs with compounds with weak C-H bonds or unsaturated groups with high electron availability. In these cases the iodine abstraction does not occur because the reaction of the aroyloxy radicals with the substrates (eqs 8 and 9) are faster than the decarboxylation (eq 10)

$$ArCOO\cdot + H\text{-}R \longrightarrow ArCOOH + R\cdot \tag{8}$$

$$ArCOO\cdot + \overset{|}{\underset{|}{C}} = \overset{|}{\underset{|}{C}} \longrightarrow Ar\text{-}COO\text{-}C\text{-}C\cdot \tag{9}$$

$$ArCOO\cdot \longrightarrow Ar\cdot + CO_2 \tag{10}$$

The main limitations with which diazonium salts are due to their low stability and to the high rate of addition of nucleophilic alkyl radicals to the diazonium group, which competes with the heterocyclic substitution and under suitable conditions leads to the "free-radical diazocoupling" (eq 11 and Scheme 3)[7]

$$R\text{-}I + 2ArN_2^+ + 2Ti^{3+} \longrightarrow R\text{-}N=N\text{-}Ar + ArI + N_2 + 2Ti^{4+} \quad (11)$$

$$ArN_2^+ + Ti^{3+} \longrightarrow Ar\cdot + N_2 + Ti^{4+}$$

$$Ar\cdot + R\text{-}I \longrightarrow Ar\text{-}I + R\cdot$$

$$R\cdot + N\equiv\overset{+}{N}\text{-}Ar \longrightarrow R\text{-}N=\underset{\bullet}{\overset{+}{N}}\text{-}Ar$$

$$R\text{-}N=\underset{\bullet}{\overset{+}{N}}\text{-}Ar + Ti^{3+} \longrightarrow R\text{-}N=N\text{-}Ar + Ti^{4+}$$

Scheme 3

This competition can be in part overcome by keeping low the stationary concentration of the diazonium salt during the reaction. A further limitation with both radical sources occurs with substrates, as quinones, very reactive towards aryl radicals: the addition of the aryl radical to the quinone competes with the iodine abstraction.

Thus more general, simple and cheap sources of alkyl radicals from alkyl iodides would have been of undoubted interest. In pursuing this aim we have found of great interest and open to important synthetic involvements the recent kinetic and thermodynamic results reported by Griller[8] for the reaction of alkyl radicals with alkyl icdide (eq 12)

$$R\cdot + R'\text{-}I \rightleftharpoons R\text{-}I + R'\cdot \quad (12)$$

The rate constants for the forward and back reactions are normally $>10^6\ M^{-1}s^{-1}$ and the equilibrium constants are strongly affected by the stability of the alkyl radicals, as the results of Table I indicate.

The strength of the C-I bonds, ranging from 56.5 kcal/mol for Me-I to 52.1 kcal/mol for t-Bu-I governs the equilibria. The methyl radical is the least stable among the alkyl radicals so that it was in principle possible to use a source of methyl radical to generate alkyl radicals of any kind according to the equilibria of Table I.

Following this idea we have developped four new general, simple and cheap sources of alkyl radicals from alkyl iodides, useful for a variety of selective reactions, which we have developped in recent years by different sources. In all cases oxy radicals were used as precursors of the methyl radical, which abstracts iodine from alkyl iodides.

TABLE I - Equilibrium constants[8], K, for the reaction

$$CH_3\cdot + R{-}I \rightleftharpoons CH_3{-}I + R\cdot$$

R•	K
Et•	20.1
n-Pr.	14.6
i-Pr.	468
s-Bu.	245
t-Bu.	1.7×10^4

Alkylation of protonated heteroaromatic bases

The substitution of protonated heteroaromatic bases by nucleophilic carbon-centered radicals is a general reaction of great synthetic interest; it reproduces most of the numerous aspects of the Friedel-Crafts aromatic substitution, but with opposite reactivity and selectivity[9]. Four new sources of alkyl radicals proved to be very effective for the heterocyclic substitution: i) Hydrogen peroxide reacts with alkyl iodides and protonated heteroaromatic bases in DMSO in the presence of catalytic amount of Fe(II) salt leading to a highly selective alkylation of the heteroaromatic ring. Eq 13 shows the stoichiometry of the reaction

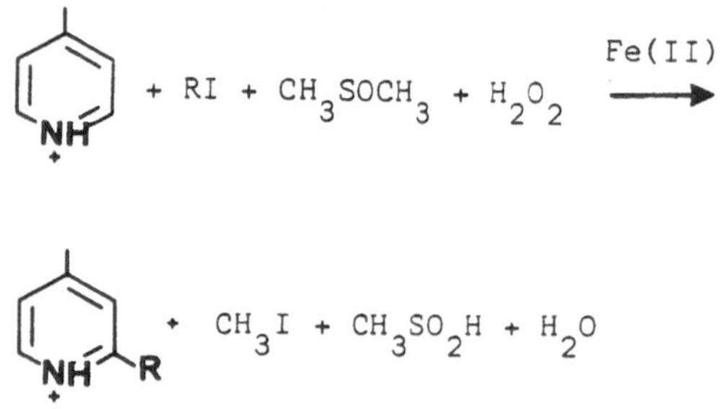

(13)

A complex, but selective, redox chain is working. The first interaction involves the well-known redox decomposition of H_2O_2 by Fe(II) salt[2] (eq 14)

$$H_2O_2 + Fe^{2+} \longrightarrow HO\cdot + HO^- + Fe^{3+} \qquad (14)$$

The high reactivity and low selectivity of the hydroxyl radicals with a large variety of organic and inorganic compounds is controlled by using DMSO as solvent. The hydroxyl radical reacts fast with DMSO (eq 15) and the possible, competitive and unselective reactions with other substrates are minimized by the excess of solvent.

$$HO\cdot + CH_3\text{-}SO\text{-}CH_3 \longrightarrow CH_3 - \overset{HO\ \ O\cdot}{\overset{\diagdown\ \diagup}{S}} - CH_3 \qquad (15)$$

The radical adduct selectively undergoes β-scission, acting thus as selective source of methyl radical (eq 16)

$$CH_3 - \overset{HO\ \ O\cdot}{\overset{\diagdown\ \diagup}{S}} - CH_3 \longrightarrow CH_3SO_2H + CH_3\cdot \qquad (16)$$

The iodine abstraction by the methyl radical (eq 17) is fast and successfully competes with other possible reactions of the same radical. It occurs according to the equilibria reported in Table I.

$$CH_3\cdot + R\text{-}I \rightleftharpoons CH_3\text{-}I + R\cdot \qquad (17)$$

However, the fact that the equilibrium of eq 17 are shifted at right is not in itself a sufficient condition to have a high selectivity because the reaction rates of the $CH_3\cdot$ and $R\cdot$ radicals can be quite different. Thus when the enthalpic factor governs the reactivity, the methyl radical is more reactive than primary, secondary, tertiary and generally α-substituted alkyl radicals; that can counterbalance the unfavourable equilibria.

The radical source can become selective when polar effects are important. In the absence of polar substituents the nucleophilic character increases from methyl to primary, secondary and tertiary alkyl radicals; polar substitutents can increase or decrease the nucleophilic character of the alkyl radicals depending on their electron-releasing or electron-withdrawing behaviour[10].

Now the addition rates of the alkyl radicals to protonated heteroaromatic bases are strongly affected by the nucleophilic character of the radicals[9,10] (eq 18)

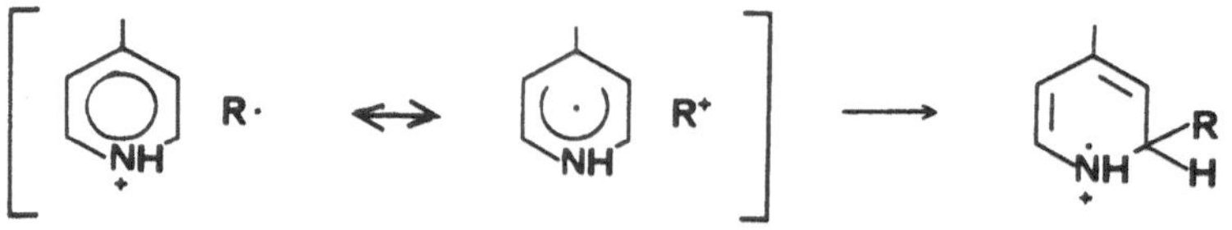 (18)

When the radical R• is more nucleophilic than the methyl radical, the faster addition to the hetorocyclic ring (eq 18) together with the favourable equilibrium (eq 17) makes the overall reaction highly selective and only the radical R. is involved in the substitution. The redox chain is particularly effective, due to the high oxidability of the pyridinyl radical intermediate (eqs 19 and 20)

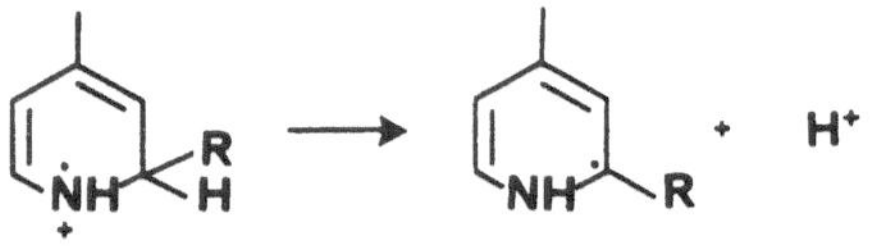 (19)

 (20)

Also in these steps (eqs 19 and 20) the thermodynamic data reported by Griller[8] for the amino radical cations and the α aminoalkyl radicals strongly contribute to understand the effectiveness of the redox chain. Griller has in fact estimated the enthalpy change for the loss of proton by trimethyl-amino radical cation (eq 21) in about 6 kcal/mol^{-1}

$$(CH_3)_3\overset{+}{N}\cdot \longrightarrow (CH_3)_2N\text{-}CH_2\cdot + H^+ \qquad (21)$$

The enthalpy change for the similar eq 19 must be even more favourable, considering the higher stability of the pyridinyl radical. Moreover, the ionization energies for the α-aminoalkyl radicals (5.4-5.7 eV) are more than 2 eV lower than those for simple alkyl radicals; they are similar to those for lithium (5.39 eV) and sodium (5.14 eV) which explain why the pyridinyl radical behaves as potent reducing agent towards the Fe(III) salt (eq 20).

Thus this new method of alkylation of heterocyclic compounds is particularly useful because of the large range of application (practically all the primary, secondary and tertiary alkyl radicals without electron-withdrawing groups bonded or conjugated to the radical center can be easily generated according to the eq 17 and successfully utilized for the heteroaromatic substitution).

The selectivity can be further or increased, particularly with primary alkyl radicals, which have less favourable

equilibria (Table I) and less marked nucleophilic character, by using an excess of alkyl iodide. Moreover, the experimental conditions are extremely simple (the reaction is completed in few minutes at room temperature), a large variety of alkyl iodides and heteroaromatic bases are easily available and utilizable, the reagents and the catalyst are cheap.
ii) The redox decomposition of t-BuOOH (eq 22) is another simple and cheap source of methyl radical, useful to generate alkyl radicals from alkyl iodides and to alkylate heteroaromatic bases.

$$Fe^{2+} + (CH_3)_3COOH \longrightarrow Fe^{3+} + OH^- + (CH_3)_3C\text{-}O\cdot \xrightarrow{k_d} CH_3\cdot + CH_3COCH_3 \qquad (22)$$

In order to make effective this source of methyl radical, two main competitive reactions of t-Bu-O. must be minimized: the hydrogen abstraction from C-H bonds in the reacting system (eq 23) and the reduction by Fe(II) salt (eq 24)

$$t\text{-BuO}\cdot + R\text{-}H \xrightarrow{k_a} t\text{-BuOH} + R\cdot \qquad (23)$$

$$t\text{-BuO}\cdot + Fe^{2+} + H^+ \longrightarrow t\text{-BuOH} + Fe^{3+} \qquad (24)$$

To reduce the incidence of the hydrogen abstraction (eq 23) we have taken advantage of solvent and temperature effects. Many years ago Walling[11] has shown the importance of the solvent and the temperature in the competition between hydrogen abstraction (k_a) (eq 23) and decomposition (k_d) (eq 22) of t-BuO• . Some data concerning the hydrogen abstraction from cyclohexane are reported in Table II.

TABLE II - k_a/k_d - Solvent effect

T °C	100°	70°	40°	25°	0°
C_2Cl_4	4.14	11.1	39.9	87.8	293
Benzene	2.82	7.62	24.7	48.6	207
MeCOOH	<1	1.34	2.9	4.87	12.6

These data suggested that refluxing acetic acid should be particularly suitable because, in addition to solvent and temperature effects (Table II), the hydrogen abstraction from acetic acid is a relatively slow process for polar reasons compared with hydrogen abstraction from cyclohexane.

To minimize the reduction of t-BuO• (eq 24) it was important to keep low the steady-state concentration of Fe(II) salt in the reacting system. That was achieved by using catalytic amounts of Fe(III) acetate. In this way a simple, cheap and effective source of methyl radical was obtained and the stoichiometry of the alkylation of the heteroaromatic bases is shown by eq 25

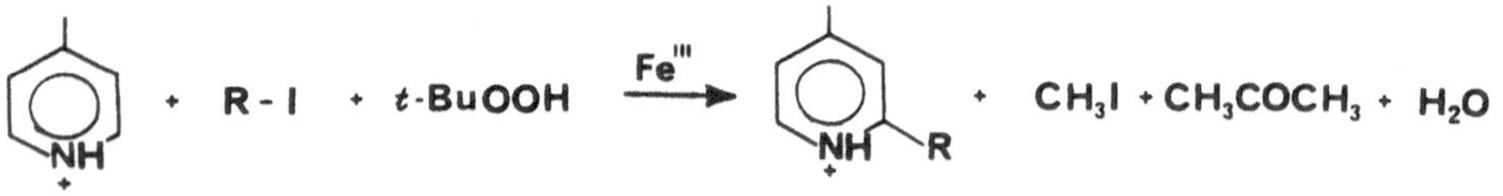
(25)

The redox chain is identical to that above described with DMSO and H_2O_2 (i); the only difference is the source of methyl radical (eq 22).

The initiation step of the redox chain appears to be due to reaction 26

$$t\text{-BuOOH} + Fe^{3+} \longrightarrow t\text{-BuOO}\cdot + Fe^{2+} + H^{+} \qquad (26)$$

This step is relatively slow compared with the steps of the redox chain leading to the heteroaromatic alkylation (eqs 17-20 and 22) so that the stationary concentration of Fe(II) salt remains very low during the reaction. No reaction occurs at room temperature in the presence of Fe(III) catalyst because the reaction 26 is too slow. No reaction occurs in refluxing acetic acid in the absence of Fe(III) salt indicating that the thermal decomposition of the peroxide has not a significant role in the initiation step because the peroxide is able to sustain a radical chain by oxidizing the pyridinyl radical intermediate (see procedure iii).

Also this radical source has a general character to generate alkyl radicals of any kind from alkyl iodides, useful for the heteroaromatic substitution.

iii) The thermal decomposition of H_2O_2 in acetone and catalytic amounts of acids proved to be another simple and cheap source of methyl radical, which has been utilized to generate alkyl radicals from alkyl iodides. Also this source is useful for the heteroaromatic substitution and the stoichiometry of the alkylation is shown in eq 27

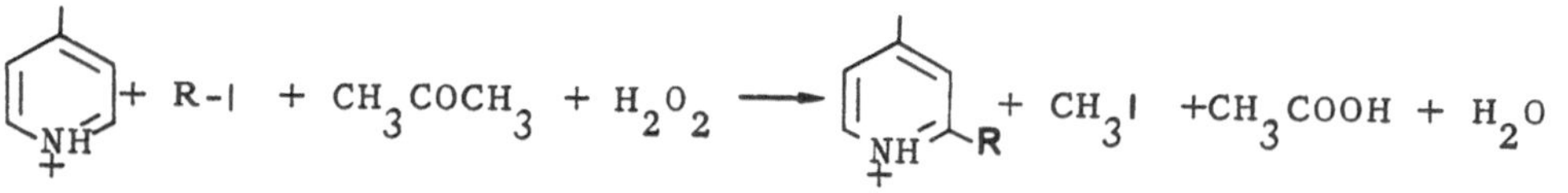
(27)

In this case the catalysis by metal salts is harmful because the decomposition of H_2O_2 takes place and no substitution occurs.

We explain this behaviour by the equilibria 28 and 29 between H_2O_2 and acetone.

$$CH_3COCH_3 + H_2O_2 \underset{}{\overset{H^+}{\rightleftharpoons}} CH_3-C(OH)(OOH)-CH_3 \qquad (28)$$

$$CH_3-C(OH)(OOH)-CH_3 + CH_3COCH_3 \overset{H^+}{\rightleftharpoons} CH_3-C(CH_3)(OH)-O-O-C(CH_3)(OH)-CH_3 \qquad (29)$$

Depending on the ratio between H_2O_2 and acetone, the amount of water and the acid catalysis other peroxides can be formed, whose structures are, however, always characterized by the $Me_2C(O-)-O-O-(-O)CMe_2$ moiety.

Now H_2O_2 is more reactive than acetone peroxide with Fe(II) salt so that the equilibrium of eq 28 is shifted to the left by the decomposition of H_2O_2 induced by the metal salt. On the contrary the thermal decomposition of the acetone peroxide (eq 30) is faster than the thermal decomposition of H_2O_2 so that the equilibria of eqs 28 and 29 are shifted to the right.

$$(CH_3)_2C(OH)-O-O-C(OH)(CH_3)_2 \longrightarrow 2\ (CH_3)_2C(O\cdot)OH \longrightarrow 2\ CH_3COOH + 2CH_3\cdot \qquad (30)$$

The overall mechanism is similar to those described in i) and ii) with the difference that a free-radical chain (Scheme 1) and not a redox chain is involved.

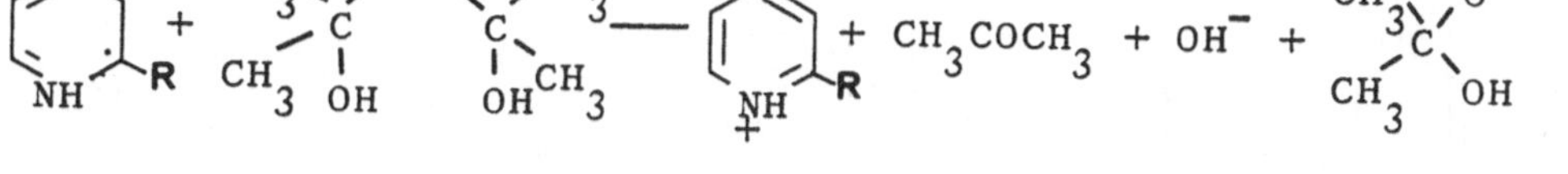

Scheme 1

The oxidation of the pyridinyl radical by acetone peroxide

is less effective than the oxidation by Fe(III) salt so that the kinetic length of the chain is shorter in this case. Moreover, in refluxing acetone the initiation step (eq 30) is rather slow. It results a slow overall reaction, which requires several hours for completion compared with few minutes with procedure i). The cheap reagents and the simple experimental conditions make, however, particularly valuable also this source of alkyl radicals.

iv) The thermal decomposition of peroxydisulphate in the presence of the acetate of the heteroaromatic bases and alkyl iodides leads to the selective alkylation of the heterocyclic ring according to the stoichiometric eq 31

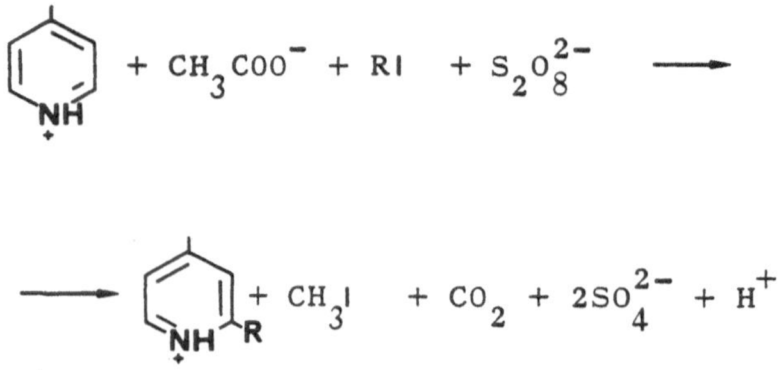

(31)

The overall mechanism is similar to those described in the previous sections. The methyl radical is generated by electron-transfer oxidation of the acetate ion (eq 33) and the rearomatization of the radical adduct occurs according to eq 34

$$S_2O_8^{2-} \longrightarrow 2\ SO_4^{-\bullet} \qquad (32)$$

$$CH_3COO^- + SO_4^{-\bullet} \longrightarrow CH_3\bullet + CO_2 + SO_4^{2-} \qquad (33)$$

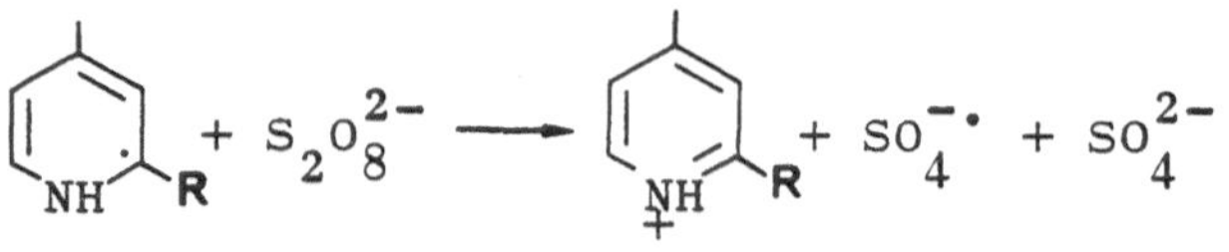

(34)

These new methods to generate alkyl radicals from alkyl iodides of any kind are much more convenient than those previously developped by us[6] and based on the use of aroyl peroxides and diazonium salts. The above discussed structural limitations with benzoyl peroxides and diazonium salts are eliminated, the reagents are much less expensive, the experimental conditions more simple.

Redox catalysis in homolytic aromatic substitution

The low selectivity of the reactions of hydroxyl and alkoxy radicals can be overcome with simple molecules, which have only one kind of C-H bond or a single particularly reactive C-H bond and therefore no problem of regioselectivity. To overcome the low chemoselectivity these simple molecules must be used in large excess in a practical way, usually as solvents.

Following this concept we have developped a variety of very simple and useful substitutions of heteroaromatic derivatives based on the general reaction 35

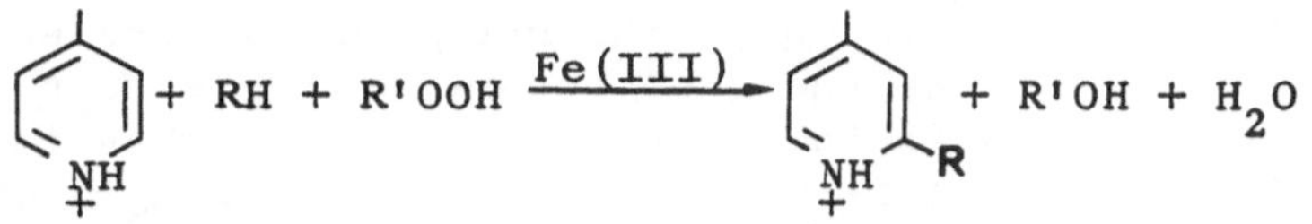

(35)

R'= H or t-Bu   R = CH_2OH, $CONH_2$, $CONMe_2$, $CHON(CH_3)CH_2$,

$CH_3CONHCH_2$, R"CO, 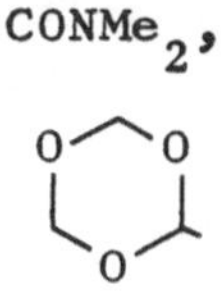 , 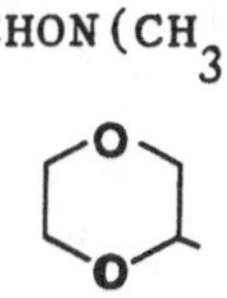

In this way important substitution reactions, such as hydroxymethylation[12], formylation[13], dioxanylation[12], carbamoylation[14], α -N-amidoalkylation[14], acylation[12] have been achieved in high yields and selectivity and simple redox catalysis by very trivial reagents, such as methanol, formaldehyde, dioxane, formamide, DMF, N-methylacetamide, aliphatic and aromatic aldehydes. All these reagents can be used without problems in large excess as solvents, thus minimizing the low chemoselectivity of the hydroxyl and alkoxy radical reactions.

The mechanism of the reaction is shown by the Scheme 2.

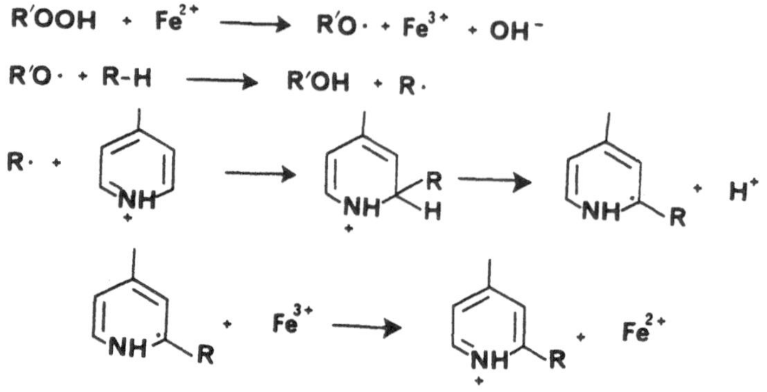

Scheme 2

The synthetic success of this redox chain is strictly connected with the polar character of the radical R•. In all the radicals indicated in eq 35 the nucleophilic character is strongly increased by the electron-releasing effect of the oxygen or nitrogen in the α-position. That makes fast the addition of the radical R• to the heterocyclic ring, but at the same time it increases the oxidation rate of the radical by Fe(III) salt (eq 36) and the reversibility of the heterocyclic addition.

$$R\bullet + Fe^{3+} \longrightarrow \text{Products} + Fe^{2+} \qquad (36)$$

Reaction 36 can be useful because it contributes to sustain the redox chain by regenerating the Fe(II) salt, but it obviously, in large occurrence, reduces the efficiency of the heteroaromatic substitution. The redox catalysis in this case has therefore more severe limitations compared with the analogous heterocyclic alkylation according to the procedure ii). High concentrations of both Fe(II) and Fe(III) salts are harmful, whereas in the procedure ii) a high concentration of Fe(III) salt is irrelevant as concerns the efficiency of the substitution because the oxidation rate of unsubstituted alkyl radicals by Fe(III) salt is much lower than the addition rate to the heterocyclic ring, but it increases the efficiency of the initiation (eq 26) and of the oxidation of the pyridinyl radical (eq 20). Thus the concentration of Fe(III) salt must be low to minimize reaction 36, but also the concentration of Fe(II) salt must be low for the same reason discussed for procedure ii), mainly the competition of reactions of the type 24. That was achieved by using small amounts of Fe(III) salt to initiate the chain (eq 26). The high reducing character of the pyridinyl radical makes very selective its oxidation also at low concentration of Fe(III) salt (eq 20). The slow initiation, due to the low concentration of Fe(III) (eq 26) is balanced by the fact that the main by-reaction (eq 36) is not a breaking step, but it contributes to increase the length of the redox chain (both reactions 20 and 36 sustain the redox chain).

Some typical results[12-14] are shown in Table III.

TABLE III - Ar-H + R-H + R'OOH ⟶ Ar-R + R'OH + H_2O

Ar-H	R-H	R'OOH	Position of attack	Conversion %	Yield[a] %
Quinoline	$NCONH_2$	HOOH	2, 4- di-substituted	100	97
Lepidine	"	"	2	82	99
"	"	t-BuOOH	2	67	93
Quinoxaline	"	HOOH	2	65	88
Isoquinoline	"	"	1	100	100
Acridine	"	"	9	68	82
Benzothiazole	"	"	2	40	68
Quinoline	Trioxane	t-BuOOH	2 (58%) 4 (42%)	81	92
Quinaldine	"	"	4	65	94
Isoquinaldine	"	"	1	83	92
Quinoxaline	"	"	2	38	94
Benzothiazole	"	"	2	37	90
Lepidine	"	"	2	89	94
"	"	HOOH	2	38	93
"	Dioxane	"	2	90	90
"	"	t-BuOOH	2	96	95
"	Methanol	"	2	93	99
"	"	HOOH	2	87	95
"	Benzaldehyde	t-BuOOH	2	80	75
"	3,4-Dimethoxy-benzaldehyde	"	2	74	85
"	Butyraldehyde	"	2	76	60
Acridine	Benzaldehyde	"	9	78	70
2-Cyano-quinoline	"	"	4	82	75

a - Based on converted heteroaromatic compound.

The yields of Table III were not optimized and the indicated conversions can be increased by increasing the amount of reagents.

These reactions for general character, the high yields and selectivity, the cheap reagents and catalyst, the simple experimental conditions, the interest of the reaction products are also suitable for industrial applications[15].

Another example of practical application of the redox catalysis with H_2O_2 concerns the hydroxylation of phenol to catechol and hydroquinone, which we[16] have developped on industrial scale about 15 years ago. The regioselectivity is good in ortho and para positions. The chemoselectivity of the hydroxyl radical has been controlled by using also in this case an excess of substrate (low conversions of phenol) in water (an innocuous solvent).

Recently, the industrial process has been modified by using a heterogeneous catalyst[17]: a special zeolite (titanium silicate). It is more convenient because it allows higher conversions without heavy loss in selectivity. Likely, that is due to the fact that, in addition to the chemical selectivity, also the absorption selectivity on the zeolite plays a significant role. The mechanism of this new catalysis is not yet made clear, as often it happens in heterogeneous catalysis, However, it appears that free radicals are involved also in this case because with toluene the same catalytic system leads to cresols, but also to significant amounts of benzyl alcohol and benzaldehyde.

Overlap aerea between free radical and ionic substitutions of heteroaromatic bases

The great synthetic interest of the substitution of protonated eteroaromatic bases by carbon-centered radicals is related to the high regio- and chemoselectivity, which are mainly due to large polar effects, with transition states similar to charge-transfer complexes[9] (eq 37)

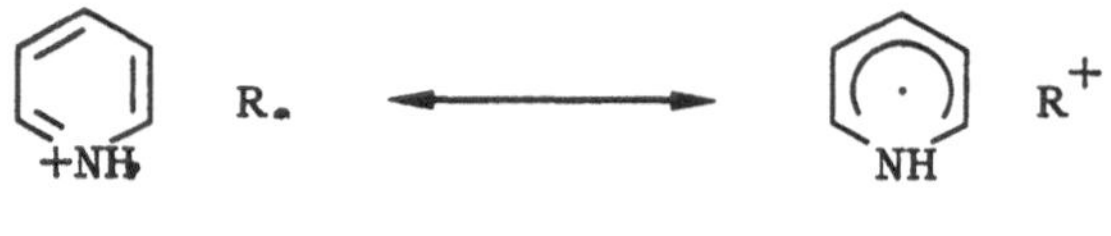

(37)

Thus, the substitution exclusively occurs in the α and γ positions of the protonated heterocyclic ring and, for example, protonated 4-cyanopyridine reacts very fast with t-Bu. (a rate constant > $10^7 M^{-1} s^{-1}$), whereas the reaction of benzene with the same radical is much slower[9] ($1-10 M^{-1} s^{-1}$). Similarly, the pyridinium cation is readily attacked by nucleophiles with very high regioselectivity in α and γ positions and high chemoselectivity.

However, these similar, general mechanistic features can be even more subtle and deep. Thus the total electron-deficiency of the pyridinium cation indicates that the charge control will lead to reaction at α position[18]. This is actually the case with nucleophiles with low-energy HOMO[19] (eq 38)

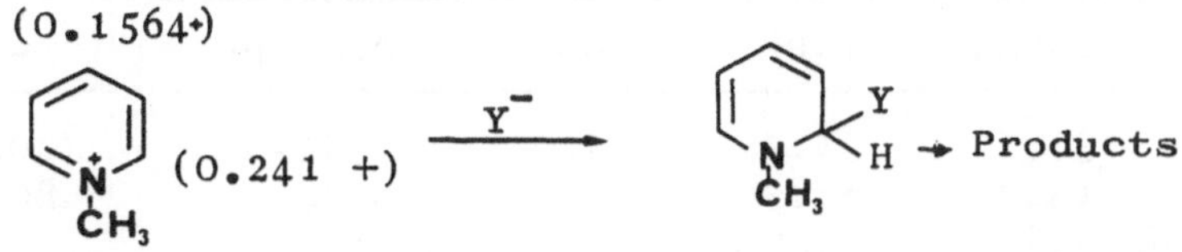

(38)

$Y^- = OH^-, NH_2^-, BH_4^-$, RMgBr

In terms of HSAB (hard and soft acids and bases) principle the hard nucleophiles (low-energy HOMO) react faster with the α-position, which is harder than the γ-position. According to the orbital interactions the hard-hard reaction is fast because of a large coulombic attraction.

However, nucleophiles with high energy HOMO attack the γ-position[20] (eq 39)

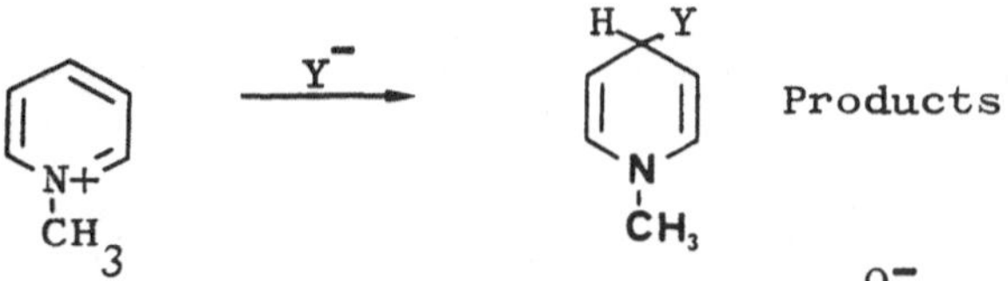

(39)

$Y^- = CN^-, CH_2 = C(O^-)-R, S_2O_4^{2-}$

That is, according to the HSAB principle, the softer nucleophiles react faster with the γ position, which is softer than the α-position. In terms of frontier orbital theory (FMO) the LUMO of the pyridinium cation has namely the form of that of benzene, but polarized by the nitrogen atom. This polarization has reduced the coefficient at C 3 and the coefficient at C 4 is larger than that at C2[21]; the frontier orbital term is largest at C4, where nucleophiles with high-energy HOMO should attack. The soft-soft reaction is fast because of a large interaction between the HOMO of the nucleophile and the LUMO of the γ-position.

Now, if we consider the regioselectivity in the substitution of protonated pyridine with a variety of carbon-centered radicals in water we observe a large change of the orientation with the nature of the alkyl radicals[9b] (Table IV).

TABLE IV - Regioselectivity for the substitution of protonated pyridine[9b]

	Me.	n-Bu.	i-Pr	t-Bu.	Dioxanyl	-THF
%	62.5	56.3	31.7	23.0	25.5	20.0
%	37.5	43.7	68.3	77.0	74.5	80.0

We believe that this behaviour has a deep connection with the regioselectivity above discussed for the substitution of the pyridinium cation with ionic nucleophiles. In terms of transition state picture this connection can be related to a charge-transfer character in the transition state of the interaction of the ionic nucleophile with the pyridinium cation (eq 40), similar to eq 37

(40)

The problem can be considered as an extension of the HSAB principle to free-radical reactions when the polar effect is the dominant factor, in the sense that the softness of a nucleophilic radical increases by decreasing the ionization potential ($\doteq$SOMO energy) (similarly, the softness of the ionic nucleophiles increases by decreasing the ionization potentials, which are roughly the energies of the HOMOs). Thus the softness of the alkyl radicals in Table IV increases in the series Me $<$ primary alkyl $<$ secondary alkyl $<$ tertiary alkyl $\simeq$ dioxanyl $<$ α-THF and therefore the attack to the softer position (γ-position of the pyridinium ion) increases according to the same sequence.

According to the FMO theory the SOMO of the radical interacts with the LUMOs of the α and γ positions of the pyridinium cation. A higher lying SOMO determines a larger interaction between the SOMO of the radical and the LUMO of the γ-

position, which has larger coefficient, and therefore a faster reaction in this position. Thus the similarity of behaviour between ionic nucleophiles and nucleophilic radicals in these reactions is related to the energies of the HOMOs and SOMOs and their interactions with the LUMOs of the pyridinium ion. Actually, the problem could be somewhat more complex because the addition of the most nucleophilic radicals to the heteroaromatic ring is reversible[9b] and that can also affect the regioselectivity of the substitution.

Free-radical diazocoupling

Nucleophilic alkyl radicals react fast with diazonium salts by two mechanisms[22]: alkyl radicals unsubstituted in α-position add fast to the diazonium group (eq 41), whereas alkyl radicals substituted in α-position with electron-releasing groups (mainly bonded to oxygen and nitrogen) react very fast by an electron-transfer process (eq 42)

$$R\cdot + N\equiv\overset{+}{N}\text{-Ar} \longrightarrow R\text{-}N=\underset{\bullet}{\overset{+}{N}}\text{-Ar} \qquad (41)$$

$$R'\cdot + N\equiv\overset{+}{N}\text{-Ar} \longrightarrow R'^{+} + \cdot N=N\text{-Ar} \longrightarrow N_2 + Ar\cdot \qquad (42)$$

No reaction occurs if the α-position of the alkyl radical is substituted by electron-withdrawing groups. Clearly polar effects play a dominant role in these interactions. The electron-transfer process (eq 42) can be considered the borderline case of a polar transition state (eq 43) of the reaction 41.

$$R\cdot \;\; N\equiv\overset{+}{N}\text{-Ar} \longleftrightarrow R^{+} \;\; \cdot N = N\text{-Ar} \qquad (43)$$

Reaction 41 is synthetically useful because it has allowed us to develop what we have called the "Free Radical Diazocoupling" reaction[22], which allows to obtain in simple way alkylarylazoderivatives. The reaction is of general synthetic interest, but particularly useful with tertiary alkylderivatives, which are difficult to prepare by conventional methods.

Taking advantage of the source i) of alkyl radicals from alkyl iodides, we have developped a new method of free-radical diazocoupling, particularly effective. The reaction is not a chain process, but stoichiometric amounts of reducing metal salts are required.

Thus the procedures iii) and iv) are not suitable in this case, because of the instability of the diazonium salts and the absence of reducing metal salt. The procedure ii) must be adjusted in order to be suitable for the diazocoupling reaction: it is necessary, obviously, to work at low temperature and with stoichiometric amounts of Fe(II) salt. Under these conditions the β -scission of the t-BuO. (eq 22) is less favoured (Table II) and its reduction by Fe(II) salt (eq 24) is exalted and an excess of t-BuOOH must be used in order to obtain good results. The stoichiometry of the reactions by the radical sources i) and ii) is shown by eqs 44 and 45

$$ArN_2^+ + R\text{-}I + CH_3SOCH_3 + H_2O_2 + 2Fe^{2+} \longrightarrow$$

$$\longrightarrow Ar\text{-}N = N\text{-}R + CH_3SO_2H + CH_3\text{-}I + 2Fe^{3+} + OH^- \quad (44)$$

$$ArN_2^+ + R\text{-}I + (CH_3)_3COOH + 2Fe^{2+} \longrightarrow$$

$$\longrightarrow Ar\text{-}N = N\text{-}R + CH_3\text{-}I + CH_3COCH_3 + 2Fe^{3+} + OH^- \quad (45)$$

The mechanism of the reaction is shown by the Scheme 3

$$H_2O_2 + Fe^{2+} \longrightarrow .OH + Fe^{3+} + OH^-$$

$$HO\bullet + CH_3SOCH_3 \longrightarrow CH_3\bullet + CH_3SO_2H$$

$$CH_3\bullet + R\text{-}I \longrightarrow CH_3\text{-}I + R\bullet$$

$$R\bullet + N \equiv \overset{+}{N}\text{-}Ar \longrightarrow R\text{-}N = \overset{+}{\underset{\bullet}{N}}\text{-}Ar$$

$$R\text{-}N = \overset{+}{\underset{\bullet}{N}}\text{-}Ar + Fe^{2+} \longrightarrow R\text{-}N = N\text{-}Ar + Fe^{3+}$$

Scheme 3

The mechanism according to procedure ii) is quite similar; the only difference is the source of the methyl radical (eq 22).

The interaction of the SOMO of the radical with the LUMO of the diazonium salt appears to govern the reactivity so that the radical become more reactive as the SOMO energy (I.P.) increases from primary to tertiary alkyl radicals (the radical becomes softer). When the SOMO is too high for the proximity of an electron-releasing group ($\bullet\overset{|}{\underset{|}{C}}\text{-O-}$, $\bullet\overset{|}{\underset{|}{C}}\text{-N}\lessdot$)

an electron-transfer process (eq 42) occurs. The similarity with the ionic diazocoupling reaction (i.e. eq 46) can be related to the interaction of the HOMO of the nucleophile with the LUMO of the diazonium salt.

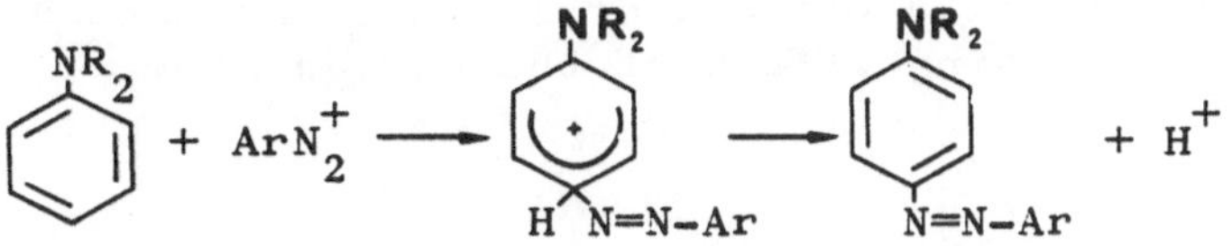

(46)

When the energy of the HOMO of the nucleophile is too high it is possible that an electron-transfer process also occurs (eq 47)

$$Y^- + ArN_2^+ \longrightarrow Y\cdot + \cdot N = N\text{-}Ar \longrightarrow Y\text{-}N = N\text{-}Ar \qquad (47)$$

The two processes 46 and 47 can be distinguished with difficulty because the two radicals formed in eq 47 can collapse in the solvent cage, before the diazenyl radical can lose N_2, and still leading to the azoderivative, in contrast with the corresponding reactions (eqs 41 and 42) of the nucleophilic alkyl radicals, which lead to different products. Actually, radical intermediates were detected in azocoupling of benzendiazonium salt and sodium phenoxide by means of ^{15}N-CIDNP [23].

We have previously[22] developped the free-radical diazocoupling starting from alkyl iodides according to eq 11 and the mechanism of Scheme 3, in which half of diazonium salt is utilized as radical source and the other half as diazocoupling reagent. The new procedure is more convenient because, in the addition to the higher yields (all the diazonium salt can be utilized for the diazocoupling reaction), it is also characterized by much cheaper reagents (H_2O_2, DMSO, Fe(II)) and simple experimental conditions.

Alkylation of pyrilium salts

The high electron affinity of pyrilium salts determines a fast and selective substitution by nucleophilic alkyl radicals from alkyl iodides. The stoichiometry with t-BuOOH is shown by eq 48

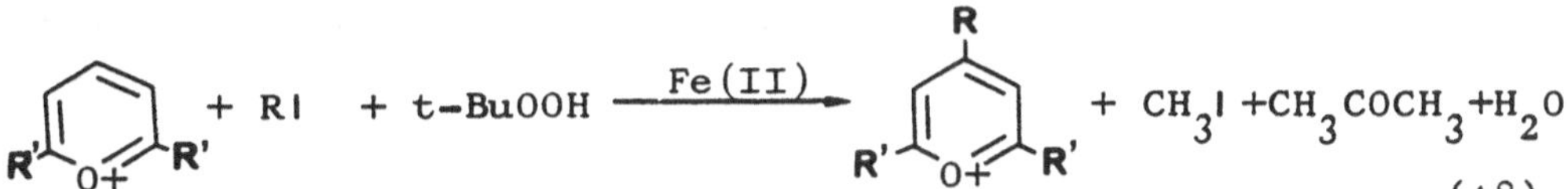

(48)

The mechanism of the reaction is quite similar to that described with protonated heteroaromatic bases; it is a redox chain process in which the radical adduct loses a proton (eq 49) and the highly nucleophilic radical intermediate is fast and selectively oxidized because of its allylic nature and of the electron-releasing effect of the oxygen atom (eq 50)

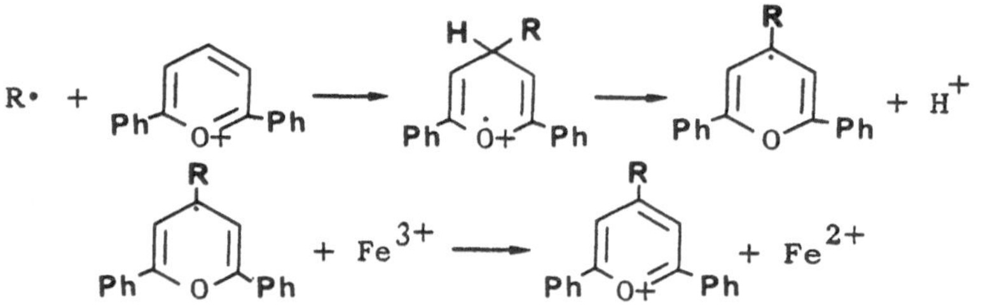

(49)

(50)

The less availability and the low stability in several media of the pyrilium salts however, does not make the substitution of the same great interest of protonated heteroaromatic bases.

<u>Alkylation of quinones</u>

Quinones are effective traps of alkyl radicals for both polar and enthalpic reasons. All the radical sources i-iv proved to be suitable for the substitution. The stoichiometries are shown by eqs 51-54

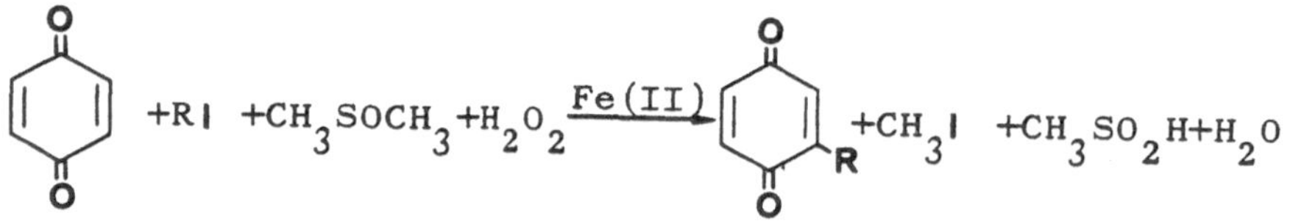

(51)

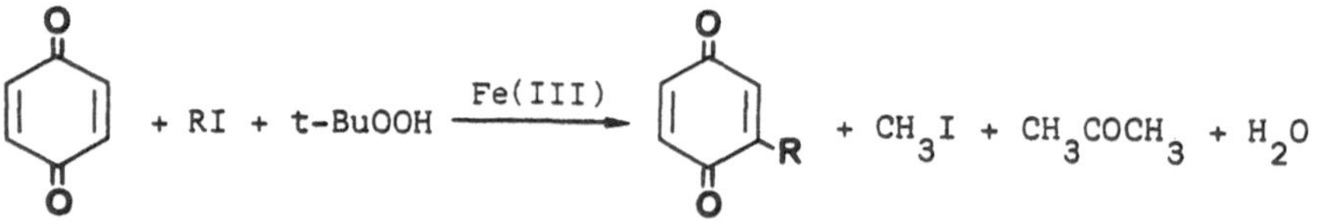

(52)

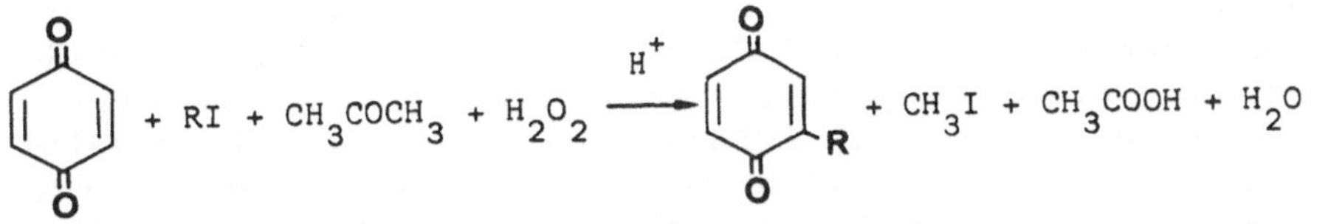

(53)

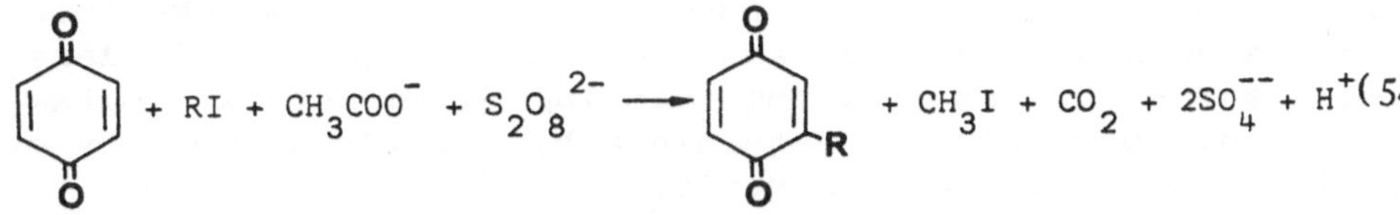

(54)

In all cases chain processes are working, involving the addition of the radical R• to the quinone (eq 55) and the oxidation of the radical adduct (eq 56), similarly with those discussed with the heteroaromatic bases

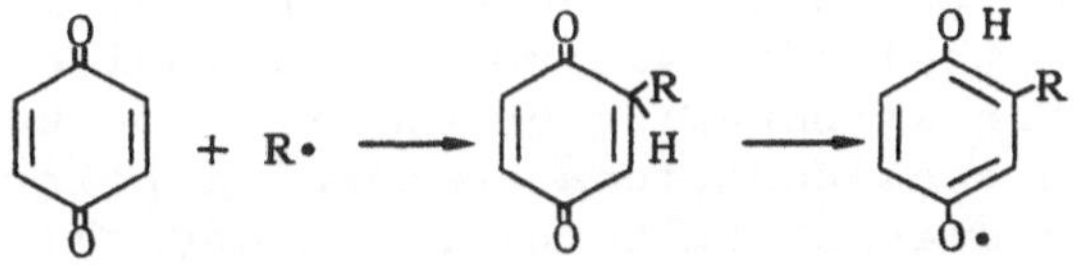

(55)

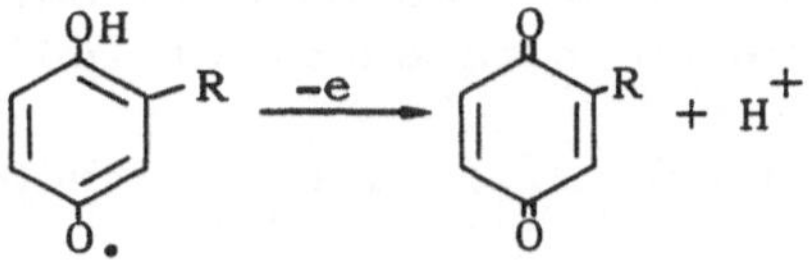

(56)

Two synthetic aspects are important in the alkylation of quinones. The methyl radical is very reactive with quinones and its addition to the quinone ring competes, at some extent, with the iodine abstraction, particularly from primary alkyl iodides. This competition can be easily overcome by using an excess of alkyl iodide.

Moreover, the introduction of an alkyl group does not substantially affect the reactivity of the quinone ring, unless steric effects are important, and polysubstitution increases with the conversions when more free position in the quinone ring are present. Thus with naphtoquinone the 2,3-

dialkylderivatives can be easily obtained in good yields with primary and secondary alkyl iodides by simply using a slight excess of alkylation reagents. Monoalkylation can be prevalently obtained at low conversions, the amount of disubstituted derivatives increasing with the conversions of the quinone. Another expedient to obtain monoalkylation concerns the use of mixtures of solvents (H_2O-DMSO, H_2O-AcOH, H_2O-MeCOMe), in which the quinone is soluble and the monoalkylquinone is less soluble and separates from the reacting mixtures. A limitation of this procedure is determined by the low solubility of the alkyl iodides in aqueous media.

With t-alkyl iodides the reaction with naphtoquinone is quite selective and good yields of monoalkyl derivatives are obtained also at high conversions. Steric effects prevent from further substitution.

Substitution of electronrich aromatics by electrophilic carbon-centered radicals

The rate and the equilibrium constants of iodine abstraction by the methyl radical from alkyl iodides leaving in α-position electronwithdrawing groups are not known. However, qualitative considerations concerning polar (eq 57) and enthalpic effects suggest that the rates and the equilibria of the abstraction should be even more favourable than those of the corresponding unsubstituted alkyl iodides[24].

$$HOOC{-}CH_2{-}I\ \ \cdot CH_3 \longleftrightarrow HOOCCH_2^{-}\cdots I\cdots \overset{+}{C}H_3 \longleftrightarrow HOOCCH_2^{\bullet}\cdots \overset{-}{I}\cdots CH_3^{+} \longleftrightarrow HOOCCH_2\cdot\ \ I{-}CH_3 \qquad (57)$$

Moreover, the resulting carbon-centered radicals have electrophilic character and react faster than the methyl radical towards substrates of high electron availability. Thus, for example, the reaction of anisole with iodoacetic acid, t-BuOOH and catalytic amount of Fe(III) salt leads to the substitution of the aromatic ring by the CH_2COOH group (eq 58)

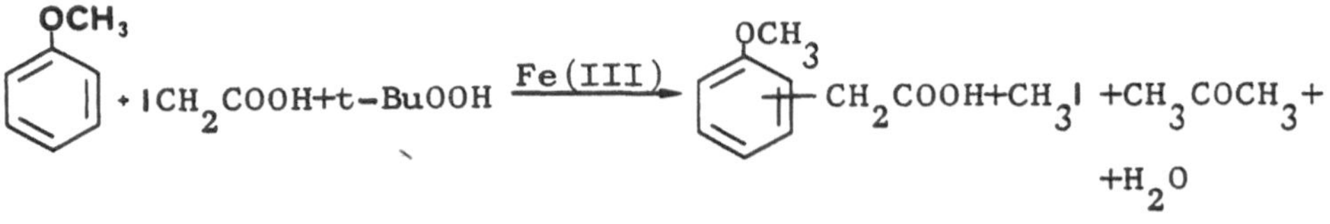

o% 79, m% 5, p% 16

A redox chain works also in this case, as in the heteroaromatic substitution; it is favoured by the reducing properties of the aromatic radical adduct (eq 59) which makes fast and selective the rearomatization step (eq 60)

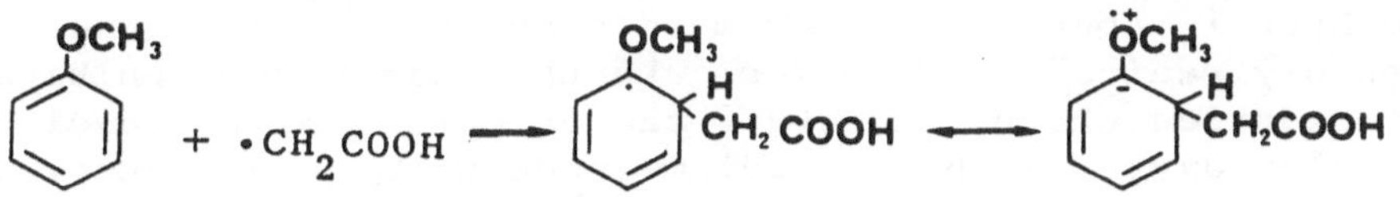

(59)

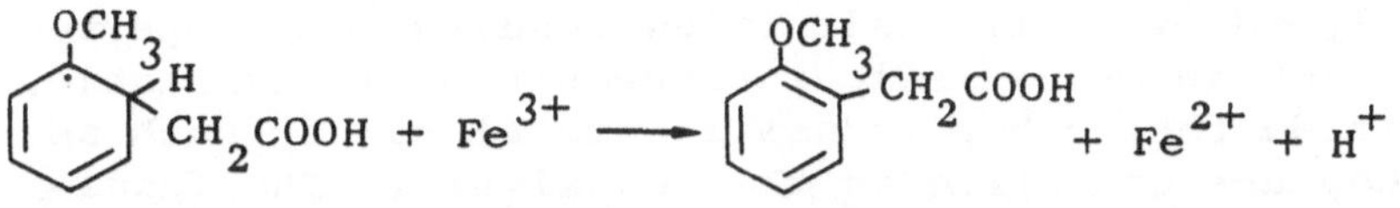

(60)

Similarly, naphthalene is substituted by the CH_2COOH group in good yields, mainly in the α position.

The easy availability of iododerivatives of general structure $I-\overset{|}{\underset{|}{C}}-X$ (X = COOH, COR, CN, NO_2, SO_2R etc.) makes this new methodology of large potential application.

Miscellaneous reactions

Preliminary results indicate that these new sources of alkyl radicals from alkyl iodides(i-iv) are suitable for synthetic purposes in other reactions already developed with different radical sources, such as alternating addition to conjugated olefins[1], the intra- or intermolecular oxidative alkylation of olefins[25], the substitution with imminium salts[9a], oximes[26], biacetyl[27].

Alkyl radicals can be generated from alkyl iodides by other reagents[28] such as metal hydrides (eq 61) or reducing metal salts (eq 62) (Cr(II), Co(II) complexes)

$$R{-}I + \cdot SnR_3 \longrightarrow R\cdot + I{-}SnR_3 \qquad (61)$$

$$R{-}I + M^{2+} \longrightarrow R\cdot + MI^{2+} \qquad (62)$$

However, these radical sources, not only are much more expensive, but above all are not compatible with the series of reactions above discussed, in which an oxidizing medium is always required.

Selective oxidation of p-methyl phenols by molecular oxygen

P-hydroxy aromatic aldehydes, such as p-hydroxybenzaldehyde, vanillin, are useful derivatives of the chemical industry. Their synthesis by oxidation of the corresponding p-hydroxymethyl benzene derivatives is of undoubted interest. The classical autoxidation is not suitable because it is difficult to stop the process at the aldehyde level; for example benzaldehyde is about 100 times more reactive than toluene towards peroxy radicals. Moreover, the phenolic group inhibits the autoxidation chain. However, the oxidation of p-cresol by molecular oxygen in basic medium leads with good selectivity to p-hydroxybenzaldehyde. In methanolic solution the reaction is catalyzed by Co(II) salts; few results were reported in patent;[29] but no mechanistic interpretation is available. We have verified that other transition metal salts, such as Fe, Mn, V, Ce, are effective in this catalysis. The fundamental role of the basic medium and the selectivity in aldehyde suggested that the metal salt catalysis was not related to the classical redox interactions (eqs 63 and 64)

$$\mathrm{ROOH} + \mathrm{M}^{n+} \longrightarrow \mathrm{RO\cdot} + \mathrm{OH}^- + \mathrm{M}^{(n+1)+} \qquad (63)$$

$$\mathrm{ROOH} + \mathrm{M}^{(n+1)+} \longrightarrow \mathrm{ROO\cdot} + \mathrm{H}^+ + \mathrm{M}^{n+} \qquad (64)$$

In methanolic solution no reaction occurs in the absence of transition metal salts and no reaction occurs in the presence of the base (NaOH, MeONa).

In an attempt to understand the role of the basic medium we have investigated the oxidation in the absence of transition metal salts, but in aprotic solvents (hydrocarbons, ethers) by using crown ethers or polyethylenglycol to solubilize the alkali phenoxide. In these conditions the reaction selectively leads to the aldehyde with oxygen at room temperature and atmospheric pressure. The stoichiometry of the reaction is shown by eq 65

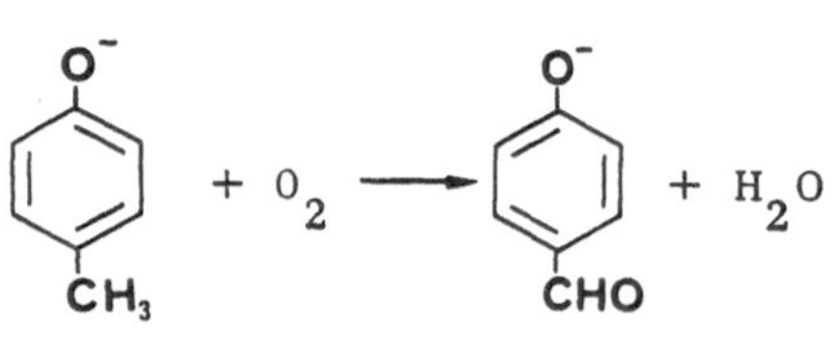

(65)

It appears that the acid-base equilibria play a fundamental role in determining the oxidation. A possible rationalization is shown in the Scheme 4

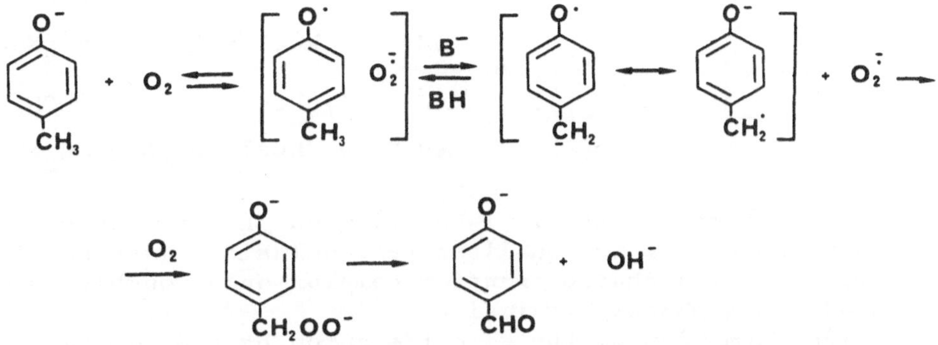

Scheme 4

Our suggestion is that the acidity of the methyl group in the charge-transfer complex is strongly increased compared with p-methylphenoxide, certainly for polar reasons, but also, perhaps prevalently, for enthalpic reasons. The delocalization of the unpaired electron in the phenoxy radical (eq 66) must significantly reduce the dissociation energy of the C-H bond in the methyl group and that should be reflected in its acidity

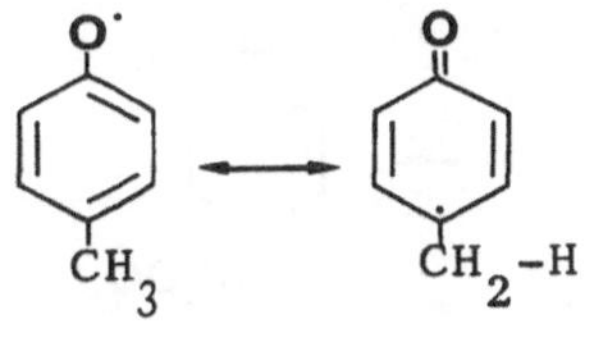

(66)

The phenomenon is quite general when a hydrogen atom is in α-position to a free-radical. Thus the pKa of the α-hydroxyalkyl radicals ($-\dot{C}-OH$) are approximately 5 units lower than those of the corresponding alcohols[30] and that can be mainly related to the lower energy of the O-H bond. Similarly the α C-H bonds of the alkoxy radicals are more acidic than those of the corresponding alcohols (eq 67)

$$-\underset{H}{\overset{|}{C}}-O\cdot + H^+ \rightleftharpoons -\underset{H}{\overset{|}{C}}-\overset{+}{\dot{O}}-H \longrightarrow -\overset{|}{\dot{C}}-OH + H^+ \qquad (67)$$

Moreover the enthalpy changes for the decomposition of the radical cations of toluene and trimethylamine certainly contribute, in addition to polar effects, to the increased acidity[8] (eqs 68 and 69)

$$(R\text{-}H)^{+} \longrightarrow R^{+} + H\cdot \qquad (68) \qquad \Delta H = 52 \text{ kcal/mol} \quad R = RhCH_2$$

$$\Delta H = 34 \text{ kcal/mol} \quad R = Me_2NCH_2$$

$$(R\text{-}H)^{+} \longrightarrow R\cdot + H^{+} \qquad (69) \qquad \Delta H = -17 \text{ kcal/mol} \quad R = PhCH_2$$

$$\Delta H = 6 \text{ kcal/mol} \quad R = Me_2NCH_2$$

Quantitative data with alkyl-substituted phenoxy radicals are not known, but the qualitative conclusions must be similar, even if a charge-transfer complex only roughly corresponds to a phenoxy radical.

The formation of the peroxide anion by radical-radical combination of $O_2^{\cdot-}$ with the benzyl radical would seem to be **ruled** out by some evidences[31]. The coupling can occur with molecular oxygen, followed by the fast reduction of the peroxy radical by $O_2^{\cdot-}$ (eqs 70 and 71)

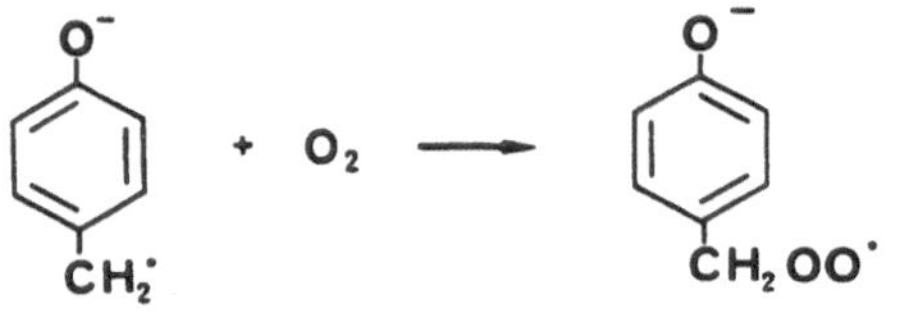

(70)

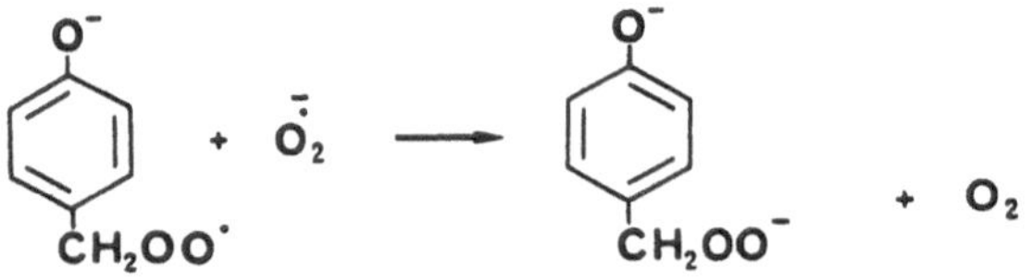

(71)

A rate constant of 10^{8} $M^{-1}s^{-1}$ has been in fact reported[32] for reaction 72

$$HOO\cdot + O_2^{\cdot-} \longrightarrow HOO^{-} + O_2 \qquad (72)$$

The basic medium has a twofold effect: it determines the acid-base equilibria of the phenoxy radical and the ionic decomposition of the peroxide anion (Scheme 4), which prevents from free-radical decomposition and therefore from further oxidation of the aldehyde.

The overall selectivity must be ascribed to polar effects: the lower reducing properties of the p-formylphenoxide compared with the corresponding p-methylderivative makes less favoured the interaction with O_2 (formation of a charge-transfer complex) and reduces the rate of oxidation. Several years ago we[33] have shown a similar behaviour in the selective oxidation of methylaromatics to aldehydes by peroxydisulphate, which occurs by an electron-transfer process.

In protic solvents, such as methanol, similar acid-base equilibria of the charge-transfer complex are less favoured; moreover, hydrogen bonds with the solvent decrease the electron availability of the phenoxide affecting the extent of the charge-transfer with oxygen and no reaction occurs in the absence of transition metal salts. We suggest that the function of the transition metal salts is a redox catalysis quite different from the classical catalysis in the autoxidation (eqs 73 and 74). A possible redox chain could be shown by eqs 73 and 74 :

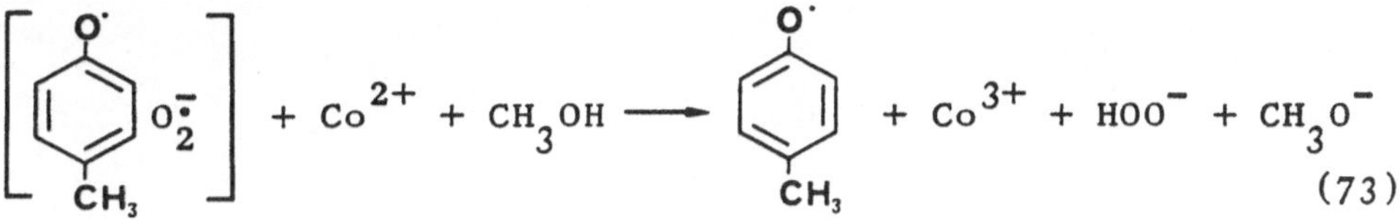

(73)

charge-transfer
complex

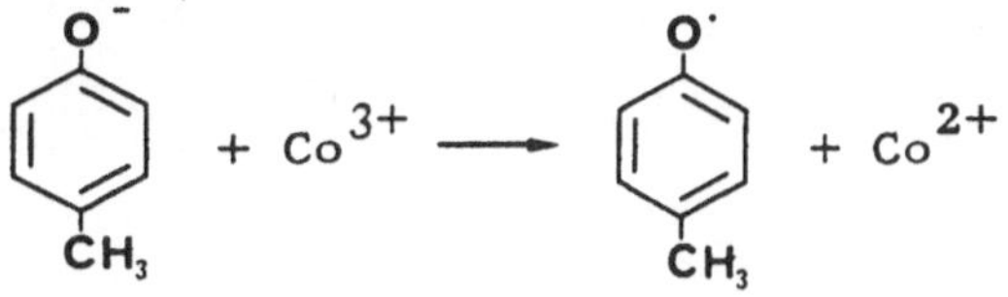

(74)

The reducing properties of $O_2^{\overline{\bullet}}$ is not consistent with the oxidation of salts such as Mn(II) or Co(II); however, its protonated form, HOO•, has much higher oxidizing character.

The acidity of the methyl group in the phenoxy radical is higher than in the charge-transfer complex and an equilibrium of the type 75 can be established also in methanol

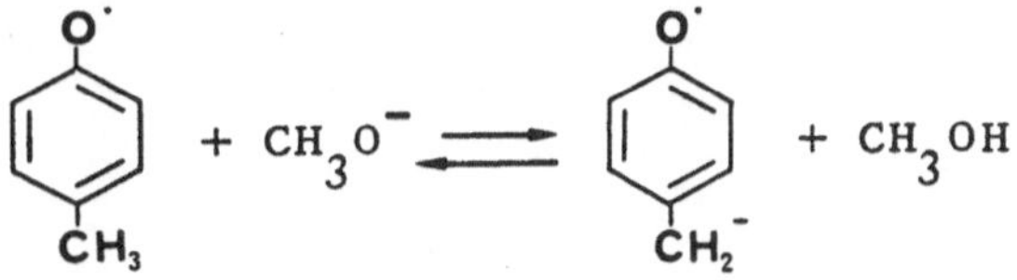

(75)

The overall interpretation is also supported by the fact that m-methylphenoxide is not oxidized under both conditions (presence of transition metal salts in methanol and absence of transition metal salt in aprotic solvent). That is very useful from practical point of view because the commercial mixtures of meta and para cresols are difficult to separate and therefore much cheaper than the pure isomers. The selective oxidation of these mixtures allows to obtain a twofold result: the synthesis of a useful product, such as p-hydroxybenzaldehyde, and the easy separation of the unchanged m-cresol.

REFERENCES

1. F. Minisci, "Fundamental Research in Homogeneous Catalysis", Plenum Publ.Corp., vol.4, pag.173-204 (1984).
2. Reviews in the subject: a) Walling, Acc.Chem.Res. 8, 125 (1975); b) F. Minisci, La Chimica e l'Industria, 65, 487 (1983).
3. T.J. Swarski, Radiation Res. 4, 483 (1965); D.M. Brown, F.S. Dainton, D.C. Walker, J.P. Keene: "Pulse radiolysis", Academic Press, New York 1965, pag 201.
4. G.E. Adams, J.W. Baag, B.D. Michael, Trans.Faraday Soc., 61, 1417 (1965); I. Kraljic, C.N. Trumbare, J.Am.Chem. Soc., 87, 2547 (1965); G.E. Adams, J.W. Baag, Proc.Chem. Soc., 112 (1964); J.F. Ward, L.S. Myers, Radiation Res., 26, 483 (1965); D. Zehavi, J. Rabani, J. Phys.Chem., 76, 312 (1972).
5. B. Giese, "Radicals in Organic Synthesis", J.E. Baldwin Ed., Pergamon Press, Oxford, 1986.
6. F. Minisci, E. Vismara, F. Fontana, G. Morini, M. Serravalle, C. Giordano, J.Org.Chem., 51, 4411 (1986).
7. A. Citterio, F. Minisci, J.Org.Chem., 47, 1759 (1982).
8. J.A. Hawari, J.M. Kanabus-Kamiuska, D.D.M. Waymer and D. Griller, "Substituent Effects in Radical Chemistry", H. G. Viehe Ed., D. Reidel Publishing Co. 1986, pag.91-105.
9a) F. Minisci, Ref.8, pag 391-433; b) F. Minisci, E. Vismara, F. Fontana, G. Morini, M. Serravalle and C. Giordano, J.Org.Chem., 52, 730 (1987).
10. F. Minisci and A. Citterio, "Advances in Free-Radical Chemistry", G.H. Williams Ed., Heyden, London 1980, pag 65-153.
11. C. Walling and P.J. Wagner, J.Am.Chem.Soc., 86, 3368 (1964).
12. F. Minisci, E. Vismara and F. Fontana, Gazz.Chim.Ital., 117, 363 (1987).
13. F. Minisci, E. Vismara, S. Levi and C. Giordano, J.Org. Chem., 51, 536 (1986).
14. F. Minisci, A. Citterio, E. Vismara and C. Giordano, Tetrahedron, 41, 4157 (1985).
15. F. Minisci, C. Giordano, E. Vismara, S. Levi and V. Tortelli, Ital.Pat. 22700 (1984); 23798 A/84 (1984).
16. P. Maggioni and F. Minisci, La Chimica e l'Industria, 59, 239 (1977).
17. G. Perego, G. Bellussi, C. Corno, M. Taramasso, P. Buonomo and A. Esposito, "Proceeding of the 7th International Zeolite Conference", Y. Murakami Ed., 1986, pag.129.

18. R.F. Hudson, Angew.Chem.Internat.Ed., 12, 36 (1973).
19. R.E. Lyle and P.S. Anderson, Adv.Heterocyclic Chem., 6, 45 (1970).
20. W. von E. Doering and W. E. McEwen, J.Am.Chem.Soc., 73, 2104 (1951); E.M. Kosower, J.Am.Chem.Soc., 78, 3497 (1956).
21. A. Streitwieser, J.I. Brauman and C.A. Coulson, "Supplemental Tables of Molecular Orbital Calculations", Pergamon Press, Oxford 1965.
22. A. Citterio and F. Minisci, J.Org.Chem., 47, 1759 (1982).
23. N.N. Bubnov, K.A. Bilevick, L.A. Poljakova and O. Yu. Okhlobystin, J.Chem.Soc.Chem.Comm., 1058 (1972).
24. Ref.10, pag.85.
25. F. Minisci, Acc.Chem.Res., 8, 165 (1975).
26. A. Citterio and L. Filippini, Synthesis, 473 (1986)
27. F. Minisci, R. Galli, M. Cecere, V. Malatesta and T. Caronna, Tetrahedron Lett., 5609 (1968).
28. Ref.5, pag.36.
29. A.T. Needham, U.S.P. 4.471.140 (1984).
30. Ref.10, pag.133.
31. A.A. Frimer, "The Chemistry of Peroxides", S. Patai Ed., Wiley, 1983, pag.454.
32. B.H.J. Bielski, J.Phys.,Chem., 81, 1048 (1977).
33. P. Maggioni and F. Minisci, La Chimica e l'Industria, 61, 101 (1979).

PARAMAGNETIC IRON COMPLEXES: THEIR ROLE IN THE TRANSFORMATION OF OLEFINS

D.Ballivet-Tkatchenko*[(1)], J. Vincent-Vaucquelin
Institut de Recherches sur la Catalyse, CNRS, conventionné à l'Université Claude Bernard
2, av. Einstein
69626-Villeurbanne Cedex
France

B. Nickel and A. Rassat
Laboratoire d'Etudes Dynamique et Structurale de la Sélectivité, U.A. 332 CNRS
B.P.68
38402-St Martin d'Hères Cedex
France

ABSTRACT. The binuclear $[Fe(NO)_2Cl]_2$ complex dissociates in solution to give a paramagnetic species. Addition of PPh_3 or Cl^- affords the tetracoordinated 17-electron complex $Fe(NO)_2PPh_3Cl$ or $[Fe(NO)_2Cl_2]^-$. None of these systems exhibits a catalytic behavior in the transformation of olefins. The catalytic activity is induced by Cl^- de-coordination. The reaction of $[Fe(NO)_2Cl]_2$ with a non-coordinating anion leads to an active species in the polymerization of olefins. On the bases or IR, ESR and EXAFS/XANES spectroscopies, the radical-cation $[Fe(NO)_2LL'_2]^{\cdot +}$ is responsible for C-C bond formation. $[Fe(NO)_2LL'_2]^{\cdot +}$ is a 19-electron complex in a trigonal-bipyramidal arrangement with the two NO being equivalent and lying in the equatorial plane. The increase of the coordination number from four to five is facilitated by the presence of a cationic charge on the complex, while the two NO ligands exhibit good π-acceptor properties by remaining linearly coordinated. Hence the complex can accomodate the 19-electron configuration.

1. INTRODUCTION

Our interest in the selective transformation of unsaturated substrates such as activated monoolefins, diolefins and acetylenic compounds has led us to examine the synthesis of catalysts based on nitrosyl complexes. Mild synthesis conditions were one of the priority in order to avoid secondary reactions (decomposition, nitrosyl transfer...).[2-9] Among the various results obtained, $[Fe(NO)_2Cl]_2$ was found to be a catalyst precursor at temperatures averaging 20 °C. A prerequisite to observe catalytic cycles is to expel the chloro ligand from the coordination sphere.

M. Chanon et al. (eds.), Paramagnetic Organometallic Species in Activation / Selectivity, Catalysis, 61–70.

Two different 'cocatalysts' were found efficient for chloro ligand abstraction. The first one implies the exchange between Cl^- and a non-coordinating anion[4] (eq 1):

$$1/2\,[Fe(NO)_2Cl]_2 \xrightarrow[-\ AgCl]{solvent,\ AgY} [Fe(NO)_2]^{\cdot +}\mathit{solvated} \qquad (1)$$

$Y = PF_6^-$, BF_4^-, ClO_4^-

The second pathway involves a one-electron reduction either by an electropositive metal[5] (Zn, Mg) or by electrochemical means[6] (eq 2):

$$1/2\,[Fe(NO)_2Cl]_2 \xrightarrow[-\ Cl^-]{solvent,\ e^-} [Fe(NO)_2]\mathit{solvated} \qquad (2)$$

Both reactions are feasible at temperatures above -40 °C and have therefore allowed the study of the reactivity of the new nitrosyl species formed over a wide range of temperatures.

This paper will focus on some of the more characteristic features of the complexes $[Fe(NO)_2Cl]_2$ and $[Fe(NO)_2]^{\cdot +}$*solvated* in relation with their paramagnetic properties.

2. MATERIALS AND METHODS

The complexes studied are air- and moisture-sensitive. All experiments were conducted with exclusion of air and water using Schlenk techniques under argon.

2.1. Materials

Acetonitrile (MeCN), tetrahydrofuran (THF), dimethylformamide (DMF), methylenechloride (CH_2Cl_2), acetone (Me_2CO), toluene (PhMe), triphenylphosphine (PPh_3), 1,2-bis(diphenyl phosphino)ethane (dppe), triphenylphosphite ($P(OPh)_3$), methyldiphenylphosphine ($PMePh_2$), dimethylphenylphosphine (PMe_2Ph) and tri-n-butylphosphine ($P(n\text{-}Bu)_3$) were purified and dried according to conventional procedures.

$AgClO_4$, $AgPF_6$ and $AgBF_4$ were purchased from Alfa Inorganics and used as received.

$[Fe(NO)_2Cl]_2$, $Fe(NO)_2PPh_3Cl$ and bis(triphenylphosphine)iminium chloride (PPNCl) were prepared by published methods described in refs 7, 10 and 11 respectively.

$[Fe(NO)_2Cl_2]^+PPN^-$ was obtained by adding PPNCl (1.72 g, 3 mmol) to $[Fe(NO)_2Cl]_2$ (0.400 g, 1.3 mmol) dissolved in CH_2Cl_2 (50 mL). After 2 h of reaction at 20 °C, the brown solution was evaporated to dryness and the residue recrystallized from CH_2Cl_2-EtOH to give $[Fe(NO)_2Cl_2]^+PPN^-$ (1.5 g, 2.1 mmol).

The cationic complexes $[Fe(NO)_2L_n]^+$ (L = THF, MeCN, Me_2CO, PPh_3, $PMePh_2$, PMe_2Ph and $P(n\text{-}Bu)_3$)) have been prepared by the general procedure described in ref 12.

2.2. Methods

Infrared (IR) spectra were obtained on a Perkin-Elmer 597 spectrometer with CaF_2 or AgCl windows.

Electron spin resonance (ESR) spectra were recorded on a Varian E9 X-band instrument equipped with a variable-temperature accessory. Diphenylpicrylhydrazyl radical (g = 2.0036) was used as the external standard in a dual-cavity arrangement. Relative errors of the g tensor are ca. ±0.0003.

3. RESULTS

3.1. The complex $[Fe(NO)_2Cl]_2$

The dimer $[Fe(NO)_2Cl]_2$ has been found to be diamagnetic. As the monomeric unit $Fe(NO)_2Cl$ contains an odd number of electrons, the existence of a metal-metal bond was proposed.[13] In polar solvents, the paramagnetic monomer $Fe(NO)_2Cl^{\bullet}$ is the only species present as evidenced by (i) one set of IR V(NO) bands (1780 and 1715 cm^{-1} in THF, 1780 and 1710 cm^{-1} in MeCN, 1805 and 1705 cm^{-1} in CH_2Cl_2 vs 1815 and 1765 cm^{-1} for $[Fe(NO)_2Cl]_2$ in nujol mull), and (ii) one ESR signal (~10 G wide; g_{av} = 2.0362 in THF, 2.0387 in MeCN and 2.0399 in CH_2Cl_2 at 20 °C). Hyperfine couplings of the unpaired electron with ^{14}NO (I = 1) and with $^{35,37}Cl$ (I = 3/2) ligands are better seen in Me_2CO-PhMe solution as reported in Figure 1.

The failure to observe hyperfine couplings other than with ^{14}NO and $^{35,37}Cl$ nuclei does not permit to determine the coordination number around the iron ion. Solvent coordination is suggested by the modifications in V(NO) and g values.

Addition of PPh_3 or $AsPh_3$ permitted to isolate a tetracoordinated complex, $Fe(NO)_2LCl^{\bullet}$, (V(NO) = 1785 and 1720 cm^{-1} L = PPh_3, and 1790 and 1730 cm^{-1} L = $AsPh_3$). The X-ray structure determination of $Fe(NO)_2PPh_3Cl^{\bullet}$ revealed a distorted trigonal pyramid geometry; the Cl, Fe, P plane may be considered as a mirror plane of the molecule with two equivalent NO ligands (Fe-N-O = 166° average).[10] Therefore, $Fe(NO)_2PPh_3Cl^{\bullet}$ is a 17-electron complex.[14] We observed an ESR spectrum (Figure 2) which shows the lines produced by $Fe(NO)_2Cl$ (Figure 1) being splitted due to coupling of the unpaired electron with the ^{31}P nucleus (I = 1/2). The spectrum simulation permitted to obtain A_N = 1.8 G and A_{Cl} = 2.4 G (^{35}Cl) and 2.8 G (^{37}Cl).[15]

The $Fe(NO)_2Cl^{\bullet}$ monomer also formed a paramagnetic adduct with the chloride anion. Addition of one equivalent of PPNCl led to an anionic 17-electron complex isolated and analysing for $[Fe(NO)_2Cl_2]^{\bullet-}PPN^+$ (V(NO) = 1775 and 1710 cm^{-1} in CH_2Cl_2). Such a complex already prepared by another method was described to present an ESR signal without any hyperfine couplings.[16] We could obtain a multi-line spectrum at low temperature (Figure 3). By comparison to that observed for $Fe(NO)_2Cl^{\bullet}$ (Figure 1) the increased number of lines is consistent with the presence of additional chloro ligand.

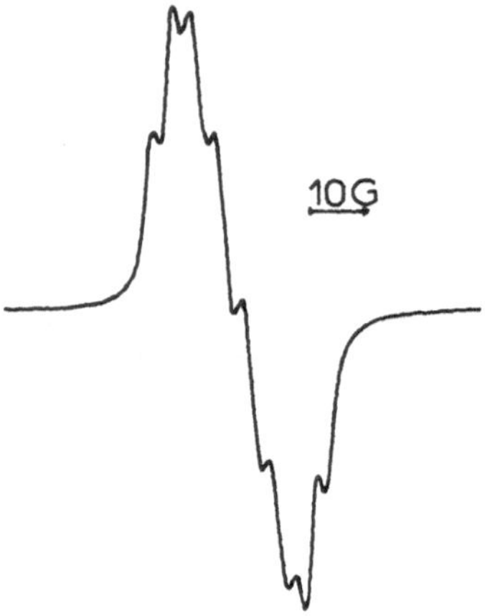

Figure 1. ESR spectrum of $Fe(NO)_2Cl^{\cdot}$ in Me_2CO-PhMe at −5 °C (g_{av} = 2.0368).

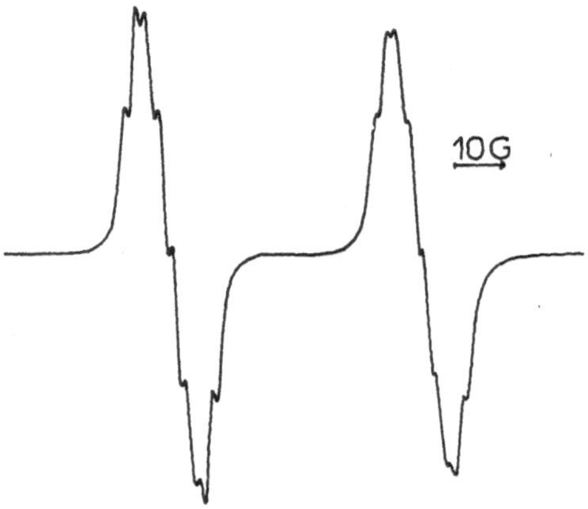

Figure 2. ESR spectrum of $Fe(NO)_2PPh_3Cl^{\cdot}$ in Me_2CO-PhMe at −10 °C (g_{av} = 2.0358, a_P = 51.5 G).

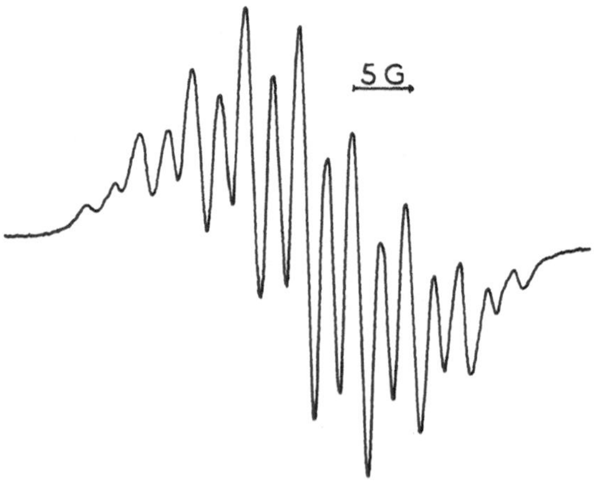

Figure 3. ESR spectrum of $[Fe(NO)_2Cl_2]^{\cdot -}PPN^+$ in CH_2Cl_2 at −50 °C (g_{av} = 2.0349).

Carbon monoxide did not react with $Fe(NO)_2Cl^{\cdot}$ even in low coordinating solvent such as CH_2Cl_2. CO coordination only occurred if a one-electron reduction reagent is added to the solution. The final isolated compound, $Fe(NO)_2(CO)_2$, is a 18-electron complex.

Neutral polydentate ligands led to ligand redistribution. For example, with 2,2'-bipyridyl and dppe, $Fe(NO)_2Cl^{\cdot}$ underwent chloro and nitrosyl ligand substitutions. No paramagnetic intermediates could be either isolated or detected.

No catalytic behavior has been detected at temperatures below 50 °C starting either with the paramagnetic $Fe(NO)_2Cl^{\cdot}$ species or with its chloro derivatives. Elimination of the chloro ligand according to eq 1 or 2 induced catalytic activity. Reaction 1 led to a new paramagnetic complex, $[Fe(NO)_2L_n]^{\cdot+}$, whose properties are discussed in the following.

3.2. Formation and paramagnetic properties of $[Fe(NO)_2L_n]^{\cdot+}$

Addition of equimolecular amounts of silver or thallium salts of non-coordinating anions to a solution of $Fe(NO)_2Cl^{\cdot}$ led to the quantitative precipitation of AgCl or TlCl (eq 1). The resulting complex is the cationic paramagnetic $[Fe(NO)_2L_n]^{\cdot+}$ species (L denotes solvent molecules or neutral two-electron donor ligands). This complex was unstable. It decomposed within a few hours at 20 °C whereas $Fe(NO)_2Cl$ was stable under reflux in THF or MeCN for hours. IR, ESR and EXAFS/XANES spectroscopies have been successfully used to characterize $[Fe(NO)_2L_n]^{\cdot+}$ for L being THF, Me_2CO, MeCN and PPh_3.[12,17] From these studies, it appears that the complex corresponds to $[Fe(NO)_2LL'_2]^+$, **1**, a pentacoordinated species with a trigonal bipyramidal geometry (TBP) with two equivalent NO's approximately linearly coordinated lying in the equatorial plane (Scheme I).

Scheme I

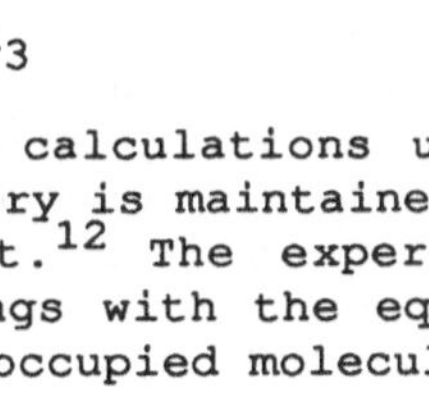

1

L = L' = MeCN, Me_2CO, THF, PPh_3

Noniterative extended Hückel calculations unequivocally demonstrated that, as far as the C_{2v} symmetry is maintained in the TBP geometry, the axial ligands are ESR silent.[12] The experimental spectra actually showed only hyperfine couplings with the equatorial ligands. In this type of situation the single-occupied molecular orbital (SOMO) lies in the equatorial plane. Such a five-coordination (19-electron configuration) is different from that observed for the chloro derivatives $[Fe(NO)_2Cl_2]^{\cdot-}PPN^+$ and $Fe(NO)_2PPh_3Cl^{\cdot}$ (tetracoordinated, 17-electron configuration). This 19- vs 17-electron structure has also been encountered on two other related $\{Fe(NO)_2\}^9$ systems formed by the reaction of

$[Fe(NO)_2I]_2$ with polydentate anionic ligands. The resulting complexes are neutral, and have been isolated and characterized by X-ray diffraction (Scheme II); they are paramagnetic but their ESR spectra have not been described.[18]

Scheme II

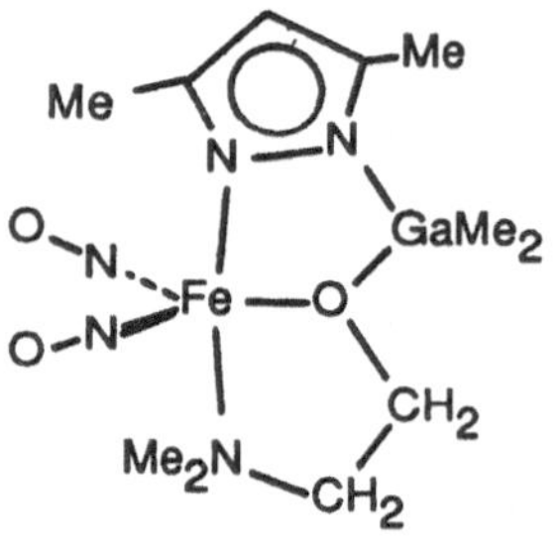

ref.18a, **2**

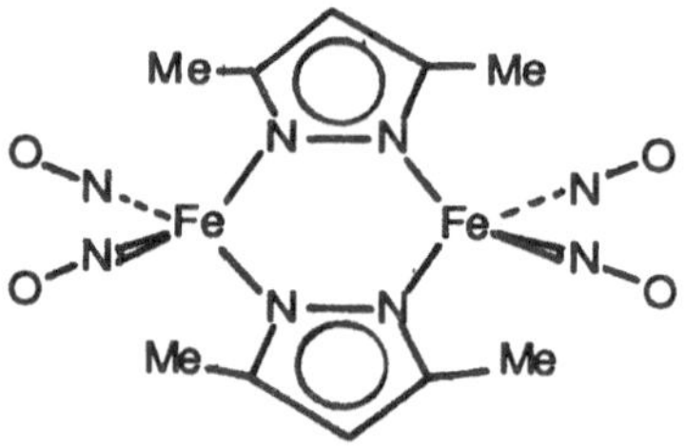

ref.18b, **3**

The tridentate chelating gallate ligand, **2**, forces a distorted TBP five-coordination as in **1** with equivalent nitrosyl ligands in the equatorial plane (V(NO) = 1750 and 1673 cm^{-1}, Fe-N-O = 158.5° average), while the bidentate anionic pyrazolyl ligand favors a dimeric structure, **3**, with no Fe-Fe interaction (Fe-Fe = 3.3359(3) Å). Each of the iron atoms is in a pseudotetrahedral environment and again the nitrosyl ligands are close to a linear arrangement.

Studying the influence of the basicity of P(III) monodentate ligands on the changes in the ESR features of $[Fe(NO)_2(MeCN)_3]^{\cdot+}$, we have observed two classes of hyperfine splitting modifications.

With PPh_3 (1 per Fe), the ESR signal represented in Figure 4 showed a doublet of resolved five-lines assigned according to computer simulation to the coupling of the unpaired electron with two equivalent ^{14}N (NO) nuclei (A_N = 3 G) and with one ^{31}P (PPh_3) nucleus (A_p = 54 G). The IR spectrum exhibited two V(NO) bands centered at 1800 and 1725 cm^{-1}. With $P(OPh)_3$, the ESR signal has the same features with the following parameters: g = 2.0268, A_P = 102 G, A_N = 3 G at -30 °C. However in this case the signal of the $[Fe(NO)_2(MeCN)_3]^{\cdot+}$ still was present. Neither a prolonged reaction time nor a large excess of phosphite (10-fold excess) increased the ESR signal of the phosphite containing species. This result shows that $P(OPh)_3$ competes with MeCN for coordination.

With more basic phosphines, $PMePh_2$, PMe_2Ph or $P(n\text{-}Bu)_3$, the IR spectra still exhibited two V(NO) stretching vibrations (1790 and 1710 cm^{-1} for $PMePh_2$, 1790 and 1720 cm^{-1} for PMe_2Ph and 1780 and 1720 cm^{-1} for $P(n\text{-}Bu)_3$) but a dramatic change occurred in the ESR features (Figure 5). A predominant five-line signal appeared. Changing either the solvent MeCN for THF or the counter-anion (ClO_4^-, BF_4^- or PF_6^-) produced no modifications on the ESR spectrum. A good fit with computer simulation spectrum was difficult to obtained. The best one was found by assuming that the coupling of the unpaired electron occurs with two equivalent ^{31}P nuclei and one ^{14}N (NO) nucleus.

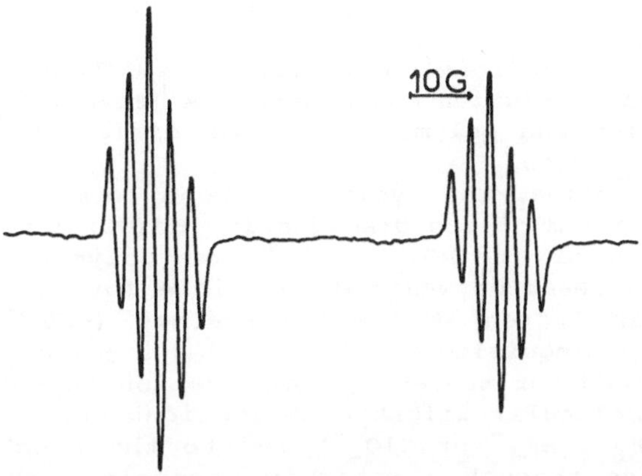

Figure 4. ESR spectrum of $[Fe(NO)_2(PPh_3)(MeCN)_2]^{\cdot+}$ in MeCN-PhMe at -10 °C (g_{av} = 2.0210, A_N = 3 G, A_P = 54 G).

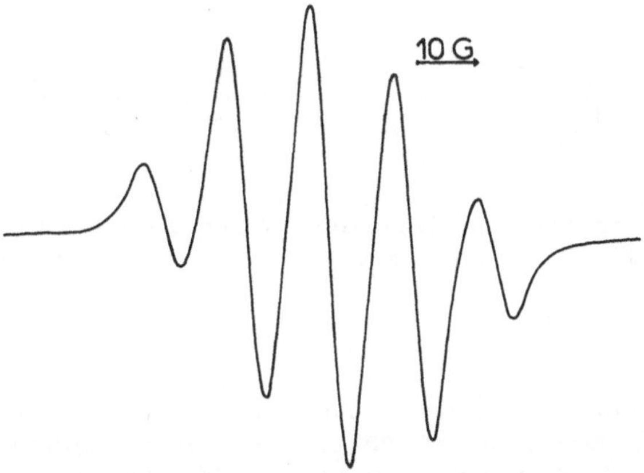

Figure 5. ESR spectrum of $[Fe(NO)_2(MeCN)_3]^{\cdot+}$ in the presence of $(P(n\text{-}Bu)_3)$ in MeCN-PhMe at -10 °C (g = 2.0185, line-separation 13 G).

Hence the NO ligands now seem to be non-equivalent on the ESR time scale. The arrangement of the ligands around the iron is no more that depicted in **1** (Scheme 1). Further work is in progress in order to get more confidence on the NO coordination modes.

3.3. Catalytic properties of $[Fe(NO)_2L_3]^{\cdot+}$ complexes

Addition of butadiene, isoprene, norbornadiene, α-methylstyrene, styrene or isobutene to a solution containing the $[Fe(NO)_2L_3]^{\cdot+}$ species led to cyclodimerization and polymerization of diolefins and only to polymerization of monoolefins.

Concerning the diolefins, the cyclic C-C bond formation is due to the formation in the medium of the diamagnetic '$Fe(NO)_2$' moiety issued from the decomposition of $[Fe(NO)_2L_3]^{\cdot+}$.[6] For polymerization, the extensive study of the reaction was made possible for monoolefins due to the activity obtained even at low temperature ($-20 \leq T \leq 40$ °C) within a few hours, a prerequisite for $[Fe(NO)_2L_3]^{\cdot+}$ stability.[7]

Kinetic data obtained for styrene polymerization together with the determination of the molecular weight distributions and the effect of the counter-anion (PF_6^-, BF_4^- or ClO_4^-) led to the conclusion that polymerization proceeds through a cationic mechanism. The system is quite active: 100 mol of styrene per mol of Fe are converted in two hours at -20 °C. Copolymerization of an equimolecular styrene-methacrylate feed led to a copolymer containing 90% styrene: a feature of a cationic mechanism. Accordingly, isobutene was polymerized at -20 °C. Blank test for styrene polymerization pointed out that the iron moiety was the active species. The proposed initiation step is depicted in Scheme III:

Scheme III

A better description of the coordination sphere is not possible at this stage of research. The fact that the cationic behavior prevails over the radicalar one for the radical-cation initiator indicates that either the SOMO has too much NO character or that styrene coordination is occurring in the axial sites of the TBP complex **1**. In both situations, the iron center behaves as a powerful electrophile, a property induced by the NO ligands. The best activity was found in low coordinating medium. Addition of PPh_3, hexamethylphosphotriamide or tetrabutylammonium iodide during the course of the reaction inhibited further polymerization.

Actually, a radicalar polymerization was found with acrylate and acrylonitrile substrates in experimental conditions where styrene was not transformed. The prerequisite to observe a radicalar behavior is to change the counter-anion to BPh_4^-. In this case the properties of the iron center are strongly modified. Acrylate and acrylonitrile are no more coordinated through the keto- or nitrile group, respectively, but through their olefinic bond as evidenced by IR.[4] Addition of PPh_3 or DMF inhibited the polymerization. Experiments conducted in order to check the reactivity of radicalar by-products arising from the decomposition of the nitrosyl complex (NO or phenyl transfer) were negative. The polymerization activity is due to the nitrosyl complex. No structural study has yet been undertaken on this system.

4. DISCUSSION

Several properties of $\{Fe(NO)_2\}^9$ compounds have been described. Both tetra- and five-coordination have been characterized according to the nature of the additional ligands stabilizing the paramagnetic dinitrosyl moiety. The increase in the coordination number from four to five is encountered with either a monoanionic tridentate chelating ligand, **3**, or neutral two-electron monodentate ones, **1**. While one can understand that the five-coordination is induced by the chelating property, its occurrence with monodentate ligands is proposed to be induced by the presence of the cationic charge on the metal.

One of the other outstanding features of these complexes is that the NO ligands remain equivalent and in a nearly linear arrangement so that tetra-coordination leads to a 17-electron configuration and five-coordination to a 19-electron one. One could have suspected that the coordination sphere expansion should have led to the bending of one NO ligand therefore keeping a 17-electron configuration. According to recent results, such M-N-O structural change seems to be favored for mononitrosyl-containing species[19] whereas for dinitrosyl complexes the 19-electron structure is stabilized, the SOMO having 2π NO character.[20-24] X-ray structural determinations of two such complexes confirm that the two M-N-O bond angles remains close to 180°.[18a, 21] In our study, the comparison of the ESR isotropic A_N(NO) values as well as the EXAFS Fe-N(NO) bond values for $Fe(NO)_2PPh_3Cl$ and $[Fe(NO)_2(MeCN)_3]^+$ confirms the increased NO ligand character of the SOMO going from the 17- to the 19-electron configuration.

Catalytic activity for the transformation of olefins is only observed with the radical-cation, **1**, in low coordinating medium. Three coordination sites are potentially available for the olefin activation instead of one for $Fe(NO)_2Cl^{\bullet}$. The polymerization-type C-C bond coupling observed is due to cationic initiation. This mechanism reflects the high electrophilicity of the iron center and its low radicalar character. The overall behavior is induced by the presence of the NO ligands. The fact that cyclodimerization did not occur with diolefins whereas it was observed with $\{M(NO)_2\}^{10}$ moieties such as the cation $[Co(NO)_2L_2]^+$ and the neutral $Fe(NO)_2L_2$ complexes [2,4] can be at least explained in structural terms although electronic factors cannot be discarded. Very recently Poliakoff et al.[25] have proposed an attractive structure for the key intermediate in the catalytic dimerization of butadiene induced by photolysis of $Fe(NO)_2(CO)_2$: a pentacoordinated complex, $Fe(NO)_2(\eta^2\text{-}C_4H_6)(\eta^4\text{-}C_4H_6)$, with one NO group changing from linear to bent mode and three coordinated C=C bonds. Hence one outstanding structural difference between **1** and the above-mentionned intermediate should be the NO bending in diamagnetic $\{Fe(NO)_2\}^{10}$ complexes and not in paramagnetic $\{Fe(NO)_2\}^9$ ones, while three coordination sites are available for olefin activation in both cases.

5. ACKNOWLEDGEMENTS

The authors wish to thank Dr. H. Faucher for helpful discussions and J. F. Dutel for technical assistance in the ESR experiments.

6. REFERENCES

(1) Present address: CNRS-LCC, 205, route de Narbonne, 31077-Toulouse Cedex, France.

(2) D. Ballivet and I. Tkatchenko, *J. Mol. Catal.*, 1975/76, **1**, 319.

(3) D. Ballivet and I. Tkatchenko, *Inorg. Chem.*, 1977, **16**, 945.

(4) D. Ballivet, C. Billard and I. Tkatchenko, *J. Organomet. Chem.*, 1977, **124**, C9.

(5) D. Ballivet, C. Billard and I. Tkatchenko, *Inorg. Chim. Acta*, 1977, **25**, L58.

(6) D. Ballivet-Tkatchenko, M. Riveccie and N. El Murr, *J. Am. Chem. Soc.*, 1979, **101**, 2763.

(7) D. Ballivet-Tkatchenko, C. Billard and A. Révillon, *J. Polym. Sci., Polym. Chem. Ed.*, 1981, **19**, 1697.

(8) D. Ballivet-Tkatchenko and C. Brémard, *J. Chem. Soc., Dalton Trans.*, 1983, 1143.

(9) D. Ballivet-Tkatchenko, A. Boughriet and C. Brémard, *J. Electroanal. Chem.*, 1985, **196**, 315.

(10) J. Kopf and J. Schmidt, *Z. Naturforsch.*, 1975, **30b**, 149.

(11) *Inorg. Synth.*, G. W. Parshall Ed., McGraw-Hill, 1974, **XV**, 85.

(12) D.Ballivet-Tkatchenko, B. Nickel, A. Rassat and J. Vincent -Vaucquelin, *Inorg. Chem.*, 1986, **25**, 3497.

(13) H. Soling and R. W. Asmussen, *Acta Chem. Scandinav.*, 1957, **11**, 1534.

(14) J. H. Enemark and R. D. Feltham, *Coord. Chem. Rev.*, 1974, **13**, 339.

(15) J. Schmidt, *Z. Naturforsch.*, 1972, **27b**, 600.

(16) N. G. Connelly and C. Gardner,*J. Chem. Soc., Dalton Trans.*,1976, 1525.

(17) D. Ballivet-Tkatchenko, C. Esselin and J. Goulon, *J. Phys.*, 1986, **47**, C8-343.

(18) (a) K. S. Chong, S. J. Rettig, A. Storr and J. Trotter, *Can. J. Chem.*, 1979, **57**, 3113; (b) *Ibidem*, 1979, **57**, 3119.

(19) W. E. Geiger, P. H. Rieger, B. Tulyathan and M. D. Rausch, *J. Am. Chem. Soc.*, 1984, **106**, 7000.

(20) J. R. Budge, J. A. Broomhead and P. D. W. Boyd, *Inorg. Chem.*, 1982, **21**, 1031.

(21) S. Yu Yeung, R. A. Jacobson and R. J. Angelici, *Inorg. Chem.*, 1982, **21**, 3106.

(22) N. M. Atherton, J. R. Morton, K. F. Preston and M.J. Vuolle, *Chem. Phys. Lett.*, 1980, **70**,4.

(23) J. R. Morton, K. F. Preston and S. J. Strach, *J. Phys. Chem.*, 1980, **84**, 2478.

(24) C. Couture, J. R. Morton, K. F. Preston and S. J. Strach, *J. Magn. Reson.*, 1980, **41**, 88.

(25) G. E. Gadd, M. Poliakoff and J. Turner, *Organometallics*, 1987,**6**, 391.

ACTIVATION OF CO-ORDINATED π-HYDROCARBONS BY ONE-ELECTRON TRANSFER

Neil G. Connelly
School of Chemistry
University of Bristol
Bristol
BS8 1TS
U.K.

ABSTRACT. General observations are made concerning the oxidative and reductive dimerisation of co-ordinated π-hydrocarbon ligands. Redox-induced C-C bond formation and cleavage reactions of a series of η^4-cyclo-octatetraene (cot) complexes and their derivatives are then described. The irreversible one-electron oxidation of $[ML_n(\eta^4-C_8H_8)]$ [ML_n = $Fe(CO)_3$, $Ru(CO)_2(PPh_3)$, RhCp (Cp = $\eta-C_5H_5$), etc.] gives 17-electron cations $[ML_n(\eta^4-C_8H_8)]^+$ which rapidly isomerise to $[ML_n(C_8H_8)]^+$ in which the monocyclic ring is η^5-dienyl bonded to M with the remaining three carbon atoms forming an unbound allyl radical fragment. Radical-radical coupling then gives the dimer $[M_2L_{2n}(\eta^5:\eta'^5-C_{16}H_{16})]^{2+}$ having two linked cyclo-octatrienyl rings. The behaviour of this dication depends on ML_n. For the iron complexes, electrocyclic ring closure gives $[Fe_2(CO)_4L_2(\eta^5:\eta'^5-C_{16}H_{16})]^{2+}$ [L = CO or $P(OPh)_3$] which contains two bicyclo[5.1.0]octadienyl groups; the hexacarbonyl is the precursor to three more $C_{16}H_{16}$ complexes sequentially formed by reduction, oxidation, and reduction processes each overall involving a two-electron change. X-Ray crystallography shows that each of these redox reactions results in stereo- and regiospecific polycyclic hydrocarbon rearrangements. The oxidation of $[Ru(CO)_2L(\eta^4-cot)]$ (L = CO or PPh_3) yields $[Ru_2(CO)_4L_2(\eta^2,\eta^3:\eta'^2,\eta'^3-C_{16}H_{16})]^{2+}$ in which each monocyclic ring is bonded to the metal *via* alkene and allyl fragments. The analogous cyclopentadienylrhodium dimer undergoes thermal isomerisation to give the asymmetric dication $[Rh_2(\eta^2,\eta^3:\sigma',\eta'^4-C_{16}H_{16})]^{2+}$.

1. INTRODUCTION

The importance of organotransition-metal chemistry derives for the most part from two sources. First, the reactivity of an organic fragment is changed on coordination to a metal. Second, the metal can readily accommodate a variation in coordination number so that the rapid interchange between 18- and 16-electron configurations can lead to the catalysis of organic reactions.

In recent years it has become clear[1] that one-electron transfer reactions can add a further dimension to organometallic chemistry. For example, oxidation or reduction provides a simple route to 17- or 19-electron complexes with chemical and structural properties very different from those of diamagnetic, 18-electron precursors.

M. Chanon et al. (eds.), Paramagnetic Organometallic Species in Activation / Selectivity, Catalysis, 71–83.

Usually such redox reactions lead to activation at the metal centre resulting, for example, in ligand substitution (both stoichiometric and catalytic), isomerisation of octahedral carbonyl derivatives, etc. However, it is rapidly becoming apparent that hydrocarbon ligand activation is also possible. Geiger et al. have studied a series of redox-induced isomerisations of cyclic polyenes[2], and $[CPh_3]^+$ oxidation of species such as $[WMe_2Cp_2]$[3], $[ReR(PPh_3)(NO)Cp]$ (R = alkyl)[4] and $[Fe(\eta^4\text{-}C_6H_7R)(\eta\text{-}C_6H_6)]$ [R = CH_2Ph, $CHS(CH_2)_3S$, etc.][5] leads to C–H activation in the resulting radical cations.

In this article we will discuss the redox-induced C–C bond formation and cleavage reactions of coordinated π-hydrocarbons. In Sections 3 and 4 we will focus on our studies of a range of η^4-cyclooctatetraene (cot) derivatives of the iron and cobalt groups. First, however, we will consider in a qualitative fashion how one-electron oxidation or reduction might induce C–C bond formation in simpler unsaturated hydrocarbons.

2. PATHWAYS FOR REDOX-INDUCED C–C BOND FORMATION

Figures 1 and 2 show schemes for the oxidative and reductive dimerisation of two of the simplest possible coordinated π-hydrocarbons. In the first example (Figure 1), oxidation of a vinyl complex gives a paramagnetic cation with a 17-electron metal

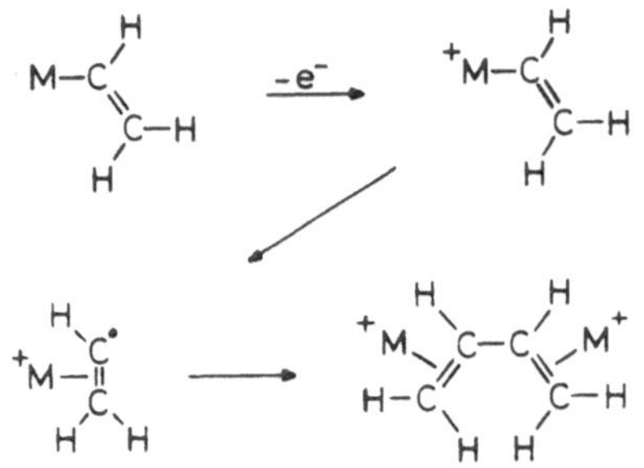

Figure 1. The oxidative dimerisation of an 18-electron metal vinyl complex.

centre. A σ–π rearrangement then allows the metal to regain the stable 18-electron configuration, and at the same time the positive charge (on the metal) is separated from the unpaired electron density (on the hydrocarbon ligand). Two carbon-based radicals then couple to give the η^2:η'^2-butadiene ligand.

This particular reaction, i.e. of a metal vinyl, is as yet hypothetical. However, the σ-allyls $[Fe(\sigma\text{-}CH_2CR^1\text{=}CR^2R^3)(CO)_2Cp]$ (R^1 = H or Me; R^2 = H, Me, or Ph; R^3 = H or Me) undergo[6] just such an oxidative dimerisation giving, for example, $[Fe_2(CO)_4(\eta^2\text{:}\eta'^2\text{-}CH_2\text{=}CHCMe_2CMe_2CH\text{=}CH_2)Cp_2]^{2+}$. Moreover, this reaction is regiospecific in that only the more highly substituted carbon atoms are linked. More recently, iodosobenzene oxidation of $[Fe(C\text{=}CHMe)(dppe)Cp]^+$ has been observed[7] to give the 2,3-dimethyl-1,3-butadien-1,4-diylidene dication $[Fe_2(\mu\text{-}C_4Me_2)(dppe)_2\text{-}Cp_2]^{2+}$. The overall reaction is more complex than that shown in Figure 1 in that oxidation of the carbene complex to $[Fe(C{=}CHMe)(dppe)Cp]^{2+}$ appears to be followed by deprotonation to the 17–electron acetylide $[Fe(C_2Me)(dppe)Cp]^+$ and then by dimerisation, again at the alkylated carbon atom. The oxidation–deprotonation step is reminiscent of the double–oxidation deprotonation mechanism for the formation of the μ–CH^+ ligand from the carbene–bridged diruthenium complex[8] $[Ru_2(\mu\text{-}CH_2)(\mu\text{-}CO)(\mu\text{-}Ph_2PCH_2PPh_2)Cp_2]$.

A reductive route to C–C bond formation, *via* a 19–electron intermediate, is outlined in Figure 2 for the simple, again as yet hypothetical, case of a cationic ethene complex. Here, the formation on reduction of the unfavourable 19–electron

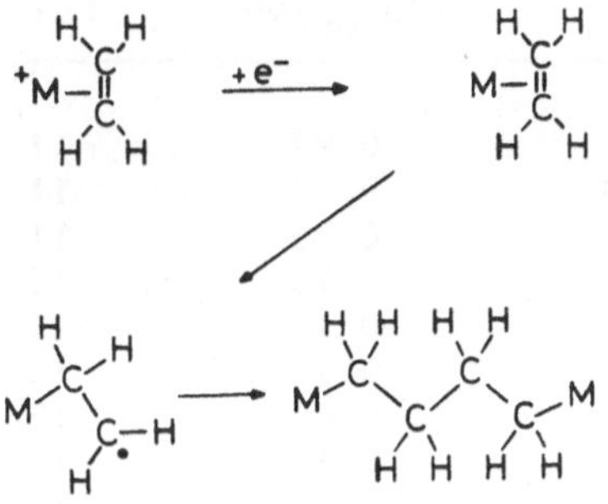

Figure 2. The reductive dimerisation of an 18–electron metal–ethene complex.

configuration is relieved by a π–σ rearrangement which again generates an 18–electron metal centre and a carbon–based radical poised for dimerisation.

Reductive coupling reactions of cationic cyclic dienyl complexes have been previously observed[1,9], and recent evidence has been presented[10] for radical intermediates in the electrochemical or zinc reduction of the cyclohexadienyl complex $[Fe(CO)_3\{\eta^5\text{-}C_6H_6(CO_2Me)\}]^+$.

Reactions of the types outlined in Figures 1 and 2 will depend upon the presence of hydrocarbons capable of variable hapticity so that, for example, an unbound site is available to supplement the bonding requirements of an electron–deficient metal generated by oxidation. The η^4–cot ligand is ideally suited to such reactions in having two unbound alkene groups in close proximity to the metal centre. We now describe how the oxidation of a range of η^4–cot complexes has led to the isolation of a series of novel polycyclic hydrocarbon complexes, formed *via* stereo– and regiospecific C–C bond making and breaking reactions.

3. THE REDOX ACTIVATION OF CO–ORDINATED CYCLO–OCTATETRAENE.

3.1. The Oxidation of η^4–Cot Complexes.

Cyclic voltammetry shows that each of the complexes $[ML_n(\eta^4\text{-}cot)]$ [ML_n = $Fe(CO)_2L$, L = CO or $P(OPh)_3$; $Fe(CO)_{3-n}\{P(OMe)_3\}_n$ (n = 1–3);[11] $Ru(CO)_2L'$, L' = CO or PPh_3;[12] $CoCp^*$ (Cp^* = $\eta^5\text{-}C_5Me_5$),[13,14] or $RhCp$[15]] undergoes irreversible one–electron oxidation at a platinum bead electrode in CH_2Cl_2 (Table 1).

The data in Table I are of note in showing (i) the large variation in $(Ep)_{ox}$ with n in $[Fe(CO)_{3-n}\{P(OMe)_3\}_n(\eta^4\text{-cot})]$, enabling the accessibility of the 17-electron cations to be tailored to the oxidising power of suitable chemical oxidants, and (ii) the relative ease with which the $CoCp^*$ and RhCp complexes are oxidised when compared with the isoelectronic iron and ruthenium tricarbonyls.

TABLE I Cyclic Voltammetric Data for the Irreversible One-Electron Oxidation of $[ML_n(\eta^4\text{-cot})]$

ML_n	$(Ep)_{ox}/V^a$	Ref.
$Fe(CO)_3$	0.97	11
$Fe(CO)_2\{P(OPh)_3\}$	0.67	11
$Fe(CO)_2\{P(OMe)_3\}$	0.55	11
$Fe(CO)\{P(OMe)_3\}_2$	0.19	11
$Fe\{P(OMe)_3\}_3$	-0.13	11
$Ru(CO)_3$	0.76	12
$Ru(CO)_2(PPh_3)$	0.46	12
$CoCp^*$	0.10	13,14
RhCp	0.55	15

[a] Oxidation peak potentials were measured at a scan rate of 200 mVs^{-1}, in CH_2Cl_2 at a Pt bead electrode, and are relative to the standard calomel electrode. Under the conditions used, the E^o value for the couple $[FeCp_2]^+/[FeCp_2]$ is 0.47 V

Further electrochemical studies of the species listed in Table I have been restricted, for example by the insolubility of the oxidation products in the case of iron. However, the results of detailed studies[15] on $[Rh(\eta^4\text{-cot})Cp]$, which have shed light on the mechanism of the chemical reaction following electron loss, are discussed in Section 3.3.

All of the complexes in Table I are chemically oxidised to give binuclear species in which hydrocarbon dimerisation has occurred. Each will be discussed in turn as the metal effect is considerable and a variety of products has been isolated.

3.2. Iron Complexes

The chemical oxidation of $[Fe(CO)_3(\eta^4\text{-cot})]$ provides a dramatic demonstration of the redox activation of a coordinated hydrocarbon. Thus, on adding a deep blue CH_2Cl_2 solution of the oxidant $[N(C_6H_4Br\text{-}p)_3][PF_6]$ to an orange-red CH_2Cl_2 solution of $[Fe(CO)_3(\eta^4\text{-cot})]$ a near quantitative yield of yellow $[Fe_2(CO)_6(\eta^5{:}\eta'^5\text{-}C_{16}H_{16})][PF_6]_2$ (1) is immediately precipitated.[11]

The X-ray crystal structure (Figure 3) of the dication $[Fe_2(CO)_4\{P(OPh)_3\}_2\text{-}(\eta^5{:}\eta'^5\text{-}C_{16}H_{16})]^{2+}$ (1-P), formed by $AgPF_6$ oxidation of $[Fe(CO)_2\{P(OPh)_3\}(\eta^4\text{-cot})]$, shows not only that dimerisation has occurred, through C–C bond formation, but also that isomerisation has led to the formation of two bicyclo[5.1.0] octadienyl fragments. In addition, it is important to note that the redox reaction

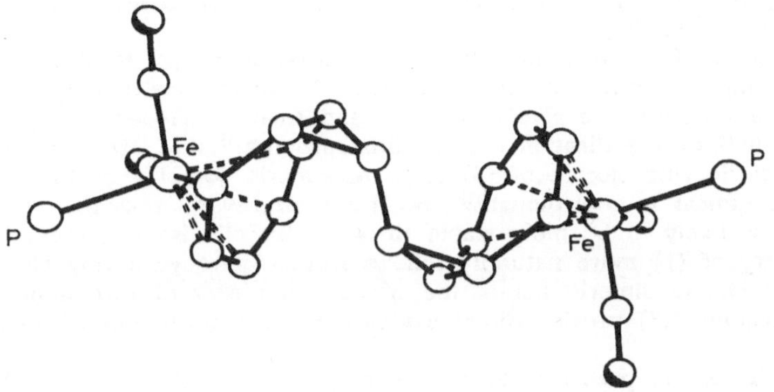

Figure 3. The X-ray structure of $[Fe_2(CO)_4\{P(OPh)_3\}_2(\eta^5:\eta'^5-C_{16}H_{16})]^{2+}$ (1–P); the phosphorus ligand substituents and the hydrogen atoms are omitted.

leads to only one of the many possible dimeric isomers.

The mechanism proposed for the formation of (1) is shown in Figure 4. It should be noted at the outset that none of the proposed radical intermediates, either formed in the oxidation reactions of the cot complexes themselves or in the subsequent reactions of the dimeric products, has been characterised. However, there is ample evidence for each of the individual steps shown.

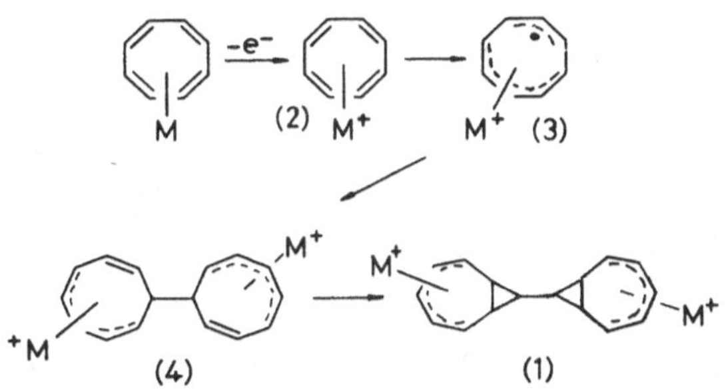

Figure 4. Proposed mechanism for the oxidative dimerisation of $[Fe(CO)_3(\eta^4-cot)]$ to (1); M=$Fe(CO)_3$.

The initial oxidation step, to form the 17–electron cation (2), is supported by photoelectron spectroscopy[16] and also by the isolation and full characterisation[17] of the tetraphenylcyclobutadiene analogue $[Fe(CO)\{P(OMe)_3\}_2(\eta^4-C_4Ph_4)][BF_4]$. The e.s.r. spectrum of this cation[17] is compatible with a metal–based radical, and a comparison[18,19] of the X-ray crystal structure with that of neutral [Fe(CO)-{P(OMe)$_3$}$_2$(η^4-C_4Ph_4)] shows little or no change in the hydrocarbon geometry on electron-transfer. In addition, the observed variations in the Fe-P and P-O bond lengths in the Fe{P(OMe)$_3$}$_2$ fragment (which, incidentally, provide evidence for the participation of P-X σ^* orbitals in metal-PX_3 bonding[19]), and the increase in $\tilde{\nu}$(CO) of *ca.* 80 cm^{-1}, also support metal- rather than hydrocarbon-based oxidation for [Fe(CO){P(OMe)$_3$}$_2$(η^4-C_4Ph_4)].

The formation of (1) then involves (i) isomerisation of (2) to the allyl-based radical (3), allowing the iron atom to regain the 18-electron configuration (c.f. Figure 1), (ii) C-C bond formation to give dimer (4), and finally (iii) electrocyclic ring closure in each half of the dication. The initial proposal[11] that dimerisation to (1) followed electrocyclic ring closure of (3) to a monomeric complex of the η^5-bicyclo [5.1.0]octadienyl radical is now thought[12] less likely for several reasons. First, the allyl radical (3) is likely to be more stable than the bicyclic isomer; second, the observed geometry of (1) more naturally follows from electrocyclic ring closure of (4); and third, the oxidative dimerisation of the η^4-cot complexes of ruthenium, cobalt, and rhodium (Section 3.3) provide direct evidence for the formation of analogues of (4).

Complex (1) is the precursor to a wide range of polycyclic hydrocarbon complexes[20-22]. Here, however, we will confine our discussion to a sequence of overall two-electron processes (Figure 5) which yields three other $C_{16}H_{16}$ complexes by reduction, oxidation, and then reduction again[20]. The reduction reactions, giving

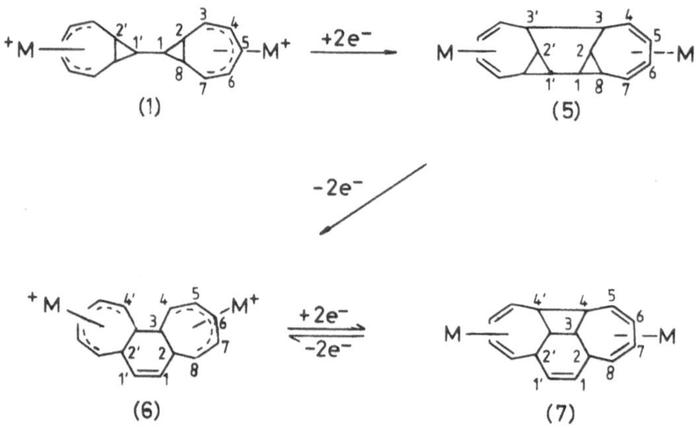

Figure 5. The reduction-oxidation sequence linking the $C_{16}H_{16}$ complexes (1), (5), (6), and (7); M = $Fe(CO)_3$

(5) and (7), can be carried out using K[BH(CHMeEt)$_3$] in thf at −78 °C; yields are only moderate but no other organometallic products are isolable. By contrast, near-quantitative yields of (6) are formed, either from (5) or (7), *via* [$FeCp_2$][PF_6] oxidation in CH_2Cl_2.

In each of the redox reactions shown in Figure 5, processes similar to those outlined in Figures 1 and 2 are envisaged. Thus, for the conversion of (1) to (5) and of (6) to (7), (i) reduction at the metal atoms generates 19-electron centres, (ii) detachment of terminal dienyl carbon atoms regenerates stable 18–electron sites, and (iii) the resulting carbon-based radicals undergo intramolecular C–C coupling to yield the neutral, diamagnetic products. For the oxidation reactions giving (6), each cationic 17–electron iron atom is able to regain the 18–electron configuration by coordination to a fifth carbon atom, supplied *via* opening of the cyclopropane ring in (5) and of the cyclobutane ring in (7).

It has been assumed, of course, that all of the reactions in Figure 5 proceed *via* converted one–electron processes at the two metal centres rather than by a stepwise mechanism. There is no evidence to support either pathway although the cyclic voltammograms of (5) and (7) each show only one, irreversible oxidation with the waveheight compatible with a two–electron process.

Although the transformations in Figure 5 can be rationalised in simplistic terms, as above, X–ray structural studies on all four of the $Fe_2(C_{16}H_{16})$ complexes have provided a more detailed insight into the stereo– and regiospecificity of the C–C bond making and breaking reactions.

Figures 6(a–d) show the metal–hydrocarbon skeletons of (1–P) [i.e. the

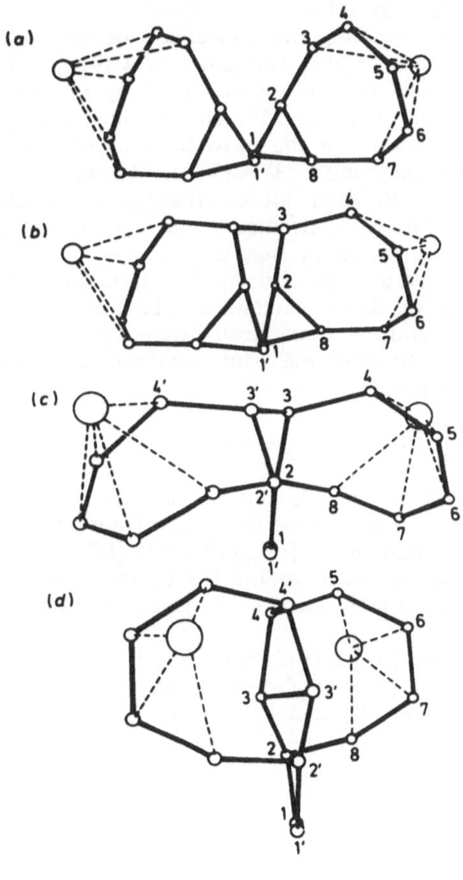

Figure 6. The Fe_2C_{16} cores of (a) (1–P), (b) (5), (c) (6), and (d) (7).

bis(triphenylphosphite)derivative of (1)][11], (5)[23], (6)[20], and (7)[20], viewed along bond C(1)–C(1'), i.e. the bond formed in the initial oxidative dimerisation of [$Fe(CO)_3$–(η^4–cot)] to (1). It should be noted first that the conformation of (1–P) is such that the two bicyclo[5.1.0]octadienyl groups are not exactly *trans* with respect to C(1)–C(1'); the C(2)–C(1)–C(1')–C(2') torsion angle is 59.7°. However, this distortion arises from unfavourable H...C non-bonded contacts between the two halves of the dication rather than from effects due to the $P(OPh)_3$ ligands. Complex (1–P), which has a ^1H n.m.r. spectrum almost identical to that of (1), can therefore be used as a model of (1), and structural comparisons with the hexacarbonyls (5) to (7) can be made with confidence.

Little conformational change is observed on reduction of (1). A simple rotation about C(1)–C(1'), which narrows the dihedral angle C(2)–C(1)–C(1')–C(2') to 27°, allows the formation of bond C(3)–C(3') in (5). The oxidation of (5) to (6) is accompanied by the opening of the cyclopropane rings at C(1)–C(8) and C(1')–C(8') [in preference to the cleavage of C(3)–C(3') which would regenerate (1)] and the formation of the cyclohexene ring C(1)–C(2)–C(3)–C(3')–C(2')–C(1'). The generation of the *cis*–alkene bond [C(1)–C(1')] results in the further closing of the dihedral angle C(2)–C(1)–C(1')–C(2') to near 0°.

The final reduction step, involving the conversion of (6) to (7), brings about the greatest conformational change in that formation of C(4)–C(4'), to give the cyclobutane ring C(3)–C(4)–C(4')–C(3'), is accompanied by a decrease in the torsion angle C(4)–C(3)–C(3')–C(4') from 160° to 25°, and by inversion of the cyclohexene ring conformation. The latter results in the movement of substituents C(4) and C(4') from *trans* to *gauche* positions on bond C(3)–C(3') i.e. from adjacent axial to equatorial sites on the C_6 ring. Perhaps not surprisingly, the oxidation of (7) leads to the rupture of the long [1.596(4)A], strained bond C(4)–C(4'), regenerating (6).

The sequence of reactions shown in Figure 6 readily demonstrates how electron–transfer can lead to stereo– and regiospecific C–C bond formation and cleavage. The high specificity of these reactions results not only from the requirement of coordinative saturation at the iron atoms, governed by the redox events, but also from the need to minimise conformational and angular strain energies in the metal–hydrocarbon skeletons.

3.3 Ruthenium, Cobalt, and Rhodium Complexes

In order to probe the effect of the metal, and perhaps of the bonding mode of the cot ligand (i.e. 1,3–bound to iron and ruthenium; 1,5–bound to cobalt and rhodium) we have begun to study the oxidation of [$ML_n(\eta^4$–cot)] [$ML_n = Ru(CO)_2(PPh_3)$, $CoCp^*$, and RhCp]. These studies have so far led to the isolation of several other binuclear species and, more important, have shed further light on the mechanism shown in Figure 4.

The chemical oxidation of [$Ru(CO)_2(PPh_3)(\eta^4$–cot)] is readily effected by [$FeCp_2]^+$ and, as in the case of [$Fe(CO)_3(\eta^4$–cot)], an excellent yield of only one product (8) is obtained. However, ^1H and ^{13}C n.m.r. spectroscopy revealed that (8) differs from (1) in containing the $\eta^2,\eta^3:\eta'^2,\eta'^3$–$C_{16}H_{16}$ ligand (Figure 7).

The behaviour of [$M(CO)_2L(\eta^4$–cot)] (M = Fe or Ru) towards one–electron oxidants mirrors their behaviour towards protons[24] in that [$Fe(CO)_2L(\eta^5$–$C_8H_9)]^+$ contains the bicyclo[5.1.0]octadienyl ligand whereas [$Ru(CO)_2L(\eta^2,\eta^3$–$C_8H_9)]^+$ has a monocyclic cyclo–octatrienyl ring bound to the metal *via* η^2–alkene and η^3–allyl fragments. There is a common intermediate in the protonation reactions, namely [$M(CO)_2L(\eta^5$–$C_8H_9)]^+$ in which the cyclo–octatrienyl group is η^5–dienyl bonded to iron or ruthenium. By analogy, it is reasonable to suggest that both (1) and (8) are formed from an intermediate containing an $\eta^5:\eta'^5$–$C_{16}H_{16}$ ligand with linked cyclo-octatrienyl rings [e.g. (4) in Figure 4].

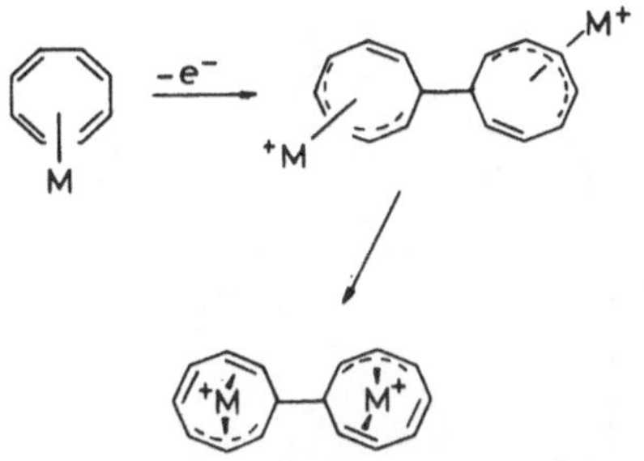

(8)

Figure 7. The oxidative dimerisation of $[Ru(CO)_2(PPh_3)(\eta^4\text{–cot})]$; M = $Ru(CO)_2(PPh_3)$

While these results provide only circumstantial evidence for dimerisation before isomerisation to (1) or (8), the oxidation of $[Co(\eta^4\text{–cot})Cp^*]$ and $[Rh(\eta^4\text{–cot})Cp]$ has led to the isolation of analogues of the bi(cyclo-octatrienyl) dication (4).

As yet only preliminary results[13] are available from the reaction of $[Co(\eta^4\text{–cot})Cp^*]$ with $[FeCp_2]^+$ although high yields of $[Co_2(\eta^5:\eta'^5\text{–}C_{16}H_{16})Cp^*_2][PF_6]_2$ have been isolated and the ^1H n.m.r. spectrum shows that the dimer has the hydrocarbon skeleton of (4) (Figure 4). In addition, cobaltocene reduction of the cobalt dication has given $[Co_2(C_{16}H_{16})Cp^*_2]$ (mass spectrum, elemental analysis, ^1H n.m.r. spectrum) as a mixture of as yet uncharacterised *asymmetric* isomers (c.f. iron).

The oxidation of $[Rh(\eta^4\text{–cot})Cp]$ is complicated, yet informative in that dimerisation is followed by a sequence of thermal isomerisation reactions not observed for any of the other metals studied. These reactions are sufficiently slow to be monitored by ^1H n.m.r. spectroscopy so that all of the binuclear species involved have been identified (Figure 8). In addition, detailed cyclic voltammetric studies have provided valuable mechanistic information[15] concerning the steps preceding dimerisation.

The scan rate and concentration dependence of the peak potential for the oxidation of $[Rh(\eta^4\text{–cot})Cp]$ shows that initial charge transfer is rapid and that the resulting radical cation undergoes a fast *first*-order reaction before dimerisation. Because the initially formed dimer has structure (11) (Figure 8), this reaction is most likely to be the rearrangement of the η^4–cot ligand of $[Rh(\eta^4\text{–cot})Cp]^+$ (9) [isoelectronic with (2), Figure 4].

The first dimeric complex shown in Figure 8, namely (11), is readily prepared by $[FeCp_2]^+$ oxidation of $[Rh(\eta^4\text{–cot})Cp]$ in CH_2Cl_2 at 0 °C. The structure, deduced from ^1H and ^{13}C n.m.r. spectroscopy, is that found for the dicobalt complex noted above with two diastereomers (11a and 11b) again observed, in an approximate 6:5 ratio. Complex (11) then isomerises to (12) [again as a diastereomeric mixture of (12a) and (12b) (*ca.* 3:1 ratio)] either slowly at room temperature or more rapidly in boiling acetone. Finally, heating (12) in CH_3NO_2 at 60 °C for 24h yields the asymmetric dirhodium complex (13).

The X-ray structure of (13), crystallised as the bis(tetraphenylborate) salt, is shown in Figure 9. One rhodium atom is η^2,η^3-bonded to the $C_{16}H_{16}$ polycyclic ring

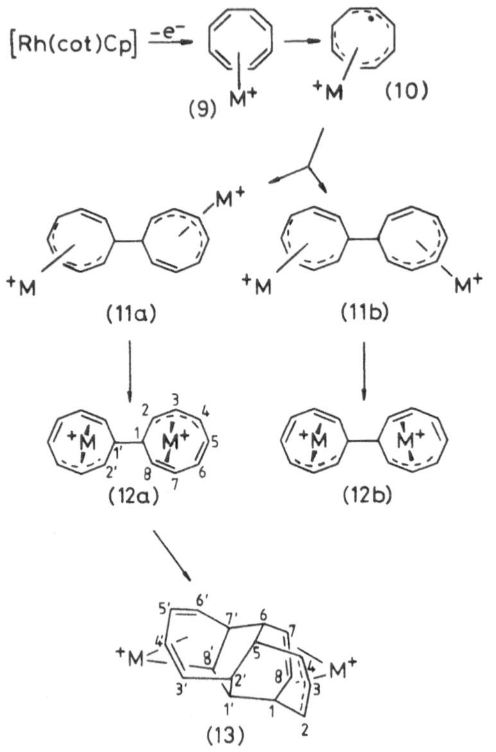

Figure 8. The oxidative dimerisation of [Rh(η^4-cot)Cp]; M = RhCp

system, rather as in (12), but the second metal atom is linked to the other half of the hydrocarbon ligand *via* one σ-bond and an η^4-butadiene fragment. In addition, two new C–C bonds [C(5)–C(2')] and C(6)–C(7')] are formed during the isomerisation of (12).

Although a complex mechanism might be envisaged for the formation of (13), only a simple conformational change in (12) is required to initiate the observed intramolecular C–C bond coupling reactions. Thus, rotation about C(1)–C(1') brings the uncoordinated double bond C(5)–C(6) into such a position that electrophilic attack by C(2'), a terminal coordinated allyl carbon atom of the second ring, can occur to generate bond C(5)–C(2'). The residual positive charge on C(6) then leads to electrophilic attack on C(7'), one of the coordinated alkene carbon atoms of the second ring, giving bond C(6)–C(7') and completing the formation of the cyclohexane ring C(1')–C(2')–C(5)–C(6)–C(7')–C(8').

It is important to note that this sequence can generate (13) only from diastereomer (12a); any similar sequence beginning with (12b) would give rise to a more highly strained polycyclic ring system. Because all of (12) is converted to only (13) we assume that (12b) isomerises to (12a).

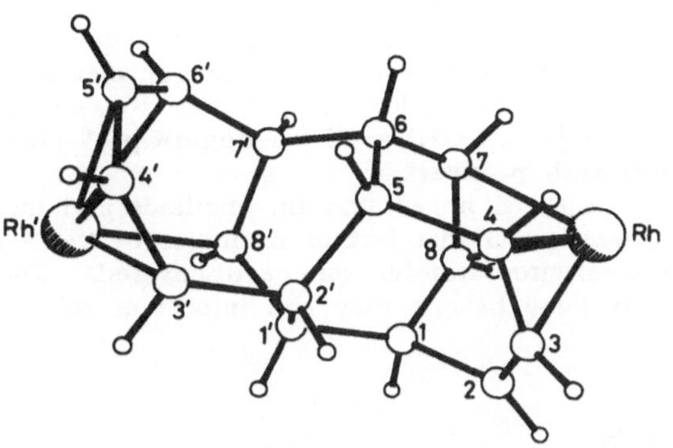

Figure 9. The $Rh_2(C_{16}H_{16})$ core of $[Rh_2(\eta^2,\eta^3{:}\sigma',\eta'^4\text{–}C_{16}H_{16})Cp_2]^{2+}$ (13).

4. CONCLUSIONS

The chemistry outlined above amply demonstrates how coordinated π–hydrocarbons can be activated *via* one–electron transfer. However, the examples given are complex, and successful partly because the cyclic polyene ligands readily display variable hapticity.

In order to carry out similar reactions with simpler π–hydrocarbons, variations of the general methods described may be necessary. For example, the oxidation of an η^4–butadiene complex, which has no additional binding site on the hydrocarbon, cannot proceed as in the case of the η^4–cot complexes described above. However, one can imagine a dimerisation mechanism such as that shown in Figure 10, where

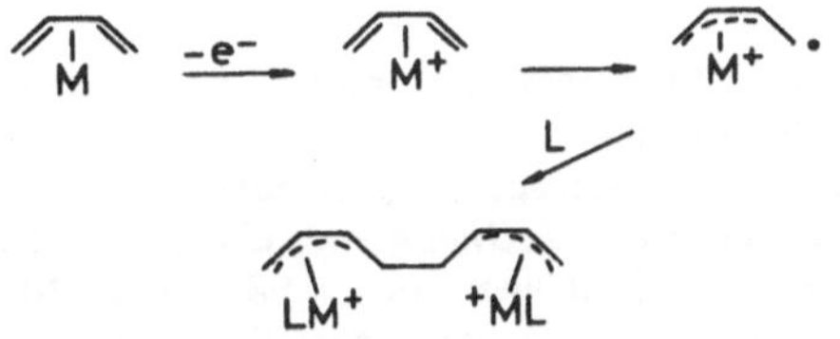

Figure 10. Proposed route for the oxidative dimerisation of η^4–coordinated butadiene; M is a 14–electron metal–ligand fragment.

coordination of an external ligand, for example CO or solvent, allows the metal to regain the favoured 18–electron configuration after the oxidation and hydrocarbon rearrangement steps have resulted in a 16–electron centre.

An alternative, and probably more promising possibility, involves the use of bi- or polynuclear species as electron-sinks where the bonding requirements of the individual metals can be accommodated by the formation or cleavage of metal-metal bonds. The example[8] referred to in Section 2, in which $[Ru_2(\mu\text{-}CH_2)(\mu\text{-}CO)(\mu\text{-}dppm)Cp_2]$ is converted to $[Ru_2(\mu\text{-}CH)(\mu\text{-}CO)(\mu\text{-}dppm)Cp_2]^+$ after double oxidation, may well involve such processes.

In summary, then, it is to be hoped that the methods and ideas outlined above will be extended and developed, and that further examples of the activation of hydrocarbon ligands by one-electron transfer will be discovered. There is no doubt that redox reactions such as these have a wide and important role to play in organometallic chemistry.

5. ACKNOWLEDGEMENTS

The work described in this article is the work of talented post-graduate and post-doctoral co-workers (Mark Gilbert, Paul Graham, Ray Kelly, Maureen Kitchen, Andy Lucy, John Sheridan, and Mark Whiteley) whose contributions I acknowledge with gratitude. Our efforts have also benefitted enormously from long-standing collaborations with the X-ray crystallographers at Bristol (Guy Orpen, Peter Woodward, and their students) and with Professor Bill Geiger at the University of Vermont, U.S.A.

6. REFERENCES

1. See, for example, N.G. Connelly and W.E. Geiger, *Adv. in Organomet. Chem.*, 1984, **23**, 1; W.E. Geiger and N.G. Connelly, *Adv. in Organomet. Chem.*, 1985, **24**, 87; J.C. Kotz in '*Topics in Organic Electrochemistry*', Ed. A.J. Fry and W.E. Britton, Plenum Press, New York, 1986; and refs. therein.
2. W.E. Geiger, T. Gennett, M. Grzeszczuk, G.A. Lane, J. Moraczewski, A. Salzer, and D.E. Smith, *J.Am.Chem.Soc.*, 1986, **108**, 7454; and refs. therein.
3. M.F. Asaro, S.R. Cooper, and N.J. Cooper, *J.Am.Chem.Soc.*, 1986, **108**, 5187; and refs. therein.
4. G.S. Bodner, J.A. Gladysz, M.F. Nielsen, and V.D. Parker, *J.Am.Chem.Soc.*, 1987, **109**, 1757; and refs. therein.
5. D. Mandon, L. Toupet, and D.Astruc, *J.Am.Chem.Soc.*, 1986, **108**, 1320.
6. P.S. Waterman and W.P. Giering, *J.Organomet.Chem.*, 1978, **155**, C47.
7. R.S. Iyer and J.P. Selegue, *J.Am.Chem.Soc.*, 1987, **109**, 910.
8. N.G. Connelly, N.J. Forrow, B.P. Gracey, S.A.R. Knox, and A.G. Orpen, *J.Chem.Soc., Chem.Commun.*, 1985, 14.
9. W.E. Geiger, T. Gennett, G.A. Lane, A. Salzer, and A.L. Rheingold, *Organometallics*, 1986, **5**, 1352; H. Ma, P. Weber, M.L. Ziegler, and R.D. Ernst, *Organometallics*, 1986, **5**, 2009, and refs. therein.
10. A.J. Pearson, Y.-S. Chen, M.L. Daroux, A.A. Tanaka, and M. Zettler, *J.Chem.Soc., Chem.Commun.*, 1987, 155.
11. N.G. Connelly, R.L. Kelly, M.D. Kitchen, R.M. Mills, R.F.D. Stansfield, M.W. Whiteley, S.M. Whiting, and P. Woodward, *J.Chem.Soc., Dalton Trans.*, 1981, 1317.
12. N.G. Connelly, P.G. Graham, and J.B. Sheridan, *J.Chem.Soc., Dalton Trans.*, 1986, 1619.

13. N.G. Connelly and M. Gilbert, unpublished results.
14. J. Moraczewski and W.E. Geiger, *Organometallics,* 1982, **1**, 1385.
15. L. Brammer, N.G. Connelly, J. Edwin, W.E. Geiger, A. G. Orpen, and J. B. Sheridan, *Organometallics,* submitted.
16. J.C. Green, P. Powell, and J. van Tilborg, *J.Chem.Soc., Dalton Trans.,* 1976, 1974; M.C. Bohm, and R. Gleiter, *Z. Naturforsch, Teil B,* 1980, **35**, 1028.
17. N.G. Connelly, R.L. Kelly, and M.W. Whiteley, *J.Chem.Soc., Dalton Trans.,* 1981, 34.
18. N.G. Connelly, A.G. Orpen, M.W. Whiteley, and P. Woodward, unpublished results.
19. A.G. Orpen and N.G. Connelly, *J.Chem.Soc., Chem.Commun.,* 1985, 1310.
20. N.G. Connelly, M.J. Freeman, A.G. Orpen, J.B. Sheridan, A.N.D. Symonds, and M.W. Whiteley, *J.Chem.Soc., Dalton Trans.,* 1985, 1027.
21. N.G. Connelly, A.R. Lucy, R.M. Mills, J.B. Sheridan, and P. Woodward, *J.Chem.Soc., Dalton Trans.,* 1985, 699.
22. N.G. Connelly, A.R. Lucy, R.M. Mills, M.W. Whiteley, and P. Woodward, *J.Chem.Soc., Dalton Trans.,* 1984, 161.
23. H.A. Bockmeulen, R.G. Holloway, A.W. Parkins, and B.R. Penfold, *J.Chem.Soc., Chem.Commun.,* 1976, 298; B.R. Penfold, Personal Communication.
24. M.A. Bennett, T.W. Matheson, G.B. Robertson, A.K. Smith, and P.A. Tucker, *Inorg.Chem.,* 1981, **20**, 2353.

REACTIONS OF PHOTOEXCITED CARBONYL COMPOUNDS WITH SILANES: MECHANISM AND REGIOSELECTIVITY

Gian Franco Pedulli
Dipartimento di Chimica Organica
Università di Bologna
Via S. Donato 15
40127 - Bologna
Italy

ABSTRACT. Silyl radicals, produced by hydrogen abstraction from silicon hydrides by photoexcited ketone triplets, add to the carbonyl group when the ketone do not contains other unsaturated reactive centers. If an olefinic double bond is also present, regiospecific or regioselective addition to the latter takes place to give the thermodynamically less stable adduct. Thermal rearrangement to the more stable adduct to oxygen, occurs only when the silyl group is attacked to a carbon atom α to the carbonyl group.

Carbonyl compounds, when photoreacted with silanes containing at least a silicon-silicon bond, are capable of bringing about displacement of a silyl radical in a fashion which is akin to the familiar homolytic substitution reaction, S_H2. Strong evidence in favour of an electron transfer mechanism for this reaction is provided by the experimental results.

1. INTRODUCTORY REMARKS

The addition of carbon centred radicals to carbon-oxygen double bonds is a practically unknown process. This is likely due to both kinetic and thermodynamic factors, since the carbon-oxygen (ca. 83 kcal/mol) or carbon-carbon bond which would be formed, depending on the site of attack, are only slightly stronger than the π bond (ca. 76 kcal/mol) which has to be cleaved. If using the thermochemical data of ref. 1, the addition of methyl radicals to acetone can be calculated to be exothermic by only 6 and 5 kcal/mol for eq.s 1 and 2, respectively. Because of the low exothermicity, these reactions are also expected to be readily reversible.

M. Chanon et al. (eds.), Paramagnetic Organometallic Species in Activation / Selectivity, Catalysis, 85–95.

$$Me\cdot + Me_2C{=}O \longrightarrow Me_2\dot{C}{-}OMe \qquad (1)$$

$$Me\cdot + Me_2C{=}O \longrightarrow Me_3C{-}O\cdot \qquad (2)$$

The analogous addition of silicon centred radicals to the oxygen atom of carbonyl groups is instead a facile and highly exothermic process since the silicon-oxygen bond being formed (111 kcal/mol) [2] is much stronger than the π bond being broken. A heat of reaction of -33 kcal/mol can be calculated for the addition of trimethylsilyl radicals to the oxygen atom of acetone (eq. 3). The potentially concurrent reaction 4, involving attack at the carbon atom, is disfavoured by ca. 28 kcal/mol with respect to 3 [3].

$$Me_2C{=}O + Me_3Si\cdot \longrightarrow Me_2\dot{C}{-}OSiMe_3 \qquad (3)$$

$$Me_2C{=}O + Me_3Si\cdot \longrightarrow Me_2(Me_3Si)C{-}O\cdot \qquad (4)$$

When reacting silicon hydrides with ketones or quinones under photolytic conditions, silyl radicals are produced *via* hydrogen abstraction from the silane by the n,π^* excited state of the carbonyl compound. Reaction 5 is as fast as the related hydrogen abstraction by *tert*-butoxyl radicals [4] (see Table I).

$$^3R_2C{=}O + R'_3SiH \longrightarrow R'_3Si\cdot + R_2\dot{C}{-}OH \qquad (5)$$

$$R'_3Si\cdot + R_2C{=}O \longrightarrow R_2\dot{C}{-}OSiR'_3 \qquad (6)$$

TABLE I - Rate constants for the reaction of benzophenone triplets and *tert*-butoxyl radicals with some silanes at 300K [4].

Substrate	$k(^3Ph_2C{=}O)/M^{-1}s^{-1}$	$k(t\text{-}BuO\cdot)/M^{-1}s^{-1}$
Et_3SiH	9.6×10^6	5.7×10^6
Cl_3SiH	7.6×10^6	4.0×10^7
$PhSiH_3$	5.0×10^6	7.5×10^6

The subsequent addition of the silyl radicals to the carbonyl group (eq. 6) proceeds with rates strongly dependent on the chemical nature of the substrate [5]. Thus, for instance, ketones containing conjugated phenyl groups exhibit higher rate constants because of the stabilization of the radical adduct by electron delocalization (see Table II). More details on the reactions of silyl radicals with variously substituted carbonyl derivatives can be found on a recently published review article [6].

Carbonyl compounds containing unsaturated substituents may undergo attack by silyl radicals either to the carbonyl group or to the substituent, or even to both. We are there-

TABLE II - Rate constants for the addition of $Et_3Si\cdot$ radicals to carbonyl compounds at 300K [5].

Substrate	$\underline{k}/M^{-1}s^{-1}$	Substrate	$\underline{k}/M^{-1}s^{-1}$
Pentan-3-one	2.8×10^{8}	Benzophenone	3.0×10^{7}
Cyclopentanone	7.2×10^{8}	Fluorenone	1.5×10^{9}
Cyclohexanone	6.6×10^{8}	Benzil	3.3×10^{8}
Acetophenone	1.2×10^{7}	Duroquinone	2.2×10^{9}

fore faced with the problem of the regiochemistry of addition to this kind of substrates. In this paper, we will deal with the reactions of silyl radicals with molecules containing both olefinic and carbonylic double bonds, which are relevant for the subsequent discussion on the photoreactions of ketones and diones with polysilanes.

2. REGIOCHEMISTRY OF THE ADDITION OF SILYL RADICALS TO UNSATURATED CARBONYL COMPOUNDS.

The simpler example of a molecule containing both C=C and C=O unsaturations is represented by acrolein. Thermodynamic, kinetic, and theoretical estimates can usually provide sensible predictions on the preferred site of radical addition to a particular substrate containing more than one reactive function.

HOMO: $O{=}C_1{-}C_2{=}C_3$ with coefficients 0.48 (C_2), 0.58 (C_3), -0.58 (O), -0.30 (C_1)

LUMO: $O{=}C_1{-}C_2{=}C_3$ with coefficients -0.39 (C_2), 0.59 (C_3), 0.51 (O), -0.48 (C_1)

The atomic orbital cefficients of the two frontier molecular orbitals of acrolein [7], shown above, suggest that the positions more susceptible to radical attack are the two terminal atoms, i.e. C_3 and the oxygen. Since however silyl radicals, due to their nucleophilic character, will mostly interact with the unoccupied orbital of lowest energy of acrolein, and being C_3 the atom bearing the larger electron density in the LUMO, a slight preference for addition to the olefinic double bond should be anticipated.

Kinetic considerations lead to the same prediction. On the basis of the Hammond postulate, the transition states for the highly exothermic reactions 7 and 8 are expected to be reactant like, and therefore radical attack to the weaker C=C bond should proceed with a lower activation energy. This point of view is consistent with the results of recent kinetic studies reported by Ingold and coworkers who showed that the room temperature rate constants for the addition of trialkylsilyl radicals to dialkyl olefines (ca. 10^6 to 10^7 $M^{-1}s^{-1}$) [8] were about an order of magnitude higher than those for the addition to structurally related dialkyl ketones (ca. 10^5 to 10^6 $M^{-1}s^{-1}$) [5].

$$Me_3Si\cdot + O{=}CH{-}CH{=}CH_2 \longrightarrow \begin{cases} Me_3SiO{-}\dot{C}H{-}CH{=}CH_2 & (7) \\ O{=}CH{-}\dot{C}H{-}CH_2SiMe_3 & (8) \end{cases}$$

The silyl adduct to carbon, on the other hand, even though kinetically favoured, is expected to be thermodynamically less stable than the adduct to oxygen. From the thermochemical data of ref.s 1, 2, and 3, the heats of reactions 7 and 8 can be calculated as ca. -41 and -28 kcal/mol, respectively. Addition to the carbonyl oxygen should then be more exothermic by ca. 13 kcal/mol, despite the fact that the cleavage of the C=O π bond (76 kcal/mol) is more costly, on energetic grounds, than that of a C=C π bond (ca. 60 kcal/mol). That is because the gain in energy associated with the formation of the very strong Si-O bond (111 against 88 kcal/mol [3] of a C-Si bond) largely exceeds the energy loss due to the cleavage of the stronger π bond.

The overall picture can be summarized as follows: kinetic control of the reaction should lead to the preferential formation of the silyl adduct to the olefinic double bond, and thermodynamic control to the adduct to the carbonyl oxygen.

Experiments performed by reacting photogenerated triphenylsilyl radicals with triphenylsilyl vinyl ketone, which is structurally similar to acrolein, led to the ESR detection of the radical adduct resulting from addition of $Ph_3Si\cdot$ to the terminal carbon atom (eq.9), in the accessible temperature range [9].

$$Ph_3Si\cdot + Ph_3SiCO{-}CH{=}CH_2 \longrightarrow \underset{\mathbf{1}}{Ph_3SiCO{-}\dot{C}H{-}CH_2SiPh_3} \quad (9)$$

Attack of the exocyclic C=C bond was also uniquely observed by reacting silyl radicals with 9-methylenanthrone (eq. 10) and with benzylidene-3,6-dimethylthieno[3,4-*b*]thiophen-2-one (eq.11). The thermodynamically more stable adduct

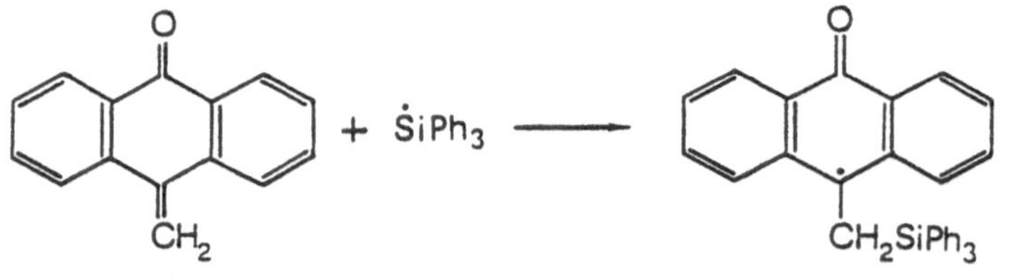

(10)

2

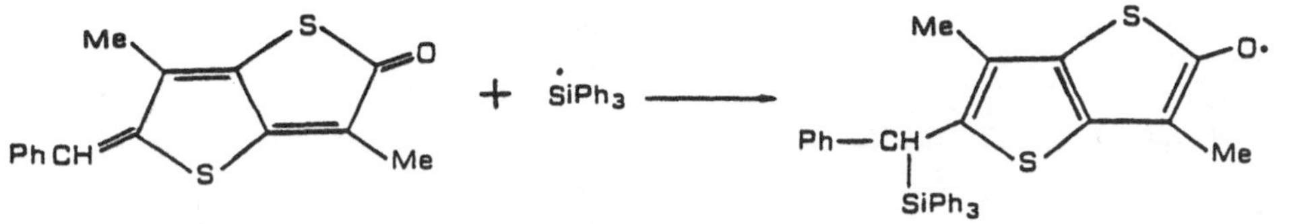

(11)

3

to oxygen was never detected even at temperatures as high as 400K, this indicating that the addition of silyl radicals to the olefinic double bond is not reversible.

Quite different results were instead obtained when reacting silyl radicals with several diones such as maleic anhydride (eq. 12) [11], 2,6-di-tert-butyl-p-benzoquinone (DBQ) (eq. 13), and 3,6-dimethylthieno[3,2-b]thiophen-2,5-dione (DMTD) (eq. 15) [10]. With the notable exception of the bulky tris(trimethylsilyl)silyl radicals which only added to the oxygen, perhaps because of steric reasons, in all other cases the nature of the radicals detected was dependent on temperature, the silyl adduct to a ring carbon being the dominant (or unique) species observed at lower temperatures and the adduct to oxygen the only species present at higher temperatures. Clean ESR spectra of radicals 7 (or 9) could also be obtained by heating, in the absence of light, solutions containing the mixture of adducts 6 and 7 (or 8 and 9) generated by low temperature

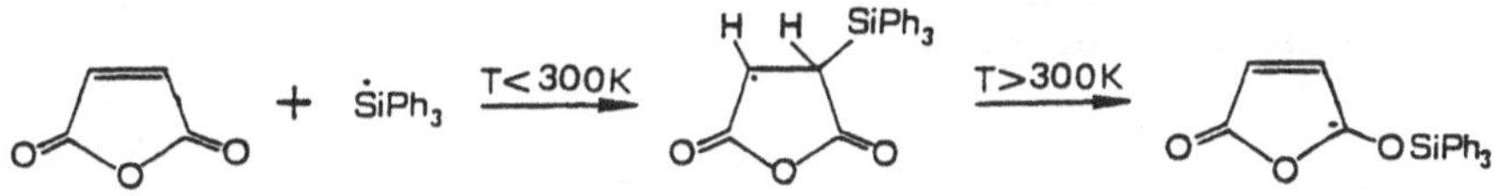

(12)

4 5

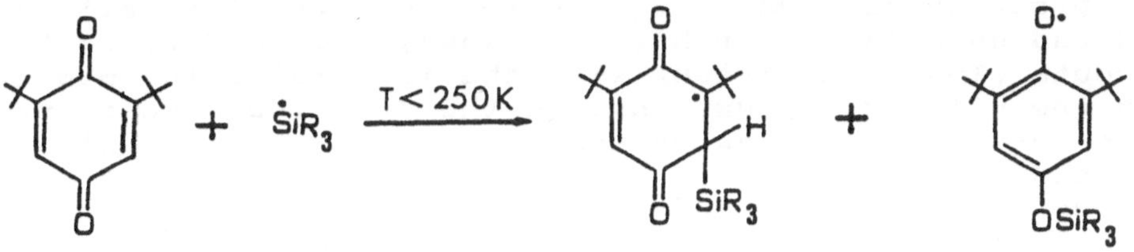

(13)

6; R=Me,Et,Ph (80%) 7 (20%)
R=Me_3Si (100%)

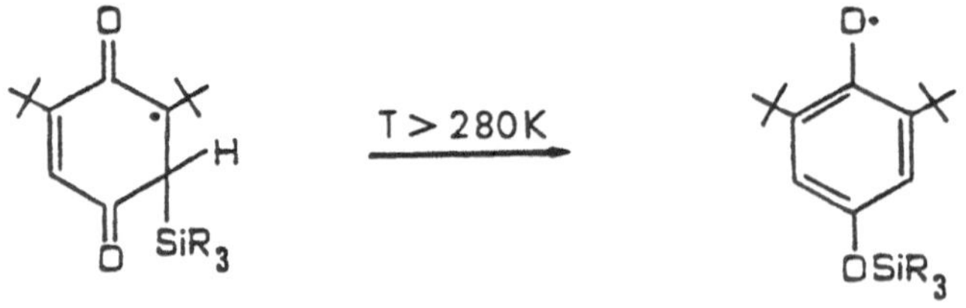

(14)

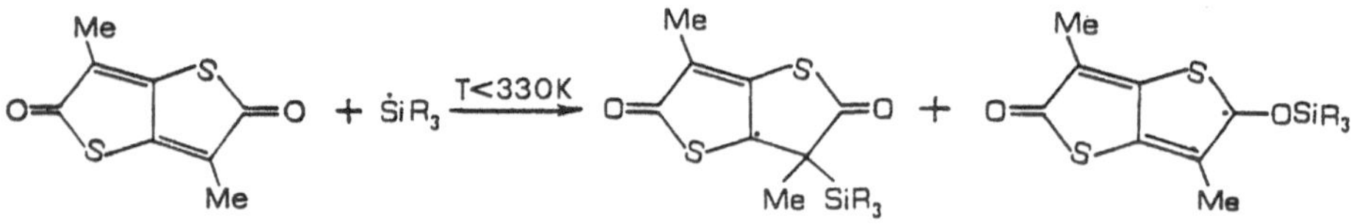

(15)

8; R=Me,Et,Ph (80%) 9 (20%)
R=Me_3Si (100%)

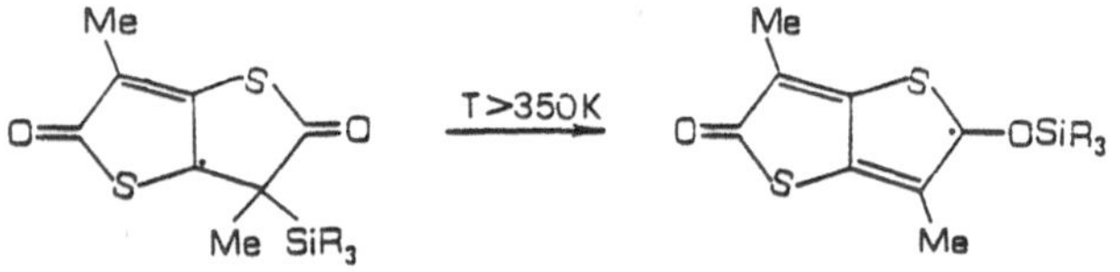

(16)

photolysis. No spectral variations were observed by decreasing the temperature again, this indicating that the conversion of the carbon to the oxygen adduct is irreversible.

In principle, isomerization might occur either by cleavage of the C-Si bond followed by readdition of the silyl radical to the carbonyl oxygen, or by an intramolecular rearrangement *via* the formation of a four membered cyclic transition state. The finding that radicals 1, 2, and 3 do not convert to the corresponding oxygen adducts, as well as experiments where the isomerization reaction was performed in the presence of a very efficient spin trap for silicon centred radicals (1,3,5-trinitrobenzene), led to discard the first route [10]. Consistent with the intramolecular migration of the silyl group was instead the fact that the isomerization of 8 into 9 was proved to be quantitative within experimental error.

Arrhenius parameters for the 6 to 7 (R=Ph) rearrangement were determined as E_a 18.2 kcal/mol and $\log(A/s^{-1})$ 13.8. The apparent activation energy for the 8 to 9 (R=Ph) conversion was found to be larger (ca. 27 kcal/mol), but the kinetic analysis was complicated in this case by a fast equilibration of the radicals with their dimers [10].

Temperature dependent reaction pathways were also observed when reacting triphenylsilyl radicals with maleic thioanhydride, maleimide [11], and 2,6-dimethoxy-p-benzoquinone [10]; at low temperature the dominant species was the ring adduct and at higher temperature the oxygen adduct.

We may therefore conclude that the attack of trialkylsilyl or triphenylsilyl radicals to the olefinic double bond of unsaturated carbonyl compounds is a more facile process than attack to the carbon-oxygen double bond, even though the adducts resulting from the former reaction are thermodynamically less stable. When addition occurs at a carbon atom α to the carbonyl group, rearrangement leading to the more stable oxygen adduct takes place via a cyclic four membered transition state, but if the attacked carbon is at a larger distance from oxygen, no isomerization occurs up to 400K.

3. PHOTOREACTIONS OF CARBONYL COMPOUNDS WITH SILANES LACKING SILICON-HYDROGEN BONDS.

In this section the photoreactions of ketones and quinones with tetramethylsilane, Me_4Si, hexamethyldisilane, $Me_3Si\text{-}SiMe_3$, and tetrakis(trimethylsilyl)silane, $(Me_3Si)_4Si$, will be examined. Due to the absence in these silanes of the relatively weak silicon-hydrogen bond (90 kcal/mol) [3], a different reactivity from that one shown by silicon hydrides is expected.

Two possible pathways can be envisaged for the first step of these reactions, as exemplified below for $Me_3Si\text{-}SiMe_3$. Eq. 17 involves hydrogen abstraction from one of the methyl groups to give a carbon centred radical, while eq. 18, implying displacement of a silyl radical from the silane, is akin to the familiar homolytic substitution reaction S_H2. The latter reaction, if exothermic, is expected to occur at elements having empty d orbitals and ,in fact, it has been reported with compounds of tervalent phosphorus [12] and of tin [13].

$$^3R_2C{=}O + Me_3SiSiMe_3 \longrightarrow R_2\dot{C}\text{-}OH + Me_3SiSiMe_2CH_2\cdot \quad (17)$$

$$^3R_2C{=}O + Me_3SiSiMe_3 \longrightarrow R_2\dot{C}\text{-}OSiMe_3 + Me_3Si\cdot \quad (18)$$

Approximate estimates of the exothermicities of reactions 17 and 18 may be obtained by by using thermochemical and photophysical data. The ΔH of each reaction can be calculated by means of eq. 19 where $\Delta H°_f$ is the standard heat of formation at 298K of a given compound and $E_T(R_2C{=}O)$ is the triplet state energy of the ketone.

$$\Delta H° = \Delta H°_f(\text{products}) - \Delta H°_f(\text{reactants}) - E_T(R_2C{=}O) \quad (19)$$

In the case of acetone, where E_T is 78 kcal/mol [14], the exothermicities of reactions 17 and 18 are calculated to be 20 and ca. 30 kcal/mol, respectively. Thus, homolytic displacement appears to be favoured by ca. 10 kcal/mol. The ease of the displacement reactions are expected to follow a trend parallel to that of the decreasing strength of the bonds to be cleaved, that is to increase along the series $Me-SiMe_3$ (89.4 kcal/mol), $Me_3Si-SiMe_3$ (80.5 kcal/mol) [3], and $Me_3Si-Si(SiMe_3)_3$. Although for the latter no data are available, a reasonable guess is that this bond is weaker than the Si-Si bond of hexamethyldisilane by ca. 10 kcal/mol [15].

The three silanes were first photoreacted with both aliphatic (acetone) and aromatic ketones (benzophenone, fluorenone, etc.) in deoxygenated benzene or tert-butylbenzene solutions. With acetone no EPR signals could be detected in the temperature range 240-330K, despite the fact that the corresponding silyl adducts are easily formed in the presence of silicon hydrides. With benzophenone the reactions were more successful and the radicals observed are shown in eq.s 20-22.

$$^3Ph_2C{=}O + Me_4Si \longrightarrow Ph_2\dot{C}{-}OH + Me_3SiCH_2\cdot \text{ (undetected)} \quad (20)$$

$$^3Ph_2C{=}O + Me_3SiSiMe_3 \longrightarrow Ph_2\dot{C}{-}OSiMe_3 \quad (21)$$

$$^3Ph_2C{=}O + (Me_3Si)_4Si \longrightarrow Ph_2\dot{C}{-}OSiMe_3 + Ph_2\dot{C}{-}OSi(SiMe_3)_3 \quad (22)$$

With Me_4Si hydrogen abstraction is therefore preferred to the displacement of a methyl radical, while with the other silanes, which contain the weaker silicon-silicon bond, homolytic substitution becomes favoured; the displaced silyl radicals could not be detected since they are easily trapped by benzophenone (see Table II). Other aromatic ketones behaved similarly to benzophenone when photoreacted with the silanes. The rate of quenching of the triplet state of benzophenone by the three silanes, i.e. $2.8x10^4$, $3.9x10^5$, and $2.6x10^6$ $M^{-1}s^{-1}$ for Me_4Si, $Me_3SiSiMe_3$, and $(Me_3Si)_4Si$, were found to increase with the decreasing strength of the bond to be cleaved in the displacement reaction [16].

When photoreacting DMTD with hexamethyldisilane in benzene at room temperature both radicals 8 and 9 (R=Me) were detected, in a ratio of ca. 1:1.8; that is radical 9 is now the dominant species contrary to what observed in the reaction with silicon hydrides. This result can be explained in terms of the displacement of trimethylsilyl radicals by triplet DMTD and concomitant formation of 9 (R=Me). The displaced $Me_3Si\cdot$ radicals then add to DMTD to afford both 8 and 9 in the proportions given in eq. 15. The photoreaction of DMTD with tetrakis(trimethylsilyl)silane afforded the three radical adducts 8 (R=Me), 9 (R=Me), and 9 ($R=SiMe_3$) in the approximate ratio 1:2:3. The unexpected formation of the ring adduct $\underline{8}$ (R=Me) indicates that along with the more predictable displacement of $(Me_3Si)_3Si\cdot$ (eq. 23), displacement of $Me_3Si\cdot$ takes also place (eq. 25).

$$^3DMTD + (Me_3Si)_4Si \longrightarrow \underline{9}\ (R{=}Me) + (Me_3Si)_3Si\cdot \qquad (23)$$

$$(Me_3Si)_3Si\cdot + DMTD \longrightarrow \underline{9}\ (R{=}SiMe_3) \qquad (24)$$

$$^3DMTD + (Me_3Si)_4Si \longrightarrow \underline{9}\ (R{=}SiMe_3) + Me_3Si\cdot \qquad (25)$$

$$Me_3Si\cdot + DMTD \longrightarrow \underline{8}\ (R{=}Me) + \underline{9}\ (R{=}Me) \qquad (26)$$

Similarly, radical $\underline{6}$ (R=Me) was detected not only in the photoreaction of DBQ with $Me_3SiSiMe_3$ but also in that with $(Me_3Si)_4Si$, this again indicating that $Me_3Si\cdot$ radicals are displaced from the latter silane.

The above results demonstrate that aromatic ketones and quinones when photoexcited in their triplet state are capable to bring about homolytic substitution reactions at the silicon atom of polysilanes. These reactions may proceed following two different routes; the initial step might be the direct attack of the triplet ketone or quinone to a silicon atom of the silane to give a biradical transition state (27$\underline{a}$), or it might involve a partial or complete electron transfer from silicon to the carbonyl compound (27$\underline{b}$).

$$^3A + M{-}M \begin{cases} \xrightarrow{\underline{a}} {}^3[A\cdots M\cdots M] \\ \xrightarrow{\underline{b}} {}^3[A^{-\cdot}\quad M{-}M^{+\cdot}] \end{cases} \longrightarrow AM\cdot + M\cdot \qquad (27)$$

If the reaction proceeds *via* route 27$\underline{a}$, the displacement of $Me_3Si\cdot$ from tetrakis(trimethylsilyl)silane when this is photoreacted with DBQ or DMTD would imply a rather unlikely attack of the triplet quinone to the highly hindered central silicon atom. Instead electron tranfer followed by fragmentation of the resulting cation according to eq. 28 could easily explain the experimental observations.

$$(Me_3Si)_4Si^{+\cdot} \longrightarrow (Me_3Si)_3Si^{+} + Me_3Si\cdot \qquad (28)$$

The ionization potentials of $(Me_3Si)_4Si$ (8.24 eV) and $Me_3SiSiMe_3$ (8.69 eV) [17] may be compatible with an electron transfer mechanism, especially in the case of the former compound. It is also remarkable that Me_4Si, whose ionization potential is much higher (10.30 eV), undergoes hydrogen abstraction by benzophenone triplets instead of homolytic substitution. Among the carbonyl derivatives, electron transfer should be favoured for those compounds having the higher electron affinity of the excited triplet state, EA_T; this can be calculated from the ground state electron affinity, EA, and the triplet state energy, E_T, by means of eq. 29 [14].

$$EA_T = EA + E_T \quad (29)$$

From the data reported in Table III, which collects the EA_T values of three representative examples, it appears that the ease of electron transfer should decrease in the order quinones > aromatic ketones > aliphatic ketones.

Table III - Electron affinities of ground (EA) and triplet states (EA_T) and triplet state energy (E_T) of carbonyl derivatives.

Substrate	EA/eV	E_T/eV[a]	EA_T/eV
Acetone	-1.51[b]	3.4	1.89
Benzophenone	0.62[c]	3.0	3.62
DBQ	1.88[c]	2.3	4.18

[a] Ref. 14. [b] Ref. 18. [c] Ref. 19.

The failure to detect any adduct radical when acetone was photoreacted with the silanes provides support in favour of route 27<u>a</u> for the analogous reactions of aromatic ketones and diones. Since solvent polarity should accelerate this reaction by stabilizing the polar transition state, we repeated the photoreaction of acetone with the silanes by using acetonitrile or neat acetone as solvent instead of aromatic hydrocarbons. In both cases the radical adducts $Me_2\dot{C}$-$OSiMe_3$ (from $Me_3SiSiMe_3$) and $Me_2\dot{C}$-$OSi(SiMe_3)_3$ (from $(Me_3Si)_4Si$) became easily detectable on a wide range of temperatures. In the latter case the intensity of the ESR signals was strong enough to allow the detection of the naturally abundant ^{29}Si satellites [16].

In conclusion, the displacement of $Me_3Si\cdot$ radicals from $(Me_3Si)_4Si$, the finding that only silanes with low ionization potentials undergo homolytic substitution, and the strong dependence on solvent polarity of the photoreactivity of acetone, provide strong evidence that reactions of triplet carbonyl compounds with polysilanes are likely to proceed <u>via</u> an electron transfer mechanism.

<u>Aknowledgements</u>. Financial support from Ministero della P.I. is gratefully aknowledged. The author wish to thank Dr. A. Alberti for invaluable assistance and Prof. M. Chanon for helpful suggestions concerning the last part of this work.

REFERENCES

1) D.F. McMillen and D.M. Golden, Ann. Rev. Phys. Chem., 1982, **33**, 492.
2) R.A. Jackson, J. Organomet. Chem., 1979, **166**, 17.
3) R. Walsh, Acc. Chem. Res., 1981, **14**, 246.
4) C. Chatgilialoglu, J.C. Scaiano, and K.U. Ingold, Organometallics, 1982, **1**, 466.
5) C. Chatgilialoglu, K.U. Ingold, and J.C. Scaiano, J. Am. Chem. Soc., 1982, **104**, 5119.
6) A. Alberti and G.F. Pedulli, Rev. Chem. Intermed., 1987, **8**, 207.
7) I. Fleming, Frontier Orbitals and Organic Chemical Reactions, John Wiley & Sons, London, 1977, p. 141.
8) C. Chatgilialoglu, K.U. Ingold, and J.C. Scaiano, J. Am. Chem. Soc., 1983, **105**, 3292.
9) A. Alberti and G.F. Pedulli, results to be published.
10) A. Alberti, C. Chatgilialoglu, G.F. Pedulli, and P. Zanirato, J. Am.Chem. Soc., 1986, **108**, 4993.
11) A. Alberti, A; Hudson, and G.F. Pedulli, Tetrahedron, 1982, **38**, 3749.
12) A. Alberti, D. Griller, A.S. Nazran, and G.F. Pedulli, J. Org. Chem., 1986, **51**, 3959.
13) A.G. Davies, J. Organomet. Chem., 1980, **200**, 87.
14) G.J. Kavarnos and N.J. Turro, Chem. Rev., 1986, **86**, 401.
15) J.M. Kanabus-Kaminska, J.H. Hawari, D. Griller, and C. Chatgilialoglu, J. Am.Chem. Soc., 1987, **109**, 5267.
16) A. Alberti, S. Dellonte, and G.F. Pedulli, results to be published.
17) H. Bock and W. Ensslin, Angew. Chem. Int. Ed. Engl., 1970, **10**, 404.
18) K.D. Jordan and P.D. Burrow, Acc. Chem. Res., 1978, **11**, 341.
19) P. Kebarle and S. Chowdhury, Chem. Rev., 1987, **87**, 513.

ALKANE FUNCTIONALIZATION BY MERCURY PHOTOSENSITIZATION

S.H. Brown, R.F. Ferguson and Robert H. Crabtree
Yale Chemistry Department
225 Prospect Street
New Haven, CT. 06511 USA

ABSTRACT. Alkanes, alcohols, ethers and silanes can be dimerised by mercury photosensitisation on a preparative scale. Alkane functionalisation can be effected by cross-dimerising an alkane with any of these substrates.

1. INTRODUCTION

The area of alkane functionalization catalysis [1] described in this lecture developed from an interest in the homogeneous catalysis of a variety of alkane conversions by phosphine complexes of the later transition metals. These are believed to involve oxidative addition as the key CH bond breaking step, and may be either thermal or photochemical. A number of catalyst deactivation processes have so far limited the number of turnovers that can be observed to <100.

All of our alkane dehydrogenation catalysts are routinely tested for homogeneity by using a mechanistic test involving metallic Hg. When we carried out this test on one of our photochemical iridium catalysts, we observed a mercury photosensitized product. This led us to take an interest in the area and we have now found several ways in which alkanes can be converted to a variety of functionalized materials on a preparative scale. These photochemical reactions lead to a much greater variety of products than the iridium catalysts, and, being a metal atom, the catalyst does not degrade significantly with time.

Our most effective photochemical alkane dehydrogenation catalyst[2] is $IrH_2(O_2CCF_3)L_2$ (**1**). Photoextrusion of H_2 and oxidative addition of alkane to the intermediate would give an alkyl hydride that could become 16e by opening of the carboxylate chelate ring. After β-elimination to give the alkene, the original dihydride is regenerated. The metal can re-enter

M. Chanon et al. (eds.), Paramagnetic Organometallic Species in Activation / Selectivity, Catalysis, 97–107.

the cycle by photoextrusion of H_2, and so the system is catalytic (eq. 1).

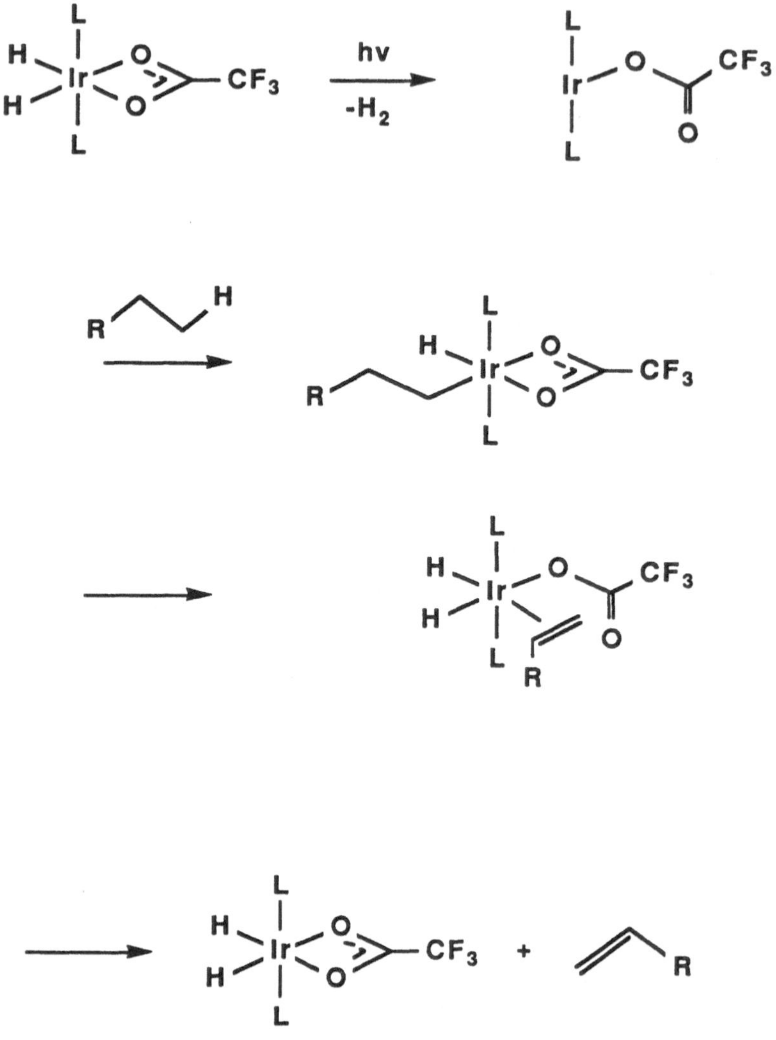

(1)

Although eq. 2 is thermodynamically unfavorable at 25°C, photochemical dehydrogenation reactions can go directly because the reaction is driven by the energy of the photons (eq. 3). A number of related catalytic[3-6] and stoichiometric[7-10] reactions have been described recently.

$$RCH_2CH_3 \xrightarrow{\text{cata}} RCH{=}CH_2 + H_2 \qquad (2)$$

$$hv + RCH_2CH_3 \xrightarrow{cata} RCH{=\!=}CH_2 + H_2 \qquad (3)$$

2. MERCURY PHOTOSENSITIZATION

As mentioned above, we needed to be sure that catalysis by metal surfaces, which might be formed by decomposition of the catalyst, was not the origin of the alkane conversions. These have been excluded by several tests including the use of metallic mercury, which poisons platinum group metal surfaces.[3e] In carrying out this test, we found that an involatile residue was formed along with the expected cyclooctene in the products. This proved to be the result of a vapor phase mercury photosensitised reaction of cyclooctane to give bicyclooctyl. No iridium is required for this, and none of the reactions I will describe below require Ir. According to the literature,[12] the general principle involved in mercury photosensitization is the homolytic cleavage of an alkane C-H bond by the $^3P_1$ excited state of mercury (Hg*) formed by irradiation at 254 nm. The equations below briefly describe the essence of the mechanistic proposal made by the early workers in the field.

$$Hg + hv \longrightarrow Hg^* \qquad (4)$$

$$Hg^* + RH \longrightarrow R\cdot + H\cdot + Hg \qquad (5)$$

$$H\cdot + RH \longrightarrow R\cdot + H_2 \qquad (6)$$

$$2R\cdot \longrightarrow R_2 \qquad (7)$$

Unlike a solution phase radical reaction, oligomers tend not to be formed even at high conversions, even though the product R_2 is intrinsically more reactive than the initial substrate, RH. The reason that this happens in our case is that the dimer has so low a vapor pressure that essentially only the original substrate, cyclooctane, is present in the vapor phase.

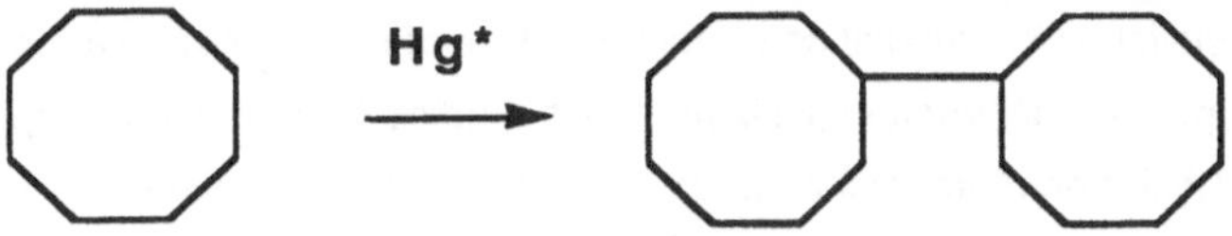

+ H_2 (8)

We find that this vapor phase selectivity effect is very useful in obtaining unusual selectivity effects,[12] and we have slightly modified the usual photochemical apparatus to take advantage of this effect. The substrate and a drop of mercury are refluxed together in a large quartz vessel and irradiated at 254 nm with four to sixteen 8 W low pressure Hg lamps. The product dimers are quickly returned to the liquid phase and the vapor phase is constantly replenished with substrate. The volume of hydrogen evolved allows us to follow the progress of the reaction. All the light is absorbed within a millimetre of the wall, as shown by running a reaction with two reactors, one held concentrically within the other. If the gap is more than a millimetre, no reaction takes place within the inner tube. The reaction rate goes up with the surface area of the reactor and not with its volume. The rate is proportional to the intensity of irradiation if the reactor geometry is held constant.

C--H bonds react in the order: $3^{o} > 2^{o} > 1^{o}$ consistent with a homolytic pathway.

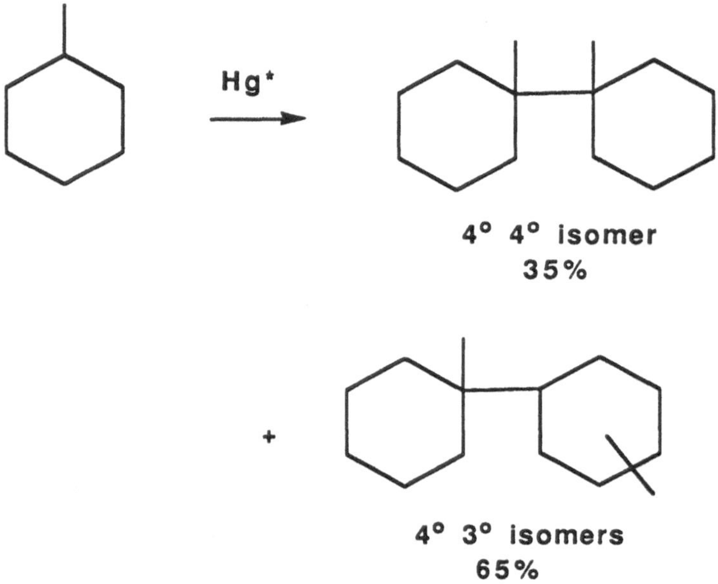

(9)

Even species with 4°-4° C-C bonds are very efficiently assembled by reaction of a precursor containing a 3° C--H bond. Some 4°-3° product is also formed and this product ratio can be altered by diluting the alkane with such gases as N_2. This leads to increased selectivity for the 4°-4° product. A possible explanation is the relaxation of the $^{3}P_{1}$ state of mercury to give the more selective $^{3}P_{0}$ state.Although mercury photosensitization was studied for many years up to 1973 (when physical chemistry based on mercury lamps rather than on lasers became unfashionable), the focus of attention was in the physical chemistry involved: energy transfer, quenching cross-sections and the like. The synthetic potential of the method was not studied. Alkanes were the chief substrates studied in the earlier work, although a few previous reports[13] deal with alcohols, ethers, amines and silanes. We find that reactions involving these and other saturated, but heteroatom-substituted species are preparatively useful when carried out under our conditions. The system is selective for C-H bonds α- to the heteroatom and for Si-H bonds. For example, Et_3SiH gives $Et_3SiSiEt_3$.

CH_3OH —> $HOCH_2CH_2OH$ (10)

$tBuOCH_3$ —> $tBuOCH_2CH_2OtBu$ (11)

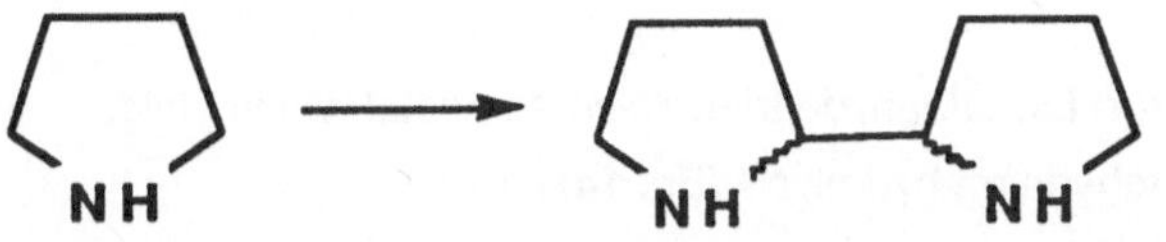

(12)

Alkanes can be cross dimerized with other species. Eq. 13 shows the results from cyclohexane and methanol. The three products are formed in approximately statistical amounts. The polarities of the three species are so different that simple solvent extraction suffices to obtain the cross dimer itself; the glycol can be removed with water and the bicyclohexyl with pentane.

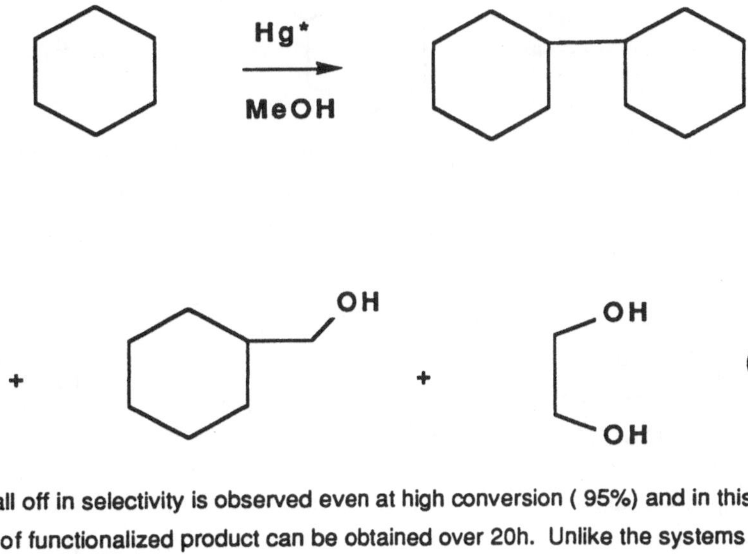

(13)

Little fall off in selectivity is observed even at high conversion (95%) and in this way tens of grams of functionalized product can be obtained over 20h. Unlike the systems mentioned in the first section of this chapter, there does not seem to be any limit to the number of catalytic cycles that can be observed. This is no doubt because mercury does not degrade under the reaction conditions, but also the refluxing of the substrate washes the walls of the vessel and prevents the build-up of colored impurities which tend to stop similar photochemical reactions in static reactors. Under certain circumstances up to 10^4 turnovers can be observed per Hg atom without a significant decrease in reaction rate.

We have recently found conditions under which even methane can be cross-dimerised with methanol to give first ethanol and then propylene glycol. The latter seems to be formed by a subsequent cross-dimerisation of the initial ethanol product with more methanol to give $MeCH(OH)CH_2(OH)$.

Cyclohexane and formaldehyde trimer give an acetal which yields cyclohexanecarboxyaldehyde on hydrolysis (Eq. 14).

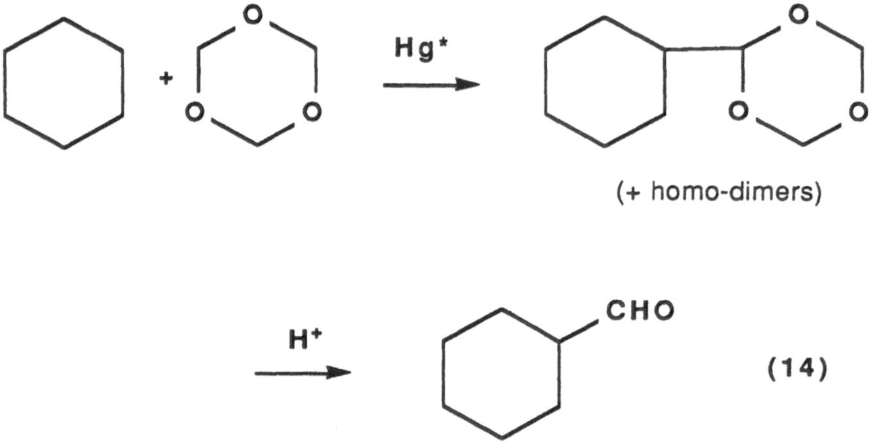

(14)

A 1-silylethanol, $R_3SiCH(CH_3)OH$, is obtained from the silane, R_3Si-H, and ethanol in spite of the much greater strength of Si-O over Si-C bonds, which normally favors the formation of a ethoxysilane (R_3SiOEt). Hydroxymethyl ethers are observed from alcohols and ethers (Eq. 15).

(15)

While only millimolar amounts of alkenes are formed with turnovers of 100 or less in the oxidative addition systems, we can now make tens of grams of functionalized product in 20h. Such a simple system may therefore have advantages over the more complicated metal phosphine catalysts such as $[IrH_2(O_2CCF_3)(PR_3)_2]$ which have many more deactivation pathways open to them. On the other hand the latter do have the very useful property of selectively attacking the least hindered C-H bond in the alkane, in contrast to the selectivity pattern observed in the Hg system and in all the 'classical' alkane functionalization reactions based on radicals and carbonium ions.

No one has been able to discover a stable metal phosphine catalyst. In none of the known examples, reported by ourselves [2], by Felkin [4,5] or by Tanaka [6], are the catalysts stable enough to give even 100 turnovers of product. For the moment, this severely limits their usefulness, but suggests that a careful study of their deactivation pathways might lead to the synthesis of a more stable catalyst.

The reason for the different selectivities of the two systems is believed to be the steric bulk of a catalyst such as $RhCl(CO)L_2$, compared with the small Hg atom, as well as the sterically demanding side-on transition state for oxidative addition (left),[11] compared with the end-on transition state (right) believed to operate for the Hg reaction.[16]

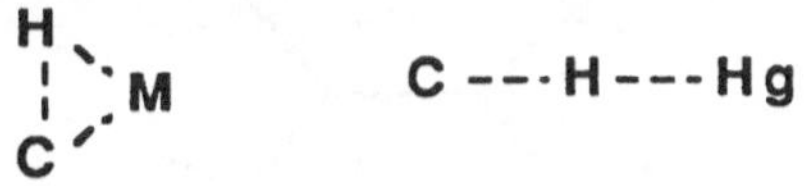

Our conditions are very different from those used in the earlier studies and so the generally accepted mechanism for the Hg photosensitisation reaction (eq. 4-7) may not apply to our version of the reaction. Our tentative conclusion is that eq. 7 is not the route by which dimer is formed. If it were, we would expect to see the disproportionation products shown in Eq. 16.

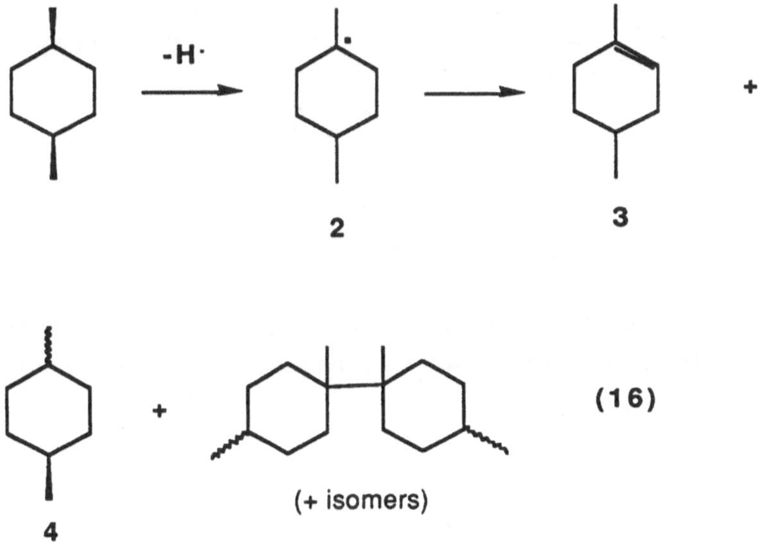

(16)

2 should give the disproportionation products **3** and **4**, which should dominate the product mixture. In particular the ratio of **3** and **4** to dimerization products should be ca. 10:1 because 3° radicals are very prone to disproportionate. Earlier workers argued that the lack of disproportionation products such as **3** is a result of its being trapped by H· to give back **2**. On the other hand **2** should also give **4**, which being a saturated compound should be much less reactive than **3** and so survive the reaction conditions. With methylcyclohexane as substrate we could not have distinguished the methylcyclohexane formed by disproportionation but cis-1,4-dimethylcyclohexane gives trans-1,4-dimethylcyclohexane on disproportionation. We do see formation of the trans alkane but at about one fifth the rate which would be expected if the dimers were formed by the combination of two free radicals.

t-Bu_2CH_2 undergoes the mercury reaction quite readily to give the 2°-2° product. On the basis of the literature mechanism, this can only arise from dimerisation of the t-Bu_2CH· radical, yet t-Bu_2CH· radical is said not to dimerize [14a].

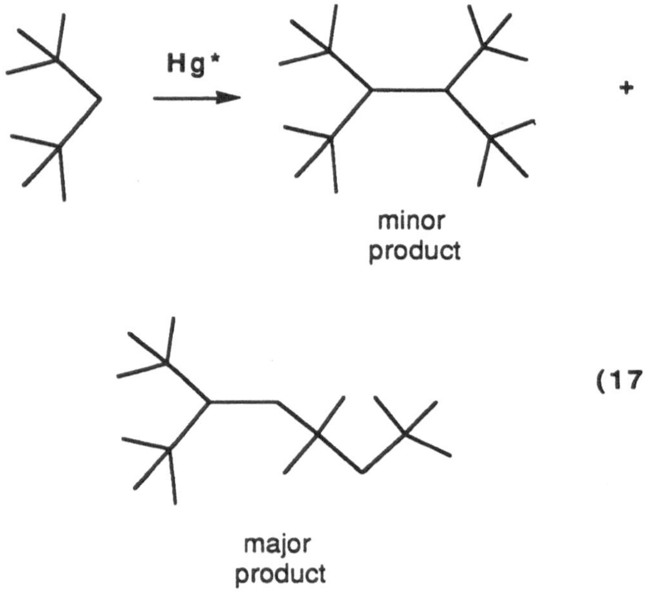

(17)

It is easy to imagine that any initially formed R· radicals might be scavenged by the large amount of Hg vapor present to give HgR·. This species could be involved in the C-C formation process as shown below.

$$R\cdot + Hg \longrightarrow RHg\cdot \quad (18)$$

$$R\cdot + RHg\cdot \longrightarrow R_2 + Hg \quad (19)$$

$$2RHg^{\cdot} \longrightarrow R_2 \qquad (20)$$

Even oxidative addition of the C--H bond to Hg* could account for the 3°>2°>1° selectivity for C--H bond attack because Hg* is so unhindered.

Photoexcitation of Hg atoms in an alkane matrix has been shown to give an EPR silent intermediate, perhaps RHgH, as the initial product.[14b] Subsequently, R· appears by decomposition of the intermediate. Other metal atoms have been studied in the matrix and it has been shown that many of them, when photoexcited, can give oxidative addition with alkane C--H bonds. For example the sequence shown in eq. 21 has been studied by Billups and Margrave[15] and by Ozin.[16] Cu(^{2}P) atoms have also been shown to oxidatively add to the C--H bond of methane, but the initial adduct, believed to be MeCuH, decomposed photolytically under the conditions of its formation to give CuH and CuMe.[17]

$$Fe + CH_4 \underset{400\ nm}{\overset{300\ nm}{\rightleftharpoons}} H{-}Fe{-}CH_3 \qquad (21)$$

Alkane activation by transition metal ions in the gas phase has been carried out by ion beam mass spectroscopy or ion cyclotron resonance (ICR). C--C, as well as C--H bonds are easily broken in these reactions.[18]

3. CONCLUSION

The mercury system is capable of rapidly assembling relatively complex molecules from simple starting materials, and doing so on a preparatively useful scale. Exxon Corp. has therefore taken an interest in possible commercial applications of the 'MERCAT' process. The catalyst resists degradation largely because it is a simple atom of a rather unreactive element.

A principle that emerges from the mercury work is that unusual selectivity patterns may be achieved by physical separation of different species. It may be possible to take advantage of this principle in other cases in which the products of a reaction are much more reactive than the starting material and therefore tend not to survive the reaction. The selective oxidation of methane to methanol may be an example.

4. ACKNOWLEDGMENT

We thank the National Science Foundation, the Petroleum Research Fund, the Department of Energy and Exxon Corp. for funding our work in this area.

5. REFERENCES CITED

1 R.H. Crabtree, Chem. Revs, 1985, 85, 245.

2 a) M.J. Burk, R.H. Crabtree and D.V. McGrath, Chem. Comm., 1985, 1829; b) M.J. Burk, R.H. Crabtree, J. Am. Chem. Soc., in press.

3 a) R.H. Crabtree, J.M. Mihelcic and J.M. Quirk, J. Am. Chem. Soc., 1979, 101, 7738; b) R.H. Crabtree, M.F. Mellea, J.M. Mihelcic and J.M. Quirk, J. Am. Chem. Soc., 1982, 104, 107; c) M.J. Burk, R.H. Crabtree, C.P. Parnell and R.J. Uriarte, Organometallics, 1984, 3, 816; d) R.H. Crabtree, R.P. Dion, D.J. Gibboni, D.V. McGrath and E.M. Holt,J. Am. Chem. Soc, 1986, 108, 7222;e) D.R. Anton and R.H. Crabtree, Organometallics, 1983, 2, 855.

4 D. Baudry, M. Ephritikine, and H. Felkin, Chem. Comm., 1982, 606, 1983, 788; D. Baudry, M. Ephritikine, H. Felkin and J. Zakrzewski, Chem. Comm., 1982, 1235, Tet. Lett., 1984, 1283.

5 a) H. Felkin, T. Fillebeen-Khan, Y. Gault, R. Holmes-Smith and J. Zakrzewski, Tet. Lett., 1984, 1279; b) C.J. Cameron, H. Felkin, T. Fillebeen-Khan, N.J. Forrow and E. Guittet, Chem. Comm., 1986, 801; b) H. Felkin, T. Fillebeen-Khan, R. Holmes-Smith, and Y. Lin, Tet. Lett. 1985, 26, 1999.

6. T. Sakakura and M. Tanaka, Chem. Comm., 1987, 758.

7. S.P. Nolan, C.D. Hoff, P.O. Stoutland, L.J. Newman, J.M. Buchanan, R.G. Bergman, G.K. Yang and K.S. Peters, J. Am. Chem. Soc., 1987, 109, 3143.

8 J. Halpern, Accts. Chem. Res., 1982, 15, 238.

9 M.J. Wax, J.M. Stryker, J.M. Buchanan, C.A. Kovac, and R.G.Bergman, J. Am. Chem. Soc., 1984, 106, 1121.

10, J.K. Hoyano and W.A.G. Graham, J. Am. Chem. Soc., 1982, 104, 3723.

11 R.J. Cvetanovic, Progr. React. Kinetics, 1964, 2, 77.

12 S.H. Brown and R.H. Crabtree, Chem. Comm., 1987, 970.

13 R.F. Porter, Inorg. Chem., 1980, 19, 447; J.S. Plotkin and L.G. Sneddon, J. Am. Chem. Soc., 1979, 101, 4155; M.A. Nay, G.N.C. Woodall, O.P. Strausz and H.E. Gunning, J. Am. Chem. Soc., 1965, 87, 179; G.J. Mains, Inorg. Chem., 1966, 5, 114.

14 a) K.U. Ingold, J. Am. Chem. Soc., 1974, 96, 2441; b) B.J. Brown, and J.E. Willard, J. Phys. Chem., 1977, 81, 977.

15 W.E. Billups, M.M. Konarski , R.H. Hauge, and J.L. Margrave, J. Am. Chem. Soc., 1982, 102, 7393.

16 G.A. Ozin and J.G. McCaffrey, J. Am. Chem. Soc., 1982, 104, 7351.

17 G.A. Ozin , D.F. McIntosh,and S.A. Mitchell, J. Am. Chem. Soc., 1981, 103, 1574.

18. J. Allison, R.B. Freas and D.P. Ridge, J. Am. Chem. Soc., 1979, 101, 1332; R.B. Freas and D.P. Ridge, J. Am. Chem. Soc., 1980, 102, 7129; 1984, 106, 825.

19 J.W. Bruno, T.J. Marks and L.R.Morss, J. Am. Chem. Soc., 1983, 105, 6824.

METAL VAPOR REACTIVITY OF COPPER TOWARD SUBSTITUTED BROMOBENZENES. COMPARISON WITH THE CLASSICAL ULLMANN COUPLING REACTION .

Jean-Claude NEGREL *, R.W. ZOELLNER **, Michel CHANON *, Françoise CHANON *

*Université d'AIX-MARSEILLE III, Centre de Saint -Jérome, 13397 MARSEILLE Cédex, France

** Dept of Chemistry,Northern Arizona University, FLAGSTAFF, 960011003, USA

ABSTRACT

The reaction of Copper vapors with a film of bromobenzenes in a solution of methylcyclohexane - tetrahydrofuran maintained at - 120°C mainly yields reduction and coupling products. The low temperature contrasts with the severe reaction conditions used when metallic copper is used as a promoter. In terms of overall conversion (percent of copper effectively used to induce chemical transformation) o-chloro; o-amino and p-RO substituents play a beneficial role. In terms of coupling /reduction ratio, unsubstituted bromobenzene yields the best results. As in the classical Ullmann reaction nitro group induces far more coupling when it is situated in ortho of bromine than when it occupies the meta or (even worse) para position. For some substrates, THF used as cosolvent reacts with the intermediates organometallic to yield the condensation of benzene with an open form of THF. Ortho substituents of apropriate structure seem able to stabilize metallic atoms of Cu.

INTRODUCTION

Metal vapor synthesis, born in the sixties (1), has now been extented to various parts of chemistry : synthesis of organometallics (2), physicochemical studies (3), catalysis (4), industrial applications (5).

Most of the studies performed in the field aim at synthetizing new organometallic compounds. In contrast, very few concentrate on the quantitative aspects of reactivity on the side of obtained organic products (6). The present report will precisely deal with this later aspect of metal vapor synthesis. We will describe the results obtained in the reaction between copper vapors (Cu°: $3d^{10}4s^{1}$) and halogenobenzene. The reaction takes place in a rotary apparatus (7) where the thin film of solution formed on the walls of glass vessel is kept at a temperature ranging from -20°C to -185°C

M. Chanon et al. (eds.), Paramagnetic Organometallic Species in Activation / Selectivity, Catalysis, 109–118.

(according to the selected solvent) ; vaporized copper metal reacts with this film. Transiently formed organometallics, whose presence is revealed by changes in coloration, are not isolated in such experiments where the work up of the final mixture is done at room temperature. In terms of organic products, the overall reaction may be summarized by :

$$Cu(at) + X\text{-}C_6H_4Br \xrightarrow[MCH + THF]{-120^\circ C} X\text{-}C_6H_4CuBr \longrightarrow X\text{-}C_6H_5 + (X\text{-}C_6H_4)_2 + CuBr$$

X = Substituent ; MCH = methylcyclohexane ; THF = tetrahydrofuran

The distribution of organic products should provide informations about the reactivity of the paramagnetic copper atoms.

Metallic copper is a well known catalyst for the Ullmann coupling to biaryls (8ab) and organocopper III have been shown to play a role in the coupling of o-bromonitrobenzene (8c). More recently Rieke used highly reactive powders of transition metals to obtain the same coupling (9). Aryl halides have been far less studied by Metal Vapor Synthesis (M.V.S.) than alkyl halides (10). Cocondensation reactions at -196°C have dealt with V, Cr, Mo and W (11), and bis (arene) metal are the products of these cocondensations. For the Nickel triad, the Klabunde products obtained result from an oxidative addition (12). For this latter case, Klabunde proposed that a π complex between the metal and the aryl halide is first formed at -196°C, this complex then evolves toward an oxidative addition product, when the temperature raises. When the organometallic intermediate is not stabilized by a phosphine ligand, it usually decomposes to yield organic compounds (coupling and reduction). Titanium atoms and C_6F_5Br yield an efficient catalyst for the polymerization of butadiene (12b).

Information about copper vapor reactivity is scarce (14,15). Klabunde has found that copper vapors react with C_6F_5Br and C_6F_6 to yield complexes (15). Murdock reported a reaction of aryl coupling when these vapors react with bromobenzene (16). Photoexcited copper atoms are able to activate methane and ethane held in a rigid matrix at 12K (17) and clusters of Cu_3, Cu_5 react with O_2 at -196°C (18). These data, added to the known rich reactivity of organocopper reagents (19) and to the catalytic effects of copper in bimetallic systems (20), give strong incentive for further studies in the field.

EXPERIMENTAL PART

The rotary metal atom reactor is a homemade adaptation of Green's reactor (21a). Several improvements have been brought to this homemade prototype

(21b). First, a 700 liter/second oil diffusion pump backed by a two stage rotary vacuum pump, a gate valve and a specially designed stainless steel liquid nitrogen-cooled trap, allow a rapid obtention and an efficient maintenance of the high vacuum needed in these experiments. Second, pitch of the vessel is adjustable from 50° to the horizontal position. Third, the built-in rotary sealing system includes four 116 mm i.d. proprietary rotary seals with differential pumping. Fourth, borosilicate reaction vessels from 6 to 20 liters are fitted on a stainless steel rotary flange and rotation is controllable from 6 to 100 rotations per minute. Fitfh, all the feedthroughs are integrated into a stainless steel cabinet-tubing system located at the center of the reaction flask ; this enables one to adjust the reaction zone on the wall of the flask and ensure easy cleaning and low outgassing rate. Sixth, a continous temperature monitoring system allows one to follow the temperature in the solution during the reaction. Seventh, twin metal vaporizations are performable. Reactants are injected into the reaction vessel as gas, liquids or solids by means of four inlet lines.

The pumping speed at the reaction flask is about 100l/s; in cocondensation experiments, a vacuum of about 10^{-6} Torr is reached in a few minutes. Solution reactions are carried out according to the following general procedure: 20 to 30 mmoles of the ligand are dissolved in a mixture of 150 ml of methylcyclohexane and 50 ml of THF. This solution is degased by freeze thaw cycles and is injected under vacuum in the reaction vessel rotating at 20 rotations per minute in a liquid nitrogen cooled methylcyclohexane bath (-120°C). Meanwhile about 1g of copper metal is vaporized in 1 hour from an alumina-coated tungsten crucible under a vacuum in the 10^{-4}-10^{-5} Torr range. At the end of the reaction, a bake out procedure of the trap under static vacuum allows the return to the reaction flask, kept cold, of any possible trap condensate. When the reaction mixture is warmed to room temperature and returned to the atmospheric pressure, quantitative recovery of the reaction products is carried out using an especially designed apparatus (27). The unreacted metal deposited on the shields and the cabinet is carefully collected and weighted to make possible a quantitative assessment of the actual amount of metal vapor having reacted with the organic substrate. Organic products from the reaction are titrated by G.C. using an internal standard, the relative response factors are calculated (22). C.G.M.S. allows the identification of products. Reproducibility was tested with p-bromotoluene and copper and also with Ti, Cr, Fe, Co, Ni. The average relative error is about 6 % with an upper limit for error of about 10 %.

RESULTS AND DISCUSSION

Table 1 contains the ratios of organic products calculated from the total mmoles of products and from the amount of copper having actually reached the thin film of solution. In this context, "conversion" means the number of millimoles of ligand converted by millimoles of reacted copper.

<u>Table 1</u>: Products obtained in the reaction between metal vapors of Cu and various bromoarenes : (c) = (a) + 2(b)

Substrate	Reduction products	(a) %	Coupling products	(b) %	Conversion % (c)
C_6H_5-Br	C_6H_6	4.5	C_6H_5-C_6H_5	21.15	46.8
p.CH_3-C_6H_4-Br	CH_3-C_6H_5	11.7	$(CH_3\text{-}C_6H_4)_2$	10.5	32.7
p. F-C_6H_4-Br	F-C_6H_5	18.9	$(F\text{-}C_6H_4)_2$	13.5	47.5
			F-C_6H_4-$CO(CH_2)_2CH_3$	0.8	
o. Cl-C_6H_4-Br	Cl-C_6H_5	38.8	$(Cl\text{-}C_6H_4)_2$	6.95	
	Br-C_6H_5	0.75	Br-C_6H_4-C_6H_4-Cl	0.95	
			Cl-C_6H_4-C_6H_5	0.9	60.7
			Cl-C_6H_4-CO-$(CH_2)_2$-CH_3	1.55	
			Cl-C_6H_4-C_6H_4-C_6H_4-Cl	0.25	
p. I-C_6H_4-Br	Br-C_6H_5	6.3	$(Br\text{-}C_6H_4)_2$	0.55	
	I-C_6H_5	1.65	Br-C_6H_4-C_6H_4I	0.3	
			C_6H_5-C_6H_5	1.15	13.6
			Br-C_6H_4-C_6H_5	0.35	
			I-C_6H_4-C_6H_5	0.25	
			C_6H_5-C_6H_4-C_6H_5	0.2	
o. NO_2-C_6H_4-Br	NO_2-C_6H_5	10	NO_2-C_6H_4-NH-C_6H_5	0.25	
			$(NO_2\text{-}C_6H_4)_2$	5.9	24.9
			Br-C_6H_4-N=NO-C_6H_4Br	1.3	
			NO_2-C_6H_4-NH-C_6H_4-Br	<0.1	
m. NO_2-C_6H_4-Br	$NO_2C_6H_5$	10.25	NO_2-C_6H_4-NH-C_6H_5	0.65	
			$(NO_2\text{-}C_6H_4)_2$	1.1	20.8
			Br-C_6H_4-N=NO-C_6H_4-Br	1.5	
			NO_2-C_6H_4-NH-C_6H_4-Br	2	
p. NO_2-C_6H_4-Br	$NO_2C_6H_5$	0.9	NO_2-C_6H_4-NH-C_6H_5	0.1	2.1
			$(NO_2\text{-}C_6H_4)_2$	0.25	
			Br-C_6H_4-N=NO-C_6H_4-Br	0.15	
			NO_2-C_6H_4-NH-C_6H_4-Br	0.1	
p. CH_3-CO-C_6H_4-Br	CH_3-CO-C_6H_5	28.5	$(CH_3\text{-}CO\text{-}C_6H_4)_2$	2.8	34.1
p. CH_3-O-C_6H_4-Br	CH_3O-C_6H_5	37.5	$(CH_3O\text{-}C_6H_4)_2$	11.7	64.3
			CH_3O-C_6H_4-$CO(CH_2)_2CH_3$	1.7	

p. $HO-C_6H_4-Br$	$HO-C_6H_5$	32.3	$(HO-C_6H_4)_2$ $HO-C_6H_4-CO(CH_2)_2CH_3$	4.1 1.5	43.5
p. $C_6H_5O-C_6H_4-Br$	$C_6H_5OC_6H_5$	35.9	$(C_6H_5-O-C_6H_4)_2$ $C_6H_5-O-C_6H_4-CO(CH_2)_2CH_3$	8.8 1.25	56.
o. $NH_2-C_6H_4-Br$	$NH_2C_6H_5$	14.3	$(NH_2-C_6H_4)_2$ $C_6H_5-NH-CO(CH_2)_2CH_3$ $Br-C_6H_4-NH-CO(CH_2)_2CH_3$	2.85 1.3 1.15	51.1
				1.2	
			Organometallic	11.9	

The values in **Table 1** show that substituent effects change the reactivity over a range of 30. There is no obvious correlation between the overall reactivity of aryl halides and Hammett substituent constants. Indeed, the most activating substituents (NH_2, OH) are, in the present reaction, less activating than phenoxy and methoxy groups. On the other hand, a methyl substituent decreases the overall reactivity and, of the three halogens considered as substituents, chlorine induces the highest reactivity. The very poor Hammett type correlation obtained provides a weak slope with a negative sign suggesting that the overall reactivity of aromatic halides toward vapors of Cu is mildly electrophilic. This result contrasts with the reactivity of Ni* toward aryl halides, where the metal behaves as a nucleophile (23). Rieke has shown that electron donating substituents on the aromatic ring decrease the reactivity (24). Poor correlations, or non linear relationships, are however usual in this kind of chemistry (23, 25).

Concerning the nature of obtained products, only for substituents Me, MeO, H the reduction-coupling reaction suffices to explain the observed products. For all the other substituents other reactions seem to intervene. When the substituent is a halogen, it may undergo reduction in competition with the reduction of the C-Br bond, except when the substituent is the fluoro group. The relative ease of reduction agrees with the one reported for the reaction between Pd and aryl halides (12ab).

For the coupling products, whatever the substituents, an important by-product appears at high conversion: it results from an opening of the THF ring to form an intermediate which then reacts with the arylhalide :

$o-Cl-C_6H_4-Br$ + Cu(at) -----------> $o-Cl-C_6H_4-CO-CH_2-CH_2-CH_3$

Besides the expected coupling product, o-chlorobromobenzene yields a small amount of 2-chloro, 2'-bromo biphenyl and two other products, probably formed through the reaction of 2,2'-chlorobiphenyl with copper atoms:

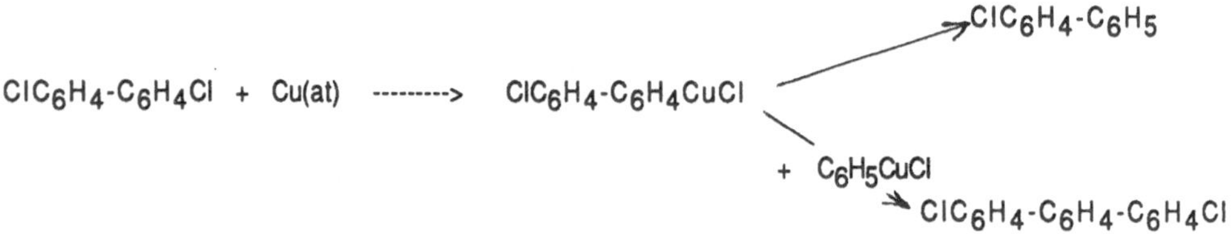

The cross coupling product :

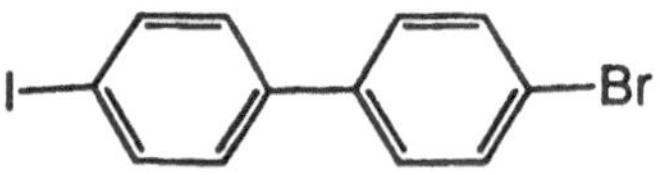

as well as the homo coupling product are obtained in smaller amounts when p-IC_6H_4Br reacts with Cu vapors. In contrast with the classical Ullmann reaction (8a), iodo aromatics are therefore not the best leaving groups to obtain the coupling reaction. The formed halo biphenyl compounds undergo further reactions of coupling and reduction as shown by the final mixture contents displayed in **Table 1**.

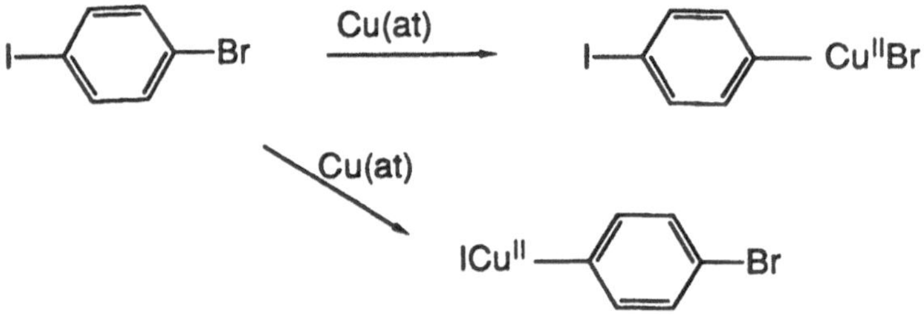

Ortho substitution by a chloro substituent induces a high reactivity and this could be due to an intramolecular stabilization of copper (II) in the complex:

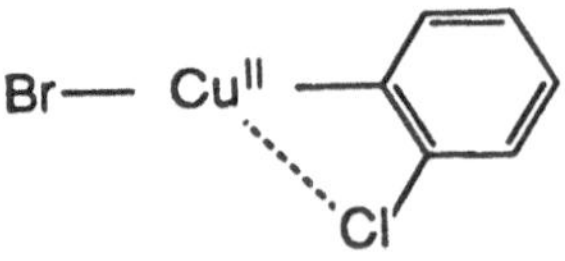

This beneficial ortho substituent effect seems to be present also for the $-NH_2$ group. It must be stressed however that this beneficial effect is in terms of overall transformation, but that reduction product dominates. This is noteworthy because in the classical Ullmann reaction, NH_2 as well as OH substituents, inhibit the coupling (28).

In the classical Ullmann coupling reaction, the nitro group is strongly activating but only in the ortho position. Results gathered in **Table 1** show that with copper metal vapor, the ortho nitro group is indeed activating: five times more coupling occurs with an ortho than with a meta nitro group. The products identified, besides coupling and reduction, are due to a specific reactivity of the nitro group, in agreement with observations reported by Gladysz, who studied the reactivity of Cr atoms ($3d^5$ $4s^1$) toward nitrobenzenes (6). Only copper and Cr atoms whose configuration contains a 4s unpaired electron seem able to promote these deoxygenation reaction of nitroarenes. On the other hand, Timms and Bashar have reported that, at -196°C, 2-nitropropane weakly complexes copper atoms (26). This type of complexation could play a role in the formation of products derived from the nitro group. The details of their formation remain unsettled at this point.

For several substituted bromobenzenes (substituents : p-F, o-Cl, p-MeO, p-C_6H_5O, p-OH, o-NH_2), the solvent THF becomes involved in the reaction. There are at least two possibilities to rationalize this involvement, but one seems to be untenable. The first supposes that a paramagnetic species may abstract one hydrogen from THF, inducing a ring opening followed by an 1-5 H migration :

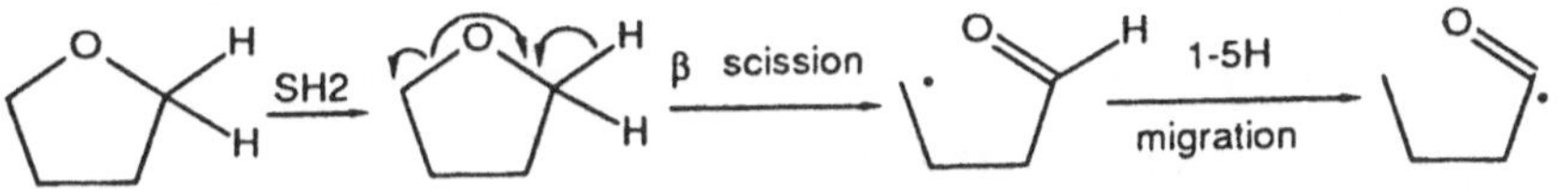

This radical R-C˙O could then add to Cu^{II} in the organometallic species

RC˙O + Br— Cu—

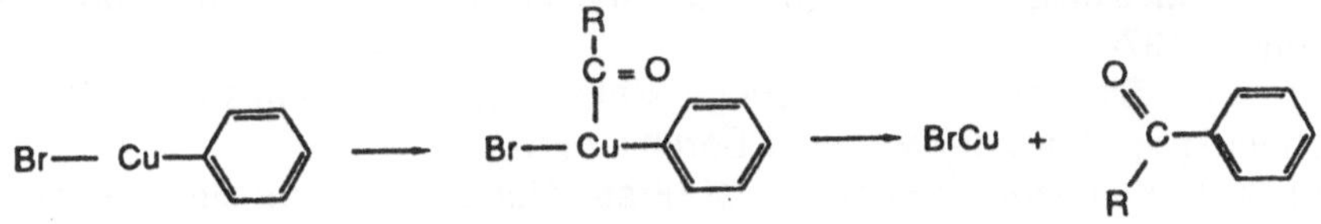

and a final reductive elimination would yield the observed $ArCOCH_2CH_2CH_3$ product. The basic flaw in such a scheme is that, to have a 1-5 H migration, the C_{sp3}-H bond should be stronger than the C_{sp2}-H bond. This does not seem to be the case [$E(C\text{-}H)_5$ = 97.5 kcal mole $^{-1}$; $E(C_{co}\text{-}H)_1$ = 101.33 kcal mole $^{-1}$].

The second possibility relies on a β elimination taking place from complexed molecules of THF :

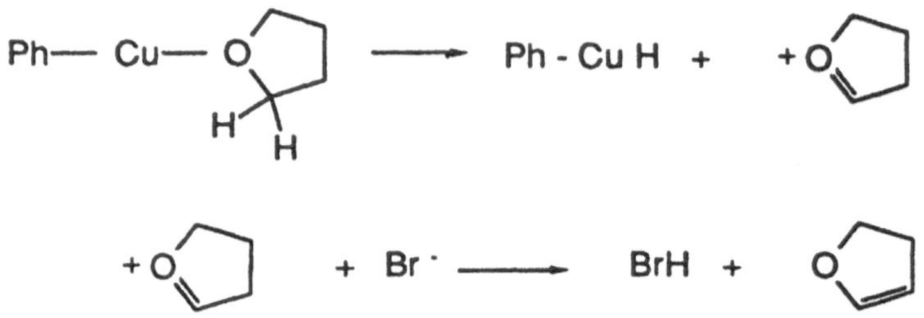

Dihydrofuran so generated would, under favorable conditions generate butanal. To our knowledge, this type of involvement of THF in organometallic reactivity has no precedent (29). We are presently studying its origin because it could possibly involve a carbon-hydrogen activation of THF by coordinatively unsaturated species of Cu (30).

In conclusion, the metal vapor chemistry of Cu is far from being competitive with Ullmann's method for coupling haloaromatics. The ortho substituent stabilizing effect is reminiscent of solvated Ni atoms discovered by Klabunde in his SMAD catalytic system, and this direction deserves further studies with Cu as solvated atom.

REFERENCES

1. P.S. Skell, L.D. Wescott Jr., J.P. Goldstein, R.R. Engel, J. Am. Chem. Soc., **87**, 1965, 2.

2. **a)** M. Moskovits, G.A. Ozin Eds, Cryochemistry, New York, John Wiley and Sons, 1976.
   b) J.R. Blackborow, D. Young, Metal Vapour Synthesis in Organometallic Chemistry, Springer-Verlag, Berlin 1979.
   c) K.J. Klabunde, Chemistry of Free Atoms and Particles, Academic Press New York, 1980.
   d) K.J. Klabunde, P.L. Timms, P.S. Skell, S.D. Ittel, in Shriver, D.F. ed. , Inorganic Syntheses, Volume 19, John Wiley and Sons, New York, 1979, Chapter 2.
   e) P.L. Timms,Proc. Roy. Soc., London, 1984, **A 396**, 1.

3. **a)** Y. Imizu, K.J. Klabunde, Inorg. Chem., 1984, **23**, 3602.
   b) P.H. Kasai, P.M. Jones, J.Am. Chem. Soc. 1984, **106**, 8018.

4. **a)** K. Matsuo, K.J. Klabunde, J. Org. Chem., 1982, **47**, 843.
b) K. Matsuo, K.J. Klabunde, J. Catal., 1982, **73**, 216.
c) K.J. Klabunde, Y. Imizu, J. Am. Chem. Soc., 1984, **106** , 2721.
d) H. Kanai, B.J. Tan, K.J. Klabunde, Langmuir, 1986, **2**,(6), 760.

5. J.C. Négrel, M. Chanon, Inform. Chim., 1984, 145.

6. S. Togashi, J.G. Fulcher, B.R. Cho, M. Hasegawa, J.A. Gladysz, J. Org. Chem., 1980, **45**, 3044.

7. R. Mackenzie, P.L. Timms, J. Chem. Soc. Chem. Comm., 1974, 650.

8. **a)** P.E. Fanta, Chem. Rev., 1964, **64**, 613 ; Synthesis, 1974, 9.
b) J.K. Kochi, Organometallic Mechanisms and Catalysis, Academic Press, New York, 1978, p. 386.
c) T. Cohen, I. Cristea, J., Am. Chem. Soc., 1976, **98**, 748.

9. **a)** R.D. Rieke, Acc. Chem. Res., 1977, **10**, 301.
b) G.W. Ebert, R.D. Rieke, J. Org. Chem., 1984, **49**, 5280.

10. Ref. 2b, p. 143 ; Ref. 2c, p. 86.

11. Ref. 2b, p. 127.

12. **a)** K.J. Klabunde, J.Y.F. Low, J. Am. Chem. Soc., 1974, **96**, 7674.
b) K.J. Klabunde, Angew. Chem. Int. Ed., 1975, **14**, 287.
c) K.J. Klabunde, Acc. Chem.; Res., 1975, **8**, 393.
d) K.J. Klabunde, B.B. Anderson, M. Bader, L.J. Radonovitch, J. Am. Chem. Soc., 1978, **100**, 1313.
e) B.B Anderson, C. Behrens, L.J. Radonovitch, K.J. Klabunde, J. Am. Chem. Soc., 1976, **98**, 5390.

13. K.J. Klabunde, J.S. Roberts, J. Organomet. Chem., 1977, **137**, 113.

14. **a)** M.T. Anthony,Thesis, Bristol, 1973.
b) Y. Tanaka, S.C. Davis, K.J. Klabunde, J. Am. Chem. Soc., 1982, **104**, 1013.
c) S.P. Kolesnikov, S.L. Povarov, S.Y. Tsvetkov, A.I. D'Yachenko, O.M. Nefedov, Bull. Acad. Sci. USSR, 1984, **33**, 4, 793.

15. **a)** K.J. Klabunde, J.Fluorine Chem., 1976, **7**, 95.
b) K.J. Klabunde, Chem. Eng. News, 1977, **24**, 23.
c) K.J. Klabunde, H.F. Efner, J. Fluorine Chem., 1976, **4**, 114.

16. T.O. Murdock, Thesis, University of North Dakota, Grand Forks, 1977.

17. **a)** J.M. Parnis, S.A. Mitchell, J. Garcia-Prieto, G.A. Ozin, J. Am. Chem. Soc., 1985, **107**, 8169.
b) G.A. Ozin, S.A. Mitchell, J. Garcia-Prieto, Angew. Chem. Int. Ed., 1982, **21**, 211.

18. J.A. Howard, R. Sutcliffe, B. Mile, J. Catal., 1984, **90**, 156.

19. J.G. Noltes and G. Van Koten, Comprehensive Organometallic Chemistry, G. Wilkinson, F.G.A Stone, E.N. Abel Ed., Pergamon,London, 1982, Vol. 2, p709.

20. **a)** J.H. Sinfelt, J.L. Carter, P.C. Yates, J. Catal. 1972, **24**, 283.
b) J.H. Sinfelt, Acc. Chem. Res. 1977, **10**, 15.

21. **a)** F.W.S. Benfield, M.L.H. Green, J.S. Ogden, D.J. Young, Chem. Comm.,1973, 866.
b) J.C. Négrel, Dossier Technique ANVAR, DAR/432.ET.01253-50574

22. J.T. Scanion,D.E. Willis, J. Chromat. Sci., 1985, **23**, 333.

23. T.T. Tsou, K.J. Kochi, J. Am. Chem. Soc., 1979, **101**, 6319.

24. H. Matsumoto, S. Inaba, R.D. Rieke, J. Org. Chem., 1983, **48**, 840.

25. M. Foa, L. Cassar, J. Chem. Soc. Dalton Trans., 1975, 2572.

26. A.B.M. Bashar, P.L. Timms, High Temp. Sci., 1984, **17**, 417.

27. A manual "solvent sweeper" apparatus with sealing system.

28 J.Forrest, J.Chem.Soc., 1960, 592.

29. A. Maaerker, Angew. Chem. Int. Ed., 1987, **26**, 972.

30. **a)** R.H. Crabtree, Chem. Rev., 1985, **85**, 245.

b) R.H. Crabtree, Adv. Organomet. Chem., 1988, **28**, 299.

RECENT ADVANCES IN THE CHEMISTRY OF SILYL RADICALS

C. Chatgilialoglu
I.Co.C.E.A.
Consiglio Nazionale delle Ricerche
40064 Ozzano Emilia (Bologna)
Italy

ABSTRACT. Absolute kinetic data for a significant number of reactions involving silyl radicals have been reported and discussed together with the available literature data of their isostructural group 14 organometallic centered radicals. Tris(trimethylsilyl)silane is found to be an effective reducing agent for organic halides and due to its efficiency would appear to offer an alternative to tin hydride.

INTRODUCTION

Silicon is the second most abundant element of earth's crust and is today the vital component of several technological revolutions; therefore the current increasing interest in silicon chemistry is not surprising. Silyl radicals, $R_3Si^{\bullet}$, have been postulated in the past as intermediates in many reactions of organometallic compounds and their structural characteristics have been the object of investigation by several scientists in the last two decades. Recently, we have been able to obtain a large body of absolute rate constants for the reactions of silyl radicals as well as many informations on their reactivity by using laser flash photolysis techniques. The rationalization of these quantitative data, in principle, can provide new approaches to the synthesis of organic compounds.

The purpose of this article is to report some recent data concerning silyl radicals and rationalize them from the point of view of the reducing ability of silanes.

STRUCTURAL CHARACTERISTICS

In the early seventies the structures of silyl radicals have been investigated in details by EPR spectroscopy(1). Thus, silyl radicals have been shown to have a pyramidal center at silicon, pyramidality depending on the electronegativity of the substituent. Furthermore, triorganosilyl radicals when generated from optically active precursors can yield products that are optically active and have retained the

M. Chanon et al. (eds.), Paramagnetic Organometallic Species in Activation / Selectivity, Catalysis, 119–129.

configuration of the starting material(2). Combination of product studies with some kinetic data allowed evaluation of the rate constant of inversion for (1-naphthyl)phenylmethylsilyl radical(3), viz.,

as $6.8x10^9 s^{-1}$ at 353K, corresponding to a half life for inversion of ca. 104ps. With the assumption that inversion at the silicon center has a normal preexponential factor (i.e., $\log(A/s^{-1})=13.3$) the inversion barrier was estimated as 5.6 kcal/mol. Similar considerations for the inversion of the analogous germyl radical indicate a barrier of 4.2 kcal/mol(4). The remarkable configurational stability of silyl radicals is further demonstrated by 1-methyl-4-*tert*-butyl-1-silacyclohexyl radical, viz.,

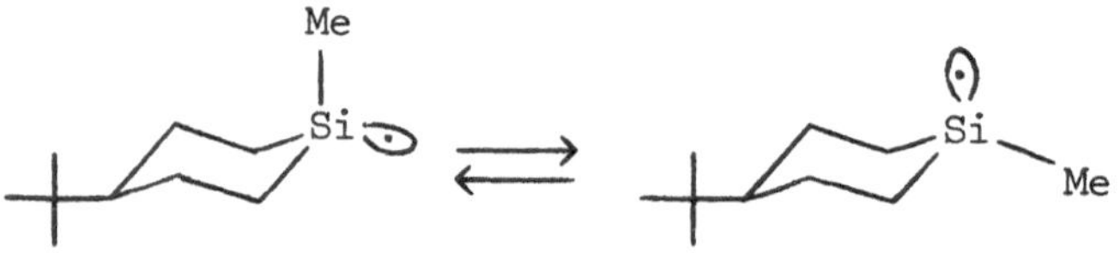

where the forward and reverse inversion rate are estimated to be $6.4x10^9$ and $2.9x10^9 s^{-1}$ respectively at 273K(3).

The optical absorption spectra in the range 280-600nm have been obtained for a variety of tri-substituted silyl radicals by using laser flash photolysis techniques(5-7). For example, trialkylsilyl radicals show a strong band at λ<300nm (λ_{max}=256nm and ε_{max}=$7200M^{-1}cm^{-1}$ in the gas phase for trimethylsilyl radical(8)) and a weaker band around 390nm. The transition assignments have been performed theoretically by using Multiple Scattering X_α calculations(9). These calculations suggest that the absorption bands in alkyl substituted silyl radicals are related to electron excitations into molecular orbitals that are predominantly Rydberg in character, the lowest band to 3s-Rydberg orbital and the second one, appearing more intense in the UV region, to the overlap of the 3d-Rydberg transition with the low-lying valence transition.

THE SUBSTITUENT EFFECT ON THE REACTIVITY OF SILANES TOWARDS FREE RADICALS

In order to form the basis of discussion, the bond dissociation energies of a variety of silane together with the absolute rate constants of their reactions with *tert*-butoxyl radicals (eq. 1) are collected in table I.

$$R_3SiH + Me_3CO^\bullet \longrightarrow R_3Si^\bullet + Me_3COH \qquad (1)$$

The factors which dominate the thermochemistry of C-H bond are

TABLE I. Bond Dissociation Energies and Related Kinetic Data for the Si-H Bonds in Silanes

Silane	$BDE(R_3Si-H)^a$, kcal/mol	k_1^b, $M^{-1}s^{-1}$ at296K
F_3Si-H	100.1	—
$(EtO)_3Si-H$	—	$<2.0x10^6$
Cl_3Si-H	91.3	$4.0x10^7$
Et_3Si-H	90.1	$4.6x10^6$
Ph_3Si-H	—	$7.7x10^6$
$Ph(H)_2Si-H$	88.2	$2.5x10^6$
$n-C_5H_{11}(H)_2Si-H$	89.6	$3.5x10^6$
H_3Si-H	90.3	—
$Me_3Si(Me)_2Si-H$	85.3	$1.5x10^7$
$(Me_3Si)_3Si-H$	79.0	$1.0x10^8$

[a]From refs 10 and 11. [b]From refs 5, 6 and 12.

essentially the importance of delocalization of the unpaired electron in the corresponding alkyl radicals and/or the relieving steric compression upon dissociation(13). From the values collected in table I it can be seen that the factors influencing the strength of carbon-hydrogen bond are essentially unimportant in the silicon cogeners, namely, (i) the absence of a weakening effect of substituent alkyl groups on Si-H bonds, (ii) the ineffectiveness of delocalization by phenyl substitution (Si-H bond weakening by phenyl substitution is very slight, ca. 1.5 kcal/mol, in contrast to the 10 kcal/mol observed for phenyl substituted C-H bonds). In other words, a striking feature of silane, alkyl silanes and phenyl silanes is the almost constant SiH bond strength at 89±1 kcal/mol. In fact, it is worth noticing that the reactivity of these silanes towards <u>tert</u>-butoxyl radicals is similar, viz., $(5\pm3)x10^6M^{-1}s^{-1}$. On the other hand, the effect produced by three neighboring fluorines or three silyl groups is profound; that is $BDE(F_3Si-H)$=100.1 and $BDE((Me_3Si)_3Si-H)$=79.0 kcal/mol. One way of rationalizing this behavior is to invoke the electronegativity character of the substituent. However, simple electronic effects do not account for the phenomenon since Si-H bond dissociation energies do not correlate with the group electronegativities of the ligands. Furthermore, alkoxy group and chlorine atom deactivate and activate, respectively, Si-H bond towards tert-butoxyl radicals although they have similar group electronegativity. These results can be interpreted by invoking a bonding interaction between d orbitals of the substituents and the semioccupied p orbital on the radical center, i.e., (d-3p)π bonding. EPR studies for $Cl_3Si^{\cdot}$ and $(Me_3Si)_3Si^{\cdot}$ radicals support such a (d-3p)π bonding interaction(12).

In summary, the reactivities of silanes containing Si-H bond towards free radicals may be of a wide range. Thus, although the importance of delocalization of the unpaired electron in the alkyl radicals is essentially unimportant in the silicon cogeners, the (d-3p)π-type bonding together with the electronegativity character of the substituents can play a very important role.

SILANES AS REDUCING AGENTS

One of the best known and most useful free-radical reaction is the reduction of alkyl halides, RX, to the hydrocarbon, RH, with tributyltin hydride(14). This reaction involves a two-step free radical chain process, viz.,

$$R^{\cdot} + Bu_3SnH \longrightarrow RH + Bu_3Sn^{\cdot} \qquad (2)$$

$$Bu_3Sn^{\cdot} + RX \longrightarrow Bu_3SnX + R^{\cdot} \qquad (3)$$

However, it is often difficult to isolate and purify the reduction products from the tin compounds. Furthermore, Giese has recently pointed out the desirability of replacing toxic hydrogen transfer agent by more ecologically acceptable compounds(15).

Alkyl halides are reduced to hydrocarbon by organosilane in good yields when catalyzed by Lewis acids(16) but this methodology has several limitations; for example, side reactions can often be predominant. On the other hand, trialkylsilanes are poor reducing agents in a free radical chain process(17). That is, although trialkylsilyl radicals are more reactive in halogen atom abstractions than tributyltin radicals(vide infra), they are rather poor H-atom donors toward alkyl radicals and tend therefore not to support chain reactions except under forcing conditions. As we have already mentioned above the silicon-hydrogen bond can be dramatically weakened by successive substitution of silyl groups at the Si-H function(11); this result suggested that tris(trimethylsilyl)silane might be a good hydrogen donor and that it would be capable of sustaining a radical chain reduction of alkyl halides analogous to reactions 2 and 3. The expectation turned out to be correct and we report some of our results in table II(18).

The reduction of alkyl halides by tris(trimethylsilyl)silane can be achieved under a variety of conditions(see table II). This reduction reaction involves a two-step free radical chain process, viz.,

$$R^{\cdot} + (Me_3Si)_3SiH \longrightarrow RH + (Me_3Si)_3Si^{\cdot} \qquad (4)$$

$$RX + (Me_3Si)_3Si^{\bullet} \longrightarrow R^{\bullet} + (Me_3Si)_3SiX \quad (5)$$

as indicated by the fact that the reaction is catalyzed by thermal or photochemical sources of free radicals, retarded by common inhibitors, and that oxygen is an excellent initiator(18).

One drawback of this method is that the product, once formed, must be separated from a full 1 equiv. of $(Me_3Si)_3SiX$ as the tris(trimethylsilyl)silane is employed in stoichiometric amounts. A method for catalytic dehalogenation via tris(trimethylsilyl)silane has also been developed(19): the organic halide is treated with an excess of sodium borohydride and catalytic amounts of $(Me_3Si)_3SiH$ or its corresponding halide and the reaction is initiated photochemically. The catalytic cycle is given in eq. 6 and 7. The reaction is carried out in realatively polar solvents,

$$(Me_3Si)_3SiX + NaBH_4 \longrightarrow (Me_3Si)_3SiH + NaX + BH_3 \quad (6)$$

$$(Me_3Si)_3SiH + RX \longrightarrow (Me_3Si)_3SiX + RH \quad (7)$$

like ethanol or ethyleneglycoldimethyl ether, in the absence of air, and preferably at room temperature. Yields are good to excellent; for example, 1-bromohexadecane has been reduced to the corresponding hydrocarbon in 98% yield(19).

TABLE II. Reduction of Organic Halides by $(Me_3Si)_3SiH$ at Room Temperature

RX	Conditions	% Yield
$C_{18}H_{37}I$	eq.M in C_6H_{12},hν,5min	100
$PhCH_2Br$	1:3 Neat, air,2.5hours	96
$C_{16}H_{33}Br$	eq.M in Monoglyme,In/hν,30min	100
$C_5H_{11}Br$	1:3 Neat, air, 30min	98
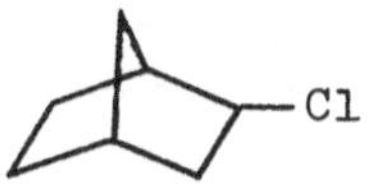	eq.M in C_6H_{12},hν,2.5hours	94
$C_{15}H_{37}Cl$	eq.M in monoglyme,In/hν,8hours	70

ABSOLUTE RATE CONSTANTS FOR THE REACTIONS OF ALKYL RADICAL WITH SILANES

The reactions of carbon-centered radicals with group 14 organometallic hydrides (see for example eqs 2 and 4) are of considerable importance in organic chemistry since they are the key steps in the free-radical chain reduction processes. In table III are listed rate constants at room temperature for hydrogen atom abstraction by primary alkyl radicals from tri-substituted tin, germanium and silicon hydrides. The order of reactivity is

$$Bu_3SnH > (Me_3Si)_3SiH > Bu_3GeH > Me_3Si(Me)_2SiH > Et_3SiH$$

with the hydrogen donating ability covering ca. 4 orders of magnitude. The difference in reactivities is expected to arise predominantly from different activation energies since the bond strengths through the series span a range of 16 kcal/mol. As a hydrogen donor to primary alkyl radicals the tris(trimethylsilyl)silane has about 1/4 the reactivity of tributyltin hydride (cf. table III). For tin hydride it has been shown that the rate constants for the reactions of primary, secondary, and terziary alkyl radicals are essentially equal at room temperature(21). This often convenient property however, is not yet confirmed for $(Me_3Si)_3SiH$ (22).

TABLE III. Rate Constants for the Reaction of Primary Alkyl Radicals with Some Group 14 Organometallic Hydrides at Room Temperature

Substrate	BDE(R_3M-H), kcal/mol	Radical	*k*, $M^{-1}s^{-1}$	Ref.
Et_3SiH	90.1	$PhCMe_2CH_2^{\bullet}$	$\sim 4 \times 10^2$	17
$Me_3Si(Me)_2SiH$	85.3	$PhCMe_2CH_2^{\bullet}$	$\sim 8 \times 10^3$	17
$(Me_3Si)_3SiH$	79.0	$CH_2{=}CH(CH_2)_3CH_2^{\bullet}$	$\sim 6 \times 10^5$	18
Bu_3GeH	~83	$CH_2{=}CH(CH_2)_3CH_2^{\bullet}$	9.3×10^4	20
Bu_3SnH	74.0	$CH_3CH_2CH_2CH_2^{\bullet}$	2.5×10^6	21

ABSOLUTE RATE CONSTANTS FOR THE REACTIONS OF SILYL RADICALS WITH ORGANIC HALIDES

Arrhenius parameters for the reactions of triethylsilyl radicals with a number of organic halides have been determined in solution by using laser flash photolysis techniques(3). These results are summarized in table IV. Comparison of these data shows that the high reactivity of CCl_4 relative to monochlorinated substrates is a consequence of its large Arrhenius preexponential factor. Such a high preexponential factor for CCl_4 has been attributed to the greater importance of polar contribution to the transition state for this chlorine atom abstraction relative to chlorine abstraction from monochlorinated alkanes(23).

$$Et_3Si^{\cdot}Cl{-}CR_3 \longleftrightarrow Et_3Si^{+}Cl^{-}\dot{C}R_3 \longleftrightarrow Et_3Si^{+}\dot{C}lCR_3^{-} \longrightarrow$$
$$\longrightarrow Et_3Si{-}Cl \;+\; {}^{\cdot}CR_3 \qquad (8)$$

That is, the greater the electron affinity of the halogen atom donor, the greater will be the contribution that canonical structures with charge separation make to the transition state. It has been recognized that charge-transfer interactions to the transition state enhance the reaction rate by reducing the activation energy, this being in excellent agreement with the present results. However, the charge-transfer may exert a greater influence on the preexponential factor than on the activation energy: when there is no polar contribution to the transition state the reaction occurs only when Si, Cl and C atoms are colinear or nearly so, whereas a strong polar contribution to the transition state will relax this restriction (see scheme I). In the limiting case of complete electron

SCHEME I

no polar contribution — strong polar contribution

transfer the resultant ion pair would not be subject to any restriction in their relative rotational motion. This gain of two rotational degree of freedom in the transition state would increase the A-factor by ca. 100. The fact that the A-factor for CCl_4 is about 2 orders of magnitude greater than that for monochlorinated alkanes means (even if there is some contribution from statistical factors) that electron transfer is extensive in the $Et_3Si^{\cdot}/CCl_4$ transition state. In a similar way, the increase in reactivity for monohalogenated compounds along the series $Cl < Br < I$ is largely due to an increase in the preexponential factors (see table IV), which has been attributed to the increased importance of polar effects along the series.

From the above discussion, charge-transfer might appear unimportant in chlorine atom abstraction from simple alkyl chlorides; however it is not so. In fact, rate constants have been measured for the reaction of the triethylsilyl radical with a series of ring-substituted benzyl chlorides(24). A Hammett plot gives a ρ=+0.64 which is the largest positive ρ value reported for a free-radical reaction. Thus, electron-withdrawing substituents increase the reaction rate, in accordance with charge separation in the transition state as indicated in eq.8.

A large body of absolute rate constants for halogen atom abstraction from RX by a variety of group 14 organometallic centered radicals is now

TABLE IV. Arrhenius Parameters for Halogen Atom Abstraction by $Et_3Si^{\cdot}$ Radical[a]

Substrate	$\log(A/M^{-1}s^{-1})$	E_a, kcal/mol	$\underline{k}$, $M^{-1}s^{-1}$ at 300K
CH_3CH_2I	10.4	1.0	4.3×10^9
$PhCH_2Br$	10.3	1.3	2.4×10^9
$(CH_3)_3CBr$	9.7	0.8_5	1.1×10^9
$CH_3(CH_2)_4Br$	9.3	0.7	5.4×10^8
CCl_4	10.2	0.8	4.6×10^9
$PhCH_2Cl$	8.9	2.1	2.0×10^7
$(CH_3)_3CCl$	8.7	3.2	2.5×10^6

[a]From ref. 3

available. Some of the data obtained mainly by laser flash photolysis techniques are listed in table V. The trends in reactivity are those which would be expected for an individual radical; that is, (i) for a particular R group the rate constants decrease along the series X = I > Br > Cl, (ii) for a particular X the rate constants decrease along the series benzyl > _tert_-alkyl > primary alkyl. For most substrates the $(Me_3Si)_3Si^{\cdot}$, $Bu_3Ge^{\cdot}$ and $Bu_3Sn^{\cdot}$ radicals have essentially similar reactivities, but are less reactive than Et_3Si radical. Table V also shows that the reactivities of silyl radicals towards a particular halide can be modified by successive substitution of silyl groups at the radical center. The difference in reactivity between tris(trimethylsilyl)silyl and triethylsilyl radicals would presumably have become more apparent if rate constants had been measured for "slow" reactions. Although the decrease of the reactivity along the series

$$Et_3Si^{\cdot} > Me_3Si(Me)_2Si^{\cdot} > (Me_3Si)_3Si^{\cdot}$$

arises probably from the different enthalpies of reaction, steric factors are also considered to be important.

Finally, the charge-transfer in the transition state proposed to explain the high reactivity of triethylsilyl radicals towards carbon tetrachloride (vide supra) can be advanced to explain also the high reactivity of $(Me_3Si)_3Si^{\cdot}$ and $Bu_3Ge^{\cdot}$ radicals towards CCl_4 and other halogen donors.

TABLE V. Absolute Rate Constants ($\underline{k}$, $M^{-1}s^{-1}$ at 297±3K) for the Abstraction of Halogen from Some Organic Halides by $R_3M^{\cdot}$ Radicals

Substrate	$Et_3Si^{\cdot}$ [a]	$Me_3Si\dot{S}iMe_2$ [b]	$(Me_3Si)_3Si^{\cdot}$ [c]	$Bu_3Ge^{\cdot}$ [d]	$Bu_3Sn^{\cdot}$ [d,e]
CH_3I	$8.1x10^9$	—	—	$>1.0x10^8$	$4.3x10^9$
$PhCH_2Br$	$2.4x10^9$	—	$9.6x10^8$	$9.7x10^8$	$1.5x10^9$
$(CH_3)_3CBr$	$1.1x10^9$	$2.6x10^8$	$1.2x10^8$	$8.6x10^7$	$1.7x10^8$
$CH_3(CH_2)_nBr$	$5.4x10^8$	$1.6x10^8$	$2.0x10^7$	$4.6x10^7$	$2.6x10^7$
CCl_4	$4.6x10^9$	—	$1.7x10^8$	$3.1x10^8$	—
$CHCl_3$	$2.5x10^8$	—	$6.8x10^6$	—	—
$PhCH_2Cl$	$2.0x10^7$	—	$<2.0x10^6$	$1.9x10^6$	$1.1x10^6$
$(CH_3)_3CCl$	$2.5x10^6$	$4.2x10^5$	—	$<5.0x10^4$	$2.7x10^4$
$CH_3(CH_2)_nCl$	$3.1x10^5$	—	—	—	$1.4x10^3$

[a] From ref. 3. [b] From ref. 17. [c] From ref. 25. [d] From ref. 4. [e] From ref. 26; original data have been normalized by multiplying by a factor of 1.7 (cf. ref. 4).

CONCLUSIONS

Tris(trimethylsilyl)silane appears to offer an alternative to tin hydride in radical chain reduction reactions of organic halides. Thus, as a hydrogen donor to alkyl radicals it is slightly less reactive than tributyltin hydride and tris(trimethylsilyl)silyl radical is reactive in halogen abstraction as much as tributyltin radical. Futhermore, tris(trimethylsilyl)silane is a far more acceptable reducing agent than tin hydride from ecological and toxicological perspectives.

ACKNOWLEDGMENT

I thank all my previous and present co-workers whose names are listed in the references.

REFERENCES

(1) A. Alberti, and G.F. Pedulli, Rev. Chem. Interm., 8, 207 (1987) and references therein.
(2) For a review on this problem, see: A.L.J. Beckwith, and K.U. Ingold, Rearrangements in Ground and Excited States (Editor P. de Mayo), Academic Press: New York, 1980, vol 1, Essay No 4.
(3) C. Chatgilialoglu, K.U. Ingold, and J.C. Scaiano, J. Am. Chem. Soc., 104, 5123 (1982).
(4) K.U. Ingold, J. Lusztyk, and J.C. Scaiano, J. Am. Chem. Soc., 106, 343 (1984).
(5) C. Chatgilialoglu, J.C. Scaiano, and K.U. Ingold, Organometallics, 1, 466 (1982).
(6) C. Chatgilialoglu, K.U. Ingold, J. Lusztyk, A.S. Nazran, and J.C. Scaiano, Organometallics, 2, 1332 (1983).
(7) C. Chatgilialoglu, Unpublished results.
(8) N. Shimo, N. Nakashima, and K. Yoshihara, Chem. Phys. Lett., 125, 303 (1986).
(9) C. Chatgilialoglu, and M. Guerra, Manuscript in preparation.
(10) R. Walsh, Acc. Chem. Res., 14, 246 (1981).
(11) J.M. Kanabus-Kaminska, J.A. Hawari, D. Griller, and C. Chatgilialoglu, J. Am. Chem. Soc., 109, 5267 (1987).
(12) C. Chatgilialoglu, and S. Rossini, Bull. Soc. Chim. Fr., in press.
(13) D.F. McMillen, and D.M. Golden, Annu. Rev. Phys. Chem., 33, 493 (1982).
(14) B. Giese, Radicals in Organic Synthesis: Formation of Carbon-Carbon Bonds, Pergamon Press: New York, 1986.
(15) B. Giese, Angew. Chem. Int. Ed. Engl., 24, 553 (1985).
(16) M.P. Doyle, C.C. McOsker, and C.T. West, J. Org. Chem., 41, 1393 (1976).
(17) J. Lusztyk, B. Maillard, and K.U. Ingold, J. Org. Chem., 51, 2457 (1986) and references therein.
(18) C. Chatgilialoglu, D. Griller, and M. Lesage, J. Org. Chem., in press.
(19) C. Chatgilialoglu, D. Griller, and M. Lesage, Italian patent: 48150A87 (1987).

(20) J. Lusztyk, B. Maillard, D.A. Lindsay, and K.U. Ingold, J. Am. Chem. Soc., **105**, 3578 (1983).
(21) C. Chatgilialoglu, K.U. Ingold, and J.C. Scaiano, J. Am. Chem. Soc., **103**, 7739 (1981).
(22) C. Chatgilialoglu, D. Griller, and M. Lesage, Studies are currently in progress.
(23) For general discussions of polar effects in radical reactions see: F. Minisci, and A. Citterio, Adv. Free-Radical Chem., **6**. 65 (1980).
(24) C. Chatgilialoglu, K.U. Ingold, and J.C. Scaiano, J. Org. Chem., **52**, 938 (1987).
(25) C. Chatgilialoglu, D. Griller, and M. Lesage, Manuscript in preparation.
(26) D.J. Carlsson, and K.U. Ingold, J. Am. Chem. Soc., **90**, 7047 (1968).

DIVALENT LANTHANIDES : A FAMILY OF REDUCING AGENTS IN ORGANIC SYNTHESIS

H.B. KAGAN and J. COLLIN
Laboratoire de Synthèse Asymétrique,
UA-CNRS n° 255
Université Paris-Sud, 91405 ORSAY (France)

ABSTRACT :

The main properties and preparations of divalent lanthanides are briefly discussed. Then organic chemistry mediated by diiodosamarium and diiodoytterbium is summarized. SmI_2 in solution in THF is a good reagent for many processes such as carbonyl reduction, pinacol formation, organic halide reduction or pseudo Barbier reaction. All these reactions were interpreted through formation of radical intermediates. Acid chlorides are coupled into α-diketones or can give a reductive coupling on ketones or aldehydes. Cp_2Sm was compared to SmI_2 in the previous reactions, it behaves sometimes differently. Especially it reacts with benzylic halides to yield stable organosamarium complexes. Some examples of reductions by Cp'_2Yb or Cp'_2Sm are also presented.

1. INTRODUCTION

Low-valent lanthanides compounds are known since a long time but remained without application in organic chemistry. The usual divalent lanthanides are based on europium, samarium and ytterbium. The main accesses to divalent lanthanides were described in recent reviews[1-3]. The introduction of divalent lanthanides (mainly Sm(II)) in organic chemistry started in 1977 when we discovered[4] an easy route to SmI_2 and YbI_2, which are soluble in THF. Most of the organic reactions mediated by divalent lanthanides was reviewed in 1985[5] and 1986[6]. We intend to discuss here some of the reactions induced by Ln(II) compounds, especially the more recent results.

2. MAIN FEATURES OF Ln(II) DERIVATIVES

Electronic distribution of Ln(II) ions can be described as a xenon core surrounded by some 4f electrons. Divalent lanthanides have a half-filled (or close to) or filled 4f orbitals. Eu(II) = [Xe] $4f^6$, Sm(II) =

M. Chanon et al. (eds.), Paramagnetic Organometallic Species in Activation / Selectivity, Catalysis, 131–147.

[Xe] $4f^7$ and Yb(II) = [Xe] $4f^{14}$. These ions behave as one-electron donors since the trivalent state is the most stable one for lanthanides. The $E°_{aq}$ values (Ln^{3+}/Ln^{2+}) are -0.35 V, -1.15 V and -1.55 V for europium, ytterbium and samarium respectively. On this basis, and by extrapolation to reactions occuring in THF solutions, one predicts that the reducing ability will follow the sequence Eu(II) < Yb(II) < Sm(II). This reactivity sequence was indeed experimentally observed[7].

In the redox couples most of the ions are paramagnetic because of an open 4f shell. The following average values of μ_D (Bohr magneton) at room temperature are taken from a recent review[8] :

Eu(II) : 1.8	Eu(III) : 4.0
Yb(II) : 0	Yb(III) : 4.3
Sm(II) : 3.6	Sm(III) : 1.7

These values are fairly constant whatever are the lanthanide derivatives.

Divalent lanthanides have some Lewis acidity as evidenced by X-ray structures of many complexes where Lewis bases such as CH_3CN orTHF, act as ligands of the metal center.

3. PREPARATIONS OF DIVALENT LANTHANIDES COMPOUNDS

Various methods for preparation of divalent lanthanides compounds have been reported. Three reactions are essentially used : oxidation of lanthanide metals, exchange of ligands on other divalent compounds, reduction of trivalent compounds (for reviews see refs. 1-3). Some of the earliest preparations of Ln(II) salts from Ln(III) precursors were based on a high-temperature dismutation (eg $SmI_3 \longrightarrow SmI_2 + I_2$) or reduction by hydrogen, again at high temperature (eg $SmI_3 + H_2 \xrightarrow{400°C} SmI_2 + HI$).

As previously stated[1] most of the interesting organic chemistry which has recently emerged using lanthanides came from Yb(II) and Sm(II) derivatives. The preparations described below are limited to some useful compounds. One key reaction which gives an easy access to divalent lanthanides of interest for organic chemists was developped in our research group in 1977[4,9] (equation [1]). It involves the attack of samarium or ytterbium metal which is commercially available (as a powder, 40 mesh) by 1,2-diiodoethane in THF, under inert atmosphere.

$$Ln + ICH_2CH_2I \xrightarrow[r.t.]{THF} LnI_2 + CH_2{=}CH_2 \qquad [1]$$

Ln = Yb or Sm

The diiodo compounds are soluble to some extent in THF, giving highly colored solutions. Thus SmI_2 solutions (solubility ≃ 0.1 M in THF) have a deep green color and can be stored under nitrogen for a very long time. We found that the stability is much increased if a small amount of Sm metal remains in contact with the solution[10]. The structure of LnI_2 (in THF solution) prepared according to equation [1] was proved by various titrations. Several variations of the original procedure of SmI_2 preparation can be found in recent literature. SmI_2 has been prepared by reaction of diiodomethane on samarium for iodomethylation reactions[11] and by reaction of iodine on samarium[12-13]. For preparative purposes it is possible to reduce the amount of solvent and to use SmI_2 as a slurry. X-ray structure of SmI_2 solvated by pivalonitrile or diglyme were recently described[14]. In the first case SmI_2 has a polymeric structure while in the second case it has a monomeric structure.

SmI_2 is a convenient starting material for several exchange reactions (equation [2]), leading to other divalent samarium complexes such as $Sm(OEt)_2$[4], $Sm(OAc)$[4], $Cp_2Sm(THF)_2$[4], $Cp'_2Sm(THF)_2$[15] $[Cp'Sm(\mu I)(THF)_2]_2$[15] ($Cp' = C_5Me_5$), C_8H_8Sm[16].

$$LnI_2 + 2\ MX \longrightarrow LnX_2 + 2\ MI \qquad [2]$$

Ln = Sm or Yb
M = Na or K

This method is easier to use than those previously described, such as reduction of Cp_3Sm[17] with potassium naphtalenide or metal vaporization technique for the synthesis of Cp'_2Sm[18] which is obtained unsolvated through this way. X-ray structures of Cp'_2Sm[19] or $Cp'_2Sm(THF)_2$[15] show a bent metallocene structure unlike metallocene complexes of transition elements.

Cp_2Yb and Cp'_2Yb have been synthesized by exchange reaction [2], the former from YbI_2[4] and the latter from $YbBr_2$[20] or $YbCl_2$[21].

Diorganolanthanides have been synthesized by Deacon from Yb metal using transmetallation reaction[22,23]:

$$(C_6F_5)_2Hg + Yb \longrightarrow (C_6F_5)_2Yb + Hg \qquad [3]$$

4. MAIN REACTIONS OF SmI_2 AND YbI_2

THF solutions of SmI_2 and YbI_2 were checked for their reducing properties toward several classes of organic compounds. This investigation was facilitated by a simple visual observation : there is an important change of color going from Ln(II) to Ln(III), namely in the case of SmI_2 from deep green to yellow. The main results obtained in the period 1977-1984 in Orsay can be found in refs[1,4-7]. Reactions were performed in THF in presence of some alcohols (which act as proton donors) or in aprotic conditions. SmI_2 soon revealed itself as much superior to YbI_2 in many reductions and was subsequently almost exclusively used in various investigations.

According to the structure of organic substrate one can expect to generate species coming from a one-electron transfer (radical anion or products arising from its cleavage) or from two one-electron transfers (dianion or anion coming from radicals in situ generated). We shall not intend to discuss the mechanistic details of these transformations in terms of outer-sphere or inner-sphere mechanisms (for a definition and significancy in organometallic chemistry see ref.[24]).

4.1. Reduction of carbonyl compounds

SmI_2 in THF is a very good reagent for the pinacol formation from aldehydes. Reaction is usually over after a few minutes at room temperature with aromatic aldehydes. Pinacols are formed from aliphatic aldehydes or aliphatic ketones after a few hours or one day respectively. For example reaction times and isolated yields in pinacols with 2 mol eq SmI_2 (10^{-1} M in THF) were found to be : PhCHO (1 min, 95%), octanal (1h, 90%), 2-octanone (1 day, 80%)[25,26].

If reaction is performed in THF in presence of small amounts of a protic source such as 1% methanol then pinacol formation is suppressed in most of the cases, the end product being an alcohol. By this way octanal was transformed quantitatively into 1-octanol (after 1 day at room temperature) while 2-octanone gave only 12% of 2-octanol. The great difference of reactivity between aliphatic aldehydes and aliphatic ketones was confirmed by competition reactions where high selectivity in aldehyde reduction was observed[7]. YbI_2 is quite inefficient in aldehyde reduction.

The above experimental data were interpreted by a mechanism involving initial formation of a ketyl arising by interaction between the aldehyde and SmI_2 :

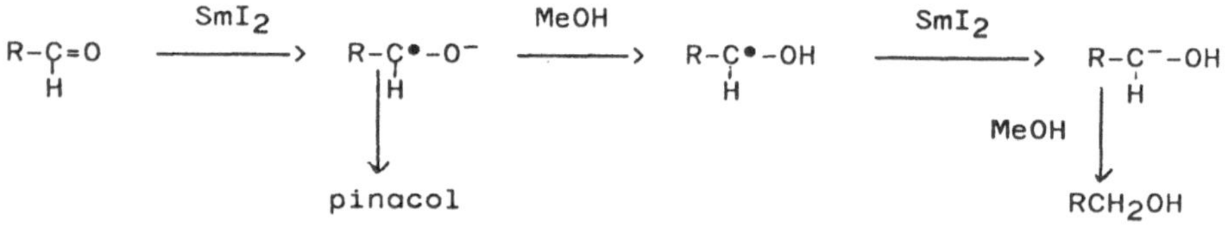

[4]

This scheme is in agreement with the preferred formation of RCDOH (90%) when MeOD replaces MeOH. However some formation of RCH_2OH (10%) indicates a minor pathway with hydrogen abstraction from THF by the hydroxyalkyl radical.

Carbonyl groups of esters or acids remain unchanged in presence of THF solutions of SmI_2 or YbI_2.

4.2. Reduction of conjugated systems

Conjugated ketones or aldehydes react with SmI_2 with competition between C=O and C=C reduction. Cinnamic acid or ester are easily reduced into 3-phenyl propionic acid or ester at room temperature by THF solutions of SmI_2 or YbI_2 containing a small amount of methanol. This reaction is not general for conjugated esters and does not occur in absence of the aromatic ring.

4.3. Deoxygenation reactions

Sulfoxides were reduced in high yield into sulfides by heating in THF a few hours with 2 eq SmI_2. However triphenylphosphine oxide could not be transformed into triphenylphosphine by this procedure[7]. Interestingly a molecular complex has been isolated (SmI_2, 2 $Ph_3P{=}O$) as a black material. This complex could be cleaved in THF and regenerated SmI_2 and triphenylphosphine oxide.

Several kinds of epoxides were transformed into the corresponding alcenes by action of SmI_2 in THF at room temperature[7]. Reaction works better in presence of small amounts of t-butanol. The mechanisms of these reactions are not well understood neither have been established conditions for stereospecific olefin formation. The oxygen removal needs two mol eq SmI_2, initially giving presumably $I_2Sm{-}O{-}SmI_2$. When an epoxide is treated with Cp'_2Sm the analogous complex $Cp'_2Sm{-}O{-}SmCp'_2$ has been isolated and its structure established by X-ray cristallography by Evans et al[27].

4.4 Organic halides

Primary or secondary alkyl halides RX upon heating a few hours in the presence of one equivalent of SmI_2 in THF are reduced to alkanes RH, without any coupling product R-R. Treatment by D_2O prior to product isolation did not give RD, which excludes the formation of an organosamarium as end product. There is no reaction at room temperature. Aromatic or vinyl halides remain unchanged apart 1-iodonaphthalene (which gives naphthalene). Interestingly (E)-$C_6H_5CH{=}CHBr$ leads to (E)-$C_6H_5CH{=}CH{-}CH$– 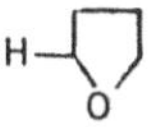
as the sole product (30% yield). Allylic or benzylic halides react in a few minutes at room temperature with 1 mol eq of SmI_2 in THF. Coupling products are isolated in high yields (eg $PhCH_2Br \longrightarrow PhCH_2CH_2Ph$). Unfortunately regioselectivity in the coupling is not very high for compounds such as cinnamyl bromide. The absence of reduction products (RH, where R = cinnamyl, benzyl, ...) was ascribed to the stability of the radical which can diffuse outside the coordination sphere of samarium (III) and combine to another radical.

The general mechanism for reaction between RX and SmI_2 is summarized as following :

$$SmI_2 + RX \longrightarrow R^\bullet \quad (+\ SmI_2X)$$

$$R^\bullet + THF \longrightarrow RH + THF^\bullet \quad \text{(when R = alkyl)}$$

R• + R• ———> R-R (when R = benzyl or alkyl)

R• + SmI_2 —?—> $RSmI_2$ (not observed)

Recently, Inanaga et al.[28] found that alkyl halides are reduced into alkanes at room temperature in THF if SmI_2 is used with prior addition of one mol eq of HMPA.

4.5. Barbier reaction

4.5.1. Basic process

Since SmI_2 was able to reduce some carbonyl groups, it was interesting to use some of the intermediates in equation [4] in order to create C-C bonds. One possibility is to trap the carbanionic intermediate by an organic halide RX or to couple the hydroxyradical with a radical produced in situ from RX according to the process described in paragraph 4.4.

It was found by Girard, Namy and Kagan[4,7] that many ketones are able to react with organic halides to lead to tertiary alcohols, according to the stoichiometry indicated in eq.[5].

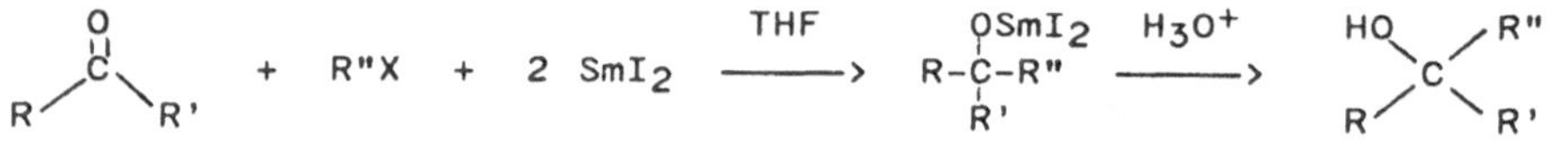

[5]

Some typical results are listed in Table I.

Barbier reaction on ketones is very general and gives good yields for various organic halides. The reactivity of R"X decreases according to the sequence R"I > R"Br > R"Cl.

The reaction on aldehydes gives a mixture of products arising from in situ oxidation of secondary samarium alcoholates of equation [5] (R = H) by the starting aldehyde. A ketone is thus produced which enters in a new Barbier reaction. It was indeed later demonstrated that diiodosamarium alcoholates are excellent catalysts for Oppenauer-Meerwein-Ponndorf-Verley reactions[29]. However aldehydes give good yields in the expected product of Barbier reaction when the organic halide is very reactive (see entries 5,6 in Table I)[30].

$ClCH_2OCH_2Ph$ easily reacts on ketones or aldehydes in presence of SmI_2, Imamoto et al[31] developed with this reagent a method for the hydroxymethylation of carbonyl compounds. Iodomethylation reactions by the combination CH_2I_2/SmI_2 were also performed[11,32].

Molander et al[33,34] showed that intermolecular Barbier reaction of ω-iodoketones is possible, with ring formation (5 or 6-member rings).

A Reformatsky-like reaction is very fast in presence of two equivalents of diiodosamarium[7]. Thus cyclohexanone and ethyl 2-bromopropionate gave ethyl 2-(1-hydroxycyclohexyl)propionate in 90% yield after 20 min at room temperature. Intramolecular Reformatsky reaction with lactone formation was recently investigated by several groups[35,36]. Similarly β- or γ-bromoesters react on ketones or aldehydes leading to lactones[37].

Mechanism of pseudo-Barbier reaction mediated by two equivalents of SmI_2 has been discussed[38]. The main routes to envisage are the following :

$$RX + SmI_2 \longrightarrow R^{\bullet} + SmI_2X \qquad [I]$$

$$R^{\bullet} + SmI_2 \longrightarrow RSmI_2 \qquad [II]$$

$$RSmI_2 + \;>C{=}O \longrightarrow R{-}C{-}OSmI_2 \qquad [III]$$

$$>C{=}O + SmI_2 \longrightarrow \;>C^{\bullet}{-}OSmI_2 \qquad [IV]$$

$$>C^{\bullet}{-}OSmI_2 + SmI_2 \longrightarrow \;>C(SmI_2){-}OSmI_2 \qquad [V]$$

$$>C(SmI_2){-}OSmI_2 + RX \longrightarrow R{-}C{-}OSmI_2 + SmI_2X \qquad [VI]$$

$$>C^{\bullet}{-}OSmI_2 + R^{\bullet} \longrightarrow R{-}C{-}OSmI_2 \qquad [VII]$$

$$>C{=}O + R^{\bullet} \longrightarrow R{-}C{-}O^{\bullet} \qquad [VIII]$$

$$R{-}C{-}O^{\bullet} + SmI_2 \longrightarrow R{-}C{-}OSmI_2 \qquad [IX]$$

Reaction [II] was discarded because of the impossibility to detect or to prepare organosamarium $RSmI_2$ in the conditions of the Barbier reaction. Reaction [VI] was also eliminated, since optically active organic halides (eg 2-bromooctyl) give a racemic product instead of inversion of configuration as expected for a SN2 type reaction. It was proposed that the Barbier reaction proceeds through equations [I], [IV] and [VII], namely by a coupling between a radical originated from RX and a ketyl. The direct addition of $R^{\bullet}$ on the carbonyl compound and further reduction (equation [VIII], [IX]) is less probable but cannot be disregarded.

Table I

Barbier reaction between ketones R-C(=O)-R' and organic halides R"X

N°	R	R'	R"X	Conditions[a]	Isolated yield in R-C(R')(OH)-R"
1	CH_3	n-C_6H_{13}	n-BuI	65°C, 8 h	76%
2	CH_3	n-C_6H_{13}	CH_3I	25°C, 1 h	95%
3	CH_3	n-C_6H_{13}	$BrCH_2CH=CH_2$	55°C, 10 min	65%
4	CH_3	n-C_6H_{13}	$PhCH_2Br$	25°C, 3 min	86%
5	t-Bu	H	$ICH_2CH=CH_2$	25°C, 5 min	78%
6	i-Pr	H	$PhCH_2Br$	25°C, 3 min	87%

a Stoichiometry [>C=O]/[SmI_2]/[R"X] = 1:1:2. For more details see the experimental section of refs 4,7,26. Solvent : THF.

6-bromo-2-hexene in presence of 2-octanone gave a mixture of tertiary alcohols :

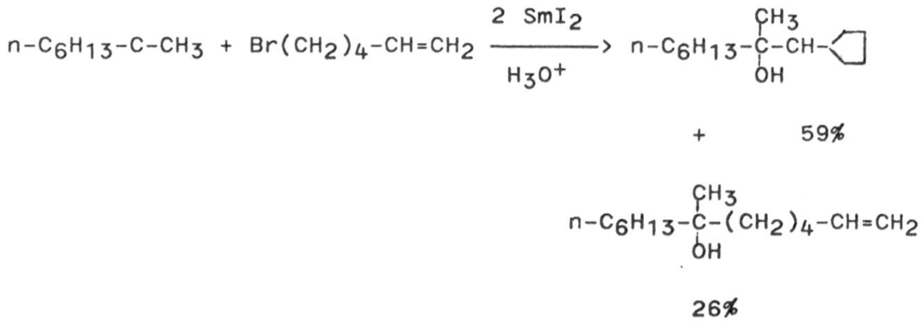

The major product has a cyclopentane structure. This is a good presumption for the intermediate formation of a radical which has time enough for cyclization before further coupling reaction to the ketone. Another piece of evidence for radical intermediates in the Barbier reaction is the identification of minor amounts of products coming from coupling with THF (eg 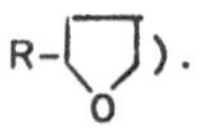).

It is interesting to point out that in the Barbier reaction with lithium metal the same type of mechanism was established[39], with the exclusion of organolithium formation.

4.6 Reaction on acid chlorides

A fast reaction occurs at room temperature when acid chlorides are added to SmI_2 in THF solution (one or two equivalents). Unexpectedly α-diketones were obtained[40] in good yields according to equation [6]. For example yield is 78% when R=Ph (reaction time : 2 minutes).

$$2\ RCOCl + 2\ SmI_2 \xrightarrow{THF} \xrightarrow{H_3O^+} 2\ RC(O)\text{-}C(O)R \qquad [6]$$

The mechanism of the reaction was investigated by Girard, Souppe, Namy and Kagan[41]. The following steps were proposed :

$$RCOCl + SmI_2 \longrightarrow RCO^\bullet + SmI_2Cl$$

$$RCO^\bullet + SmI_2 \longrightarrow RCOSmI_2$$

$$RCOSmI_2 + RCOCl \longrightarrow RC(O)\text{-}C(O)R + SmI_2Cl$$

An argument for a fast acyl radical reduction by SmI_2 is the good yield of $PhCH_2CO\text{-}COCH_2Ph$ (75%) obtained from $PhCH_2COCl$. Since neither toluene or $PhCH_2CH_2Ph$ was detected it means that $PhCH_2CO$ is reduced at a rate much faster than the decarbonylation rate ($k = 5.2 \times 10^7 s^{-1}$)[42]. In the inverse addition, where SmI_2 solution was slowly added to acid chloride solution, some dibenzyl was observed. The acyl anion $RCOSmI_2$ has not been isolated. Its structure could tentatively be represented in resonance with a samarium alkoxycarbene structure, by analogy with known Cp_2LuCO-<u>t</u>-Bu[43].

Trapping of intermediate $RCOSmI_2$ by electrophiles other than RCOCl was possible, by the following method[41]. RCOCl and an electrophile (D_2O, H_2O or RR'C=O) are mixed in THF. This solution is poured into a THF solution of SmI_2 at room temperature. The method allows to prepare in good yields many mixed ketols from acid chlorides and aldehydes or ketones, according to equation [7].

$$RCOCl + R'R''C{=}O + 2\ SmI_2 \longrightarrow \xrightarrow{H_3O^+} R\text{-}C(O)\text{-}C(OH)R'R'' \qquad [7]$$

D_2O leads to formation of deuterated aldehydes RCOD.

4.7 Miscellaneous

Diiodosamarium was found to be a versatile reagent for many additional types of reduction. A review article gave the main results till 1986[6]. Fragmentation reactions[44], cleavage of isoxazoles[45] and reductions of R-C-CHR' into R-C-CH$_2$R'[46-48] are some examples of transformations in-
$\quad\quad\overset{\|}{O}\ \overset{|}{X}\quad\quad\quad\quad\overset{\|}{O}$
duced by SmI$_2$. The X group is an heteroatom (e.g. OAc, halide, vicinal epoxide). Another interesting reaction is the reductive coupling by SmI$_2$ between a ketone and an electrophilic olefin such as ethyl acrylate[49]. The creation of rings is also possible by an intramolecular reaction[50]. The intramolecular reductive coupling of unsaturated ketones between a carbonyl group and a terminal non activated C=C double bond was obtained at low temperature in presence of 2 mol eq of SmI$_2$ and t-BuOH[51]. By this process polysubstituted β-hydroxycyclopentanecarboxylates are stereoselectively created with three contiguous chiral centers.

Diiodomethane in presence of SmI$_2$ leads to unusual chemistry, presumably through transient organometallics. Thus allylic alcohols give cyclopropyl alcohols, cyclopropanation is very selective and does not occur on isolate double bonds[52]. Cyclopropanation also occurs on lithium enolates, giving cyclopropanols after hydrolysis[53].

5. REACTIVITY OF Cp$_2$Sm, Cp'$_2$Sm, Cp'$_2$Yb

Cp$_2$Sm, Cp'$_2$Sm and to a minor extent Cp'$_2$Yb behave as powerful one electron donors, as does SmI$_2$. This reducing property has been used for organic synthesis, but also for organometallic chemistry, owing to the presence of the stabilizing cyclopentadienyl or pentamethylcyclopenta-dienyl ligands. These divalent complexes are presently investigated in two major fields of organometallic chemistry : reactions with halogenated derivatives leading to "Grignard-like" complexes and reactions on unsaturated molecules producing new types of complexes.

5.1 Barbier-like reactions

Cp$_2$Sm promotes a Barbier-like reaction[54] between alkyl iodides or various activated halogeno compounds and aldehydes or ketones.

$$\begin{array}{c}R_1 \\ R_2\end{array}\!\!\!>\!\!C = O + RX \xrightarrow[\text{2) H}_3\text{O}^+]{\text{1) Cp}_2\text{Sm/THF}} \begin{array}{c}R_1 \\ HO\end{array}\!\!\!>\!\!C\!\!<\!\!\begin{array}{c}R_2 \\ R\end{array}$$

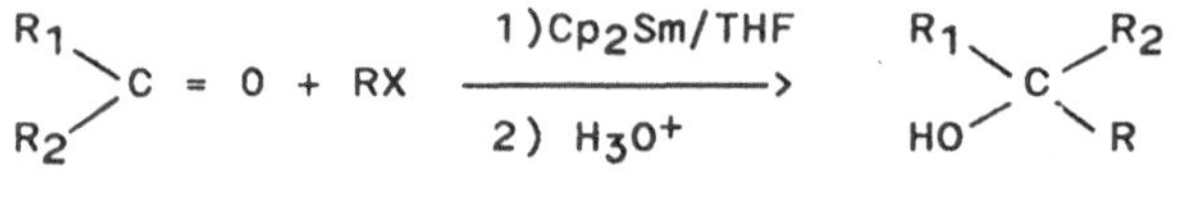

RX = nBuI,	R$_1$ = C$_6$H$_{13}$,	R$_2$ = CH$_3$	69%
RX = CH$_2$=CH-CH$_2$I,	R$_1$ = C$_6$H$_{13}$,	R$_2$ = H	86%
RX = iPrI,	R$_1$ = C$_6$H$_{13}$,	R$_2$ = H	50%
RX = nC$_{12}$H$_{25}$I	R$_1$ = C$_6$H$_{13}$,	R$_2$ = H	71%
RX = BrCH$_2$COOEt,	R$_1$ = tBu,	R$_2$ = H	80%

It proves to be a better coupling agent than SmI_2. With ketones yields are as high as those obtained with SmI_2 however the reaction works at room temperature instead of refluxing THF , with aldehydes yields are moderate although higher than with SmI_2. Cp_2Sm does not catalyze MPVO reactions unlike SmI_2[55] (Oppenauer reactions decrease yields in Barbier reactions between alkyl iodides and aldehydes mediated by SmI_2).

5.2 Acyl chlorides

Acyl chlorides react with SmI_2 yielding α diketones[40] or can be coupled with aldehydes or ketones to give ketols[41]. Study of mechanism leads to the hypothesis that reaction is mediated by a transient acyl anion (see § 4.6). Since bis cyclopentadienyl acyl lutetium was already isolated by Evans[43], use of cyclopentadienyl as ligand could be envisaged as a way to stabilize acyl samarium species. Reaction of acyl chlorides with Cp_2Sm depends on the substitution of the carbon atom in α position of the carbonyl group and on the reaction temperature. Preliminary results indicate that acyl samarium complexes can be generated from acyl chlorides RCOCl. Complexes where R has a tertiary atom adjacent to the carbonyl group are stable at low temperature (-30°C)[56].

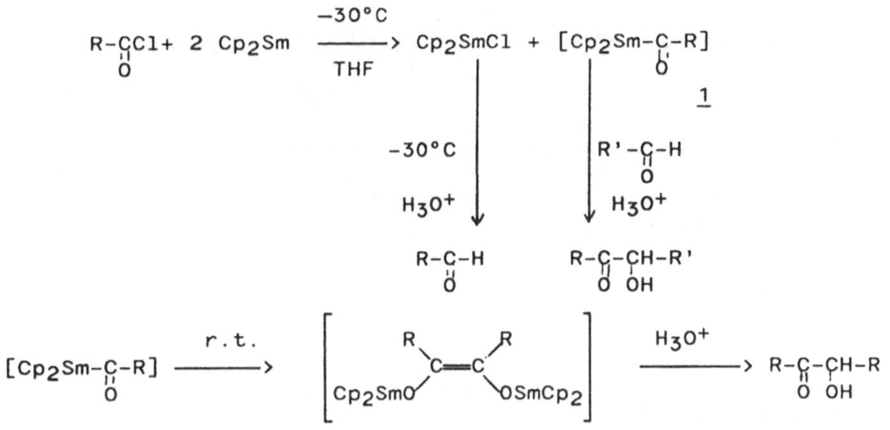

1 2

At low temperature intermediate 1 reacts with water or aldehydes and affords respectively aldehydes or ketols. When temperature raises acyl samarium dimerizes giving an enediolate intermediate 2 which after hydrolysis yields a ketol. X-ray structure of similar enediolate samarium complexes obtained by dimerization of a formyl intermediate has been published by Evans[57].

5.3 Organic halides

Reaction of divalent lanthanides with organic halides first yields to an organic radical and then to a "Grignard-like" lanthanide complex, after oxidative addition of the radical on a second molecule of the divalent lanthanide complex. The scheme is believed to be as follows :

$$Ln^{II} + RX \longrightarrow Ln^{III} - X + R^{\bullet}$$

$$R^{\bullet} + Ln^{II} \longrightarrow Ln^{III} - R$$

$$2\ R^{\bullet} \longrightarrow R - R$$

$$2\ R^{\bullet} \longrightarrow RH + R(-H)$$

Study of such a route to trivalent organolanthanide compounds was attractive for preparation of the well known Cp_2LnR or Cp'_2LnR type complexes[58,59].

With Cp_2Sm, Cp'_2Sm and Cp'_2Yb the nature of the products obtained depends on the metal and the ligands and on the conditions of reaction. Cp_2Sm reacts with benzylic halides to give products 3 which display reactivity pattern of organometallic compounds[60]. They react with deuterated water, aldehydes or ketones and acyl chlorides to give respectively $ArCH_2D$, alcohols or ketones.

$$2\ Cp_2Sm + ArCH_2X \longrightarrow \underset{\underline{3}}{Cp_2SmCH_2Ar} + Cp_2SmX$$

$$Cp_2SmCH_2Ar \xrightarrow{D_2O} ArCH_2D,$$

$$Cp_2SmCH_2Ar + R_1-\underset{\|}{\overset{}{C}}(O)-R_2 \xrightarrow{H_3O^+} ArCH_2-C(R_1)(OH)-R_2$$

$$Cp_2SmCH_2Ar + R-C(=O)-Cl \longrightarrow ArCH_2-C(=O)-R$$

Reactions of various non benzylic organohalides with Cp_2Sm give organometallics as well[61].

Reactivity of $Cp'_2Sm\ OEt_2$ and $Cp'_2Yb\ OEt_2$ toward alkyl and aryl halides[62,63] was studied by nmr methods by Finke and Watson. With both derivatives a mixture of products was obtained. Trivalent lanthanide compounds and organic products are formed according to generalized equations [8] and [9].

[8] $$(a + b)\ Cp'_2Sm,\ OEt_2 + (a + c)\ RX \longrightarrow (a-2c)\ Cp'_2SmX + b\ Cp'_2SmR + c/z\ [Cp'_3Sm_2X_3]_2 + c\ C'pR + (a-b)\ R^{\bullet} + (a+b)\ Et_2O$$

[9] $$Cp'_2Yb,\ OEt_2 + (1+a)\ RX \longrightarrow (1-a)\ Cp'_2YbX + a\ Cp'YbX_2 + a\ Cp'R + 0.5\ [R-R,\ RH,\ R(-H)] + Et_2O$$

There is no evidence of a transient organometallic species through reaction of Cp'_2Sm,OEt_2 with benzyl chloride[62]. On the other hand when Cp'_2Yb,OEt_2 reacts with benzyl chloride an Yb^{III} "Grignard-like" complex is intermediately formed. This complex is reactive towards an excess of benzyl chloride yielding bibenzyl and $Cp'CH_2Ph$. Organosamarium complexes Cp'_2Sm-R were observed by Finke with phenyl or neopentyl as R group.

Thus the major differences between reactivities of divalent $Cp_2Sm,(THF)_2$ and Cp'_2Sm,OEt_2 towards benzyl halides are following : unexpectedly Grignard-like products seem more stable with cyclopentadienyl ligand and no coupling of cyclopentadienyl ligand with benzyl group has been observed.

5.4 <u>Reaction of Cp'_2Sm with unsaturated molecules</u>

Evans found that Cp'_2Sm reduces some organic molecules yielding trivalent complexes which themselves are very reactive, towards CO insertion for instance, and have new reactivity patterns such as cleavage of N=N double bond, CH bond activation ... Structure of the complexes were determined by X-ray studies. These reactions meantimes permit easy building of skeletons difficult to obtain by standard organic synthesis. So they are of interest for organometallic and organic chemistry.

Cp'_2Sm and CO form complex <u>4</u> $[Cp'_4Sm_2(O_2CCCO)THF]$[64]. It has a ketene dicarboxylate structure derived from three CO molecules and two one-electron reductions. Formation of this complex requires cleavage of a carbon-oxygen bond of a CO molecule.

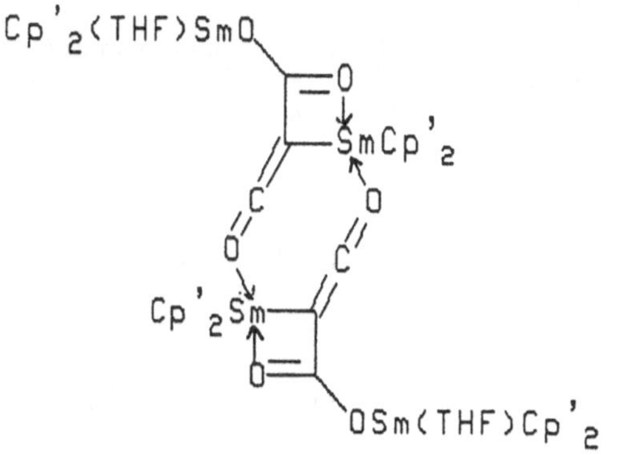

<u>4</u>

Addition of Cp'_2Sm on the triple bond of hexyne leads to the formation of a vinyl complex <u>5</u>[65].

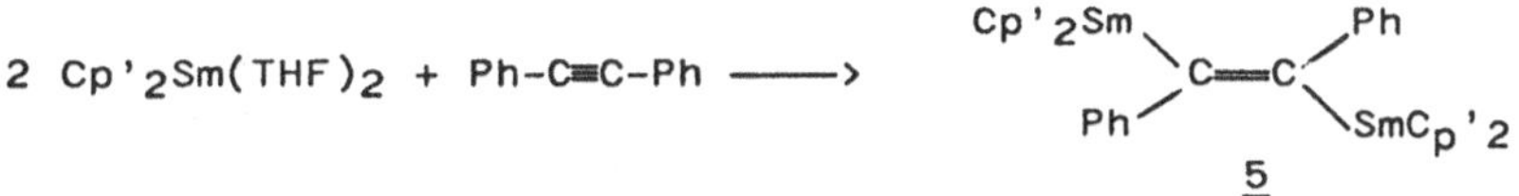

<u>5</u>

Complex 5 was transformed under H_2 into the hydride $[Cp'_2SmH]_2$[65], which was characterized by X-ray analysis. Complex 5 and the hydride are catalysts for hydrogenation of alkynes. Complex 5 reacts with CO and gives a complex 6 of formula $[Cp'_2Sm]_2$ $(O_2C_{16}H_{10})$ which presents a dihydroxyindenoindene unit, formed by CO insertion followed by C-H activation[66].

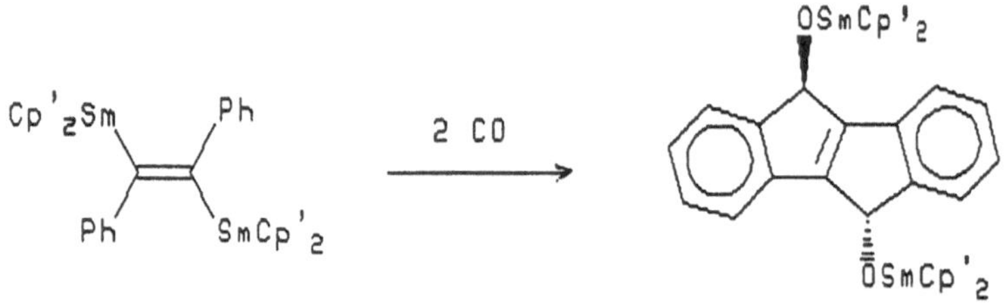

5 6

The diyne Ph-C≡C-C≡C-Ph and $Cp'_2Sm\ (THF)_2$ afford a complex 7 of formula $[Cp'_2Sm]_2\ C_4Ph_4$[67] with a reaction pathway different from that of diphenylacetylene. In the structure of complex 7 the trivalent samarium atoms are η_2 coordinated to the triple bonds. It is the first report on a samarium η_2 coordination to carbon-carbon multiple bond. It is important to note that until recently no η_2 complex of lanthanide has been isolated. Complexes Cp'_2Yb (η_2MeC≡CMe) and $Cp'_2Yb(\mu C_2H_4)Pt(PPh_3)_2$ have been prepared by Andersen from divalent Cp'_2Yb[68,69].

$Cp'_2Sm(THF)_2$ reacts with azobenzene with addition of two Cp'_2Sm units on the N=N double bond[70]. This complex 8 is transformed by CO into 9 resulting of a cleavage of the N=N double bond and insertion of two CO molecules[71].

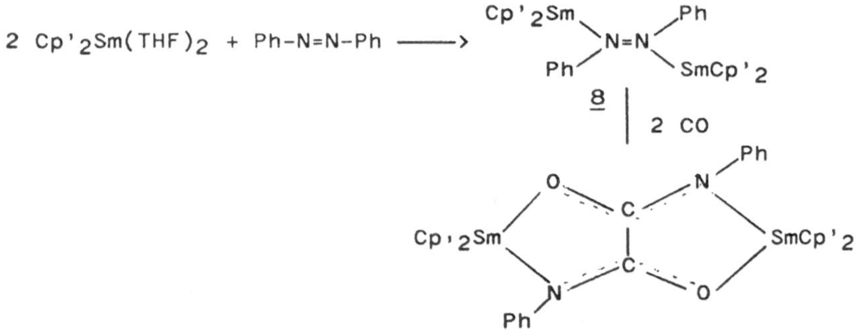

9

$Cp'_2Sm(THF)_2$ reacts with C_5H_6 to form $Cp'_2Sm\ Cp$[72] and a mixed-valent complex $Cp'_2Sm\ (\mu Cp)\ Sm\ Cp'_2$ containing one trivalent and one divalent samarium. X-ray structure of this mixed-valent complex shows that the samarium(III) atom is symmetrically bonded to the five atoms of the cyclopentadienyl ring while samarium(II) is close to two cyclopentadienyl carbon atoms.

$$Cp'_2Sm(THF)_2 + C_5H_6 \longrightarrow Cp'_2Sm\ C_5H_5 + 1/2\ H_2 + 2\ THF$$

$$Cp'_2SmC_5H_5 + Cp'_2Sm \longrightarrow Cp'_2Sm\ \mu(C_5H_5)SmCp'_2$$

6. CONCLUSION

Amongst the divalent lanthanide ions samarium appears to be the most useful one electron donor. Diiodosamarium is easily available and establishes as a versatile reagent in organic synthesis. Many transformations of organic compound have been observed : deoxygenation reactions, reduction of carbonyl groups, hydrogenolysis in α position of a carbonyl, C-C bond formation arising from pinacolizations, pseudo-Barbier reactions or reductive coupling involving acid chlorides. Cp_2Sm and Cp'_2Sm are also interesting reducing agents.

Besides their uses in organic chemistry, divalent lanthanide compounds allow the development of new routes in organometallic chemistry and the synthesis of new types of complexes. They are starting materials for synthesis of trivalent complexes which themselves are powerful catalysts for polymerisation[73] and hydrogenation of olefins[74,75].

REFERENCES

1) Kagan,H.B. ; Namy,J.L., in Handbook on the Physics and Chemistry of Rare Earths (Edited by K.A. Gschneidner Jr and Eyring,L.), Chapter 50, North Holland Publishing Co, Amsterdam,1984.
2) Eick,H.A. ; J. of the Less-Common Metals, 1987, *127*,7.
3) Evans,W.J., Polyhedron, 1987,*6*,803.
4) Namy,J.L. ; Girard,P. ; Kagan,H.B., Nouv.J.Chim., 1977,*1*,5.
5) Kagan,H.B. in Fundamental and Technological Aspects of Organo-f Elements Chemistry.(Edited by T.J.Marks and I.L.Fragola) p.49 D.Reidel, New York 1985.
6) Kagan,H.B. ; Namy,J.L., Tetrahedron, 1986,*42*,6573.
7) Girard,P. ; Namy,J.L. ; Kagan,H.B., J.Am.Chem.Soc., 1980,*102*,2693.
8) Evans,W.J.,J.Organomet.Chem., 1987,*326*,299.
9) Namy,J.L. ; Girard,P. ; Kagan,H.B. ; Caro,P.E., Nouv.J.Chim., 1981,*5*,479.
10) This observation was used to stabilize THF solutions of SmI_2 commercially available since 1986 (Alfa Chemicals).
11) Imamoto,T. ; Takayama,T. ; Koto,H., Tetrahedron Lett., 1986,*27*,3243.
12) Souppe,J., Thesis 1983, Université Paris-Sud, Orsay.
13) Imamoto,T. ; Ono,M., Chem.Lett., 1987,501.
14) Chebolu,V. ; Whittle,R.R ; Sen,A., Inorg.Chem.,1985,*24*,3082.
15) Evans,W.J ; Grate,J.W. ; Choi,H.W. ; Bloom,I. ; Hunter,W.E. ; Atwood,J.L., J.Am.Chem.Soc., 1985,,*107*,941.

16) Wayda,A.L. ; Cheng,S. ; Mukerji,I., J.Organometal.Chem., 1987, *330*, C 19.
17) Watt,G ; Gillow,J., J.Am.Chem.Soc., 1969,*91*,775.
18) Evans,W.J. ; Hughes,L.A. ; Hanusa,T.P., J.Am.Chem.Soc., 1984,*106*, 4270.
19) Evans,W.J. ; Bloom,I. ; Hunter,W.E. ; Atwood,J.L., J.Am.Chem.Soc., 1981, *103*,6507.
20) Watson,P.L., J.Chem.Soc.Chem.Comm., 1980, 652.
21) Don Tilley,T. ; Andersen,R.A. ; Spencer,B. ; Ruber,H. ; Zolkin,A. ; Templeton,D.H.,Inorg.Chem., 1980, *19*, 2999.
22) Deacon,G.B.; Vince,V.G., J.Organometal.Chem., 1976,*112*, C1
23) Deacon,G.B. ; Roverty,W.D. ; Vince,D.G., J.Organometal.Chem., 1977, *135*,103.
24) Kochi,J.K., Organometallic Mechanisms and Catalysis, Academic Press, New York, 1978.
25) Namy,J.L. ; Souppe,J. ; Kagan,H.B., Tetrahedron Lett., 1983, *24*, 765.
26) Souppe,J. ; Danon,L. ; Namy,J.L. ; Kagan,H.B., J.Organometal.Chem., 1983, *250*, 227.
27) Evans,W.J. ; Grate,J.W. ; Bloom,I. ; Hunter,W.E. ; Atwood,J.L., J.Am.Chem.Soc., 1985,*107*,405.
28) Inanaga,J. ; Ishikawa,M. ; Yamaguchi,M., Chem.Lett., 1987, 1425.
29) Namy,J.L. ; Souppe,J. ; Collin,J. ; Kagan,H.B., J.Org.Chem., 1984, *49*,2045.
30) Souppe,J. ; Namy,J.L. ; Kagan,H.B., Tetrahedron Lett., 1982, *23*, 3497.
31) Imamoto,T. ; Takeyama,T. ; Yokoyama,M., Tetrahedron Lett., 1984, *25*,3225.
32) Tabuchi,T. ; Inanaga,J. ; Yamaguchi,M., Tetrahedron Lett., 1986,*27*,3891.
33) Molander,G.A. ; Etter,J.B., J.Org.Chem., 1986,*51*,1778.
34) Molander,G.A. ; Etter,J.B., J.Am.Chem.Soc.,1987,*109*,453.
35) Tabuchi,T. ; Kawamura,K. ; Inanaga,J. ; Yamaguchi,M., Tetrahedron Lett.,1986,*27*,3889.
36) Molander,G.A. ; Etter,J.B., J.Am.Chem.Soc.,1987,*109*,4556.
37) Otsubo,K. ; Kawamura,K. ; Inanaga,J. ; Yamaguchi,M., Chem. Lett.,1987,1487.
38) Kagan,H.B. ; Namy,J.L. ; Girard,P. ; Tetrahedron Supplement n° 1,1981,*37*,175.
39) Dubois,J.E. ; Bauer,P. ; Kaddari,B., Tetrahedron Lett.,1985, *26*,57.
40) Girard,P ; Couffignal,R. ; Kagan,H.B., Tetrahedron,1981,*22*,3959.
41) Souppe,J. ; Namy,J.L. ; Kagan,H.B., Tetrahedron Lett.,1984,*25*,2869.
42) Lunazzi,L. ; Ingold,K.U. ; Scalano,J.C. ; J.Phys.Chem.,1983,*87*,529.
43) Evans,W.J. ; Wayda,A.L. ; Hunter,W.E. ; Atwood,J.L., J.C.S.Chem.Comm., 1981,706.
44) Ananthanarayan,P. ; Gallagher,T. ; Magnus,M., J.C.S.Chem.Comm.,1982, 709.
45) Natale,N.R., Tetrahedron Lett.,1982,*23*,5009.
46) Molander,G.A. ; Hahn,G., J.Org.Chem.,1986,*51*,1135.
47) Molander,G.A. ; Hahn,G., J.Org.Chem.,1986,*51*,2596.
48) Otsubo,K. ; Inanaga,J. ; Yamaguchi,M., Tetrahedron Lett.,1987,*28*, 4437.
49) Otsubo,K. ; Inanaga,J. ; Yamaguchi,M., Tetrahedron Lett.,1986,*27*, 5763.

50) Fukuzawa,S. ; Iida,M. ; Nakanishi,A. ; Fujinami,t. ; Sakai,S., J.C.S.Chem.Comm., 1987,*920*,
51) Molander,G.A. ; Kenny,C., Tetrahedron Lett., 1987,*28*,4367.
52) Molander,G.A. ; Etter,J.B., J.Org.Chem., 1987,*52*,3942.
53) Imamoto,T. ; Takiyama,N., Tetrahedron Lett., 1987, *28*,1307.
54) Namy,J.L. ; Collin,J. ; Zhang,J. ; Kagan,H.B., J.Organomet.Chem., 1987,*328*,81.
55) Collin,J. ; Namy,J.L. ; Kagan,H.B., Nouv.J.Chim., 1986,*10*,229.
56) Collin,J. ; Namy,J.L. ; Kagan,H.B., unpublished results.
57) Evans,W.J. ; Grate,J.W. ; Daedens,R.J., J.Am.Chem.Soc., 1985,*104*,1671.
58) Evans,W.J., Advances in Organometallic Chemistry, 1985,*24*,131.
59) Schumann,H., Angew.Chem.Int.Ed.Engl., 1984,*23*,474.
60) Collin,J. ; Namy,J.L. ; Bied,C. ; Kagan,J.L., Inorg.Chim.Acta, 1987,*140*,29.
61) Collin,J. ; Bied,C. ; Kagan,H.B., unpublished results.
62) Finke,R.G. ; Keenan,S.R. ; Schiraldi,D.A. ; Watson,P.L., Organometallics, 1987,*6*, 1356.
63) Finke,R.G. ; Keenan,S.R. ; Schiraldi,D.A. ; Watson,P.L., Organometallics, 1986,*5*, 598.
64) Evans,W.J. ; Grate,J.W. ; Hughes,L.A. ; Zhang,H. ; Atwood,J.L., J.Am.Chem.Soc., 1985,*107*, 3278.
65) Evans,W.J. ; Bloom,I. ; Hunter,W.E. ; Atwood,J.L., J.Am.Chem.Soc., 1983,*105*,1401.
66) Evans,W.J. ; Hughes,L.A. ; Drummond,D.K. ; Zhang,H. ; Atwood,J.L., J.Am.Chem.Soc., 1986,*108*, 1722.
67) Evans,W.J. ; Keyen,R.A. ; Zhang,H. ; Atwood,J.L., J.Chem.Soc. Chem.Comm., 1987,837.
68) Burns,C.J. ; Andersen,R.A., J.Am.Chem.Soc., 1987;*109*,941.
69) Burns,C.J. ; Andersen,R.A., J.Am.Chem.Soc., 1987,*109*, 915.
70) Evans,W.J. ; Drummond,D.K., J.Am.Chem.Soc., 1986,*108*,7440.
71) Evans,W.J. ; Drummond,D.K. ; Bott,S.G. ; Atwood,J.L., Organometallics, 1986,*5*,2389.
72) Evans,W.J. ; Ulibarri,T.A., J.Am.Chem.Soc., 1987,*109*,4292.
73) Watson,P.L. ; Parshall,G.W., Acc.Chem.Res., 1985,1851.
74) Jeske,G. ; Lauke,H. ; Mauermann,H. ; Swepson,P.N. ; Schumann,H. ; Marks,T.J., J.Am.Chem.Soc., 1985,*107*,8091.
75) Jeske,G. ; Lauke,H. ; Mauermann,H. ; Schumann,H. ; Marks,T.J., J.Am.Chem.Soc., 1985,*107*,8111.

ELECTRON-TRANSFER ACTIVATION OF METAL CARBONYLS

J. K. Kochi
Department of Chemistry
University of Houston
Houston, Texas 77004
U.S.A.

ABSTRACT. The availability of various carbonylmetal species is used to underscore the importance of electron transfer, transient radicals and ion radicals in the reactions of both mononuclear analogues and polynuclear clusters. Thus the coupling of carbonylmetallate anions with carbonylmetal cations to dimetal carbonyls is chosen as a prototypical system to emphasize the role of 17- and 19-electron carbonylmetal radicals in the formation of metal-metal bonds. The chemical properties of these reactive intermediates are examined independently by electrochemical generation from the oxidation and reduction of diamagnetic carbonylmetal anions and cations, respectively. These carbonylmetal radicals show unusually enhanced substitutional lability of the ligands. The formation of the hydrido and formyl derivatives of metal carbonyls is also shown to proceed readily via such metastable paramagnetic intermediates.

The heterolytic coupling of carbonylmetal cations and anions is indicated in the treatment of tetracarbonylcobaltate(-I) with hexacarbonylrhenium(I) to afford the mixed dimetal carbonyl,[1] i.e.

$$Co(CO)_4^- + Re(CO)_6^+ \rightarrow ReCo(CO)_9 + CO \quad (1)$$

Similarly the treatment of $Co(CO)_4^-$ with the phenanthroline-substituted tetracarbonylmanganese(I) cation leads to the substituted heterobimetallic carbonyl,[2] viz.

$$Co(CO)_4^- + (\eta^2\text{-phen})Mn(CO)_4^+ \rightarrow (\eta^2\text{-phen})MnCo(CO)_7 + CO \quad (2)$$

It is noteworthy that the coupling of metal centers in eq 2 is equivalent to the microscopic reverse of the well-known base-induced disproportionation of metal carbonyls,[3] e.g.

$$Co_2(CO)_8 + 2\,B \rightarrow [Co(CO)_4^-][Co(CO)_3B_2^+] + CO \quad (3)$$

M. Chanon et al. (eds.), Paramagnetic Organometallic Species in Activation / Selectivity, Catalysis, 149–169.

where B = phosphorus, nitrogen and oxygen-centered bases.[4] In the course of our recent electrochemical redox studies of metal carbonyls,[5] we observed the highly selective coupling of the carbonylmanganese anion and cation, i.e.

$$Mn(CO)_5^- + Mn(CO)_6^+ \rightarrow Mn_2(CO)_{10} + CO \quad (4)$$

Furthermore in the reduction of substituted-carbonylmanganese cations, cyclic voltammetry establishes that the anion $Mn(CO)_5^-$ forms at the cathode in the presence of the cationic precursor $Mn(CO)_5L^+$ and affords both $Mn_2(CO)_{10}$ and $Mn_2(CO)_8L_2$, the relative amounts of each varying considerably with the nature of L = MeCN, pyridine, phosphine, etc. The unusual selectivities prompts us to examine the coupling processes in detail, especially with regard to the products and stoichiometry in the formation of manganese-manganese bonds from various substituted-carbonylmanganese cations and anions, i.e.

$$Mn(CO)_5L^+ + Mn(CO)_4P^- \rightarrow Mn_2(CO)_8(L)(P) + CO \quad (5)$$

with an emphasis placed on phosphorus and nitrogen ligands. In particular, the carbonylmanganese cations $(OC)_5MnL^+$ in eq 5 can be placed into three groups: I, L = CO; II, L = the nitrogen-centered ligands acetonitrile and pyridine; III, L = phosphorus(III) ligand consisting of a series of aryl and alkylphosphines with different base strengths[6] and steric properties (cone angles).[7] In turn, three different types of carbonylmanganate(-I) anions are useful in the treatment of each of the carbonylmanganese(I) cations, as described individually below.

Pentacarbonylmanganate(-I) A and hexacarbonylmanganese(I) react upon mixing (< 5 min) in tetrahydrofuran solutions and afford dimanganese decacarbonyl as the sole carbonyl product.[8] Essentially the same results are obtained with the acetonitrile and pyridine derivatives, IIa and IIb, respectively, i.e.

$$Mn(CO)_5^- + Mn(CO)_5L^+ \rightarrow Mn_2(CO)_{10} + L \quad (6)$$

where L = CO, MeCN, and pyridine. When pentacarbonylmanganate(-I) is treated with the triarylphosphine-substituted cations IIIa and IIIb, high yields of $Mn_2(CO)_{10}$ are formed together with minor amounts of the bis-substituted dimer $Mn_2(CO)_8L_2$ and the substituted hydride $HMn(CO)_4L$, i.e.

$$Mn(CO)_5^- + Mn(CO)_5L^+ \rightarrow \underset{(60\%)}{Mn_2(CO)_{10}} + \underset{(14\%)}{Mn_2(CO)_8L} + \underset{(10\%)}{HMn(CO)_4L} \quad (7)$$

where L = phosphines (Table I). Treatment of pentacarbonylmanganate with the alkyl and arylphosphine derivatives IIIa-IIIg also afford variable amounts of $Mn_2(CO)_{10}$ and the analogous carbonylmanganese products $Mn_2(CO)_8L_2$ and $HMn(CO)_4L$. In addition, small but distinctive amounts of the mono-substituted dimanganese carbonyl $Mn_2(CO)_9L$ are detected. Furthermore, the coupling processes of $Mn(CO)_5^-$ with IIIa-IIIg often proceed at significantly attenuated rates, as qualitatively indicated by the reaction times listed in column 5, Table I. The material balance

TABLE I. Products and Stoichiometry from the Coupling of $Mn(CO)_5^-$ and $Mn(CO)_5L^+$.

	$Mn(CO)_5L^+$ L	$Mn_2(CO)_{10}$ %	$Mn_2(CO)_8L_2$ %	$HMn(CO)_4L$ %	Time (min)	Material Balance
I	CO	80			< 5	80
II a	CH_3CN	88			< 5	88
b	py	84			< 5	84
III a	PPh_3	66	14	8	< 5	88
b	$P(p\text{-}Tol)_3$	56	14	15	< 5	85
c	PPh_2Et	46	16	18	30	76
d	$PPhEt_2$	38	12	22	200	72
e	PEt_3	28	12	17	300	67
f	PPh_2Me	36	8	17	5	61
g	$PPhMe_2$	40	20	16	200	81

in Table I indicates that all of the phosphine (L) included in $Mn(CO)_5L^+$ is not retained in the carbonylmanganese products. The examination of the ^{31}P NMR spectrum of the reaction mixture indicates that the deficit is present as free phosphine.

Tetracarbonyl(triphenyl phosphite)manganate(-I) B and the cationic hexacarbonylmanganese(I) rapidly yield two manganese dimers, i.e.

$$Mn(CO)_4P^- + Mn(CO)_6^+ \rightarrow [\underset{(40\%)}{Mn_2(CO)_{10}} + \underset{(30\%)}{Mn_2(CO)_8P_2}] + CO \quad (8)$$

where $P = P(OPh)_3$. When the acetonitrile and pyridine-substituted cations, IIa and IIb are employed, the same pair of dimanganese carbonyls is obtained upon mixing, together with significant amounts of the mono-substituted analog, i.e.

$$Mn(CO)_4P^- + Mn(CO)_5L^+ \rightarrow [\underset{(30\%)}{Mn_2(CO)_{10}} + \underset{(30\%)}{Mn_2(CO)_9P} + \underset{(20\%)}{Mn_2(CO)_8P_2}] + L \quad (9)$$

where $P = P(OPh)_3$ and L = MeCN and py. The absence of any other carbonyl-containing product in the reaction mixture indicates that the nitrogen ligands (MeCN and py) are rapidly replaced, as given by the approximate stoichiometry presented in eq 9. In marked contrast, the reaction of the phosphine-substituted cations IIIa-IIIg with the carbonylmanganate B affords no $Mn_2(CO)_{10}$ or $Mn_2(CO)_9P$; the most important products being the other dimanganese carbonyls, i.e.

$$Mn(CO)_4P^- + Mn(CO)_5L^+ \rightarrow \underset{(20\%)}{Mn_2(CO)_8P_2} + \underset{(30\%)}{Mn(CO)_8L_2} + \underset{(20\%)}{Mn_2(CO)_8(L)(P)} \quad (10)$$

In addition, minor amounts of the manganese hydrides $HMn(CO)_4L$ and $HMN(CO)_4P$ are observed (Table II). With $Mn(CO)_5PEt_3^+$ and $Mn(CO)_5PEt_2Ph^+$, the IR spectrum of the solution taken immediately (∿ 5 min) after mixing

TABLE II. Cross Coupling Products in Ion-Pair Annihilation.

				$Mn_2(CO)_{10}$	$Mn_2(CO)_9P$	$Mn_2(CO)_8P_2$	
$Mn(CO)_4[P(OPh)_3]^-$	+	$Mn(CO)_6^+$	→	40	10	30	
"	+	$Mn(CO)_5py^+$	→	40	20	26	
"	+	$Mn(CO)_5NCMe^+$	→	20	40	16	
$Mn(CO)_4[PPh_3]^-$	+	$Mn(CO)_6^+$	→	32	20	30	
"	+	$Mn(CO)_5py^+$	→	28	14	58	
"	+	$Mn(CO)_5NCMe^+$	→	22	18	52	
				$Mn_2(CO)_{10}$	$Mn_2(CO)_9L$	$Mn_2(CO)_8L_2$	HMn
$Mn(CO)_5^-$	+	$Mn(CO)_5[PPh_3]^+$	→	66	1	14	8
"	+	$Mn(CO)_5[PEt_3]^+$	→	28	10	12	17
				$Mn_2(CO)_8P_2$	$Mn_2(CO)_8PL$	$Mn_2(CO)_8L_2$	HMn
$Mn(CO)_4[P(OPh)_3]^-$	+	$Mn(CO)_5[PPh_3]^+$	→	20	24	22	(+)
"	+	$Mn(CO)_5[PEt_3]^+$	→	18	30	36	(+)
$Mn(CO)_4[PPh_3]^-$	+	$Mn(CO)_5[PPh_3]^+$	→		78		(+)
"	+	$Mn(CO)_5[PEt_3]^+$	→	8	12	5	38
"	+	$Mn(CO)_5[PPh_2Et]^+$	→	26	46	18	(+)

with $Mn(CO)_4P(OPh)_3^-$ shows the presence (∿ 20%) of the reactant ion pair, the complete disappearance of which requires approximately an hour.

Tetracarbonyl(triphenylphosphine)manganate(-I) C reacts upon mixing with hexacarbonylmanganese(I) and its acetonitrile and pyridine derivatives IIa and IIb to produce three dimanganese carbonyls, i.e.

$$Mn(CO)_4P^- + Mn(CO)_5L^+ \rightarrow \underset{(30\%)}{[Mn_2(CO)_{10}} + \underset{(20\%)}{Mn_2(CO)_9P} + \underset{(40\%)}{Mn_2(CO)_8P_2]} + L \quad (11)$$

where P = PPh_3 and L = CO, MeCN and py. The complete replacement of acetonitrile and pyridine in IIa and IIb, respectively, according to eq 11 is supported by the singular absence of other carbonyl-containing products. However with the phosphine-substituted cations IIIa-IIIg, the ligand L is completely retained to afford a mixture of the homo- and cross-coupled dimers, similar to that described in eq 10, where P = PPh_3 and L = triaryl- and alkylphosphines. Of the phosphine-substituted cations, $Mn(CO)_5PEt_3^+$ is somewhat unusual in that it also affords significant amounts of the hydrides $HMn(CO)_4PEt_3$ and $HMn(CO)_4PPh_3$.

Effect of Added Phosphine on the Ion-Pair Couplings.

All of the anionic carbonylmanganates(-I) and carbonylmanganese(I) cations are substitution stable. Thus separate solutions of these ions in acetonitrile or tetrahydrofuran remain unchanged in the presence of various amounts of added phosphine--certainly for periods far exceeding the time required for the ion-pair coupling (vide supra) to proceed to completion. The results in Table II show three important changes accompanying the presence of triphenylphosphine on the ion-pair coupling of pentacarbonylmanganate A and the acetonitrile-substituted cation IIa, as represented in eq 6. First, with increasing amounts of PPh_3, the yield of $Mn_2(CO)_{10}$ drops precipitously, as indicated by the trend in column 2, Table III. The deficit is made by a pair of new products $Mn_2(CO)_8(PPh_3)_2$ in column 4 and $HMn(CO)_4PPh_3$ in column 5.

$$Mn(CO)_5^- + Mn(CO)_5(NCMe)^+ \xrightarrow{PPh_3} Mn_2(CO)_8(PPh_3)_2 + HMn(CO)_4PPh_3, \text{ etc.} \quad (12)$$

Small amounts of the mono-substituted dimer $Mn_2(CO)_9PPh_3$ are observed only at the highest concentrations of added PPh_3. The effect of added triphenylphosphine on the ion-pair coupling of the PPh_3-substituted anion C and the MeCN-substituted cation IIa in eq 11 is more subtle. Thus the increasing amounts of added PPh_3 leads to a gradual diminution in $Mn_2(CO)_{10}$ yield, which is solely made up by an increase in the yield of $Mn_2(CO)_8(PPh_3)_2$. The amount of the mono-substituted $Mn_2(CO)_9PPh_3$ remains relatively invariant. Hydridomanganese carbonyl $HMn(CO)_4PPh_3$ is found only at the highest concentrations of added triphenylphosphine. Added phosphine also affects the ion-pair reaction of the phosphine-substituted cations. For example, when $Mn(CO)_5PEt_3^+$ is treated with pentacarbonylmanganate (compare entry 8, Table I) in the presence of two equiv. of added triethylphosphine, the formation of $Mn_2(CO)_{10}$ is quenched with a

TABLE III. Effect of Added Phosphine on Ion-Pair Annihilation.[a]

Added PPh_3 mmol	$Mn_2(CO)_{10}$ %	$Mn_2(CO)_9P$ %	$Mn_2(CO)_8P_2$ %	$HMn(CO)_4P$ %	Material Balance
0	88	-	-	-	88
0.1	70	-	8	8	86
0.3	64	-	14	12	90
0.5	42	-	18	20	80
0	22	18	52	-	92
0.05	16	20	54	-	90
0.15	12	18	64	-	94
0.25	6	16	66	-	88
0.10	-	-	40	40	-
0.10	-	-	70	30	-

[a]Conditions same as Table I. P = PPh_3

concomitant increase in the yield of $Mn_2(CO)_8(PEt_3)_2$ to 40%, i.e.

$$Mn(CO)_5^- + Mn(CO)_5PEt_3^+ \xrightarrow{PEt_3} Mn_2(CO)_8(PEt_3)_2 + 2CO \quad (13)$$

together with $HMn(CO)_4PEt_3$ (20%) and $HMn(CO)_3(PEt_3)_2$ (20%).

Effect of Solvent and Added Salt on the Ion-Pair Coupling.

Acetonitrile and tetrahydrofuran represent aprotic media of different properties, especially as evaluated by most measures of solvent polarity.[9] The large solvent effect on the ion-pair coupling of the pentacarbonylmanganate anion A is indicated by the more than 100-fold difference which separates its reactivity with hexacarbonylmanganese(I) cation in tetrahydrofuran and in acetonitrile (compare entries 1 and 2, Table IV.) The prolonged reaction times in the more polar acetonitrile can be related to the stabilization of the separate ions. The latter is consistent with

TABLE IV. Solvent and Salt Effects on the Ion-Pair Coupling of Pentacarbonylmanganate(-I) with Various Carbonylmanganese(I) Cations.

$Mn(CO)_5L^+$ L	Solvent (M)	Salt (M)	Time (min)
CO	THF	0	< 5
CO	MeCN	0	> 1000
CO	THF	0.3	∼ 90
CO	MeCN	0.2	> 2000
CH_3CN	THF	0	< 5
CH_3CN	MeCN	0	∼ 500
CH_3CN	THF	0.3	∼ 30
Pyridine	THF	0	< 5
Pyridine	THF	0.3	∼ 30
PEt_3	THF	0	∼ 10^2
PEt_3	THF	0.3	> 10^3

the decreased reactivity of the ions in tetrahydrofuran containing 0.3 M tetra-n-butylammonium perchlorates (see entries 3 and 9, Table IV). In both cases, the change in reactivity can be attributed to the alteration in the concentration of ion pairs as a result of the medium effect on the association constant K, i.e.

$$Mn(CO)_5^- + Mn(CO)_5L^+ \overset{K}{\rightleftharpoons} [Mn(CO)_5^- \; Mn(CO)_5L^+]$$

Such ion pairs can be monitored directly by the growth of the charge-transfer absorption band ($h\nu_{CT}$), as shown in Figure 1 for a related system:[10]

$$Cp_2Co^+ + Co(CO)_4^- \rightleftharpoons [Cp_2Co^+ Co(CO)_4^-] \xrightarrow{h\nu_{CT}} [Cp_2Co\cdot \; Co(CO)_4\cdot]$$

Thus the absorbance decrease at $\lambda \sim 500$ nm in Figure 1 relates directly to ion-pair separation arising from an increase in solvent polarity and by the added salt (tetra-n-butylammonium perchlorate).[11]

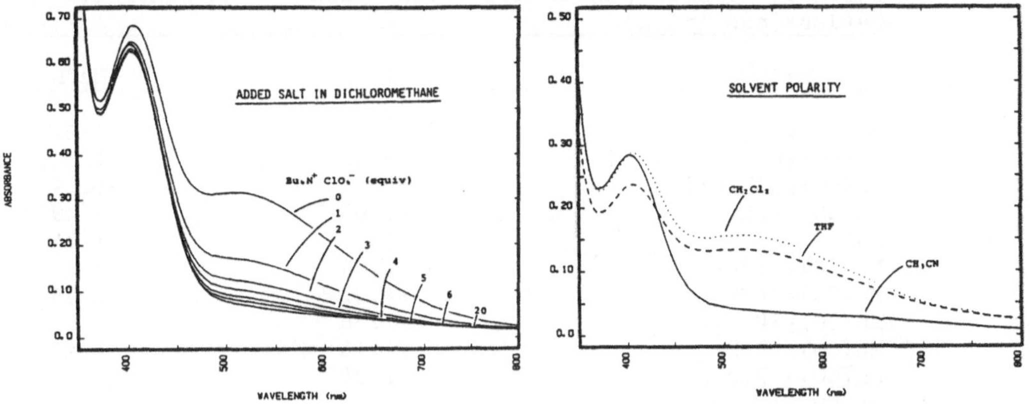

Figure 1. Charge-transfer spectrum of the electron donor-acceptor ion pair $[Cp_2Co^+ Co(CO)_4^-]$ as a function of the added salt $n\text{-}Bu_4N^+ ClO_4^-$ (left) and solvent polarity (right).

Electrochemical Properties of Anionic Carbonylmanganates(-I) and Carbonylmanganese(I) Cations.

The energetics of the 1-electron oxidation of the anionic pentacarbonylmanganates(-I) and reduction of carbonylmanganese(I) cations are provided by cyclic voltammetry in either acetonitrile solution containing 0.1 M tetraethylammonium or tetrahydrofuran containing 0.3 M tetra-n-butylammonium perchlorate (TBAP). The initial negative scan cyclic voltammogram of hexacarbonylmanganese(I) in acetonitrile shows a 1-electron cathodic wave with a peak potential $E_p^{red} = -1.27$ V vs SCE at 500 mV s^{-1} by comparison with a ferrocene standard.[12] The anodic wave on the return scan is not observed until the CV scan rate is increased to v > 70,000 V s^{-1} using a Pt microelectrode.[13] Accordingly the reversible $E_{1/2} = -1.0$ V for

$$Mn^I(CO)_6^+ + e \rightleftharpoons Mn^0(CO)_6\cdot \qquad (14)$$

is taken as the average of the cathodic and anodic peak potentials.[14] The cyclic voltammograms of the other carbonylmetal cations show only a single cathodic wave, the accompanying anodic wave being absent even at the very fastest sweep rates. The irreversible peak potentials E_p^{red} listed in Table V for the cations I-III are obtained at a standard sweep rate of 500 mv s^{-1} in acetonitrile and in tetrahydrofuran. The initial positive scan cyclic voltammogram of the PPh_3-substituted carbonylmanganate(-I) C is reversible only at sweep rates v > 1500 V s^{-1}. The reversible $E_{1/2}$ = -0.33 V vs SCE for $Mn(CO)_4PPh_3^-$ in acetonitrile cannot obtain

$$Mn^{-I}(CO)_4PPh_3^- \rightleftharpoons Mn^0(CO)_4PPh_3\cdot + e \qquad (15)$$

for either the parent carbonylmanganate(-I) $Mn(CO)_5^-$ or the phosphite analog B owing to the irreversible cyclic voltammograms. Accordingly, the irreversible anodic peak potentials E_p^{ox} are measured at a standard sweep rate of v = 500 mV s^{-1} (Table V).

TABLE V. Cyclic Voltammetric Potentials for Carbonylmetal Cations and Anions.

$Mn(CO)_5L^+$	THF	MeCN
$Mn(CO)_6^+$		-1.27
$Mn(CO)_5py^+$	-0.81	-1.12
$Mn(CO)_5[CH_3CN]^+$	-0.84	-1.19
$Mn(CO)_5PPh_3^+$	-1.05	-1.29
$Mn(CO)_5[P(p\text{-}Tol)_3]^+$	-1.19	-1.34
$Mn(CO)_5(PPh_2Et)^+$	-1.24	-1.39
$Mn(CO)_5(PPhEt_2)^+$	-1.32	-1.55
$Mn(CO)_5(PEt_3)^+$	-1.36	-1.67
$Mn(CO)_5(PPh_2Me)^+$	-1.16	-1.41
$Mn(CO)_5(PPhMe_2)^+$	-1.34	-1.54
$Mn(CO)_5(\eta^2\text{-DPPE})_2^+$	-1.84	
$Mn(CO)_4P^-$	**THF**	**MeCN**
$Mn(CO)_5^-$	-0.03	-0.11
$Mn(CO)_4]P(OPh)_3]^-$	-0.11	
$Mn(CO)_4PPh_3^-$	-0.52	-0.50
$Mn(CO)_4(\eta^1\text{-DPPE})^-$	-0.48	

Routes to Dimetal Carbonyls via Ion-Pair Annihilation.

The ion-pair interaction of an anionic carbonylmanganate with a carbonylmanganese cation to afford various dimanganese carbonyls formally represents an annihilation of -1 and +1 oxidation states to a pair of metal (0). As such the material balance for the coupling of manganese centers can be represented from among the three possible combinations of ligand attachments, viz.

Scheme I:

$$Mn^{-I}(CO)_4P^- + Mn^{I}(CO)_5L^+ \longrightarrow \begin{cases} Mn^0_2(CO)_8(P)_2 & (16) \\ Mn^0_2(CO)_8(P)(L) & (17) \\ Mn^0_2(CO)_8(L)_2 & (18) \end{cases}$$

The most direct transformation of an ion pair is included in the cross-coupled combination in eq 17. The corresponding stoichiometry may be expressed in eq 6 when P = CO and L = CO, MeCN or pyridine (see entries 1-3 in Table I). On the other hand, the interactions of the other ion pairs can lead to the homo-coupled products $Mn_2(CO)_8(P)_2$ and $Mn_2(CO)_8(L)_2$. For example, when P = CO and L = phosphine, the primary dimanganese carbonyls are the homo-coupled $Mn_2(CO)_{10}$ and $Mn_2(CO)_8(L)_2$ with only minor amounts of the cross-coupled $Mn_2(CO)_9L$ (see entries 4-7, Table I). Moreover, when P and L are both phosphines, all three combinations, viz., the homo-coupled $Mn_2(CO)_8P_2$ and $Mn_2(CO)_8(L)_2$ as well as the cross-coupled $Mn_2(CO)_8(P)(L)$, are obtained (see Table II).

In order to account for this apparent ligand-dependent selectivity in the ion-pair couplings, we classify them in terms of the relative amounts of each combination (i.e., eq 16, 17 or 18) which is formed. In doing so however we take cognizance of only the major changes in the product compositions listed in Tables I and II, owing to the limited quantification of all the products, especially in complex mixtures (vide supra). Nonetheless, the general trends in the product variation with changes in the ligands P and L are instructive. Two clear-cut categories emerge as extremes from the data in Tables I and II--namely those in which the carbonylmanganese ions labelled with ligands P and L either (A) maintain their original identity or (B) they are randomly scrambled, as described individually below.

(A) The cross-coupled combination of carbonylmanganese moieties given by the stoichiometry

$$Mn(CO)_5^- + Mn(CO)_5L^+ \rightarrow Mn_2(CO)_{10} + L \qquad (19)$$

corresponds to the replacement of the ligand L in the cation by a carbonylmanganese group. The extent to which the latter is an electron-rich anion then represents a nucleophilic substitution at a cationic manganese center. Such a transformation obtains in eqs 5 and 6 when L is CO and a nitrogen-centered ligand such as pyridine and acetonitrile, respectively. However the situation is somewhat ambiguous since the dimanganese decacarbonyl can also derive from a pair of $Mn(CO)_5^-$ anions (via oxidation) and/or a pair of $Mn(CO)_5L^+$ cations (via reductive loss of L).

(B) The homo-coupled combination of carbonylmanganese moieties given by the stoichiometry

$$Mn(CO)_4P^- + Mn(CO)_5L^+ \rightarrow [\underset{(25\%)}{Mn_2(CO)_8P_2} + \underset{(25\%)}{Mn_2(CO)_8L_2} + \underset{(50\%)}{Mn_2(CO)_8(P)(L)}] + CO \qquad (20)$$

represents the complete loss of anionic and cationic identities in the ion-pair precursor and the subsequent statistical reassemblage of the carbonylmanganese moieties as dimers. Such a randomization is unambiguously represented in eq 10 owing to the retention of either P or L (or both) in the dimanganese products. The same basic stoichiometry obtains in eqs 8 and 9, but in a somewhat less obvious manner owing to the ligand loss of either CO or L. The statistical mixture of manganese dimers as exemplified in eq 20 does provide unusual insight as to how the manganese-manganese bond can be formed by ion-pair annihilation, as described below.

Formation of the Manganese-Manganese Bond via Radical Coupling.

The electron balance for the ion-pair annihilation leading to the homo-coupled products $Mn_2(CO)_8(P)_2$ and $Mn_2(CO)_8(L)_2$ formally requires the oxidation of the anionic $Mn(CO)_4P^-$ and the reduction of the cationic $Mn(CO)_5L^+$, respectively. As such, the most direct pathway for the formation of these dimeric products is via the well-known coupling of carbonylmanganese radicals,[15] i.e.

Scheme II:

$$Mn(CO)_4P^- \xrightarrow{-e} Mn(CO)_4P\cdot \longrightarrow \tfrac{1}{2}Mn_2(CO)_8P_2 \qquad (21a,b)$$

$$Mn(CO)_5L^+ \xrightarrow{+e} Mn(CO)_5L\cdot \xrightarrow{-CO} \tfrac{1}{2}Mn_2(CO)_8L_2 \qquad (22a,b)$$

These radicals are the direct products of electron transfer within the ion pair, i.e.

$$Mn(CO)_4P^- + Mn(CO)_5L^+ \longrightarrow Mn(CO)_4P\cdot + Mn(CO)_5L\cdot \qquad (23)$$

Furthermore if the CO loss from $Mn(CO)_5L\cdot$ were rapid in eq 22b, the electron transfer in eq 23 would be tantamount to the simultaneous production of a pair of similar radicals, i.e.

$$Mn(CO)_4P^- + Mn(CO)_5L^+ \xrightarrow{-CO} Mn(CO)_4P\cdot + Mn(CO)_4L\cdot$$

which would afford a more or less random mixture of dimanganese carbonyls. Accordingly, the homo-coupled combination found in eq 20 points to an electron-transfer mechanism for the ion-pair annihilation leading to the manganese-manganese couplings. By the same token, the cross-coupling combination observed in eqs 5 and 6 seems to disfavor the same carbonylmanganese radicals as intermediates.

In order to resolve this conundrum we recognize that the two carbonylmanganese radicals arising directly from the electron transfer in eq 23 are inherently different.

Mechanistic Distinction between 19- and 17-Electron Carbonylmanganese Radicals.

The carbonylmanganese radical $Mn(CO)_5L\cdot$ is a supersaturated 19-electron species, whereas $Mn(CO)_4P\cdot$ an unsaturated 17-electron species.[5] The difference in the coordinative saturation in these two types of carbonylmanganese radicals immediately serves as a point of differentiation which is modulated by the rate of ligand loss from the 19-electron species. Accordingly it is worthwhile to review the properties and behavior of 19- and 17-electron carbonylmanganese radicals when they are separately generated by independent methods.

A. The 19-electron radical $Mn(CO)_5L\cdot$ is a transient species which is conveniently produced from the cathodic reduction of the carbonylmanganese(I) cations.[5] For example, Figure 2 shows the initial negative scan cyclic voltammograms (CV) of a series of carbonylmanganese cations undergoing 1e reduction to the 19-electron radicals. Ligand dissociation of the 19-electron $Mn(CO)_5L\cdot$ is rapid and can involve the loss of either L or CO to produce the 17-electron carbonylmanganese radicals, i.e.

$$Mn(CO)_5L\cdot \xrightarrow{k_L} Mn(CO)_5\cdot + L \qquad (24)$$

$$Mn(CO)_5L\cdot \xrightarrow{k_{CO}} Mn(CO)_4L\cdot + CO \qquad (25)$$

In the case of the nitrogen-substituted radicals $Mn(CO)_5(NCMe)\cdot$ and $Mn(CO)_5(py)\cdot$, the loss of CO is not competitive with the expulsion of either MeCN or py since $Mn_2(CO)_{10}$ is obtained as the sole dimer. This selectivity is in accord with the expected difference in leaving group abilities of CO relative to MeCN and py. On the other hand, for the phosphine derivatives of $Mn(CO)_5L\cdot$ there is a competition between the loss of carbon monoxide and phosphine. Moreover the logarithm of the relative rate constants k_L/k_{CO} in eqs 24 and 25 shows a linear correlation with the pK_a value of the phosphine ligand, (Figure 3). It is

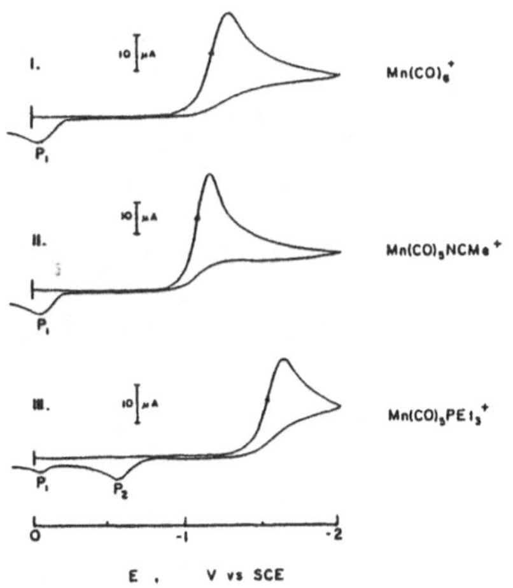

Figure 2. Initial negative scan cyclic voltammograms of carbonylmanganese cations at 500 mV s^{-1} in acetonitrile containing 0.1 M $Bu_4N^+ ClO_4^-$.

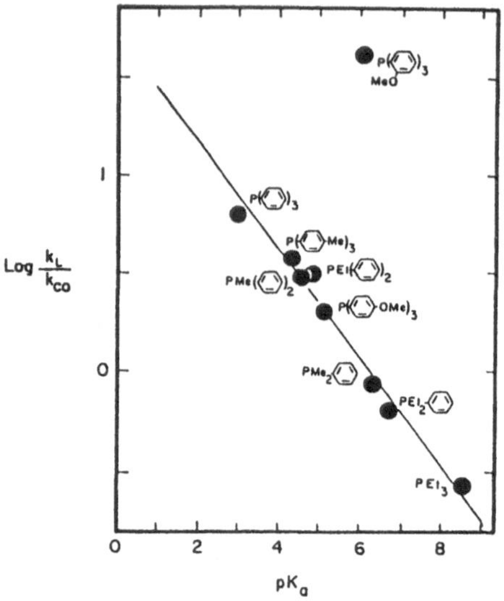

Figure 3. Competitive decomposition of the 19e $Mn(CO)_5L^{\cdot}$ according to eqs 24/25 as a function of the pk_a of LH^+.

noteworthy that the ratio k_L/k_{CO} = 6.7 for L = PPh_3 is more than a factor of 20 larger than k_L/k_{CO} = 0.3 for the more basic PEt_3. Note that the anodic waves P_1 and P_2 in Figure 2 represent the oxidation of the anions formed by the subsequent reduction of the 17-electron radicals, i.e.

$$Mn(CO)_5\cdot \xrightarrow{+e} Mn(CO)_5^- \qquad \text{(CV wave } P_1)$$

$$Mn(CO)_4L\cdot \xrightarrow{+e} Mn(CO)_4L^- \qquad \text{(CV wave } P_2)$$

B. The 17-electron radical $Mn(CO)_4P\cdot$ can be independently generated by the oxidation of the carbonylmanganate anion by electron acceptors such as tropylium cation (T^+), i.e.[16]

$$Mn(CO)_5^- + T^+ \longrightarrow Mn_2(CO)_{10} + T_2 \ (?)$$

More conveniently, the oxidation of carbonylmanganates can be carried out anodically, i.e.

$$Mn(CO)_5^- \xrightarrow{0.04V} \tfrac{1}{2}Mn_2(CO)_{10}$$

$$Mn(CO)_4[PPh_3]^- \xrightarrow{-0.39V} \tfrac{1}{2}Mn_2(CO)_8[PPh_3]_2$$

The persistence of the 17e radical is considerably enhanced by the presence of phosphine ligands. Thus Figure 4 shows that successively faster scan rates are required for cyclic voltammetry.[13] The ease of oxidation of $Mn(CO)_4P^-$ follows the σ-donor properties of the ligands

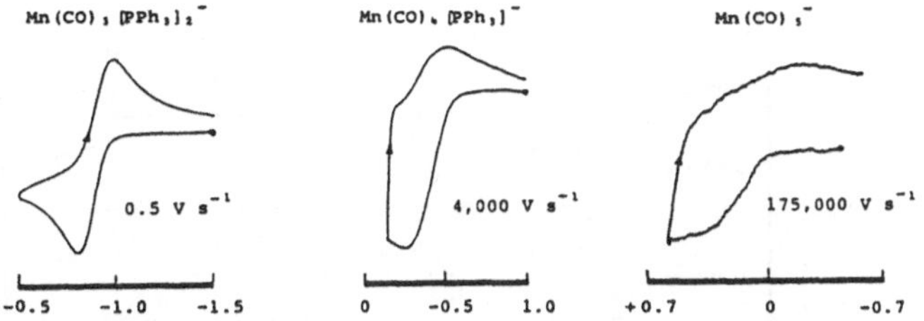

Figure 4. Variation of E^0 and the reversibility of carbonylmanganate oxidation in acetonitrile with increasing phosphine substitution, as indicated by the CV sweep rate.

in the order: $P = PPh_3 >> P(OPh)_3 > CO$; and it is an indication of the reactivity of the resulting radical $Mn(CO)_4P\cdot$, as reflected in the rates of self-dimerization.[15]

$$2\ Mn(CO)_4P\cdot \xrightarrow{k_d} Mn_2(CO)_8P_2 \qquad (26)$$

For example, the second-order rate constant for $Mn(CO)_5\cdot$ is $k_d = 9 \times 10^8\ M^{-1}\ s^{-1}$, and for $Mn(CO)_4PPh_3\cdot$ it is roughly two orders of magnitude slower ($k_d = 1 \times 10^7\ M^{-1}\ s^{-1}$).[15,17] These 17-electron species are also labile and undergo rapid ligand substitution by an associative mechanism[17,18] e.g.

$$Mn(CO)_5\cdot + PPh_3 \xrightarrow{k_s} Mn(CO)_4PPh_3\cdot + CO$$

where $k_s = 1.7 \times 10^7\ M^{-1}\ s^{-1}$. The 19-electron intermediate above is the same as that in eq 23 where L = phosphine. Thus the presence of added phosphine during the ion-pair annihilation would serve to equilibrate all carbonylmanganese radicals to a common species, as observed in eqs 12 and 13. The reversible CV of the multiply substituted $Mn(CO)_3[PPh_3]_2^+$ at the low scan rates indicated in Figure 4 is associated with the persistent 17e radical, the dimerization of which is not favored,[19] i.e.

$$2\ Mn(CO)_3[PPh_3]_2 \rightleftarrows Mn_2(CO)_6[PPh_3]_4$$

As a result of the ready reversibility, the primary fate of $Mn(CO)_3[PPh_3]_2\cdot$ is the formation of the hydride derivative i.e.

$$Mn(CO)_3[PPh_3]_2\cdot \longrightarrow H\,Mn(CO)_3[PPh_3]_2$$

This transformation can be examined quantitatively in the presence of tri-n-butyltin hydride as the hydrogen atom donor,[20] i.e.

$$Mn(CO)_3P_2^- \underset{}{\overset{-e}{\rightleftharpoons}} Mn(CO)_3P_2\cdot \qquad (27)$$

$$Mn(CO)_3P_2\cdot + Bu_3SnH \xrightarrow{k_2} HMn(CO)_3P_2 + Bu_3Sn\cdot \qquad (28)$$

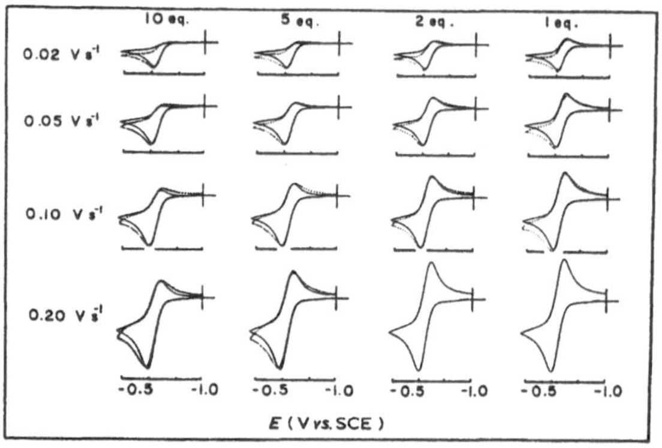

Figure 5. Effect of tri-n-butyltin hydride concentration on the reversible cyclic voltammogram of $Mn(CO)_3(PPh_3)_2^-$ in THF containing 0.3 M TBAP at various swap rates. The computer simulated cyclic voltammograms (·······) based on $k_2 = 7.3\ M^{-1}s^{-1}$.

The matrix of cyclic voltammograms in Figure 5 illustrate the effects of stannane concentration (horizontal) and scan rates (vertical) on the chemical reversibility of eq 27. The quantitative analyses of these cyclic voltammetric alterations lead to the second order rate constant k_2 listed in Table VI for hydrogen atom transfer in eq 28 as a dependence on the ligand parameters θ and the cone angle.[7]

Table VI. The Effect of Phosphine Structure P on the Reactivity of 17e Radicals $Mn(CO)_3P_2\cdot$

Ligand P	k_2 ($M^{-1}\ s^{-1}$)	θ (cm^{-1})	Cone Angle (deg)
$(PhO)_3P$	8.1	2085.3	128
$(o\text{-}CH_3C_6H_4O)_3P$	2.1	2084.1	141
$(Ph_2PCH_2)_2$	25.1	2066.7	125
$(n\text{-}Bu)_3P$	13.2	2060.3	132
$(i\text{-}PrO)_3P$	5.0	2075.9	130
Ph_3P	< 0.5	2068.9	145
$i\text{-}Pr_3P$	< 0.01	2059.7	160

C. Hydrogen atom transfer as described for the persistent di-substituted radicals $Mn(CO)_3P_2\cdot$ in Table VI also applies to $Mn(CO)_5\cdot$ and $Mn(CO)_4P\cdot$ in the form of competition with dimerization (eq 26). Thus hydrogen abstraction by $Mn(CO)_5L\cdot$ from solvent (SH) occurs as a side reaction and yields the hydridomanganese complexes $HMn(CO)_4L$ (Table II) when followed by the rapid extrusion of carbon monoxide,[21] i.e.

$$Mn(CO)_5L\cdot \xrightarrow[(SH)]{k_H} Mn(CO)_4(CHO)L \xrightarrow{-CO} HMn(CO)_4L \qquad (29)$$

Competition from this homolytic process is most severe with L = phosphines. For the more labile 19-electron radicals derived from L = CO, MeCN and py, the rate constant k_L in eq 24 appears to be sufficiently greater than k_H in eq 29 to obviate this competition, and hydridomanganese carbonyls are not important products.

With this background established for the properties of relative 19- and 17-electron carbonylmanganese radicals, let us now reconsider the distribution among the diverse products obtained from the various ion-pair annihilations in Tables I and II.

Radical-Pair Interactions of Transient 19- and 17-Electron Carbonylmanganese Species.

Several pathways are conceivable for the formation of $Mn_2(CO)_{10}$ in high yields from $Mn(CO)_5^-$ and $Mn(CO)_5(NCMe)^+$ in Table I. On one hand, the most economical route for this ion-pair annihilation involves cross-coupling accompanied by the displacement of MeCN, i.e.

$$Mn(CO)_5^- + Mn(CO)_5(NCMe)^+ \longrightarrow Mn_2(CO)_{10} + MeCN$$

On the other hand, a prior electron transfer (eq 23) involves a multistep process in which the 19- and 17-electron radical pair are involved in ligand loss (eq 24) and dimerization (eq 26), i.e.

Scheme III:

$$Mn(CO)_5^- + Mn(CO)_5(NCMe)^+ \longrightarrow Mn(CO)_5^{\cdot} + Mn(CO)_5(NCMe)^{\cdot} \qquad (30)$$

$$Mn(CO)_5(NCMe)^{\cdot} \xrightarrow{fast} Mn(CO)_5^{\cdot} + MeCN \qquad (31)$$

$$2\ Mn(CO)_5^{\cdot} \xrightarrow{fast} Mn_2(CO)_{10} \qquad (32)$$

The marked effect of added triphenylphosphine strongly supports the latter mechanism. Thus the presence of PPh_3 effectively diverts the course of reaction from $Mn_2(CO)_{10}$ to a pair of new products $Mn_2(CO)_8(PPh_3)_2$ and $HMn(CO)_4PPh_3$ without materially affecting the rate (see Table IV). Both of the products are readily attributed to the efficient interception of the carbonylmanganese radical, i.e.

Scheme IV:

$$Mn(CO)_5^{\cdot} + PPh_3 \rightleftharpoons Mn(CO)_5PPh_3^{\cdot} \qquad (33)$$

$$Mn(CO)_5PPh_3^{\cdot} \xrightarrow{-CO} Mn(CO)_4PPh_3 \qquad (34)$$

$$Mn(CO)_5PPh_3^{\cdot} \xrightarrow{SH} HMn(CO)_4PPh_3 + CO \qquad (35)$$

followed by rapid dimerization. Essentially the same results obtain when the ion-pair annihilation of $Mn(CO)_4PPh_3^- / Mn(CO)_5(NCMe)^+$ is carried out with added PPh_3, and $Mn(CO)_5^- / Mn(CO)_5PEt_3^+$ with added PEt_3 (eq 13). At this juncture, it is important to re-emphasize that the control experiments establish both the precursor carbonylmanganate(-I) anion and carbonylmanganese(I) cation to be substitution stable, and no ligand exchange of the type described above can occur within the time span of these experiments.

The principal byproduct of the ion-pair coupling of $Mn(CO)_4P^-$/ $Mn(CO)_5L^+$ is the hydridomanganese carbonyl $HMn(CO)_4L$, which most commonly arises when L = phosphines. On the other hand when L = CO, MeCN or py, the corresponding hydride is absent. These results accord with hydride formation occurring via the 19-electron radical $Mn(CO)_5L\cdot$, as described in eq 29. Furthermore, hydrogen atom transfer to phosphorus-substituted radicals with L = phosphine should occur more readily than that with L = CO, MeCN or py owing to the dissociation rate constant $k_L > k_{CO}$ in eqs 24 and 25. It is possible that the small amounts of $HMn(CO)_4P$ observed in some cases arise by protonation of the carbonylmanganate anion by adventitious moisture[22] or by multiple ligand substitutions of the 17-electron radical, i.e.

Scheme V:

$$Mn(CO)_4P\cdot \ + \ L \longrightarrow Mn(CO)_4L\cdot \ + \ P \tag{36}$$

$$Mn(CO)_5\cdot \ + \ P \longrightarrow Mn(CO)_5P\cdot, \text{ etc.} \tag{37}$$

Indeed, the ion-pair annihilation of $Mn(CO)_5^-/Mn(CO)_5PEt_3^+$ carried out in the presence of two equiv. PEt_3 afforded $Mn_2(CO)_8(PEt_3)_2$, $HMn(CO)_4PEt_3$ and the bis-substituted hydride $HMn(CO)_3(PEt_3)_2$ as the principal products, with no evidence of either $Mn_2(CO)_{10}$ or $Mn_2(CO)_9PEt_3$. These results accord with efficient interception of $Mn(CO)_5\cdot$ (compare eqs 33-35). Moreover, the appearance of $HMn(CO)_3(PEt_3)_2$ indicates the facility with which further homolytic substitution occurs.[18d] i.e.

$$Mn(CO)_4PEt_3\cdot + PEt_3 \xrightarrow{-CO} Mn(CO)_3(PEt_3)_2\cdot, \text{ etc.} \tag{38}$$

According to the electron-transfer formulation, the relative amounts of products are largely determined by the magnitudes of k_L, k_{CO} and k_H since they regulate the steady-state concentration of the 19-electron radical. The extent to which k_d for the coupling of the 17-electron radical is dependent on X = CO, P and L also affects the distribution among the various dimanganese carbonyls.[18] The availability of free phosphine is responsible for raising the steady-state concentration of the 19-electron species by the microscopic reverse of eq 24. This results in increased hydride (eq 29) and phosphine incorporation into the dimanganese carbonyls (eqs 36 and 38). Most importantly, the diversion of $Mn(CO)_5^-$ and $Mn(CO)_5(NCMe)^+$ by added PPh_3 (Table III) demonstrates that a substantial portion, if not all, of the ion-pair annihilation occurs via scavengable radicals according to Scheme III and IV.

Mechanism of Mn-Mn Bond Formation by Ion-Pair Annihilation.

Taken together, the foregoing analyses show that the diverse products derived from the ion-pair annihilations in Tables I-III can be completely accounted for by the known behavior of 19- and 17-electron carbonylmanganese radicals. Accordingly, the initiation by electron-transfer is included in the generalized formulation below for Mn-Mn bond formation.

Scheme VI:

$$Mn(CO)_4P^- \; + \; Mn(CO)_5L^+ \longrightarrow [Mn(CO)_4P\cdot \; Mn(CO)_5L\cdot] \quad (39)$$

$$[Mn(CO)_4P\cdot \; Mn(CO)_5L\cdot] \xrightarrow{\text{collapse}} Mn_2(CO)_8(P)(L) + CO \quad (40)$$

$$[Mn(CO)_4P\cdot \; Mn(CO)_5L\cdot] \xrightarrow{\text{diffuse}} Mn(CO)_4P\cdot + Mn(CO)_5L\cdot \quad (41)$$

$$Mn(CO)_5L\cdot \xrightarrow{k_L, k_{CO}} Mn(CO)_4X\cdot + L(CO) \quad (42)$$

$$2\,Mn(CO)_4X\cdot \xrightarrow{k_d} Mn_2(CO)_8X_2 \quad (43)$$

where X = CO, P, L

The first-order rate constants k_L and k_{CO} represent the ligand dissociation from the 19-electron radical, and k_d is the second-order rate constant for the couplings of pairs of 17-electron radicals.

According to Scheme VI, electron-transfer is the first step (eq 39) in the annihilation of the carbonylmanganese cations by anions. The nonstatistical distribution among the homo-coupled and cross-coupled dimers in Scheme I derive primarily from the competition between cage collapse (eq 40) and diffusive separation (eq 41) of the initially formed radical pair. The extent to which cage collapse occurs faster than diffusive separation will determine the amounts of cross-coupled products obtained. Alternatively, the efficiency with which added phosphines diverts the products to $Mn_2(CO)_8P_2$ (Table III) reflects the extent to which 17-electron carbonylmanganese radicals have escaped and consequently are subject to ligand substitution. In the same way the trapping of 19-electron $Mn(CO)_5L\cdot$ by hydrogen transfer reflects cage escape. Indeed the multiple substitution of phosphine in the hydride products (see eq 38) suggests that the 17-electron carbonylmanganese species are longer lived then their 19-electron precursors. A further indication of this difference is shown in the carbonylmanganate $Mn(CO)_4(\eta^1\text{-DPPE})^-$ in which the corresponding 17-electron radical will be susceptible to intramolecular trapping, i.e.

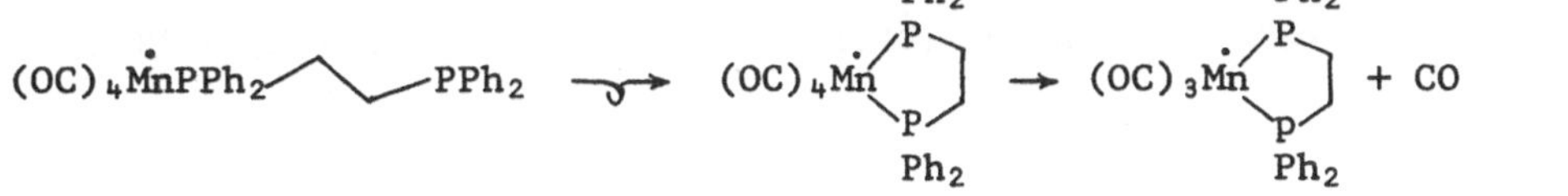

(44)

The facile annihilation of $Mn(CO)_4(\eta^1\text{-DPPE})^-$ by an equivalent amount of either $Mn(CO)_5(NCMe)^+$ or $Mn(CO)_5py^+$ afforded three principal products, viz., $Mn_2(CO)_{10}$ (30%), $Mn_2(CO)_9(\eta^1\text{-DPPE})$ (15%), and $HMn(CO)_3(\eta^2\text{-DPPE})$ (35%).

The formation of the chelated $HMn(CO)_3(\eta^2\text{-DPPE})$ is consistent with the trapping of the 19-electron intermediate in eq 44, as indicated by the mechanism for hydride formation in eq 29. The observation of the monohapto dimanganese carbonyl $Mn_2(CO)_9(\eta^1\text{-DPPE})$ suggests that the cross-coupling either involves a concerted process or a cage collapse in eq 40 which is faster than intramolecular trapping (eq 44). Be that as it may, from a purely operational point of view, there are two pathways for the formation of Mn-Mn carbonyls by ion-pair annihilation, namely, (i) a process in which pairs of carbonylmanganese radicals behave more or less independently as discrete species as a result of diffusive separation and (ii) a minor pathway in which the cation and anion more or less maintain their identity by either the cage collapse of the radical pair (eq 40) or by concerted action.

Driving Force for the Formation of Mn-Mn Bonds by Ion-Pair Annihilation.

Since the radical-pair mechanism in Scheme VI commences with an electron-transfer initiation, a critical part of the driving force for Mn-Mn bond formation is the oxidation/reduction of the anion/cation pair. As such, the energetics for the production of the 19- and 17-electron carbonylmanganese radical pair derives from the reversible electrode potentials for $Mn(CO)_5L^+$ and $Mn(CO)_4P^-$.

(a) For the reduction of the parent cation $Mn(CO)_6^+$ in eq 14, a value of $E_{1/2} \cong -1.0$ V vs SCE is obtained from the partially reversible cyclic voltammogram at very high sweep rates. Although we are unable to measure directly the reversible electrode potentials for the other carbonylmanganese(I) cations, they can be reliably estimated from the peak potentials listed in Table V by an additive correction of $\sim$ 0.3 V. The wide span in potentials indicates a large ligand dependence on the ease of reduction of $Mn(CO)_5L^+$ in the order: L = py > MeCN > CO > PPh_3 > PPh_2Et > $PPhEt_2$ > PEt_3 >> η^2-DPPE.

(b) For the oxidation of the anionic $Mn(CO)_4PPh_3^-$ in eq 15, $E_{1/2} \cong -0.33$ V vs. SCE. The reversible potentials for the other carbonylmanganates can be estimated from the peak potentials in Table V by an additive correction of $\sim$ 0.2 V. The ligand effect on the ease of oxidation is opposite to that in the cation in the order: L = PPh_3 >> $P(OPh)_3$ > CO.

On this basis, the driving force for electron transfer will be the most favorable for the ion pair $Mn(CO)_4PPh_3^-/Mn(CO)_5py^+$ with $\Delta E \sim 0.5$ V and the least favorable for $Mn(CO)_5^-/Mn(CO)_2(DPPE)_2^+$ with $\Delta E \sim 1.6$ V. Indeed the latter is an unreactive ion pair, and it can be readily isolated as a simple salt from THF solution, i.e.

$$Na^+ Mn(CO)_5^- + Mn(CO)_2(DPPE)_2^+ BF_4^- \rightarrow [Mn(CO)_5^- Mn(CO)_2(DPPE)_2^+] + NaBF_4 \qquad (45)$$

owing to the insolubility of $NaBF_4$. Although we have not yet measured the reaction rates quantitatively, the ion-pair reactivities in Tables I-III qualitatively follow the trend in the driving forces. Thus we find the ion pairs $Mn(CO)_4PPh_3^-/Mn(CO)_5py^+$ and $Mn(CO)_4PPh_3^-/Mn(CO)_5(NCMe)^+$

in Table III to be the most reactive in this study. At the other extreme, the couplings of the parent anion $Mn(CO)_5^-$ with the phosphine-substituted cations in Table I require the longest times. Since $\Delta E \sim 1.2$ V for $Mn(CO)_5^-/Mn(CO)_5PEt_3^+$, it appears that the threshold in the driving force for ion-pair annihilation lies somewhere between 1.2 and 1.6 V in THF solution. If the rate of electron transfer in eq 39 is taken in the outer-sphere context of Marcus theory,[23] two other factors must also be considered. Thus the interaction of oppositely charged ions will be aided considerably by the electrostatics.[24] Indeed such a positive work term accords with the strong solvent dependence and the negative salt effect described in Table IV. The contribution from the reorganization energies of $Mn(CO)_4P^-$ and $Mn(CO)_5L^+$ cannot be evaluated at this juncture. However these large highly polarizable, 5- and 6-coordinate ions are likely to show only minor differences in basic structural change. If so, the driving force ΔE will be the dominant factor in determining the ease with which ion pairs are annihilated to form metal-metal bonds.

SUMMARY AND CONCLUSIONS

The carbonylmanganese(I) cation $Mn(CO)_6^+$ and the carbonylmanganate(-I) $Mn(CO)_5^-$ react upon mixing in THF solution to afford high yields of dimanganese decacarbonyl. Similarly the substituted cations $Mn(CO)_5L^+$ with L = py, MeCN, aryl and alkylphosphines and the substituted anions $Mn(CO)_4P^-$ with P = PPh_3 and $P(OPh)_3$ lead to mixtures of dimanganese carbonyls labelled with the P and L tracers. The extensive (if not complete) scrambling of the carbonylmanganese moieties during ion-pair annihilation is ascribed to the 17- and 19-electron radicals $Mn(CO)_4L\cdot$ and $Mn(CO)_5P\cdot$, respectively, as the reactive intermediates. This conclusion is strongly supported by the known behavior of both types of radicals when they are independently generated by the anodic oxidation of $Mn(CO)_4P^-$ and the cathodic reduction of $Mn(CO)_5L^+$. The reversible addition of ligands to 17-electron radicals provides a ready means for interconversion with their 19-electron counterparts. Thus the effect of added phosphine in altering the course of Mn-Mn coupling via ligand substitution and the formation of the hydrido byproducts $HMn(CO)_4L$ via hydrogen transfer (eq 29) provide compelling evidence for carbonylmanganese radicals since neither the cation nor the anion is susceptible to additives on the timescale of the coupling experiments. These experiments also rule out any type of concerted displacement mechanism as a dominant route in the formation of $Mn_2(CO)_{10}$ from ion pairs such as $Mn(CO)_5^-$ / $Mn(CO)_5(NCMe)^+$ or $Mn(CO)_5^-/Mn(CO)_5py^+$ in which the carbonylmanganese moieties are not labelled. The participation of specific cross-coupling is included in the unified mechanism as a minor pathway stemming from cage collapse (eq 40) of radicals immediately following an initial electron-transfer. The latter is consistent with the reactivity trends for carbonylmanganese ion pairs which qualitatively parallel the differences ΔE in the oxidation and reduction potentials of $Mn(CO)_4P^-$ and $Mn(CO)_5L^+$.

The annihilation of carbonylmanganese cation-anion pairs thus underscores the importance of electron-transfer activation of otherwise stable species. Particularly germane is the enhanced substitutional lability of the carbonylmanganese radicals. Indeed the latter is a general phenomenon applicable to the remarkably efficient (ETC) catalysis of a wide variety of reactions of metal carbonyls both as mononuclear species as well as polynuclear clusters.[25]

Acknowledgment: I thank my coworkers particularly D. J. Kuchynka, K. Y. Lee, B. Narayanan, H. Ohst and T. M. Bockman, for their enthusiastic participation in this research. I am also deeply indebted to Dr. C. Amatore (École Normal Supérieure, Paris) for help with his invaluable electrochemical insights, and the National Science Foundation and Robert A. Welch Foundation for financial support.

References

1. (a) Kruck,T.; Hofler, M. Angew. Chem. Int. Eng. Ed. 1964, 3, 701. (b) Joshi, K. K.; Pauson, P. L. Z. Naturforsch 1962, 17b, 565. (c) Nesmeyanov, A. N.; Anisimov, K. N.; Kolobova, N.E.;Kolomnokov, I.S. Izvest Akad. Nauk SSSR, Otdel, Khim, Nauk 1963, 194. (d) Szabo, P.; Fekete, L.; Bor, G.; Nagy-Magos, Z.; Marko, L. J. Organomet. Chem. 1968, 12, 245.
2. Kruck, T.; Hofler, M. Chem. Ber. 1964, 97, 2289.
3. Stiegman, A. E.; Tyler, D. R. Coord. Chem. Rev. 1985, 63, 217.
4. (a) Allen, D. M.; Cox, A.; Kemp, T. J.; Sultana, Q.; Pitts, R. B. J. Chem. Soc. Dalton Trans. 1976, 1189. (b) Hieber, W.; Schropp, W. Z. Naturforsch, Tiel B., 1960, 15, 271. (c) McCullen, S. B.; Brown, T. L. Inorg. Chem. 1981, 20, 3528. (d) Hudson, A.; Lappert, M. F.; Nicholson, B. K. J. Organomet. Chem. 1975, 92, C11. (e) Hudson, A.; Lappert, M. F.; Macquitty, J. J.; Nicholson, B. K.; Zainal, H.; Luckhurst, G. R.; Zannoni, C.; Bratt, S. W.; Symons, M. C. R. J. Organomet. Chem. 1976, 110, C5 (f) Stiegman, A. E.; Stieglitz, M.; Tyler, D. R. J. Am. Chem. Soc. 1983, 105, 6032. (g) Stiegman, A. E.; Goldman, A. S.; Philbin, C. E.; Tyler, D. R. Inorg. Chem. 1986, 25, 2976.
5. Kuchynka, D. J.; Amatore, C.; Kochi, J. K. Inorg. Chem. 1986, 25, 4087.
6. (a) Streuli, C. A. Anal. Chem. 1960, 32, 985. (b) Henderson, Jr., W. A.; Streuli, W. A. J. Am. Chem. Soc. 1960, 82, 5791. (c) Zizelman, P. M.; Amatore, C.; Kochi, J. K. J. Am. Chem. Soc. 1984, 106, 3771.
7. Tolman, C. A. Chem. Rev. 1977, 77, 313.
8. Lee, K. Y.; Kuchynka, D. J.; Kochi, J. K. Organometallics 1987, 6, 0000.
9. (a) Kosower, E. M. Introduction to Physical Organic Chemistry 1968, Wiley: New York. (b) Reichardt, C. Liebig Ann. 1971, 752, 64.
10. Bockman, T. M. unpublished results.
11. Szwarc, M., Ed. Ions and Ion Pairs in Organic Reactions, Vols. 1 and 2, Wiley, New York 1972 and 1974.
12. Gagne, R. R.; Koval, C. A.; Lisensky, G. C. Inorg. Chem. 1980, 19, 2854.

13. D. J. Kuchynka, unpublished results. See also Howell, et al.[14]
14. Howell, J. O.; Goncalves, J. M.; Amatore, C.; Klasinc, L.; Wightman, R. M.; Kochi, J. K. J. Am. Chem. Soc. 1984, **106**, 3968.
15. (a) Hughey, J. L.; Anderson, C. P.; Meyer, T. J. J. Organomet. Chem. 1977, **125**, C49. (b) Wegman, R. W.; Olsen, R. J.; Gard, D. R.; Faulkner, L. R.; Brown, T. L. J. Am. Chem. Soc. 1981, **103**, 6089. (c) Walker, H. W.; Herrick, R. S.; Olsen, R. J.; Brown, T. L. Inorg. Chem. 1984, **23**, 3748. (d) Yesaka, H.; Kobayashi, T.; Yasufuku, K.; Nagakura, S. J. Am. Chem. Soc. 1983, **105**, 6249. (e) Rothberg, L. J.; Cooper, N. J.; Peters, K. S.; Vaida, V. J. Am. Chem. Soc. 1982, **104**, 3536. (f) Waltz, W. L.; Hackelberg, O.; Dorfman, L. M.; Wojcicki, A. J. Am. Chem. Soc. 1978, **100**, 7259. (g) Meyer, T. J.; Caspar, J. V. Chem. Rev. 1985, **85**, 187.
16. Armstead, J. A.; Cox, D. J.; Davis, R. J. Organometal. Chem. 1982, **236**, 213.
17. For three recent summaries see (a) Herrinton, T. R.; Brown, T. L. J. Am. Chem. Soc. 1985, **107**, 5700. (b) Turaki, N. N.; Huggins, J. M. Organometallics, 1986, **5**, 1703. (c) Meyer, et al in ref 15g and related references therein.
18. (a) Fox, A.; Malito, J.; Poe, A. J. Chem. Soc. Chem. Commun. 1981, 1052. (b) Wrighton, M. S.; Ginley, D. S. J. Am. Chem. Soc. 1975, **97**, 2065. (c) Kidd, D. R.; Brown, T. L. J. Am. Chem. Soc. 1978, **100**, 4095. (d) McCullen, S. B.; Walker, H. W.; Brown, T. L. J. Am. Chem. Soc. 1982, **104**, 4007. (e) Byers, B. H.; Brown, T. L. J. Am. Chem. Soc. 1977, **99**, 2527. (f) Hoffman, N. W.; Brown, T. L. Inorg. Chem. 1978, **17**, 613. (g) Absi-Halabi, M.; Brown, T. L. J. Am. Chem. Soc. 1977, **99**, 2982. (h) Shi, Q.-Z.; Richmond, T. G.; Trogler, W. C.; Basolo, F. J. Am. Chem. Soc. 1982, **104**, 4032.
19. (a) Kidd, D. R.; Cheng, C. P.; Brown, T. L. J. Am. Chem. Soc. 1978, **100**, 4103. (b) Brown, T. L.; McCullen, S. B. J. Am. Chem. Soc. 1982, **104**, 7496.
20. Kuchynka, D. J.; Amatore, C.; Kochi, J. K. J. Organometal. Chem. 1987, **328**, 133.
21. (a) Narayanan, B. A.; Amatore, C. A.; Kochi, J. K. Organometallics, 1984, **3**, 802. (b) Narayanan, B. A.; Kochi, J. K. J. Organomet. Chem. 1983, **272**, C49. (c) Narayanan, B. A.; Amatore, C.; Kochi, J. K. Organometallics, 1986, **5**, 926. (d) See also Hanckel, J. M.; Lee, K.-W.; Rushman, P.; Brown, T. L. Inorg. Chem. 1986, **25**, 1852.
22. Zotti, G.; Zeechin, S.; Pilloni, G. J. Organomet. Chem. 1983, **246**, 61.
23. Marcus, R. A. J. Chem. Phys. 1956, **24**, 966; 1957, **26**; 867; 1965, **43**, 679. For a review, see Cannon, R. D. Electron Transfer Mechanisms; Butterworths: London, 1980.
24. For the strong ion-pairing of carbonylmetal ions in solvents such as THF, see (a) Darensbourg, D. J.; Darensbourg, M. Y. Inorg. Chim. Acta. 1971, **5**, 247. (b) Darensbourg, M. Y.; Jimenez, P.; Sackett, J. R. J. Organomet. Chem. 1980, **202**, C68. (c) Darensbourg, M. Y.; Hanckel, J. M. J. Organomet. Chem. 1981, **217**, C9. (d) Drew, D.; Darensbourg, M. Y.; Darensbourg, D. J. J. Organomet. Chem. 1975, **85**, 73. (e) Kao, S. C.; Darensbourg, M. Y.; Schenk, W. Organometallics, 1984, **3**, 871.
25. Kochi, J. K. J. Organometal. Chem. 1986, **300**, 139.

OXIDATIVELY INDUCED SUBSTITUTION REACTIONS OF TRANSITION METAL ORGANOMETALLIC COMPOUNDS

John C. Kotz
Department of Chemistry
State University of New York
Oneonta, New York, 13820 USA

ABSTRACT. Oxidation, both chemical and anodic, can induce substitutions, bond-breaking and -making, rearrangements, and coupling reactions of organometallic compounds. In some cases the processes are catalytic. This article reviews the recent literature on oxidatively induced substitutions and briefly describes a novel case of induced halogen substitution in bis(cyclopentadienyl)molybdenum dihalides (Cp_2MoX_2) by solvent as well as the importance of the 17-electron radical cation $[Cp_2MoX_2]^{+\cdot}$ as an intermediate in the catalytic oxidation of triphenylphosphine.

The inducement of reactions by electron transfer has been an active area of investigation in organometallic chemistry,[1] as in other fields of chemistry.[2] In organometallic chemistry, electron transfer-induced reactions include (a) the substitution of neutral donor ligands or anionic two-electron donor ligands; (b) structural changes such as the isomerization of the primary ligand sphere or the rearrangement of donors; or (c) reactions at coordinated ligands. We have recently reviewed the electrochemistry of transition metal organometallic compounds, and described these various types of reactions in that review.[1] The present article will cover new developments for **reactions induced by oxidation**, particularly **anodic oxidation.** The literature is covered to September, 1987, and discussion is limited to oxidatively induced substitution of anionic, σ-bonded ligands (halides and alkyl or aryl groups) or neutral, σ-bonded ligands such as CO.[3,4]

Anodically or cathodically induced reactions usually involve odd-electron intermediates, since most such reactions are one-electron processes. The study of these reactions is important, since the "participation of odd-electron intermediates in 'traditional' organometallic reactions has become a matter of growing speculation and concern."[5] Cathodically induced reactions have been particularly well studied,[6] so their anodic counterparts offer many opportunities for further work.

M. Chanon et al. (eds.), Paramagnetic Organometallic Species in Activation / Selectivity, Catalysis, 171–185.

1. ANODICALLY INDUCED SUBSTITUTION OF NEUTRAL DONOR LIGANDS

1.1. Arene Metal Carbonyls.

A recent study of the electrochemical behavior of some arene metal carbonyls has shed light on many other studies of oxidatively induced substitution reactions. In general, arene metal carbonyls are first oxidized to a 17-electron cation, but this is followed by decomposition of the compound, presumably with loss of arene, and a further two-electron oxidation.[7] However, the anodic behavior in CH_3CN of substituted arene metal carbonyls such as (hexamethylbenzene)$W(CO)_3$ has recently been reexamined.[5] This complex undergoes one-electron oxidation at +0.64 V, but the coupled reduction wave appears nearer 0 V in CH_3CN or -0.6 V in DMF. To account for these observations the authors proposed that the cation radical incorporated solvent in some manner, and that this solvento complex was more difficult to reduce than the solvent-free cation. To further test this notion, various phosphines and phosphites were added to $(C_6Me_6)W(CO)_3$ in CH_3CN. In each case the one-electron oxidation wave was associated with a reduction wave at a negative potential, and, for PBu_3, there was evidence for a phosphine substitution product $(C_6Me_6)W(CO)_2(PBu_3)$. These results have been interpreted as follows: (a) On adding a phosphine to a solution of the cation radical $[(C_6Me_6)W(CO)_3]^+$, ligand sphere expansion occurs with the added phosphine attached to the metal to give a 19-electron complex. (b) The 19-electron complex does not dissociate CO to give the observed substitution product; instead, the latter is formed only upon reduction of the 19-electron species.

There may be an alternative explanation for the results above. Specifically, cationic arene metal complexes are well known to form complexes by attack of the nucleophile at the ring.[8] If such complexes were formed between $[areneM(CO)_3]^+$ (M = Cr, W) and phosphines, 19-electron complexes would not be required, and the reduction of $[(arene\text{-}PR_3)M(CO)_3]^+$ would indeed occur at a different potential.[9] In any event, the question of the interaction of solvent or other added bases with organometallic substrates, and the effect of such interactions, bears further study.[10]

1.2. Cyclopentadienyl Metal Carbonyls

Among the best documented studies of oxidatively induced substitutions involving metal carbonyls, particularly by **electron transfer catalysis (ETC)**, are those of Kochi and co-workers.[11] Such reactions occur by the **general ETC sequence**

(1) chain initiation: $L\text{-}M \rightleftharpoons [L\text{-}M]^+ + e^-$

(2) substitution: $[L\text{-}M]^+ + N \underset{k_{-1}}{\overset{k_1}{\rightleftharpoons}} [N\text{-}M]^+ + L$

(3) chain propagation: $[N\text{-}M]^+ + L\text{-}M \xrightarrow{k_2} N\text{-}M + [L\text{-}M]^+$

(4) chain termination: $[N\text{-}M]^+ + e^- \rightleftharpoons N\text{-}M$

where, in Kochi's work, L-M is a metal carbonyl complex (e.g., MeCpMn-$(CO)_2$(pyridine), $(MeCN)_3W(CO)_3$, etc.) and N is a nucleophile such as a phosphine. At least for substitution reactions of $MeCpMn(CO)_2L$, step (2) is rate determining. This, the observed dependence on the electronic properties of L and N and the steric properties of the incoming phosphines (N), and other observations, imply an irreversible associative ligand exchange mechanism for the substitution reaction.

An associative mechanism has also been observed for substitution reactions of the 17-electron molecule $V(CO)_6$,[12] and such mechanisms are increasingly found for 18-electron cyclopentadienyl metal complexes.[13,14] However, the 17-electron molecules and ions studied thus far have substitution rates many times greater than the 18-electron molecules, with much lower enthalpies of activation. Kochi and co-workers believe that the difference in rates for 18- and 17-electron molecules "may be ascribed to the ability of the Cp ligand in the 18-electron [complexes] to accommodate the '20-electron' transition state by converting from η^5 to η^3 ligation and essentially neutralize ... the electronic effects of the entering nucleophile. ... such a change is expected to be less necessary in the electron-deficient 17-electron system ... ."[11c] Finally, these same authors have called attention to the analogy between the mechanism of substitution observed for $[MeCpMn(CO)_2L]^+$ and nucleophilic S_N2 substitution at a tetrahedral carbon center.

The electrochemistry of cyclopentadienyl carbonyl derivatives of Group 9 metals (Co and Rh) of the type $(C_5R_5)M(CO)L$ (L = phosphine, arsine, phosphite; R = H, Me) has been the subject of several studies.[15] All these compounds undergo oxidatively induced substitution of CO for a phosphine or phosphite. For example, the CO ligand of $(C_5H_5)Co(CO)L$ (L = PPh_3) cannot be substituted by PPh_3 either thermally or photochemically. However, the bis-phosphine complex $[(C_5H_5)CoL_2]^+$ is readily obtained in an oxidatively induced reaction, either anodically or chemically. Even though the potential for the oxidation of $(C_5H_5)Co(CO)_2$ is estimated to be at least 1 V (see below), a value far greater than the reduction potential of $[Cp_2Fe]^+$, both CO ligands are substituted by PPh_3 when ferricenium ion is added to $(C_5H_5)Co(CO)_2$ in the presence of PPh_3.

$$(C_5H_5)Co(CO)_2 + PPh_3 \rightleftharpoons (C_5H_5)Co(CO)(PPh_3) + CO$$
$$(C_5H_5)Co(CO)(PPh_3) \rightleftharpoons [(C_5H_5)Co(CO)(PPh_3)]^+ + e^-$$
$$[(C_5H_5)Co(CO)(PPh_3)]^+ + PPh_3 \rightleftharpoons [(C_5H_5)Co(PPh_3)_2]^+ + CO$$

The authors speculated that a small concentration of $(C_5H_5)Co(CO)(PPh_3)$ was formed initially. Since this compound has a much lower oxidation potential than $(C_5H_5)Co(CO)_2$, the 17-electron species was rapidly formed by chemical oxidation or at the anode, and this latter material underwent rapid, oxidatively induced substitution. The 17-electron cationic products of these reactions were found readily to undergo "radical coupling reactions." For example, reaction of $[(C_5H_5)Co(CO)L]^+$ with Br_2 or I_2 gave $[(C_5H_5)Co(CO)XL]^+$, where the cobalt is formally +3.

The electrochemistry of $(C_5H_5)Co(CO)(PPh_3)$ in THF at a Pt electrode exhibited two waves, a reversible process at E^o = +0.21 V and another, irreversible wave, not observed in CH_2Cl_2, at 0.75 V. The authors attributed this new wave to the oxidation of a 19-electron cationic complex $[(C_5H_5)Co(CO)(PPh_3)(THF)]^+$ to an 18-electron dication, a result reminiscent of the solvent interactions with (arene)$M(CO)_3$ (Section 1.1).

Remarkable differences are often observed between analogous complexes of the cyclopentadienyl and pentamethylcyclopentadienyl ligands. Therefore, Connelly and Raven have studied complexes of the pentaphenylcyclopentadienyl ligand, $(C_5Ph_5)M(CO)L$ (M = Co, Rh), largely because of the ligand's "apparent ability ... to stabilize reactive organotransition metal radicals generated in one-electron transfer reactions."[15b] As outlined below, their studies successfully demonstrated this.

The complex $(C_5Ph_5)Co(CO)_2$ does not give a PPh_3 substitution product even after lengthy irradiation in toluene, nor does the rhodium analog undergo substitution after several days in refluxing heptane. However, substitution was rapid on forming the 17-electron radical cations $[(C_5Ph_5)M(CO)_2]^+$. Similar results are obtained for the cobalt complexes $(C_5R_5)M(CO)_2$ when R = H or Me, but the 17-electron rhodium-containing complexes are too unstable when R = H or Me to allow their subsequent reduction to the 18-electron, neutral substitution product.

When the ferricenium ion was used as a chemical oxidant [E^o = 0.47 V vs. SCE], oxidatively driven substitution of CO for N in $(C_5Ph_5)Rh(CO)_2$ was completely successful only for N = $P(OPh)_3$. Although there is a large and unfavorable difference between the potentials for the oxidation of L-M = $(C_5Ph_5)Rh(CO)_2$ and the reduction of ferricenium ion, the irreversibility of step 2 in the ETC scheme above (L = CO) allows substitution to produce the 17-electron radical cation $[(C_5Ph_5)Rh(N)(CO)]^+$. Once formed, the substitution product [0.67 V for N = $P(OPh)_3$, step 4] is a strong enough oxidant that ferricenium ion is regenerated, thereby renewing the catalytic cycle. Unfortunately, when the ligands L are more basic (PPh_3, $AsPh_3$), the overall process leads only to an inseparable equilibrium mixture of starting materials and products.

1.3. Phosphine-Substituted Metal Carbonyls

The 17-electron cation radical $[Fe(CO)_3(PPh_3)_2]^+$, generated anodically or chemically, is about 10^9 more reactive to CO substitution by pyridine than its 18-electron precursor. In the most recent paper on the electrochemistry of $Fe(CO)_3(PPh_3)_2$, it was proposed that CO substitution by a base such as pyridine in $[Fe(CO)_3(PPh_3)_2]^+$ occurs according to the first two steps in the ETC scheme above (where L = CO and N = pyridine).[16] However, a possible ETC reaction is intercepted by a third step,

$$[Fe(CO)_2(py)(PPh_3)_2]^+ + [Fe(CO)_3(PPh_3)_2]^+ \xrightarrow{k_2} Fe(CO)_3(PPh_3)_2 + [Fe(CO)_2(py)(PPh_3)_2]^{2+}$$

the disproportionation of 17-electron Fe(I) ions. A product of this reaction, the 16-electron dication, is not stable and rapidly forms $[Fe(py)_6]^{2+}$.

The reactivity of $[Fe(CO)_3(PPh_3)_2]^+$ has been compared with that of other 17-electron, metal carbonyl radicals. In general, such radicals react more slowly as CO is replaced by phosphines and as the bulk of the phosphines is increased. These observations lead to the conclusion that substitution is associatively activated, just as were the substitution reactions of other 17-electron compounds described in Section 1.2.

The 17-electron cation generated chemically or electrochemically from trans-$Cr(CO)_4(PPh_3)_2$ is light and moisture sensitive and, unlike the 18-electron parent, the radical reacts rapidly with acetonitrile, acetone, iodide or bromide ion, and water.[17] In every case at least one of the products is the neutral complex trans-$Cr(CO)_4(PPh_3)_2$. In the case of I^- or Br^-, the reaction is a simple redox process, the deep blue cation rapidly reverting to the yellow 18-electron, neutral complex.

$$2\,[Cr(CO)_4(PPh_3)_2]^+ + 2\,X^- \longrightarrow 2\,Cr(CO)_4(PPh_3)_2 + X_2$$

In contrast, the neutral complex is produced in other cases by a disproportionation reaction that, the authors argue, is catalyzed by potentially coordinating ligands.

$$2\,[Cr(CO)_4(PPh_3)_2]^+ \longrightarrow Cr(CO)_4(PPh_3)_2 + Cr^{2+} + 2\,PPh_3 + 4\,CO$$

There was no evidence for electron-transfer catalyzed substitution of the Cr(I) cation.

1.4 Dimeric Compounds.

The anodic electrochemistry of numerous dimeric organometallic compounds has been studied recently.[18] Although some of these dimeric complexes can undergo oxidatively induced substitution, $Cp_2Mo_2(CO)_4(\mu\text{-}SR)_2$ is unique in that it can lead to a radical cation that is the reactive species in a substitution reaction that can be either oxidatively induced or catalyzed by an ETC mechanism (**Scheme I**, R = Me, Ph; L = MeCN).[18k] Like other thiolate bridged dimers, the neutral compound undergoes a two-electron oxidation where both electrons are transferred at the same potential, presumably owing to the formation of a metal-metal bond.

SCHEME I

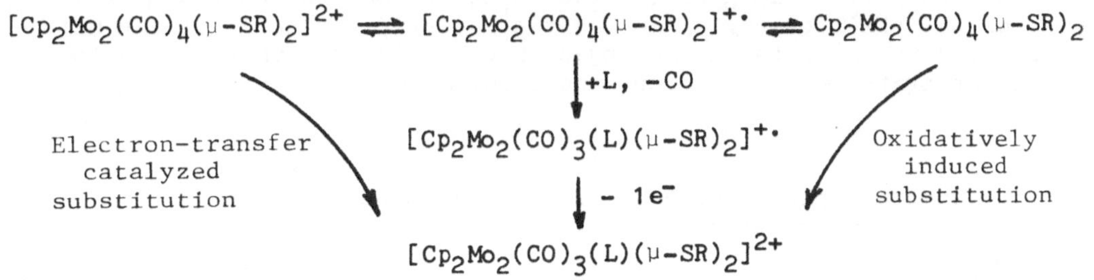

However, substitution of CO by solvent also occurs, and the origin of the product is thought to be an intermediate radical cation. Neither the neutral dimer nor the dication $[Cp_2Mo_2(CO)_4(\mu\text{-}SR)_2]^{2+}$ in **Scheme I**

undergoes thermal or photochemically activated substitution. However, chemical oxidation of $Cp_2Mo_2(CO)_4(\mu\text{-}SR)_2$ with two equivalents of Ag^+ ion leads to the solvent-substituted dication $[Cp_2Mo_2(CO)_3(L)(\mu\text{-}SR)_2]^{2+}$, a product of an oxidatively-induced substitution reaction. On the other hand, electrochemical reduction of $[Cp_2Mo_2(CO)_4(\mu\text{-}SR)_2]^{2+}$ leads to an approximately equal mixture of $[Cp_2Mo_2(CO)_3(L)(\mu\text{-}SR)_2]^{2+}$ and the parent complex $Cp_2Mo_2(CO)_4(\mu\text{-}SR)_2$ in an electron-transfer catalyzed process.

Scheme I, and the redox potentials of the species involved, imply that the following reactions can occur in solution once the intermediate radical cation $[Cp_2Mo_2(CO)_4(\mu\text{-}SR)_2]^{+\bullet}$ is formed:

(a) Reduction to the neutral parent.

$$[Cp_2Mo_2(CO)_4(\mu\text{-}SR)_2]^{+\bullet} + e^- \longrightarrow Cp_2Mo_2(CO)_4(\mu\text{-}SR)_2$$

(b) Reduction by the solvent-substituted radical cation, a better reducing agent than the unsubstituted radical cation.

$$[Cp_2Mo_2(CO)_4(\mu\text{-}SR)_2]^{+\bullet} + [Cp_2Mo_2(CO)_3(L)(\mu\text{-}SR)_2]^{+\bullet} \longrightarrow Cp_2Mo_2(CO)_4(\mu\text{-}SR)_2 + [Cp_2Mo_2(CO)_3(L)(\mu\text{-}SR)_2]^{2+}$$

(c) Chain-carrying electron transfer in the ETC mechanism.

$$[Cp_2Mo_2(CO)_4(\mu\text{-}SR)_2]^{2+} + [Cp_2Mo_2(CO)_3(L)(\mu\text{-}SR)_2]^{+\bullet} \longrightarrow [Cp_2Mo_2(CO)_4(\mu\text{-}SR)_2]^{+\bullet} + [Cp_2Mo_2(CO)_3(L)(\mu\text{-}SR)_2]^{2+}$$

(d) Oxidation of the intermediate, substituted radical cation.

$$[Cp_2Mo_2(CO)_3(L)(\mu\text{-}SR)_2]^{+\bullet} \longrightarrow [Cp_2Mo_2(CO)_3(L)(\mu\text{-}SR)_2]^{2+} + e^-$$

Reactions (a) and (b) above are responsible for formation of the neutral, unsubstituted dimer, but only (b) can be operative when the electrode is turned off; both steps are responsible for chain-termination in the ETC mechanism. Step (d) was not found to contribute significantly.

2. OXIDATIVELY INDUCED SUBSTITUTION OF ANIONIC DONOR LIGANDS

The oxidative electrochemistry of a number of transition metal alkyls and aryls have been investigated, and the earlier papers are outlined in our previous review.[1] In general, the oxidation of such compounds leads to loss of the alkyl or aryl group as the radical, the metal in the metal-containing fragment being unchanged in oxidation number (or in a lower oxidation state if two R groups are eliminated).

$$M\text{-}R \longrightarrow M^+ + R^\bullet + e^-$$

We have used the phrase **OXIRE** (oxidatively induced reductive elimination) to describe such processes.

2.1. Metal Alkyls and Aryls with Nitrogen Ligands.

One of the studies we described in our previous review concerned the oxidation of the series of compounds *cis*-$(bipy)_2FeR_2$.[19] Oxidation of the iron to either the +2 or +3 state led to the R groups appearing as dialkyls (R-R), alkanes (RH), or olefins (R-H), a result best interpreted in terms of the production of R initially as a free radical. More recently it has been found that chemical oxidation of a similar compound, the Co(III) complex *cis*-$[(bipy)_2CoR_2]^+$ (R = Me, Et),[20] gives *exclusively* R_2 and $[(bipy)_2Co]^{2+}$.

$$cis\text{-}[(bipy)_2CoR_2]^+ \longrightarrow [(bipy)_2Co]^{2+} + 1/2\ R\text{-}R + e^-$$

This parallels the oxidatively induced reductive elimination of R groups from $[cis\text{-}(bipy)_2FeR_2]^{2+}$, where iron has an oxidation number of +3. On the other hand, it is in contrast with other work on cobalt dialkyl complexes where the alkyl groups were located *trans* to one another.[21] For example, when Co(III) complexes of the type *trans*-L_4CoMe_2 were oxidized by one electron (L_4 is a macrocyclic ligand), only one alkyl group is lost to give a good yield of ethane, the hydrocarbon being formed by an *inter*molecular process in which methyl radicals from two separate organocobalt complexes are dimerized. In a good hydrogen-donor solvent, however, methane is formed by H atom abstraction from the solvent.

The usual mechanism for an outer-sphere electron transfer reaction was suggested for the electron transfer reactions of the *cis*-dialkyl cobalt(III) complex:[21,22] (a) Formation of a precursor complex ({*cis*-$[(bipy)_2CoR_2]^+$/oxidant}); (b) Activation of the precursor, electron transfer, and relaxation of the successor complex ({*cis*-$[(bipy)_2CoR_2]^{2+}/ox^{-\bullet}$}; (c) Dissociation of the successor complex to give separated products. Step (c) was suggested as the rate determining step in the process. In this regard, the calculated rate constant for the step ($\log k_d > 6.2$) was found to be significantly greater than that for *trans*-L_4CoMe_2 when oxidized by $[IrCl_6]^{2-}$ ($\log k_d = 5.0$). Thus, a pair of alkyl groups is eliminated much more rapidly than the homolytic cleavage of one alkyl group in oxidized *trans*-L_4CoMe_2.

A recent examination of the oxidative dealkylation of alkylcobalamins with I_2 strongly supports an electron transfer mechanism.[23] Using aqueous iodine, the reaction stoichiometry for methylcobalamin was as follows,

$$Me\text{-}B_{12} + I_2 + H_2O \longrightarrow [H_2O\text{-}B_{12}]^+ + Me\text{-}I + I^-$$

while, in the presence of excess Cl^-, the reaction gave exclusively methyl chloride instead of methyl iodide.

$$Me\text{-}B_{12} + I_2 + H_2O + Cl^- \longrightarrow [H_2O\text{-}B_{12}]^+ + Me\text{-}Cl + 2\ I^-$$

The rate law for the oxidative Co-C cleavage reaction was first order in both $Me\text{-}B_{12}$ and I_2. To account for this, and other experimental observations, the following reaction scheme, which owes much to the earlier work of Halpern and co-workers,[24] was proposed:

$$Me\text{-}B_{12} + I_2 \underset{k_{-1}}{\overset{k_1}{\rightleftharpoons}} [Me\text{-}B_{12}]^{+\cdot} + I_2^{+\cdot}$$

$$[Me\text{-}B_{12}]^{+\cdot} + I^- \xrightarrow{k_2} Me\text{-}B_{12} + I^\cdot$$

$$[Me\text{-}B_{12}]^{+\cdot} + X^- \xrightarrow{k_3} B_{12r} + Me\text{-}X$$

$$B_{12r} + I_2 + H_2O \xrightarrow{rapid} [H_2O\text{-}B_{12}]^+ + I_2^{-\cdot}$$

$$I^\cdot + I^- \xrightarrow{rapid} I_2^{-\cdot} \xrightarrow{rapid} 1/2\ I_2 + I^-$$

The one study of the anodic oxidation of $Me\text{-}B_{12}$ that has been done showed that the oxidation of the compound occurs in a highly irreversible manner with $E_{p,anodic} = +0.87$ V (vs. SCE).[25] Since the reduction potential for I_2 is +0.08 V (vs. SCE), the net potential for the oxidation of $Me\text{-}B_{12}$ is very negative. Therefore, why is an electron transfer mechanism apparently operative in the demethylation of alkyl cobalamins? To rationalize this, a charge transfer complex between I_2 and $Me\text{-}B_{12}$ is postulated.[26,27] Electron transfer occurs within this complex, the driving force being the formation of the ion pair.

$$Me\text{-}B_{12} + I_2 \xrightarrow{K_{CT}} \{Me\text{-}B_{12}:I_2\} \xrightarrow{k_{et}} \{Me\text{-}B_{12}^{+}, I_2^{-\cdot}\} \longrightarrow \text{products}$$

The electron transfer step is irreversible because it is followed by the rapid, nucleophilic displacement by halide ions or by the reduction of $[Me\text{-}B_{12}]^+$ by I^- ion.

The oxidative behavior of methylcobalamin contrasts with aryl-iron porphyrin complexes.[28] For either (OEP)FePh or (TPP)FePh [OEP = octaethylporphyrin; TPP = tetraphenylporphyrin], it is suggested that the first step in the process is oxidation of Fe(III) to Fe(IV). This is apparently followed by cleavage of the Fe-C bond and migration of the phenyl group to a nitrogen atom of the porphyrin ligand; the iron atom in the resulting complex is thus assigned an oxidation number of +2. This cationic complex can then undergo a reversible oxidation to the dication containing Fe(III).

2.2. Metal Alkyls and Aryls with Cyclopentadienyl Ligands

The oxidative cleavage of Fe-R bonds in CpFe(CO)LR (Fp-R) depends on the nature of the oxidant and the presence of nucleophiles in solution.[29,30] Using copper salts as oxidants, or in electrochemical oxidation, the 17-electron species $[Fp\text{-}R]^{+\cdot}$ was most surely formed in the first, one-electron step. If the alkyl group R is susceptible to nucleophilic attack (Me, $PhCH_2$), and a good nucleophile is present (Cl^-, Br^-), an S_N2

process occurs with concomitant formation of the alkyl halide and Fp• radical. For example, using Cu(II) halides, the overall process is

$$CpFeLL'R + 2\ CuX_2 \longrightarrow CpFeLL'X + RX + 2\ CuX$$

It is thought that the Fe-C bond is cleaved in a one-electron oxidation reaction to give a 17-electron intermediate.

$$CpFeLL'R + CuX_2 \longrightarrow [CpFeLL'R]^{+\bullet} + X^- + CuX$$

Nucleophilic attack by X^- at the alpha carbon of R (where possible) apparently gives R-X and the radical species [CpFeLL'•].

$$[CpFeLL'R]^+ + X^- \longrightarrow [CpFeLL'^\bullet] + R\text{-}X$$

The latter then abstracts a halogen atom from a second molecule of CuX_2 to complete the process.

$$[CpFeLL'^\bullet] + CuX_2 \longrightarrow CpFeLL'X + CuX$$

On the other hand, when the alkyl group is less susceptible to nucleophilic attack, or in the absence of a good nucleophile, homolysis of the M-C bond to give an alkyl radical in a reductive elimination process is a competitive reaction path. Halogens, however, were thought to be two-electron oxidants, with the transfer occurring in one-electron steps. The first step is a one-electron transfer to give a solvent-caged ion-pair. The $[X_2]^-$ ion is still a good oxidant and transfers a second electron within the solvent cage.

$$FpR + X_2 \longrightarrow [FpR]^+[X_2]^- \longrightarrow [FpRX]^+[X]^- \longrightarrow \text{products}$$

When $CpFe(CO)LCH_3$ is oxidized in the presence CO or pyridine as a nucleophile, the results differ from those above: migratory insertion of CO into the M-alkyl bond occurs to give the metal acyl in an ETC process.[31] Further, there is a trillionfold increase in the equilibrium constant for the reaction of the 17-electron radical cation relative to the neutral, 18-electron parent. Quite recently, good evidence has been found that the reaction proceeds through a 19-electron intermediate.[32]

$$[CpFe(CO)(PPh_3)CH_3]^{+\bullet}\ (17\text{-e}) + Nu \longrightarrow [CpFe(CO)(PPh_3)(Nu)CH_3]^{+\bullet}\ (19\text{-e})$$

$$[CpFe(CO)(PPh_3)(Nu)CH_3]^{+\bullet}\ (19\text{-e}) \longrightarrow [CpFe(Nu)(PPh_3)(COCH_3)]^{+\bullet}\ (17\text{-e})$$

As is now clear, we find that many reactions of organometallic compounds actually involve electron transfer. One such reaction, and one that has led to many other studies, is the reaction of the trityl cation, Ph_3C^+, with the neutral tungsten(IV) complex $Cp_2W(CH_3)_2$.[33] Although the trityl cation is well-known as a hydride abstraction reagent, it behaved instead as a one-electron oxidizing agent in this case. Reaction of Cp_2WMe_2 and Ph_3CBF_4 in CH_2Cl_2 at 25°C gave the final product, the ethylene hydride, in 96% yield. That the reaction proceeds through the 17-electron cation was suggested when addition of Ph_3CPF_6 to the neutral dimethyl compound at -78°C gave a good yield of oxidized starting material $[Cp_2WMe_2]^+$.

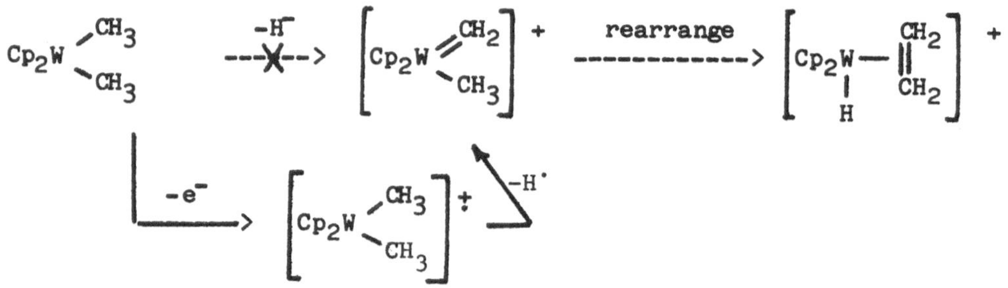

An unusual feature of the reaction scheme above is the abstraction of H from an α-carbon atom. Therefore, the compound $Cp_2W(CH_3)(C_2H_5)$, in which either α- or β-H atoms may be abstracted, was prepared. Reaction of the methyl-ethyl derivative with the trityl radical, $Ph_3C^{\cdot}$, in the presence of a phosphine gave, in good yield, a phosphonium ylid complex, $[Cp_2W(C_2H_5)(CH_2PR_3)]^+$. This product indicates that α-hydrogen abstraction occurs preferentially since the compound must arise as a result of the trapping of an intermediate methylidene complex $[Cp_2W(C_2H_5)CH_2]^+$ by phosphine.[34]

One method for the preparation of transition metal olefin complexes has been by β-hydride abstraction with trityl cation from an appropriate metal alkyl. However, when $CpRe(NO)L(CH_2CH_2R)$ is treated with Ph_3C^+, α-hydride abstraction occurs to give alkylidene complexes.[35]

$$CpRe(NO)(L)\text{-}CH_2CH_2R + Ph_3C^+ \longrightarrow CpRe(NO)(L)\text{=}CHCH_2R^+ + Ph_3CH$$

A thorough study of the mechanism of this process has revealed that the initial step involves a one-electron transfer followed by a rate-determining H-atom transfer between the radical species produced in the electron transfer step.[36] For example, for the benzyl complex,

$$(Re)\text{-}CH_2Ph + Ph_3C^+ \rightleftharpoons [(Re)\text{-}CH_2R]^{+\cdot} + Ph_3C^{\cdot}$$

$$[(Re)\text{-}CH_2R]^{+\cdot} + Ph_3C^{\cdot} \longrightarrow [(Re)\text{=}CHR]^+ + Ph_3C\text{-}H$$

the equilibrium constant for the first step was found to be 2.5 x 10^{-5}, a value 2000 times smaller than for the complex $CpRe(NO)L[CH_2C(Me)Ph]$.

It would now appear to be a general phenomenon that the trityl cation can act as a one-electron oxidant in its reactions with organometallic alkyls. Indeed, other examples include the α-hydride abstraction reaction of Ph_3C^+ with $(C_5Me_5)Fe(CO)_2(CH_2OH)$ and reactions of Ph_3C^+ with $Me_3M(CH_2)_3M'Me_3$ (where M is a Group 14 element).[37,38] Finally, both oxidatively catalyzed migratory CO insertion and β-elimination occur on reaction of Ph_3C^+ with CpFe(CO)2(2-norbornyl) complexes.[39]

Oxidation of a metal-benzyl complex, $Cp_2Zr(CH_2Ph)_2$, has led to <u>very</u> interesting results. The complex contains the metal in a formal oxidation state of 0, but oxidation clearly leads to Zr-C bond cleavage.[40] When ferricenium was used as the oxidant, benzylferrocene was observed as a minor product, the latter assumed to arise from coupling between ejected benzyl radicals and $[Cp_2Fe]^+$.

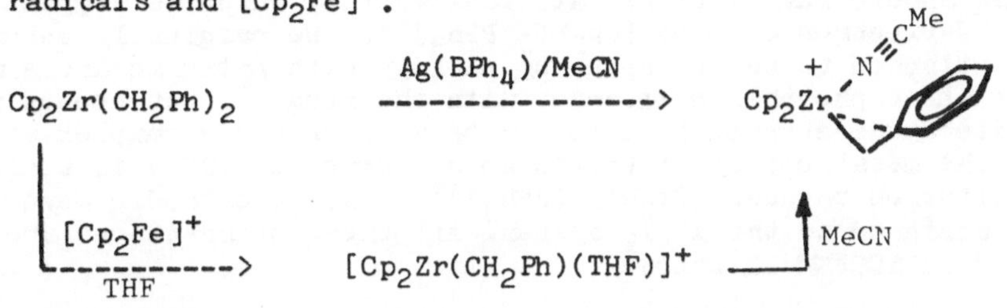

Although this was only a preliminary report, it would be interesting to investigate the reaction using Ph_3C^+ or anodic oxidation. Further, in view of the fact that $(bipy)_2MR_2$ (M = Fe, Co) complexes generally give rise to a coupling of R-R groups (see above), it is interesting that oxidation of $Cp_2Zr(CH_2Ph)_2$ does not give bibenzyl. (In contrast, we have observed biferrocene as the product of the oxidation of a Ti(0) complex, Cp_2TiFc_2.[41])

2.3. Cyclopentadienyl Metal Halides

Cyclic voltammetry shows that compounds of the type Cp_2MX_2 (M = Mo, W; X = Cl, Br, I) are oxidized in a quasi-reversible manner at 0.5-0.55 V (vs. SCE) in CH_3CN.[41,42] Electrolysis at more anodic potentials, however, indicates a net 1-electron oxidation, where the products for X = I are $[Cp_2M(X)(MeCN)]^+$ and I_2. Since $[Cp_2M(X)(MeCN)]^{2+}$ has a reduction potential of about 1.1 V, one way in which the observed products can be formed is the following ETC process:

Chain initiation: $Cp_2MX_2 \longrightarrow [Cp_2MX_2]^{+\cdot} + e^-$

Substitution: $[Cp_2MX_2]^{+\cdot} + MeCN \longrightarrow [Cp_2M(X)(MeCN)]^{2+} + X^-$

Chain propagation: $[Cp_2M(X)(MeCN)]^{2+} + Cp_2MX_2 \longrightarrow [Cp_2M(X)(MeCN)]^+ + [Cp_2MX_2]^{+\cdot}$

X^- oxidation: $X^- \longrightarrow 1/2\ X_2 + e^-$

To study this mechanism,[43] we have used $[(bipy)_3Ru]^{3+}$ as a chemical oxidant. With a stoichiometric deficiency of the Ru(III) complex [Ru(III):Cp_2MX_2 = 1:5], we find a rapid color change occurs when Cp_2MX_2 and $[(bipy)_3Ru]^{3+}$ are mixed; this is followed by a slower substitution reaction, where approximately one-half of the starting material is transformed into $[Cp_2M(X)(MeCN)]^+$ in 25 minutes.

To further probe the reaction, we studied CH_2Cl_2 solutions of Cp_2MX_2 and PPh_3. For X = I, the cyclic voltammogram of such solutions showed the following features: (a) The anodic wave for Cp_2MX_2 was elongated, shifted anodically, and showed reduced current for the cathodic portion. Such features indicate the participation of $[Cp_2MX_2]^{+\bullet}$ in a catalytic process (presumably the oxidation of PPh_3 to give $[PPh_3]^{+\bullet}$, which reacts with traces of water to give the observed products $Ph_3P{=}O$ and Ph_3PH^+.)[44] (b) A new anodic wave appeared at about +0.7 V, which we suggest arises from the 19-electron complex $[Cp_2MX_2 \cdot PPh_3]^{+\bullet}$. We originally assumed the PPh_3 was attached to the Cp ring, by analogy with cationic arene metal complexes where phosphines interact with the ring,[45] but recent results in the literature suggest that it may be a 19-electron complex with PPh_3 bound to the metal center.[46] (c) An anodic wave at 1.05 V is assigned to the substitution product $[Cp_2M(X)(PPh_3)]^+$. (d) A cathodic wave at about 0.2 V is assigned to the I^-/I_2 system. All these observations are summarized in **SCHEME II** below.

SCHEME II

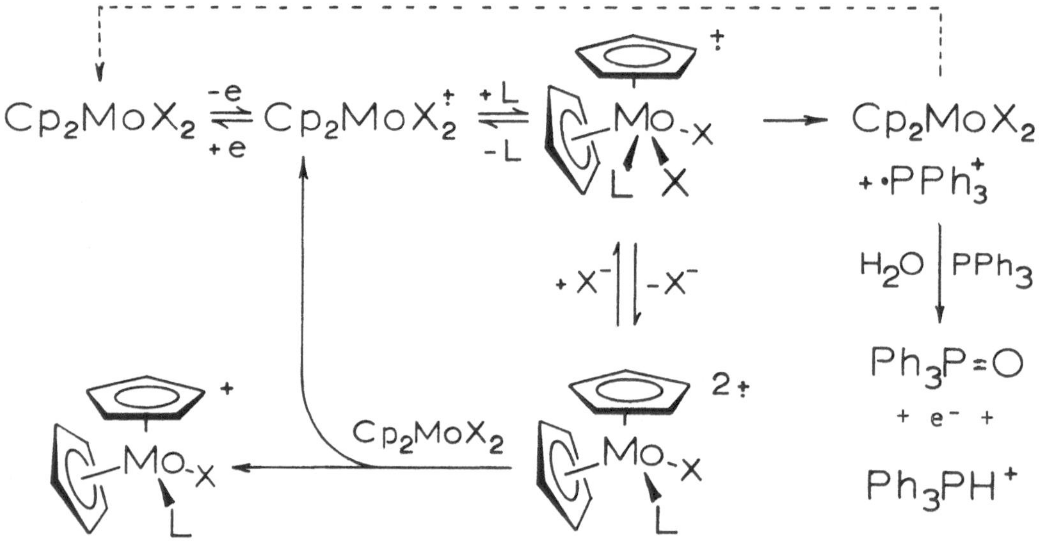

Finally, it must be noted that the presence of PPh_3 and I_2 (the latter from anodic oxidation of I^-, itself a product of the substitution reaction; Scheme II) in the same solution leads to further complications. We have observed that the phosphine severely affects the electrochemical behavior of I^-, presumably because PPh_3 and I_2 give $(PPh_3I)I_3$.[47] Thus, we believe the catalytic current we observe in solutions of Cp_2MI_2 and PPh_3 may be due in part to the decomposition of such phosphine/iodine complexes in the presence of water.

3. ACKNOWLEDGEMENTS

We wish to thank the American Chemical Society-Petroleum Research Fund for the support of our research. The United States Information Agency and the Fulbright Commission also provided some travel funds. Finally, I thank my co-workers, especially Rute Costa, Helena Garcia, Alberto Dias, Emmanual Quaye, and Laura Pence for their efforts in this area.

REFERENCES

1. Kotz, J.C. in Topics in Organic Electrochemistry, Fry, A.J.; Britton, W.E. (eds.) Plenum Press, New York, 1986.
2. Chanon, M. Acc. Chem. Res. **1987**, 20, 214.
3. The following references refer to examples of anodically induced reactions that are not covered in the present article: **Isomerization:** (a) Bond, A.M.; Colton, R.; Kevekordes, J.E.; Panagiotidou, P. Inorg. Chem. **1987**, 26, 1430. (b) Vallat, A.; Person, M.; Roullier, L.; Laviron, E. Inorg. Chem. **1987**, 26, 332. (c) Connelly, N.G.; Raven, S.J.; Carriedo, G.A.; Riera, V. J. Chem. Soc. Chem. Commun. **1986**, 992. **Rearrangement:** Samuel, E.G.; Kochi, J.K. J. Am. Chem. Soc. **1986**, 108, 4790. **Reactions of coordinated ligands:** Freeman, M.J.; Orpen, A.G.; Connelly, N.G.; Raven, S.J. J. Chem. Soc. Dalton Trans. **1985**, 2283.
4. Other reviews on oxidatively induced reactions are: (a) Daub, G. Prog. Inorg. Chem. **1977**, 22, 409. (b) Halpern, J. Angew. Chem. Int. Ed. Engl. **1985**, 24, 274.
5. Doxsee, K.M.; Grubbs, R.H.; Anson, F.C. J. Am. Chem. Soc. **1984**, 106, 7819.
6. For example see: (a) Narayanan, B.A.; Amatore, C.; Kochi, J.K. Organometallics **1987**, 6, 129. (b) Lane, G.A.; Geiger, W.E.; Connelly, N.G. J. Am. Chem. Soc. **1987**, 109, 402. (c) Pearson, A.J.; Chen, Y-S.; Daroux, M.L.; Tanaka, A.A.; Zettler, M. J. Chem. Soc. Chem. Commun. **1987**, 155. (d) Richmond, M.G.; Kochi, J.K. Inorg. Chem. **1986**, 25, 656.
7. (a) Connelly, N.G.; Demidowicz, Z.; Kelly, R.L. J. Chem. Soc. Dalton Trans. **1975**, 2335. (b) Degrand, C.; Radecki-Sudre, A.; Besancon, J. Organometallics **1982**, 1, 1311. (c) Rieke, R.D.; Tucker, I.; Milligan, S.N.; Wright, D.R.; Willeford, B.R.; Radonovich, L.J.; Eyring, M.W. Organometallics **1982**, 1, 938.
8. Domaille, P.J.; Ittel, S.D.; Jesson, J.P.; Sweigart, D.A. J. Organometal. Chem. **1980**, 202, 191.
9. Relevant to this is a paper describing the substitution of arene in $[(arene)FeCp]^+$ by PR_3 by reductive electron transfer catalysis. Darchen, A. J. Chem. Soc. Chem. Commun. **1983**, 768.
10. In this connection, it is worth noting that electrochemical observations suggesting solvent interactions with the ligands in metal carbonyls have been reported: Chadwick, I.; Diaz, C.; Gonzalez, G.; Santa Ana, M.A.; Yutronic, N. J. Chem. Soc. Dalton Trans. **1986**, 1867.
11. (a) Hershberger, J.W.; Amatore, C.; Kochi, J.K. J. Organometal. Chem. **1983**, 250, 345; and references therein. (b) Hershberger, J.W.; Klinger, R.J.; Kochi, J.K. J. Am. Chem. Soc. **1983**, 105, 61. (c) Zizelman, P.M.; Amatore, C.; Kochi, J.K. J. Am. Chem. Soc. **1984**, 106, 3771.

12. (a) Shi, Q-Z.; Richmond, T.G.; Trogler, W.C.; Basolo, F. J. Am. Chem. Soc. **1984**, 106, 71. (b) Kowalski, R.M.; Basolo, Trogler, W.C.; Gedridge, R.W.; Newbound, T.D.; Ernst, R.D. ibid., **1987**, 109, 4860.
13. Rerek, M.E.; Basolo, F. J. Am. Chem. Soc. **1984**, 106, 5908; and references therein.
14. Associatively activated substitution of 17-electron molecules and ions contrasts with 18-electron metal carbonyls such as $Ni(CO)_4$, where the evidence favors an I_d mechanism for CO substitution. Atwood, J.D. Inorganic and Organometallic Reaction Mechanisms, Brooks/Cole Publishing Company, Monterey, CA, 1985, pp.106-118.
15. (a) Broadley, K.; Connelly, N.G.; Geiger, W.E. J. Chem. Soc. Dalton Trans. **1983**, 121. (b) Connelly, N.G.; Raven, S.J. ibid. **1986**, 1613. (c) Gennett, T.; Grzeszczyk, E.; Jefferson, A.; Sidur, K.M. Inorg. Chem. **1987**, 26, 1856.
16. Therien, M.J.; Ni, C-L.; Anson, F.C.; Osteryoung, J.G.; Trogler, W.C. J. Am. Chem. Soc. **1986**, 108, 4037.
17. Bagchi, R.N.; Bond, A.M.; Brain, G.; Colton, R.; Henderson, T.L.E.; Kevekordes, J.E. Organometallics **1984**, 3, 4.
18. (a) Moulton, R.; Weidman, T.W.; Vollhardt, K.P.C.; Bard, A.J. Inorg. Chem. **1986**, 25, 1846. (b) Lacombe, D.A.; Anderson, J.E.; Kadish, K.M. Inorg. Chem. **1986**, 25, 2074. (c) Kadish, K.M.; Lacombe, D.A.; Anderson, J.E. Inorg. Chem. **1986**, 25, 2246. (d) Gross, R.; Kaim, W. Inorg. Chem. **1986**, 25, 498. (e) Connelly, N.G.; Finn, C.J.; Freeman, M.J.; Orpen, A.G.; Stirling, J. J. Chem. Soc. Chem. Commun. **1984**, 1025. (f) Legzdins, P.; Wassink, B. Organometallics, **1984**, 3, 1811. (g) Connelly, N.G.; Payne, J.D.; Geiger, W.E. J. Chem. Soc. Dalton Trans. **1983**, 295. (h) Connelly, N.G.; Raven, S.J.; Geiger, W.E. J. Chem. Soc. Dalton Trans. **1987**, 467. (i) Zhuang, B.; McDonald, J.W.; Schultz, F.A.; Newton, W.E. Organometallics **1984**, 3, 943. (j) Rosenhein, L.D.; Newton, W.E.; McDonald, J.W. Inorg. Chem. **1987**, 26, 1695. (k) Gueguen, M.; Guerchais, J.E.; Petillon, F.Y.; Talarmin, J. Chem. Soc. Chem. Commun. **1987**, 557; and references therein. (l) McDonald, J.W. Inorg. Chem. **1985**, 24, 1734. (m) Broadley, K.; Connelly, N.G.; Lane, G.A.; Geiger, W.E. J. Chem. Soc. Dalton Trans. **1986**, 373. (n) Connelly, N.G.; Garcia, G.; Gilbert, M.; Stirling, J.S. J. Chem. Soc. Dalton Trans. **1987**, 1403.
19. Lau, W.; Huffman, J.C.; Kochi, J.K. Organometallics, **1982**, 1, 155
20. Fukuzumi, S.; Ishikawa, K.; Tanaka, T. J. Chem. Soc. Dalton Trans. **1985**, 899.
21. Tamblyn, W.H.; Klinger, R.J.; Hwang, W.S.; Kochi, J.K. J. Am. Chem. Soc. **1981**, 103, 3161.
22. Purcell, K.F.; Kotz, J.C. Inorganic Chemistry, Saunders College Publishing, Philadelphia, 1977, p. 660.
23. Fanchiang, Y-T. Organometallics, **1985**, 4, 1515.
24. Topich, J.A.; Halpern, J. Inorg. Chem. **1979**, 18, 1339 and references therein.
25. Rubinson, K.A.; Itabashi, E.; Mark, H.B., Jr. Inorg. Chem. **1982**, 21, 3571.
26. Fukuzumi, S.; Kochi, J.K. J. Am. Chem. Soc. **1980**, 102, 2141.
27. Kochi, J.K. Pure Appl. Chem. **1980**, 52, 571.

28. Lancon, D.; Cocolios, P.; Guilard, R.; Kadish, K. J. Am. Chem. Soc. **1984**, 106, 4472; Organometallics **1984**, 3, 1164.
29. Rogers, W.N.; Page, J.A.; Baird, M.C. Inorg. Chem. **1981**, 20, 3521.
30. Since ruthenium compounds often accommodate higher oxidation states, Fp-R and its ruthenium analog were compared. (Joseph, M.F.; Page, J.A.; Baird, M.C. Organometallics, **1984**, 3, 1749.) However, aside from the fact that the Ru(II) complexes were more difficult to oxidize than their iron-containing analogs, oxidative reactions of the Ru(II) complexes gave results very similar to the iron compounds.
31. Magnuson, R.H.; Meirowitz, R.; Zulu, S.J.; Giering, W.P. Organometallics, **1983**, 2, 460, and references therein.
32. Therien, M.J.; Trogler, W.C. J. Am. Chem. Soc. **1987**, 109, 5127.
33. Hayes, J.C.; Cooper, N.J. J. Am. Chem. Soc. **1982**, 104, 5570; Hayes, J.C.; Cooper, N.J. Organometallic Compounds: Synthesis, Structure, and Theory, Texas A and M University Press, College Station, Texas, 1983, p.353.
34. For similar results on another Cp_2WR_2 complex, see Chong, K.S.; Green, M.L.H. Organometallics, **1982**, 1, 1586.
35. Bodner, G.S.; Gladysz, J.A.; Nielsen, M.F.; Parker, V.D. J. Am. Chem. Soc. **1987**, 109, 1757, and references therein.
36. Bodner, G. S.; Gladysz, J.A.; Nielsen, M.F.; Parker, V.D. Organometallics **1987**, 6, 1628. See also Asaro, M.F.; Bodner, G.S.; Gladysz, J.A.; Cooper, S.R.; Cooper, N.J. ibid, **1985**, 4, 1020.
37. Guerchais, V.; Lapinte, C. J. Chem. Soc. Chem. Commun. **1986**, 663.
38. Traylor, T.G.; Koermer, G.S. J. Org. Chem. **1981**, 46, 3651.
39. Bly, R.S.; Silverman, G.S.; Bly, R.K. Organometallics **1985**, 4, 374.
40. Jordon, R.F.; LaPointe, R.E.; Bajgur, C.S.; Echols, S.F.; Willett, R. J. Am. Chem. Soc. **1987**, 109, 4111.
41. Kotz, J.C.; Vining, W.; Coco, W.; Rosen, R.; Dias, A.R.; Garcia, M.H. Organometallics, **1983**, 2, 68.
42. Asaro, M.F.; Cooper, S.R.; Cooper, N.J. J. Am. Chem. Soc. **1986**, 108, 5187.
43. Costa, R.; Dias, A.R.; Kotz, J.C.; Quaye, E. unpublished research.
44. The production of $Ph_3P{=}O$ has been observed by others in solutions containing oxidized organometallics, but the explanation has been that the oxide is formed from traces of oxygen (Bagchi, R.N.; Bond, A.M.; Heggie, C.L.; Henderson, T.L.; Mocellin, E.; Seikel, R.A. Inorg. Chem. **1983**, 22, 3007). We believe the following reaction is more likely:

$$3\ PPh_3 + H_2O \longrightarrow Ph_3P{=}O + 2\ Ph_3PH^+ + 2\ e^-$$

See Schiavon, G.; Zecchin, S.; Cogoni, G; Bontempelli, G. Electroanal. Chem. Interfac. Electrochem. **1973**, 48, 425.
45. See ref. 8 and Aviles, T.; Royo, P. J. Organometal. Chem. **1981**, 221, 333.
46. 19-electron complexes were observed by Doxsee, et al. (ref.5); Broadley, et al. (ref. 15a); and Therien and Trogler (ref. 32).
47. Cotton, F.A.; Kibala, P.A. J. Am. Chem. Soc. **1987**, 109, 3308.

COMPETITIVE REACTIVITIES OF PARAMAGNETIC ORGANOMETALLIC INTERMEDIATES

Theodore L. Brown* and Richard J. Sullivan
School of Chemical Sciences
University of Illinois, Urbana-Champaign
505 S. Mathews Avenue
Urbana, Illinois 61801 USA

ABSTRACT. The competitive reactivities of 17-e and 19-e metal carbonyl radicals in atom transfer reactions are analyzed in terms of the Marcus/Agmon-Levine equation. The intrinsic barriers to atom transfer are low for rhenium carbonyl radicals in halogen atom transfer reactions. Assuming similarly low barriers for other radical species, the analysis suggests that, where 17-e and 19-e species are in equilibrium, the reactivity of the (usually) more abundant 17-e radical will dominate. However, interconversion between 17-e and 19-e radicals seldom obtains, and the mode of radical formation is usually critical.

1. INTRODUCTION

During the past several years a great deal of information and new insights have accumulated regarding the reactivities of paramagnetic organometallic intermediates. It has become clear that these species are potentially important in a variety of reaction settings. Their importance stems from their high reactivities for several different kinds of reactions, including substitution, atom and electron transfer. It has become increasingly evident that species with differing coordination environment at the metal center may be involved in these kinds of reactions. It is common to speak of 17-electron (17-e) and 19-electron (19-e) radicals in reference to a formal valence shell electron count at the metal. (For example, $Mn(CO)_4P(Me)_3$ is a 17-e radical, $Mn(CO)_4[P(Me)_3]_2$ is 19-e.) Under a given set of reaction conditions one or the other of these inter-related species is likely to be kinetically dominant. In this contribution we present an analysis of the relative reactivities of 17-e and 19-e species in reactions of metal carbonyl radicals, with application to atom transfer processes.

M. Chanon et al. (eds.), Paramagnetic Organometallic Species in Activation / Selectivity, Catalysis, 187–200.

2. FORMATION AND PROPERTIES OF METAL CARBONYL RADICALS

2.1. Formation Processes

Metal carbonyl radicals can be formed in several kinds of reactions; the most important are electrochemical or chemical reduction or oxidation of 18-electron species, photochemical or thermal homolysis of metal-metal bonds in dinuclear metal carbonyls or atom abstraction processes. Depending on the mode of formation, the radical initially produced may be 17-e or 19-e in character. Some examples follow:

17-electron:

(a) $Mn_2(CO)_{10} \xrightarrow{h\nu} 2Mn(CO)_5\cdot$ [1-4]

(b) $Cr(CO)_6 \longrightarrow Cr(CO)_6^{+}\cdot + e^-$ [5]

(c) $HRe(CO)_5 + Q \longrightarrow Re(CO)_5\cdot + QH$ [6]

(d) $\eta^5\text{-}CH_3C_5H_4Mn(CO)_2L \longrightarrow \eta^5\text{-}CH_3C_5H_4Mn(CO)_2L^{+}\cdot + e^-$ [7]

19-electron:

(a) $Cr(CO)_6 + e^- \longrightarrow Cr(CO)_6^{-}\cdot$ [5,8,9]

(b) $Mn(CO)_2(CN\text{-}t\text{-}Bu)_2(N\text{-}N)^+ + Na/Hg \longrightarrow Mn(CO)_2\text{-}$
$(CN\text{-}t\text{-}Bu)_2(N\text{-}N)\cdot$ [10]

(c) $HCo(CO)_4 + e^- \longrightarrow HCo(CO)_4^-$ (low temp. matrix) [11]

The radicals, once formed, are generally quite reactive, and rapidly react via any one of several pathways, depending on solution conditions. However, examples exist of reasonably stable, persistent radicals. For the most part these are of the 17-e variety (L = phosphine or phosphite):

$Mn(CO)_3L_2\cdot$ [12] $Re(CO)_3[P(C_6H_{11})_3]_2\cdot$ [13] $CpCr(CO)_2PPh_3\cdot$ [14]

$CpCoL_2^{+}\cdot$ [15]

These 17-e radicals are stabilized by steric crowding at the metal center, which prevents dimerization or other reaction with the medium, or by a delocalization of the SOMO, thereby reducing reactivity.

Persistent radical species which can be viewed formally as 19-e are known.[16] However, for the most part these are best viewed as 18-e at the metal center; the unpaired spin density is largely distributed in the ligand system. Examples of this type include the ortho-quinone complexes of manganese and rhenium carbonyls.[17]

In the absence of an extraordinary stabilizing factor of the kind just described, the radicals are capable of undergoing reaction via recombination, substitution, atom transfer or electron transfer[4,18,19] Of particular importance to an understanding of the reactivities of the radicals is knowledge of the comparative reactivities of 17-e and 19-e species, which are related via the addition or elimination of a two-electron donor group;[16] e.g.,

$$C_5H_5Mo(CO)_3\cdot + L \longrightarrow C_5H_5Mo(CO)_3L\cdot$$ [20]

$$(arene)W(CO)_3{}^{+}\cdot + L \longrightarrow (arene)W(CO)_3L^{+}\cdot$$ [21]

$$Mn(CO)_5\cdot + PPh_3 \longrightarrow Mn(CO)_5PPh_3\cdot$$ [22]

In many reaction situations, both the 17-e and 19-e radicals are capable of undergoing reaction. These reactions may be different; for example, $Mn(CO)_5$ might be formed under conditions under which it could undergo reduction or dimerization, or add a phosphine to form the 19-e species which might undergo a fast hydrogen atom abstraction.[22] In other circumstances the 17-e and 19-e species may be in competition for reaction with a single substrate, e.g., a halogen atom donor, or an electron acceptor. The relative rate constants for reaction of the two radicals can be expected to vary, depending on the kind of reaction involved. Of critical importance to the behavior of the system is the rapidity with which the 17-e and 19-e species interconvert, relative to the rates of any other reactions the radicals might undergo.

2.2. Substitution Processes

It has been evident since the earliest work involving metal carbonyl radicals that the 17-e species can be very labile toward substitution. In the first demonstrated example of a radical chain substitution process, in the reaction $HRe(CO)_5 + L \rightarrow HRe(CO)_4L + HRe(CO)_3L_2$, rapid substitution of the $Re(CO)_5\cdot$ radical intermediates by L was postulated as a key step.[6] Since then the lability of metal carbonyl radicals has been shown to be quite general. Poë was the first to provide direct evidence that substitution of the 17-e radicals generally occurs via an associative process.[23] Examples of some bimolecular rate constants are shown in Table I. The results of the various studies show that (a) the substitution process can be very rapid; for the binary carbonyl radicals, the rate constants approach the diffusion limit, and (b) in all cases where such effects have been evaluated the substitution process is sensitive to the steric and electronic characteristics of the entering ligand.

From the fact that substitution is an associative process we can infer that there exists a 19-e transition state or intermediate. Until recently, the evidence for intermediates has been based largely on the nature of the products formed in electron transfer processes involving radical intermediates.[27,28] The instances in which the evidence was most convincing were of two kinds: (a) One of the

TABLE I
Examples of rate constants for substitution at 17-e metal centers

Reaction	Rate Constant ($M^{-1}\ s^{-1}$)	Ref.
$Re(CO)_5\cdot + PPh_3 \longrightarrow Re(CO)_4PPh_3\cdot + CO$	1.1×10^9	23
$Mn(CO)_5\cdot + P(n\text{-}Bu)_3 \longrightarrow Mn(CO)_4P(n\text{-}Bu)_3\cdot + CO$	1.0×10^9	24
$Mn(CO)_5\cdot + AsPh_3 \longrightarrow Mn(CO)_4AsPh_3\cdot + CO$	6.5×10^4	24
$Mn(CO)_3[P(i\text{-}Bu)_3]_2\cdot + CO \longrightarrow Mn(CO)_4P(i\text{-}Bu)_3\cdot + CO$	0.32	25
$\eta^5\text{-}C_5H_4CH_3Mn(CO)_2(4\text{-}NO_2C_5H_4N)^+\cdot + PPh_3 \longrightarrow \eta^5\text{-}C_5H_4CH_3Mn(CO)_2PPh_3{}^+\cdot + 4\text{-}NO_2C_5H_4N$	1.6×10^1	7
$V(CO)_6\cdot + P(n\text{-}Bu)_3 \longrightarrow V(CO)_5P(n\text{-}Bu)_3\cdot + CO$	50	26

ligands in the 17-e radical is η^5-C_5H_5 or a related type, capable of variable hapticity at the metal center. The capacity of such ligands to affect the course of substitution processes by changing hapticity has been well documented for substitution at 18-e metal carbonyl centers.[29,30] (b) A multidentate ligand capable of intramolecular interaction to form a 19-e species is involved.[16,28a] In other than these two circumstances any equilibria between 17-e and 19-e forms of a metal carbonyl radical are likely to heavily favor the 17-e form. Nevertheless the 19-e radical can be kinetically dominant when reaction occurs immediately upon formation of the 19-e radical, and before ligand dissociation can occur, or when the rate constant for reaction of the 19-e form is sufficiently greater than that for the 17-e form. In this contribution we address the question of the competition between the 17-e and 19-e radicals for atom transfer reactions. The general principles delineated below apply also to electron transfer reactions, but the detailed considerations are sufficiently different to warrant a separate treatment.

3. THE MARCUS/AGMON-LEVINE MODEL FOR ATOM TRANSFER

3.1. Application to Competitive Atom Transfer Reactions

We have shown that the model originally put forward by Marcus,[31] and developed as a general phenomenological model by Agmon and Levine,[32] serves to relate a substantial body of experimental data for halogen atom transfer reactions of $Re(CO)_4L\cdot$ radicals.[33] This model should be applicable to both atom and electron transfer reactions of the radicals. Consider the atom transfer reaction in terms of the following equilibria:

$$ML_n\cdot + HX \underset{k_{21}}{\overset{k_{12}}{\rightleftharpoons}} \{L_nM\cdot\cdot HX\} \underset{k_{32}}{\overset{k_{23}}{\rightleftharpoons}} \{L_nMH\cdot\cdot X\} \xrightarrow{k_{30}} L_nMH + X\cdot \quad (1)$$

The overall rate constant for atom transfer, k_T, is expressed in the M/A-L model in terms of four parameters: $A = k_{21}/k^o_{23}$, which measures the fraction of precursor complex that undergoes atom transfer (k^o_{23} is the frequency factor for the atom transfer step); $B = k_{21}/k_{30}$, a measure of the efficiency of final product formation; the overall free energy change in the atom transfer step, ΔG_{23}, and the intrinsic barrier to the reaction, $\Delta G^\dagger(0)$, which measures the free energy barrier for the hypothetical ergoneutral atom transfer reaction. The free energy barrier for hydrogen atom transfer within the encounter complex, $\Delta G^\dagger_{23}$, is given by

$$\Delta G^*_{23} = \Delta G_{23} + \frac{\Delta G^*(0)}{\ln 2}\ln\left\{1 + \exp\left[\frac{-\Delta G_{23}\ln 2}{\Delta G^*(0)}\right]\right\} \quad (2)$$

This in turn leads to the following expression for the atom transfer rate constant:

$$\log k_T = \log k_{12} - \log\left[1 + A\exp(\Delta G_{23}/RT)\exp\left\{\frac{\Delta G^*(0)}{RT\ln 2}\ln\left[1 + \exp(-\Delta G_{23}\ln 2/\Delta G^*(0))\right]\right\} + B\exp(\Delta G_{23}/RT)\right] \quad (3)$$

(k_{12} is the rate constant for formation of the encounter complex.)

3.2. 17-e/19-e Competition

Our present purpose is to consider the competition between the 17-e $ML_n\cdot$ radical and the 19-e $ML_nL'\cdot$ radical for the same atom transfer substrate. It is useful to think in terms of a reaction series, in which a given radical reactant undergoes reactions with a series of substrates for which a common set of values of A, B and $\Delta G^\dagger(0)$ apply, and in which k_T varies as a function of ΔG_{23}. In the reactions of a 17-e species $ML_n\cdot$ with an atom transfer reagent, the transition state will involve metal-to-atom bond formation to an extent that varies with the overall free energy change in the reaction.[32] The 17-e radical may also be involved in an equilibrium with a nucleophile L' present in solution (L' may be the solvent itself), to form a formally 19-e species $ML_nL'\cdot$. This 19-e radical may also undergo reaction with the atom transfer reagent; in that reaction the transition state may involve the metal center in the same way as for the 17-e radical, or it may involve some other point of attack in the molecule.

The reaction system of interest is described by the reactions:

$$ML_n\cdot + L' \xrightarrow{k_1} ML_nL'\cdot \tag{4}$$

$$ML_nL'\cdot \xrightarrow{k_2} ML_n\cdot + L' \tag{5}$$

$$ML_n\cdot + HX \xrightarrow{k_3} HML_n + X\cdot \tag{6}$$

$$ML_nL'\cdot + HX \xrightarrow{k_4} HML_n + L' + X\cdot \tag{7}$$

It is assumed that the observed product of reaction of $ML_nL'\cdot$ is the same as that for reaction of $ML_n\cdot$. In fact the initial product of H atom transfer may be a formyl compound which decarbonylates to form the hydride. Let us assume that $ML_n\cdot$ and $ML_nL'\cdot$ are in equilibrium. Then,

$$K = k_1/k_2 = \frac{[ML_nL'\cdot]}{[ML_n\cdot][L']} \tag{8}$$

The overall rate of reaction is given by

$$\frac{d[HML_n\cdot]}{dt} = -\left\{\frac{d[ML_n\cdot]}{dt} + \frac{d[ML_nL'\cdot]}{dt}\right\}$$

$$= -[k_3 + K[L']k_4][HX][ML_n\cdot] \tag{9}$$

The terms k_3 and $K[L']k_4$ represent the relative contributions of $ML_n\cdot$ and $ML_nL'\cdot$, respectively, to the formation of hydride product.

Figure 1 shows graphs of k_3 or $K[L']k_4$ vs. ΔG_{23} for several assumed values of the relevant parameters (Table II). These graphs correspond to the assumption that the 19-e radical is in lower equilibrium concentration than the 17-e, and that the atom transfer process for a given atom donor is more exergonic for the 19-e radicals than for the 17-e. In all cases we have assumed that A = 1, B = 2.0. Variation in the relative value of A for the 19-e radical as compared with the 17-e has much the same effect as variation in the K[L'] term. The plots are not sensitive to the assumed value for B in the energy range of interest here.

An important and obvious conclusion to be drawn from Figure 1 is that, in the exergonic region, the 17-e radical is the kinetically dominant species, when it is present in significantly higher concentration. We do not know the relative values of $\Delta G^{\dagger}(0)$ for the 17-e and 19-e radicals. However, for reaction of $Re(CO)_4L\cdot$ radicals with a series of organic halides, $\Delta G^{\dagger}(0)$ is 5.0 kcal mol^{-1} for L = $P(i\text{-}O\text{-}Pr)_3$ as compared with 3.5 kcal mol^{-1} for $P(CH_3)_3$,[33] probably because of the larger steric requirement at the metal center for the former ligand. Further, in reactions of a series of $Re(CO)_4L\cdot$ radicals with a given halogen atom donor or with $HSn(Bu)_3$, the rate constants exhibit a dependence on the cone angle of L.[33,34] It is therefore likely that $\Delta G^{\dagger}(0)$ will be larger for the 19-e radical than for the 17-e when the atom transfer is directly to the metal for both radicals, Figure 1C

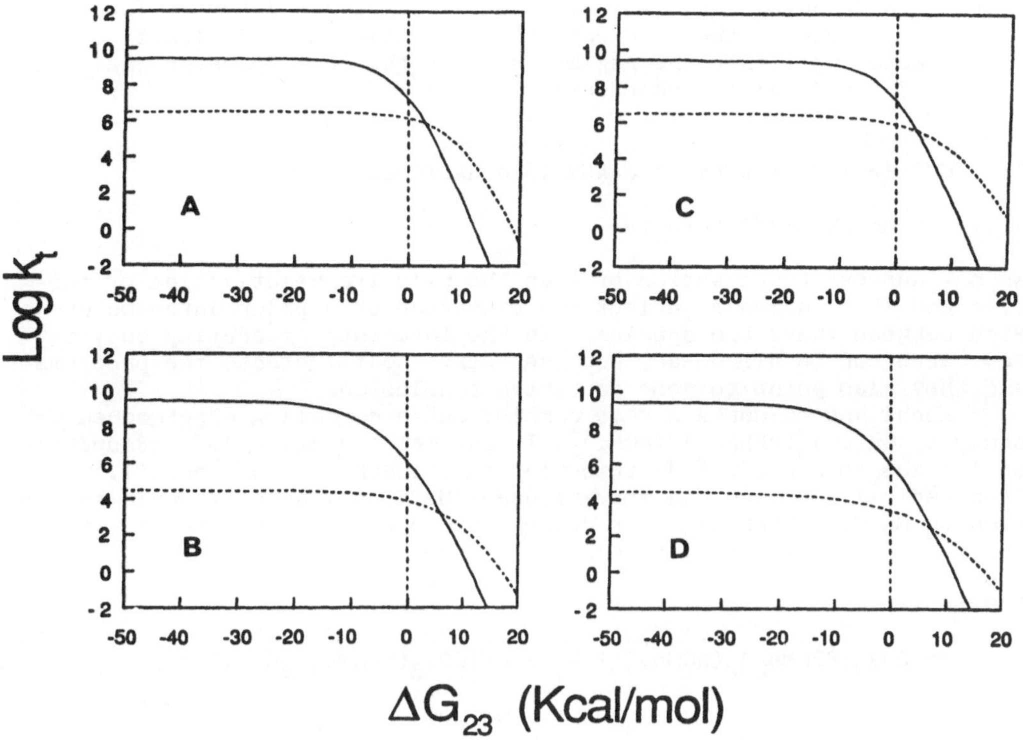

Figure 1. Relative reactivies of 17-e and 19-e radicals toward an atom transfer agent, as expressed by the Marcus/Agmon-Levine equation. For the 17-e radical $k_t = k_3$; for the 19-e radical $k_t = K[L']k_4$. The values chosen for the parameters of equation (3) are listed in Table II. The ΔG_{23} scale is that for the 17-e radical. The scale for the 19-e radical is obtained by adding $\Delta\Delta G$.

TABLE II
Atom transfer reaction parameters assumed for 17-e and 19-e radicals in Figure 1

Figure	$\Delta G^{\dagger}(0)$(17-e)[a]	$\Delta G^{\dagger}(0)$(19-e)[a]	K[L']	$\Delta\Delta G_{23}$[a,b]
A	3.5	3.5	1×10^{-3}	-11
B	5.0	5.0	1×10^{-5}	-15
C	3.5	5.0	1×10^{-3}	-15
D	5.0	7.0	1×10^{-5}	-20

a. kcal mol^{-1}
b. ΔG_{23}(19-e)-ΔG_{23}(17-e)

and 1D. We conclude that for reactions involving relatively low intrinsic barriers, the 17-e radical can remain the kinetically dominant species well into the regime in which the atom transfer reaction to the 17-e radical is endergonic.

4. EQUILIBRIUM BETWEEN 17-e AND 19-e RADICALS

4.1. Electrochemical Results

We now address the question of when the relative reactivities of the 17-e and 19-e radicals reflect the existence of a prior interconversion between these two species. In the interests of brevity only a few cases can be discussed, but they serve to illustrate the problems and they also point to some tentative conclusions.

Kochi and co-workers have carried out a revealing electrochemical study of $Mn(CO)_3(PPhMe_2)_2(NCMe)^+$, I, and related cations.[35] Reduction of I leads to a product distribution as illustrated in Scheme I. The 19-e radical can lose the solvent base, MeCN, followed by rapid reduction to $Mn(CO)_3(PPhMe_2)_2^-$, or undergo hydrogen atom transfer to carbon, yielding a formyl intermediate. The formyl rapidly

Scheme I:

$$Mn(CO)_3(PPhMe_2)_2(NCMe)^+ + e^- \rightarrow Mn(CO)_3(PPhMe_2)_2H \ (80\%) \tag{10}$$

$$\rightarrow Mn(CO)_3(PPhMe_2)_2^- \ (20\%) \tag{11}$$

decarbonylates to form the observed hydride product, observed in about 70-80 percent yield. In this case it appears that (a) hydrogen atom transfer to the 19-e radical dominates over loss of the most labile ligand as the first step in establishing a 19-e/17-e equilibrium. The 17-e radical, once formed, undergoes rapid reduction under electrochemical conditions; this process obviates any other chemistry that the 17-e species might exhibit. One might ask whether there is not a fast equilibrium between the 19-e and 17-e species involving loss and re-entry of MeCN, inasmuch as MeCN is the solvent medium. However, Kochi shows that the $Mn(CO)_3(PPhMe_2)_2^-$ anion undergoes reversible electrooxidation in acetonitrile.[35] If reaction (12) were facile,

$$Mn(CO)_3L_2\cdot + MeCN \xrightarrow{k_{12}} Mn(CO)_3L_2(NCMe)\cdot \tag{12}$$

the 19-e product should have undergone rapid hydrogen atom transfer, contrary to observation. One can estimate that k_{12} must be less than about 1 $M^{-1}\ s^{-1}$.

Examples of rapid hydrogen atom transfer reactions that follow immediately upon formation of 19-e $Mn(CO)_5L\cdot$ radicals (L = phosphine) have also been demonstrated.[36] On the other hand, Kochi and co-workers have argued for prior 17-e/19-e equilibria in advance of hydrogen atom transfer to a 19-e radical, e.g., in the constant current electroreduction of $Mn(CO)_6^+$, conducted in the presence of $PPhMe_2$:[36]

$$Mn(CO)_6^+ + PPhMe_2 \xrightarrow{e^-} HMn(CO)_3(PPhMe_2)_2 \quad (13)$$

The proposed steps in this process are shown in Scheme II:

Scheme II:

$$Mn(CO)_6^+ + e^- \longrightarrow Mn(CO)_6\cdot \quad (14)$$

$$Mn(CO)_6\cdot \longrightarrow Mn(CO)_5\cdot + CO \quad (15)$$

$$Mn(CO)_5\cdot + L \longrightarrow Mn(CO)_5L\cdot \quad (16)$$

$$Mn(CO)_5L\cdot \longrightarrow Mn(CO)_4L\cdot + CO \quad (17)$$

$$Mn(CO)_4L\cdot + L \longrightarrow Mn(CO)_4L_2\cdot \quad (18)$$

$$Mn(CO)_4L_2\cdot + HB \longrightarrow \{Mn(CO)_3L_2CHO\} \longrightarrow HMn(CO)_3L_2 + CO \quad (19)$$

$$2\ Mn(CO)_4L\cdot \longrightarrow Mn_2(CO)_8L_2 \quad (20)$$

$$Mn(CO)_4L\cdot + e^- \longrightarrow Mn(CO)_4L^- \quad (21)$$

A difficulty in interpreting these results is that the hydrogen atom source is not identified. It is probably $PPhMe_2$, present in 0.15M concentration. It is not clear that $Mn(CO)_6\cdot$ itself could not undergo fast atom transfer with an active donor. Reaction 22 would in that case be competitive with reaction 15:

$$Mn(CO)_6\cdot + HB \longrightarrow \{Mn(CO)_5CHO\} \longrightarrow HMn(CO)_5 + CO \quad (22)$$

Kochi et al. have observed an anodic return wave on CV reduction of $Mn(CO)_6^+$ at high scan rates,[22] from which one can estimate a half-life for CO loss from $Mn(CO)_6\cdot$ to be on the order of 4×10^{-5} s^{-1}. A rate constant of 10^6 M^{-1} s^{-1}[16] for H atom transfer to $Mn(CO)_6\cdot$ would make formation of hydride dominant over CO loss. However, assuming that CO loss, equation 15, dominates over atom transfer, addition of L to $Mn(CO)_5\cdot$ must compete with alternative pathways for loss of $Mn(CO)_5\cdot$, notably formation of $Mn_2(CO)_{10}$ or $Mn(CO)_5^-$. Reaction 16 is at least as rapid as the overall substitution of L for CO, for which rate constants are $1 \times 10^9 M^{-1}$ s^{-1} for $P(n\text{-}Bu)_3$ and 2×10^7 M^{-1} s^{-1} for PPh_3[24] (Table I). Thus, reaction 16 should indeed be competitive with the alternative reaction pathways for $Mn(CO)_5\cdot$.

To account for the quantitative formation of $HMn(CO)_3[PPhMe_2]_2$, Kochi et al. propose reaction of the 17-e species, $Mn(CO)_4L\cdot$, with a second molecule of L to form $Mn(CO)_4L_2\cdot$, reaction 18.[36] This species then undergoes hydrogen atom atom transfer. However, the limited data available suggest that a $Mn(CO)_4L_2\cdot$ 19-e radical will not undergo very fast hydrogen atom transfer. Evidence for this is seen in the behavior of $Mn(CO)_4[PPh_3]_2\cdot$, generated by reduction of the corresponding cation. Even in the presence of 0.08M $HSn(Bu)_3$, a very active hydrogen atom donor, loss of phosphine, followed by reduction of the $Mn(CO)_4PPh_3\cdot$ to form $Mn(CO)_4PPh_3^-$, accounts for a significant portion of the overall reaction under cv conditions.[9] This result is likely to be the consequence of both a rapid loss of L from $Mn(CO)_4L_2$ to form the 17-e species, and a low rate of hydrogen atom transfer to $Mn(CO)_4L_2\cdot$. In the reaction of the 17-e $Mn(CO)_3L_2$ radicals with $HSn(Bu)_3$, the rate constants for H atom transfer were found to decline rapidly with increasing cone angle of the phosphine L.[12b] As an alternative to portions of Scheme II we propose that formation of the disubstituted hydride occurs via a radical chain substitution process involving the monohydride, scheme III.

Scheme III:

$$Mn(CO)_6\cdot \longrightarrow Mn(CO)_5\cdot + CO \quad (15)$$

$$Mn(CO)_5\cdot + L \longrightarrow Mn(CO)_5L\cdot \quad (16)$$

$$Mn(CO)_5L\cdot + HB \longrightarrow \{Mn(CO)_4LCHO\} \longrightarrow HMn(CO)_4L + CO \quad (23)$$

$$Mn(CO)_4L\cdot + L \longrightarrow Mn(CO)_3L_2\cdot + CO \quad (25)$$

$$Mn(CO)_3L_2\cdot + HMn(CO)_4L \longrightarrow HMn(CO)_3L_2 + Mn(CO)_4L\cdot \quad (26)$$

Scheme III does not include all potential atom transfer and other reactions that might occur. The main idea is that radical chain substitution processes account for conversion of all hydrides formed to the disubstituted hydride. Thus, $HMn(CO)_5$, if formed initially by hydrogen atom transfer to $Mn(CO)_6\cdot$, would also be converted to $HMn(CO)_3L_2$.

4.2. Other Examples

It seems likely that 19-e species are involved in several interesting examples of redox-catalyzed processes that can be thought of as intramolecular atom or group transfers.[37-39] For example, the oxidatively catalyzed migratory insertion of CO in $(\eta^5\text{-}C_5H_5)(PPh_3)(CO)FeCH_3^+$ [37] has been shown to proceed via formation of a 19-e radical $(\eta^5\text{-}C_5H_5)(PPh_3)(CO)(Nu)FeCH_3^+$ by addition of a nitrogen base, Nu, to the 17-e radical cation.[40] The 19-e radical undergoes facile migration of CH_3 onto the CO group. In this case the reactivity of the 19-e radical dominates the observed chemistry. The driving force, as in formation of formyl species upon hydrogen atom transfer to the 19-e radical, derives from the shift of unpaired spin density from the metal to the carbonyl carbon, forming a bent, acyl-like group.[41]

An example drawn from photochemistry involves the formation of persistent manganese radicals, $Mn(CO)_3L_2$. Photolysis of $Mn_2(CO)_{10}$ in the presence of excess L, with periodic removal of liberated CO, leads to formation of the persistent radical.[12] The IR spectra of the product solutions reveals the presence of $HMn(CO)_3L_2$, in amounts that vary with L: $P(n\text{-}Bu)_3 > P(i\text{-}Bu)_3 > P(i\text{-}Pr)_3$. Inasmuch as the solutions of persistent radicals in the presence of excess ligand are stable for extended periods of time following the photochemical formation process, the hydrides are evidently not formed via a thermal hydrogen atom transfer reaction to $Mn(CO)_3L_2$.

The source of hydrogen atom in such reactions is probably the excess ligand. Hydrogen atom transfer could involve a 19-e species such as $Mn(CO)_5Ln$, which must be present as substitution occurs, or it might involve the 17-e species, $Mn(CO)_4Ln$. In either case, a monosubstituted hydride formed in the atom transfer reaction would eventually undergo atom transfer to the persistent radical to form the observed product, $HMn(CO)_3L_2$.

In an attempt to address this question we have photolyzed 5 x 10^{-4}M $Mn_2(CO)_8[PMe_3]_2$ in hexane solution under various conditions, using 366 nm radiation. In the absence of any other reagents, the starting dimer under photolysis for several hours yields small amounts of $Mn_2(CO)_9PMe_3$ and $Mn_2(CO)_{10}$. In the presence of 0.01 M $HSiEt_3$ under Ar, the starting dimer is converted under photolysis almost completely to $HMn(CO)_4PMe_3$ in about 90 min. When the same reaction is carried out under slightly more than 1 atm CO, the rate of formation of $HMn(CO)_4$ PMe_3 is decreased, and $Mn_2(CO)_9PMe_3$ and $Mn_2(CO)_{10}$ are seen. The overall rate of disappearance of $Mn_2(CO)_8[PMe_3]_2$ is lower. Under 3.8 atm CO the rate of formation of $HMn(CO)_4PMe_3$ is still further decreased, and the overall rate of reaction is further reduced. Extended photolysis of $Mn_2(CO)_{10}$ in hexane with excess $HSiEt_3$ and under 3.8 atm CO leads to formation of $HMn(CO)_5$.

These results suggest that the metal carbonyl hydride is not being formed via hydrogen atom transfer from $HSiEt_3$ to a 19-e radical formed by addition of CO to the $Mn(CO)_4PMe_3$ radical. If that were the case, more hydride should have been seen under higher CO pressure, whereas the rate of hydride formation actually declines with increasing [CO]. At the same time, the fact that CO does interact with the $Mn(CO)_4PMe_3$ radical is evidenced by the increased rate of formation of $Mn_2(CO)_9L$ at higher CO pressure. It appears most likely that hydride formation occurs as a result of a reaction involving the 17-e Mn-$(CO)_4PMe_3$ radical. $HSiEt_3$ was chosen as the hydrogen atom donor in these studies because the Si-H bond energy is comparatively high, 90 kcal mol^{-1}.[42] Thus hydrogen atom transfer to $Mn(CO)_4L$ is substantially endergonic. The quantum yield for formation of $HMn(CO)_4PMe_3$ in photolysis of $Mn_2(CO)_8[PMe_3]_2$ under Ar is on the order of 0.1. Given what is known about the rates of competing processes, this suggests a value for k_T for the reaction, $HSiEt_3$ + $Mn(CO)_4PMe_3n$: $HMn(CO)_4PMe_3$ + $SiEt_3n$, on the order of 10^2 to 10^3 M^{-1} s^{-1}. In spite of this comparatively slow hydrogen atom transfer rate, it appears that the 17-e species first formed in metal-metal bond homolysis reacts to form the hydride more rapidly than the 19-e radical that forms via addition of CO.

5. SOME CONCLUSIONS

Few experiments have been carried out with the aim of distinguishing the reactivities of the 17-e and 19-e radicals with respect to atom transfer processes. Kochi's work and the recent study by Therien and Trogler[40] provide the best evidences available, but they apply to a restricted set of compounds under particular conditions. Additional experimental work based on electrochemical, flash photolysis and thermal methods is needed. The present analysis suggests that:

(a) It will often be the case that the reactivity of either a 17-e or 19-e radical with respect to some reaction, be it electron transfer, atom transfer or recombination, will dominate over reactions that interconvert the 17-e and 19-e radicals. In these cases the mode of formation of the radical determines the subsequent reaction pathway.
(b) When a 17-e/19-e equilibrium is established, it is likely to be heavily weighted toward the 17-e radical unless there is present a ligand of variable hapticity such as C_5H_5. This means that the 17-e radical will dominate in atom transfer processes with low activation barriers. The existing counter-examples involve Mn radicals in which the energy of the M-H bond, formed via H atom transfer to the 17-e radical, is much lower than the energy of the acyl C-H bond, formed via H atom transfer to the 19-e radical. The rate constant for atom transfer to metal or carbon is determined not only by the energetics of bond rupture and formation, but by the relative spin densities on the two sites. The fact that atom transfer occurs to the carbonyl carbon in the 19-e radical is consistent with what is known from epr spectroscopy about the electron distribution in such species. For example, Preston and co-workers have shown that the unpaired spin density is quite high on the carbon atom of a bent carbonyl in $Fe(CO)_5^-$.[41] The difference in reactivities of the 17-e and 19-e radicals will be smaller for second and third row metal carbonyl radicals, because the metal-hydrogen bond energies are comparatively higher.
(c) The 19-e species might in some cases be the kinetically dominant species in halogen atom transfer. We have shown that the rate constants for halogen atom transfer can be related to the reduction potentials for organic halogen atom donors.[33] Because the 19-e species is more readily oxidized than the 17-e, halogen atom transfer should be more exergonic for the 19-e radical. However, as shown in Figure 1, this advantage does not produce much effect for processes of low intrinsic barrier until the atom transfer process for the 17-e species has become significantly endergonic. It should be possible to explore systems in which the appropriate conditions apply, using either electrochemical or flash photolysis techniques.
(d) Highly substituted 19-e radicals are unlikely to undergo fast atom transfer reactions because of the steric impediment represented by the ligands, and because the 17-e/19-e equilibrium is shifted heavily toward the 17-e radical.
(e) Radical chain substitution processes involving 17-e and 19-e species must be taken into account, particularly in systems containing excess ligand and in which radicals are being generated.

Acknowledgement: This research has been supported by the National Science Foundation through research grant CHE-86 08839.

References

1. Wrighton, M. S.; Ginley, D. S. J. Am. Chem. Soc. 1975, **97**, 2065.
2. Morse, D. L.; Wrighton, M. S. J. Am. Chem. Soc. 1976, **98**, 3931.

3. Walker, H. W.; Herrick, R. S.; Olsen, R. J.; Brown, T. L. Inorg. Chem. 1984, **23**, 3748.
4. Meyer, T. J.; Casper, J. V. Chem. Rev. 1985, **85**, 187.
5. (a) Pickett, C. J.; Pletcher, D. J. Chem. Soc., Dalton Trans. 1975, 879.
   (b) Pickett, C. J.; Pletcher, D. J. Chem. Soc., Chem. Commun. 1974, 660.
   (c) ibid., J. Chem. Soc., Dalton Trans. 1976, 636.
   (d) ibid. J. Chem. Soc., Dalton Trans. 1976, 749.
6. (a) Beyers, B. H.; Brown, T. L. J. Am. Chem. Soc. 1975, **97**, 947.
   (b) Beyers, B. H.; Brown, T. L. J. Am. Chem. Soc. 1977, **99**, 2527.
7. Zizelman, P. M.; Amatore, C.; Kochi, G. K. J. Am. Chem. Soc. 1984, **106**, 3771.
8. Seurat, A.; Lemoine, P.; Gross, M. Electrochim. Acta, 1978, **23**, 1219.
9. Narayanan, B. A.; Amatore, C.; Kochi, J. K. Organometallics 1986, **5**, 926.
10. Garcia Alonzo, F. J.; Riera, V.; Valia, M. L.; Morieras, D.; Vivanco, M.; Solons, X. J. Organomet. Chem. 1987, **326**, C71.
11. Fairhurst, S. A.; Morton, J. R.; Preston, K. F. J. Magn. Res. 1983, **55**, 453.
12. (a) Kidd, D. R.; Cheng, C. P.; Brown, T. L. J. Am. Chem. Soc. 1978, **100**, 4103.
    (b) McCullen, S. B.; Brown, T. L. J. Am. Chem. Soc. 1982, **104**, 7496.
13. Walker, H. W.; Rattinger, G. B.; Belford, R. L.; Brown, T. L. Organometallics 1983, **2**, 775.
14. (a) Adams, R. D.; Collins, D. E.; Cotton, F. A. J. Am. Chem. Soc. 1974, **96**, 749.
    (b) Madack, T.; Vahrenkamp, H. Z. Naturforsh. B.: Anorg. Chem. Org. Chem. 1978, **33B**, 1301.
    (c) Cooley, N. A.; Watson, K. A.; Fortier, S.; Baird, M. C. Organometallics, 1986, **5**, 2563.
15. McKinney, R. J. Inorg. Chem. 1982, **21**, 2051.
16. Stiegman, A. E.; Tyler, D. R. Comments on Inorg. Chem. 1986, **5**, 215.
17. (a) Creber, K. A. M.; Chen, K. S.; Wan, J. K. S. Revs. Chem. Intermed. 1984, **5**, 37.
    (b) Kabachnik, M. I.; Bubnov, N. N.; Solodovnikov, S. P.; Prokof'ev, A. I. Russ. Chem. Revs. 1984, **53**, 288.
    (c) Kaim, W. Coord. Chem. Revs. 1987, **76**, 187.
18. Brown, T. L. Ann. N. Y. Acad. Sci. 1980, **330**, 80.
19. Wrighton, M. S.; Graff, J. L.; Luong, J. C.; Reichel, C. L.; Robbins, J. L. in Reactivity of Metal-Metal Bonds, Chisholm, M. H. ed.; ACS Symposium Series 155; American Chemical Society: Washington, DC 1981, p. 85.
20. Stiegman, A. E.; Stieglitz, M.; Tyler, D. J. Am. Chem. Soc. 1983, **105**, 6032.
21. Doxsee, K. M.; Grubbs, R. H.; Anson, F. C. J. Am. Chem. Soc. 1984, **106**, 7819.

22. Lee, K. Y.; Kuchynka, D. J.; Kochi, J. K. Organometallics, 1987, **6**, 1886.
23. Fox, A.; Malito, J.; Poë, A. J. Chem. Soc., Chem. Commun. 1981, 1062.
24. Herrinton, T. R.; Brown, T. L. J. Am. Chem. Soc. 1985, **107**, 5700.
25. McCullen, S. B.; Walker, H. W.; Brown, T. L. J. Am. Chem. Soc. 1982, **104**, 4007.
26. Shi, Q. Z.; Richmond, T. G.; Trogler, W. C.; Basolo, F. J. Am. Chem. Soc. 1982, **104**, 4032; ibid., 1984, **106**. 71.
27. Jones, W. D.; Huggins, J. M.; Bergman, R. G. J. Am. Chem. Soc. 1981, **103**, 4415.
28. (a) Stiegman, A. E.; Tyler, D. R. Inorg. Chem. 1984, **23**, 527.
    (b) Stiegman, A. E.; Goldman, A. S.; Philbin, C. E.; Tyler, D. R. Inorg. Chem. 1986, **25**, 2976.
    (c) Stiegman, A. E.; Tyler, D. R. J. Am. Chem. Soc. 1985, **107**, 967.
    (d) Goldman, A. S.; Tyler, D. R. Inorg. Chem. 1987, **26**, 253.
29. (a) Basolo, F. Inorg. Chim. Acta 1985, **100**, 33.
    (b) O'Connor, J. M.; Casey, C. P. Chem. Rev. 1987, **87**, 307.
30. Kershner, D. L.; Rheingold, A. L.; Basolo, F. Organometallics 1987, **6**, 196 and references therein.
31. Marcus, R. A. J. Phys. Chem. 1968, **72**, 891.
32. (a) Agmon, N.; Levine, R. D. Chem. Phys. Lett. 1977, **52**, 197.
    (b) Agmon, N. Int. J. Chem. Kinet. 1981, **13**, 333.
33. Lee, K. W.; Brown, T. L. J. Am. Chem. Soc. 1987, **109**, 3269.
34. Hanckel, J. M.; Lee, K. W.; Rushman, P.; Brown, T. L. Inorg. Chem. 1986, **25**, 1852.
35. Narayanan, B. A.; Amatore, C.; Kochi, J. K. Organometallics 1987, **6**, 129.
36. Kuchynka, D. J.; Amatore, C.; Kochi, J. K. Inorg. Chem. 1986, **25**, 4087.
37. Magnuson, R. H.; Meirowitz, R.; Zulu, S. J.; Giering, W. P. Organometallics 1983, **2**, 460.
38. (a) Reger, D. L.; Mintz, E. Organometallics 1984, **3**, 1759.
    (b) Reger, D. L.; Mintz, E.; Lebioda, L. J. Am. Chem. Soc. 1986, **108**, 1940.
39. Davies, S. G.; Simpson, S. J. J. Organomet. Chem. 1982, **240**, C48.
40. Therien, M. J.; Trogler, W. C. J. Am. Chem. Soc. 1987, **109**, 5127.
41. Fairhurst, S. A.; Morton, J. a.; Preston, K. F. J. Chem. Phys. 1982, **77**, 5872.
42. Kanabus-Kaminska, J. M.; Howari, J. A.; Griller, D.; Chatgilialoglu, C. J. Am. Chem. Soc. 1987, **109**, 5266.

STRUCTURE AND SUBSTITUTION REACTIONS OF NINETEEN-ELECTRON ORGANOMETALLIC COMPLEXES

David R. Tyler,* Cecelia Philbin, and Mao Fei
Department of Chemistry
University of Oregon
Eugene, Oregon 97403
USA

ABSTRACT. Seventeen-electron organometallic radicals react with 2-electron ligands to yield 19-electron adducts. Little is known about the structure and reactivity of these species, and so we have begun a systematic study of their properties. The substitution reactions of the 19-electron $Co(CO)_3L_2$ complex (L_2 = 2,3-bis(diphenylphosphino)maleic anhydride) were investigated. The complex reacts with phosphines and phosphites in first-order reactions to yield $Co(CO)_2L_2L'$. A dissociatively-activated pathway involving loss of CO is proposed. Irradiation ($\lambda > 500$ nm) of $Cp_2*Fe_2(CO)_4$ and PR_3 yields $(C_5Me_4H)Fe(CO)_2CH_3$. The 19-electron $Cp*Fe(CO)_2PR_3$ complex is the proposed intermediate. This species likely contains an η^4-Cp ring.

Reactions of 17-electron organometallic radicals with 2-electron ligands yield 19-electron adducts.[1] The 19-electron complexes are in many cases simply transition states, as for example in the substitution reactions of $V(CO)_6$:[2]

$$V(CO)_6 + L \longrightarrow [V(CO)_6L] \longrightarrow V(CO)_5L + CO \qquad (1)$$

In other cases, the 19-electron complexes are genuine intermediates with finite lifetimes and characteristic chemical properties and reactivities. Examples of this type of adduct include the 19-electron complexes formed by reaction of $CpMo(CO)_3$, $CpFe(CO)_2$, or $Mn(CO)_5$ with various L, e.g.:[1,3]

$$Cp_2Mo_2(CO)_6 \xrightarrow{h\nu} 2\ \underset{17}{CpMo(CO)_3} \xrightarrow{L} \underset{19}{CpMp(CO)_3L} \qquad (2)$$

Because 19-electron adducts represent a new class of organometallic complexes and because these molecules are important intermediates whenever metal radicals are formed,[1] we are systematically studying their structure and reactivity.

M. Chanon et al. (eds.), Paramagnetic Organometallic Species in Activation / Selectivity, Catalysis, 201–209.

Four basic reaction types have been established for 19-electron complexes:[1] outer-sphere electron-transfer (e.g., eq 3), ligand dissociation (e.g., eq 4), ligand coupling reactions (e.g., eq 5), and substitution.

$$CpFe(CO)dppe + Cp_2Co^+ \longrightarrow CpFe(CO)dppe^+ + Cp_2Co \qquad (3)^4$$

$$Fe(CO)_5^- \longrightarrow Fe(CO)_4^- + CO \qquad (4)^5$$

$$2\ (C_7H_7)Mo(CO)_3 \longrightarrow (CO)_3Mo(\eta^6\text{-}C_7H_7\text{-}\eta^6\text{-}C_7H_7)Mo(CO)_3 \qquad (5)^6$$

The first three of these reaction types are reasonably well understood and have been discussed in a recent review of 19-electron adducts.[1] Considerably less is known, however, about the substitution reactions of 19-electron species, and in this paper we discuss recent experimental results on the substitution reactions of the 19-electron $Co(CO)_3L_2$ complex (L_2 = 2,3-bis(diphenylphosphino)maleic anhydride).

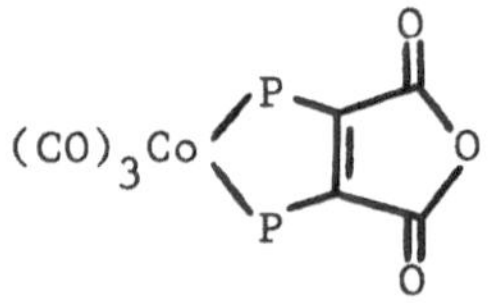

In this paper we also discuss recent results that give insights into the structures (electronic and geometric) of 19-electron complexes. For this purpose, it is conceptually useful to define three limiting classes of structures for these complexes. Class I: σ* complexes; the 19th electron is in a M–L antibonding orbital, e.g., $Mn(CO)_5Br^-$,[7] $CpMo(CO)_3I^-$.[8] Class II: complexes that have a metal valence electron count lower than 19 due to a geometry change such as a slipped Cp ring, bent CO ligand, or a "five-coordinate" phosphoranyl radical structure:

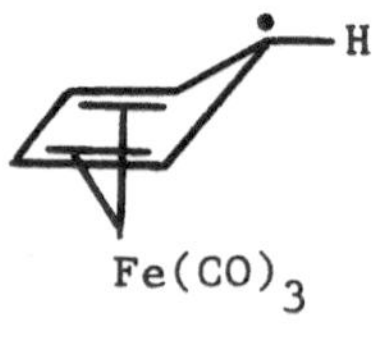

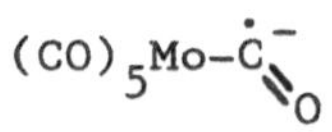

Class III: π* complexes; complexes where the 19th electron is primarily in a low-energy ligand orbital (generally a π* orbital), e.g., $Mo(CO)_4(bpy)^-$,[9] $CpMo(CO)_2(c\text{-hex-DAB})$[10] (c-hex-DAB = c-$C_6H_{11}$N=CHCH=N-c-$C_6H_{11}$). The class III complexes are generally the most stable complexes, followed by class II, then class I.

Results and Discussion

Substitution reactions of $Co(CO)_3L_2$ (L_2 = 2,3-bis(diphenylphosphino)-maleic anhydride. The $Co(CO)_3L_2$ complex was first synthesized by Fenske,[11] who showed that the odd electron is delocalized over the Co atom and the π^* system of the L_2 ligand. The molecule clearly belongs to Class III as defined above, and the relative stability of the complex can be attributed to the ability of the L_2 ligand to delocalize the extra electron.[11]

$Co(CO)_3L_2$ reacts thermally with phosphine and phosphite ligands to form monosubstituted products of the type $Co(CO)_2L_2L'$ (eq 6).[12]

$$Co(CO)_3L_2 + L' \longrightarrow Co(CO)_2L_2L' + CO \tag{6}$$

$$L' = PR_3,\ P(OR)_3$$

In the case of L' = PPh_3, the crystal structure of the product was determined by X-ray diffraction. A simplified structure, omitting the phenyl rings on the phosphorus atoms, is shown in Figure 1. Note that while the geometry of the $Co(CO)_3L_2$ complex is a distorted square pyramid, the $Co(CO)_2L_2PPh_3$ complex assumes a distorted trigonal bipyramidal structure with two CO ligands in equatorial sites.

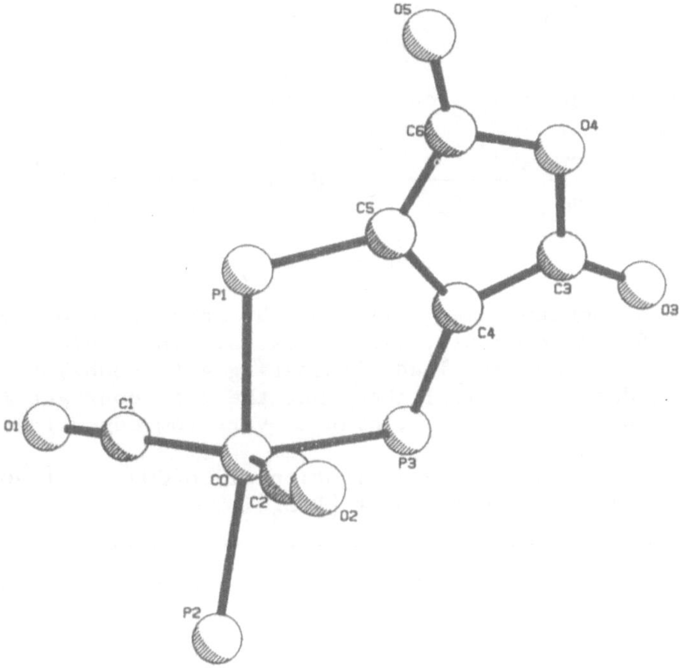

Figure 1. Crystal structure of the $Co(CO)_2L_2PPh_3$ complex (L_2 = 2,3-bis(diphenylphosphino)maleic anhydride).

Kinetics and mechanism of the substitution reaction. Three possible pathways for the substitution reaction are outlined in Scheme I: Pathway (i) is a dissociatively-activated mechanism in which CO dissociates reversibly to give a four-coordinate intermediate, followed by reaction with L' to give the products. Pathway (ii) is an associatively-activated pathway, and pathway (iii) is a dissociatively-activated mechanism involving dissociation of one end of the chelate ligand followed by subsequent steps.

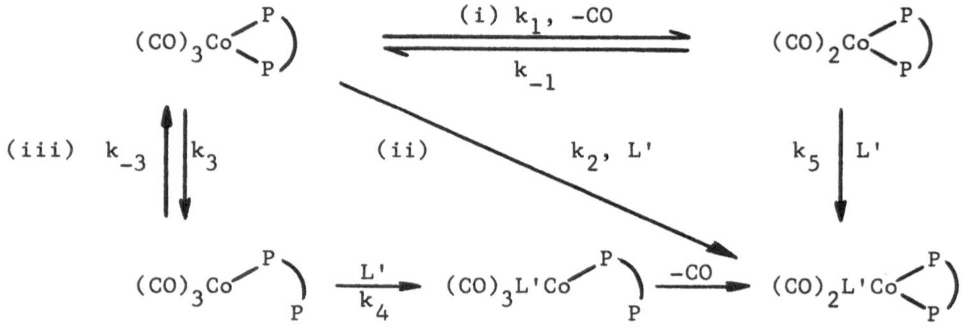

Scheme I

The total rate can be expressed as follows:

$$-\frac{d[S]}{dt} = \frac{k_1k_5[S][L']}{k_{-1} + k_5[L']} + k_2[S][L'] + \frac{k_3k_4[S][L']}{k_{-3} + k_4[L']} \qquad (7)$$

where $S = Co(CO)_3L_2$

The reaction was monitored by following the absorbance decay of $Co(CO)_3L_2$ at 758 nm in CH_2Cl_2. In all cases, linear plots of $-\ln(A_t - A_\infty)$ vs. t were obtained, indicating a (pseudo-) first-order reaction. The data in Table I show that the rate constant is independent of ligand concentration, even at very low concentrations of L'.

TABLE I. Rate Constants for Substitution of $Co(CO)_3L_2$ (1.40×10^{-3} M) at 25 ^oC in CH_2Cl_2 as a Function of $[PPh_3]$.

$[PPh_3]$ M	0.148	0.296	0.371	0.445	0.556	0.704
k (sec^{-1}) x 10^4	54.8	54.5	54.7	55.2	54.2	55.0

This result eliminates the associatively-activated mechanism [pathway (ii)]. Note, however, that these data cannot differentiate between the ring-opening pathway [(iii)] and pathway (i) because the third term in the above rate expression can be reduced to the first-order expression rate = $k_3[S]$ if $k_4[L] >> k_{-3}$. However, this inequality is probably not valid in this reaction (or many other related reactions). In terms of electron donating ability and steric size, the bidentate ligand and the entering ligand are comparable to each other. In addition, the rigidity of the chelate ligand (caused by the carbon-carbon double bond) assures that the dissociated phosphorus of the bidentate ligand has a high "effective concentration" in the vicinity of the metal center. Thus, a strong argument can be made that $k_4[L']$ is much smaller than k_{-3}. In fact, in numerous studies it has been shown that the "competition ratio," k_{-3}/k_4 is generally <u>much</u> greater than one. The conclusion is that if pathway (iii) were followed then the rate would be first-order in [L'].

Assume, however, that pathway (iii) is occurring but that k_{-3} is anomolously small so that the reaction rate is simply first-order in [S], as found. This assumption was tested by lowering the concentration of both the entering ligand (10^{-3} M) and the complex (10^{-3} M) in order to decrease the magnitude of $k_4[L']$. Under these conditions, the dependence of the rate on the ligand concentration would be more obvious. However, first-order rate constants essentially identical to those in Table I were obtained in these experiments. The ring-opening mechanism can, therefore, be reasonably excluded. The data is thus consistent only with the dissociatively-activated route in pathway (i).

Table II lists the rate constants obtained at different temperatures.

TABLE II. Rate Constants for the Substitution of $Co(CO)_3L_2$ by PPh_3 in CH_2Cl_2 as a Function of Temperature.

t (^oC)	k (sec^{-1}) x 10^4
10.0	5.98 ± 0.17
14.9	12.6 ± 0.4
19.9	26.7 ± 0.3
24.9	59.7 ± 0.3
29.9	102 ± 1

A plot of $-\ln(k/T)$ vs. $(1/T)$ yielded the following activation parameters: $\Delta H^{\ddagger} = 23.8 \pm 0.6$ kcal/mole; $\Delta S^{\ddagger} = 11.1 \pm 2.2$ cal mole^{-1} K^{-1}. The positive entropy is consistent with the dissociative pathway; the enthalpy can be roughly viewed as the Co-CO bond energy.

As a further test of the proposed dissociative pathway, other ligands were reacted with $Co(CO)_3L_2$. Similar rate constants were obtained with all of the ligands, a result consistent with a dissociatively-activated mechanism.

In conclusion, the 19-electron $Co(CO)_3L_2$ complex undergoes substitution via a dissociatively-activated pathway involving loss of CO. Other 19-electron complexes also probably substitute via dissociatively-activated pathways. Examples are shown in eqs 8-11.

$$Mn(CO)_4(DBOQ) + t\text{-}Bu\text{-}DAB \xrightarrow{RT,\ dark} Mn(CO)_3(DBOQ)(t\text{-}Bu\text{-}DAB\text{-}N') + CO \qquad (8)$$ [13]

DBOQ = 3,5-di-t-butyl-ortho-benzoquinone; the -N' nomenclature for the DAB ligand indicates only one N atom is coordinated to the Mn)

$$Mo(CO)_4(bpy)^- + PPh_3 \xrightarrow{RT} Mo(CO)_3PPh_3(bpy)^- + CO \qquad (9)$$ [9]

$$Mn(CO)_3(CH_3CN)_3 \xrightarrow[RT]{L} Mn(CO)_3(CH_3CN)_2L \xrightarrow{L} Mn(CO)_3(CH_3CN)L_2 \qquad (10)$$ [14]

$L = PPh_3$ or PMe_2Ph

It is logical to propose that these reactions also proceed via a dissociatively-activated mechanism. However, some caution is necessary. For example, the 19-electron complex Cp_2MnPR_3 is known, as is the "21-electron" $Cp_2Mn(dppm)$ (dppm = 1,2-bis(diphenylphosphino)-methane).[15] The existence of the latter complex suggests that the phosphine substitution reactions of the former could proceed via an associatively-activated mechanism. Admittedly, this is an extreme case because the Cp_2MnPR_3 complex is decidedly ionic; nevertheless, it serves to caution us about the possibility of associatively-activated reactions.

<u>Structure of the 19-electron $(C_5Me_5)Fe(CO)_2PPh_3$ complex.</u> Irradiation ($\lambda > 500$ nm) of the $Cp_2{}^*Fe_2(CO)_4$ complex ($Cp^* = \eta^5\text{-}C_5Me_5$) and PPh_3 in benzene gives $(C_5Me_4H)Fe(CO)_2CH_3$:

$$Cp_2{}^*Fe_2(CO)_4 \xrightarrow[PPh_3]{\lambda > 500\ nm} (C_5Me_4H)Fe(CO)_2CH_3 \qquad (11)$$

9,10-dihydroanthracene

(Continued irradiation produces the CO-insertion product $(C_5Me_4H)Fe(CO)(PPh_3)(COCH_3)$ as the final product of the reaction.) An important observation is that no $(C_5Me_4H)Fe(CO)_2CH_2$ forms if the ligand and H atom donor (9,10-dihydroanthracene) are not present. The quantum yield for reaction 11 is very low ($\Phi < 10^{-3}$).

Recent work by Blaha and Wrighton[16] was instrumental in our interpretation of reaction 11. They reported the photochemical reaction of $CpFe(CO)_2R$ ($R = CH_2C_6H_5$) with CO:

$Cp^*Fe(CO)_2R + CO \xrightarrow{h\nu}$ 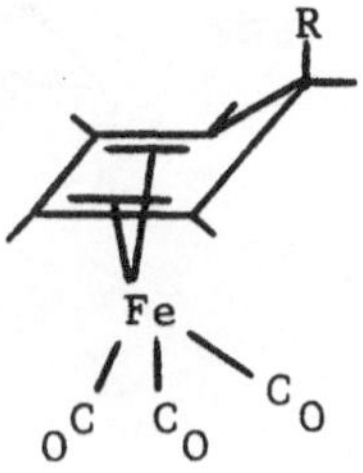 (12)

The following mechanism was suggested:

$Cp^*Fe(CO)_2R \xrightarrow{h\nu} Cp^*Fe(CO)_2 + R$

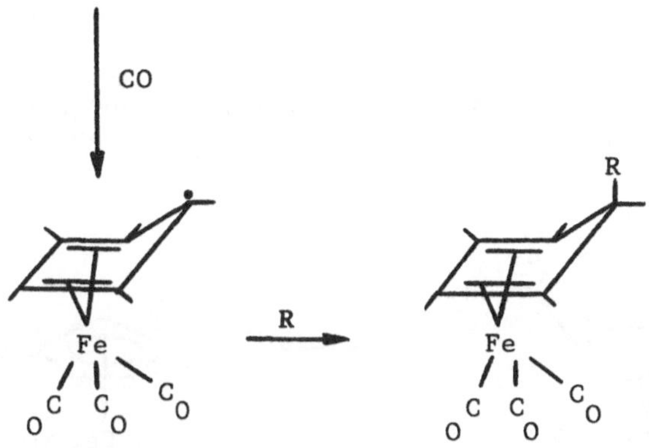 (13)

The key point here[5] is that the 19-electron $Cp^*Fe(CO)_3$ intermediate does not have an η^5-Cp ring but rather an η^4-ring. Apparently, this structure is more favorable because of the 18-valence electron configuration at the metal.

We suggest that a similar structural change is responsible for the products in reaction 11. The following mechanism is suggested:

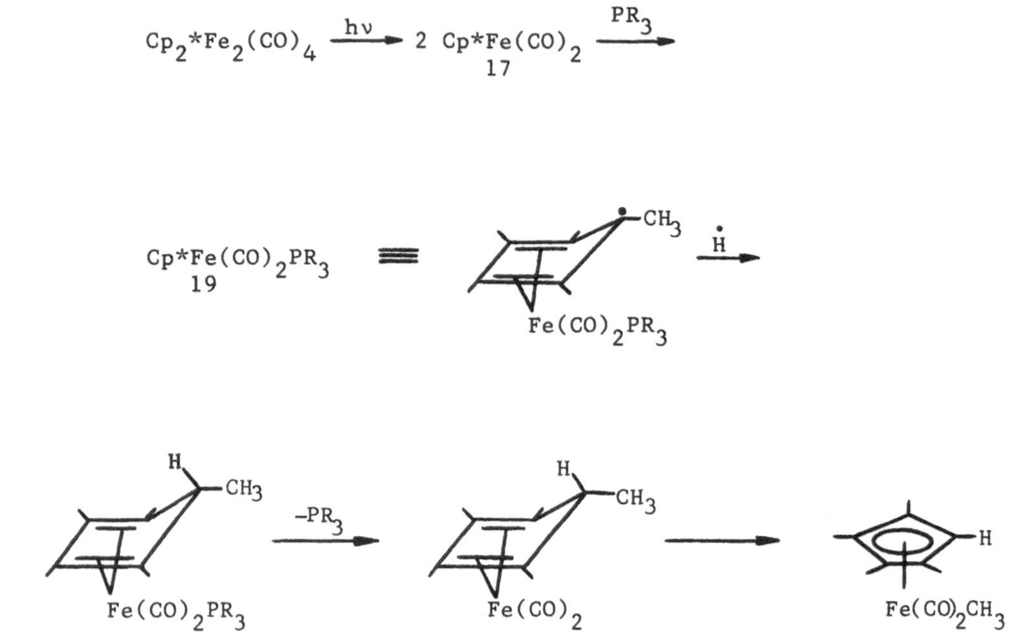

Scheme II

The conversion of species II to III has numerous precedents in the literature. Thus, for example, Eilbracht and Dahler reported[17] the following reaction:

(14)

In conclusion, reactions 11 and 12 strongly suggest that the 19-electron $CpFe(CO)_2L$ complexes have an η^4-Cp ring. (Of course, the η^4-Cp structure may be in equilibrium with an η^5-Cp structure.) In these reactions, the relationship between the structure of the radicals and the reactivity of the radicals is clearly evident.

One final comment is appropriate. Our study above (as well as that of Wrighton and Blaha) provides information only on the structure of the $CpFe(CO)_2L$ 19-electron complexes. It would be premature to conclude that all 19-electron species with Cp rings are η^4 complexes. As discussed previously, structural examples are known for 19-electron complexes in each of the three classes. At present, the factors that determine the structures of 19-electron adducts are not known, and it is therefore not possible yet to predict a priori the electronic or geometric structures of these species.

Acknowledgment. Our research at the University of Oregon is supported by the National Science Foundation, the Air Force Office of Scientific Research, and the Donors of the Petroleum Research Fund, administered by the American Chemical Society.

References

1. Stiegman, A. E.; Tyler, D. R. Comments Inorg. Chem. 1986, 5, 215.
2. (a) Shi, Q.-Z.; Richmond, T. G.; Trogler, W. C.; Basolo, F. J. Am. Chem. Soc. 1982, **104**, 4032. (b) Shi, Q.-Z.; Richmond, T. G.; Trogler, W. C.; Basolo, F. J. Am. Chem. Soc. 1984, **106**, 71.
3. Stiegman, A. E.; tyler, D. R. Coord. Chem. Res. 1985, **63**, 217.
4. Goldman, A. S.; Tyler, D. R. Inorg. Chem. 1987, **26**, 253.
5. Pickett, C. J.; Pletcher, D. J. Chem. Soc., Dalton Trans. 1975, 879, and references therein.
6. Armstead, J. A.; Cox, D. J.; Davis, R. J. Organomet. Chem. 1982, **236**, 213.
7. Anderson, O. P.; Symons, M. C. R. J. Chem. Soc., Chem. Commun. 1972, 1020.
8. Symons, M. C. R.; Bratt, S. W.; Wyatt, J. L. J. Chem. Soc., Dalton Trans. 1983, 1377.
9. (a) Conner, J. A.; Overton, C.; Murr, N. E. J. Organomet. Chem. 1984, **277**, 277. (b) Alegria, A. E.; Lozada, O.; Rivera, H.; Sanchez, J. J. Organomet. Chem. 1985, **281**, 229. (c) Miholova, D.; Vlcek, A. A. J. Organomet. Chem. 1985, **279**, 317.
10. Bruce, A. E.; Tyler, D. R., unpublished work.
11. Fenske, D. Chem. Ber. 1979, **112**, 363.
12. Mao, F.; Tyler, D. R., manuscript in preparation.
13. (a) Alberti, A.; Hudson, A. J. Organomet. Chem. 1983, **241**, 313. (b) Alberti, A.; Hudson, A. J. Organomet. Chem. 1983, 248, 197.
14. Narayanan, B. A.; Amatore, C.; Kochi, J. K. J. Chem. Soc., Chem. Commun. 1983, 397.
15. Howard, C. G.; Girolami, G. S.; Wilkinson, G.; Pett, M. T.; Hursthouse, M. B. J. Am. Chem. Soc. 1984, **106**, 2033.
16. Blaha, J. P.; Wrighton, M. S. J. Am. Chem. Soc. 1985, 107, 2694.
17. Eilbracht, P.; Dahler, P. Chem. Ber. 1980, **113**, 542.

ROLE OF TERMINATION STEPS ON THE EFFICIENCY OF ELECTRON TRANSFER CHAIN CATALYSIS

C. AMATORE, A. JUTAND and J.N. VERPEAUX

Ecole Normale Supérieure
Laboratoire de Chimie, UA CNRS 1110
24, rue Lhomond
75231 PARIS CEDEX 05 - FRANCE.

ABSTRACT : The role of activation/termination sequences on the efficiency of electron transfer chain catalysis is discussed and illustrated by two examples related to transition metal catalysis of organic reactions.

1. INTRODUCTION AND PRESENTATION

1.1 Situation of the problem.

A number of propagation sequences involved in metal catalysis are now thought in terms of various combinations of radical formation and radical consumption steps. Thus electron transfer chain catalysis appears to be as pervasive as e.g. proton transfer catalysis in chemistry (1). However the mechanisms of most of these catalytic processes are poorly understood, and their radical nature often only established through the use of specific scavengers, or derived from inferential evidences related to thought analogous reaction mechanisms. Indeed proper identification of the intimate mechanism of electron transfer catalytic processes has been so far extremely scarce in the literature and most of the time relative to experimental situations which complexity is by far beyond that of actual catalytic systems operative in chemistry. For example the $S_{RN}1$ mechanism for aromatic nucleophilic substitution has been proposed by Bunnett (2) as an extension of an original proposal by Kornblum (3) nearly fifteen years before its intricacies could be established on solid kinetic grounds (1e,4). Nevertheless, in the time interval the great potential of the $S_{RN}1$ reaction in synthetic chemistry has been revealed and developped to a large extent (5).

Scheme I

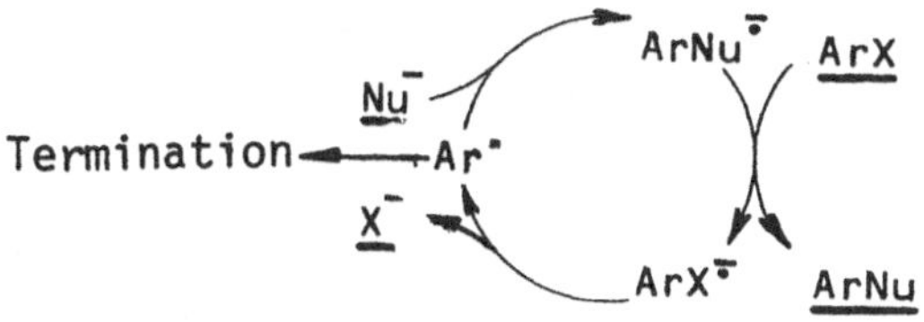

M. Chanon et al. (eds.), Paramagnetic Organometallic Species in Activation / Selectivity, Catalysis, 211–224.

The difficulties in establishing the exact mechanism of a chain reaction are mainly related to the fact that their kinetic behavior is by far more complex than that of usual reactions. This originates from the simultaneous involvement of the activation and termination sequences in addition with the propagation of the chain. Indeed the activation/termination competition controls the amount of the catalytic species responsible for the chain propagation. Thus the overall kinetics, defined e.g. by the overall rate of conversion of the substrate, depends greatly on the activation/termination sequence (4). This is a well known problem in other chain mechanisms as e.g. in combustion reactions. Indeed it is intuitively understood that a "simple" reaction such as alcanes combustion obeys different kinetic regimes ran-

$$C_nH_{2n+2} + (3n+1)/2\ O_2 \longrightarrow nCO_2 + (n+1)\ H_2O \qquad (1)$$

ging from controlled combustion (heaters) to explosive behavior, depending on the exact experimental conditions. Similarly the role of many fireextinguishers consists in adding a large variety of termination steps to a fast propagating chain sequence, which results in rapid deactivation of an otherwise efficient chain reaction.

A further difficulty in the interpretation of the observed kinetics of chain reactions is then obviously related to the identification of the activation and of the termination processes. Indeed for efficient chain reactions those processes concern only a negligible amount of material, most of the reactant being converted by the catalytic cycle. This obviously results in considerable experimental difficulties in the delineation of the activation/termination sequence.

1.2 Role of the Activation/Termination sequences.

In $S_{RN}1$ reactions it has been postulated for a long time that the main deactivation route consisted in the duplication of the σ-aryl radical in eqn 2 (2,5). In the practice such a termination step,

$$Ar^{\bullet} + Ar^{\bullet} \longrightarrow Ar — Ar \qquad (2)$$

although extremely fast, does not play a significant role owing to the too small quantities of σ-aryl radical present during the reaction (4). Detailled kinetic studies have shown years after that for most of the situations of interest, the main deactivation route involves the reduction of the σ-aryl radical by the electron rich anion radicals, $ArX^{\bar{\bullet}}$ and $ArNu^{\bar{\bullet}}$, of the substrate and of the product, to afford after

$$Ar^{\bullet} + ArNu^{\bar{\bullet}} \longrightarrow Ar^- + ArNu \qquad (3)$$

$$Ar^{\bullet} + ArX^{\bar{\bullet}} \longrightarrow Ar^- + ArX \qquad (4)$$

$$Ar^- \xrightarrow{H^+} ArH \qquad (5)$$

protonation the corresponding aromatic. Both reactions overcome reaction 2 in the deactivation of the $S_{RN}1$ chain propagation, owing to their diffusion controlled rate constants as well as to the fact that both anion radicals are present in considerably larger concentrations than the σ-arylradical (4,6).

Particularly crucial in the efficiency of a chain conversion is the role of the activation process. Indeed for efficient chains, the overall amount of the catalytic species involved in the chain propagation, is fixed by the instant balance between their production (activation sequence) and their consumption (termination steps). Thus the rate and efficiency of the chain conversion is intimately controlled by the activation rate for a given set of termination steps, inherent to the reaction considered.

For bimolecular termination steps, such as those described above for the $S_{RN}1$ reaction, the amount of material escaping the chain through the termination sequences increases when the activation rate increases. Noteworthy it is of importance in this context to recall that electrochemical activation of $S_{RN}1$ reactions usually leads to less efficient conversions than photochemical activation. Indeed the importance of the bimolecular termination steps is considerably less for the latter owing to the extremely small amount of chain carriers resulting from a poor activation rate (5). Conversely under reductive conditions (electrochemistry, dissolved metals,...) the activation is much more rapid which leads to short conversion times but is generally associated with poorer conversion yields in substituted products. The role of the activation rate is well examplified by the cyclic voltammograms in figure 1A. Thus when the electrode potential is made cathodic enough to reduce significantly the aromatic halide a very poor substitution yield is obtained, as evidenced by the small magnitude of the substituted product observed upon scan reversal, in figure 1Aa. On the

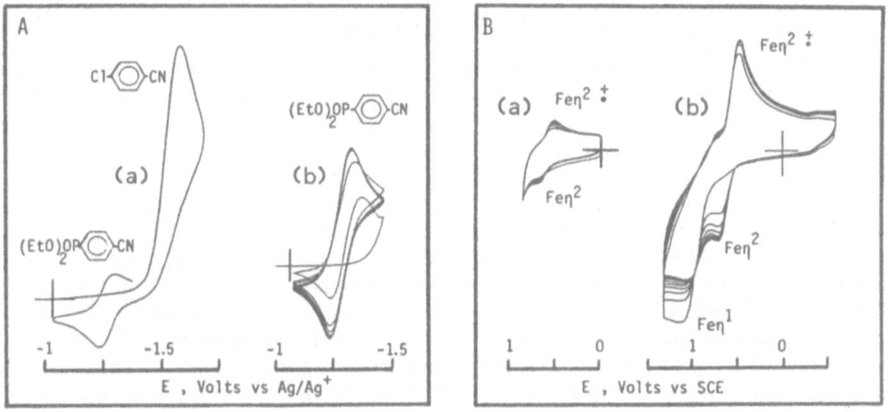

<u>Figure 1</u> : Effect of the activation rate on the overall efficiency of electron transfer chain catalysis as evidenced by cyclic voltammetry. (A) Aromatic nucleophilic substitution (eqn.6) in liquid ammonia, from ref.7. (B) Chelation of N,N-dialkyldithiocarbamate (dtc); compare eqn.11 and Scheme II. $Fe\eta^1 = (Me_5Cp)Fe(\eta^1\text{-dtc})(CO)_2$, $Fe\eta^2 = (Me_5Cp)Fe(\eta^2\text{-dtc})(CO)$, in THF at 20°C and 20 $V.s^{-1}$. See text for both systems.

other hand, when only the foot of the halide wave is scanned, which

$$Cl{-}\langle O\rangle{-}CN \; + \; {}^{-}OP(OEt)_2 \xrightarrow{[+e]} Cl^- \; + \; (EtO)_2P(O){-}\langle O\rangle{-}CN \qquad (6)$$

amounts to create very small quantities of the chain initiator, $ArX^{\bullet}$, a nearly quantitative conversion is obtained, although the reaction needs a more important delay to be completed as shown by the gradual increase of the product wave upon scan cycling in figure 1Ab (7).

Conversely when the termination steps are unimolecular or pseudo first order in chain carriers, their rate does not depend on the amount of chain carriers. In such a situation, the faster the activation, the faster the chain conversion and the better the yields. For example in the electrocatalytic chelation of dithiocarbamate iron carbonyl complex in Scheme II (8), figure 1B demonstrates that the

Scheme II : (dtc = N,N-dialkyldithiocarbamate; Cp'= Me_5Cp)

$$Cp'Fe(\eta^1dtc)(CO)_2 \underset{}{\overset{-e}{\rightleftharpoons}} Cp'Fe(\eta^1dtc)(CO)_2^{+\bullet} \qquad (7)$$

$$Cp'Fe(\eta^1dtc)(CO)_2^{+\bullet} \longrightarrow Cp'Fe(\eta^2dtc)(CO)^{+\bullet} + CO \qquad (8)$$

$$Cp'Fe(\eta^2dtc)(CO)^{+\bullet} \overset{+e}{\rightleftharpoons} Cp'Fe(\eta^2dtc)(CO) \qquad (9)$$

Termination :

$$Fe(\eta^2dtc)(CO)^{+\bullet} \longrightarrow Fe^{3+} + \ldots \qquad (10)$$

Overall :

$$Fe\,(\eta^1dtc)(CO)_2 \xrightarrow{[-e]} Fe(\eta^2dtc)(CO) + CO \qquad (11)$$

the larger the oxidative driving force, i.e. the more anodic the electrode potential, the more efficient is the overall electron catalyzed conversion in eqn.11, as evidenced by the magnitude of the increase of the oxidation wave of the chelated species with the number of cycles.

From the above presentation it is infered that most of the experimental problems arising in the kinetic characterization of a chain process are related to the ability of clearly indentifying and controlling the activation/termination sequences. This is particularly difficult for electron transfer catalysis, especially for reductive activation owing to the nature of chemical reductant. Thus it is not surprising that electrochemistry which allows a facile and precise tuning of the reductive or oxidative driving force (see above figs 1A and 1B) had played an important role in the investigation of electron transfer catalysis of organic and organometallic reactions. Indeed since the early work by Feldberg and Jeftic on electrocatalysis of ligand exchange in 1972, a large number of papers dealing with electron transfer catalysis has appeared in the literature (9).

As explained above an important feature of the electrochemical approach in the investigation of rates and mechanisms of electron transfer catalysis is related to the exact control of the activation sequence by the electrode potential. A less apparent but obvious other advantage is that the current flow from or to the electrode constitutes a precise determination of the electron consumption and thus of the termination/chain propagation competition kinetics. In this respect it is important to emphasize that electrochemistry is one of the only experimental techniques in which the same device (the electrode) is used to control the energy injected into the system (potential) and the way this energy is used (current).

In the following we want to present two examples illustrating the use of electrochemical techniques in the investigation of the mechanism of electron transfer chain catalysis involving transition metals.

2. OXIDATION PROMOTED COUPLING OF α-SULFONYL CARBANIONS TO OLEFINS

The benzenesulfonyl group ($PhSO_2$-noted Σ- in the following) allows an easy stabilization of α-carbanions. Thus α-sulfonyl carbanions are readily obtained in aprotic solvents such e.g. as THF. The synthetic applications of their catalytic or stoichiometric oxidation by transition metal salts have been widely examined by Julia and coll. (10).

Three main products may be obtained as a function of experimental conditions, i.e. as a function of the nature and relative amount of the transition metal oxidant.

When the latter is present in catalytic amounts (2% or less) the reaction results in the formation of olefins (10a) as in eqn.12. Conversely in the presence of stoichiometric quantities of the metal

$$2\ \Sigma\text{-}\bar{C}H - R \xrightarrow[\text{THF, reflux}]{[Ni(acac)_2, 2\%]} RCH = CHR + 2\,\Sigma^- \qquad (12)$$

salt, a disulfone (10b-10d) formed by oxidative coupling of two anions is generally obtained as in eqn.13, where X=halide, methylsulfonate,

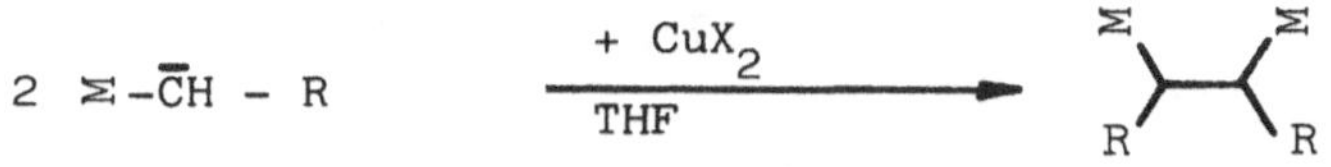

(13)

trifluoroacetate,... Yet for particular ligands, such as acetate, the stoichiometric oxidation results in dehydrogenation (10c-10d) (eqn.-14),

$$\Sigma\text{-}\bar{C}H - CHRR' \xrightarrow[\text{THF}]{+\ Cu(CH_3CO_2)_2} \Sigma - CH = CRR' \qquad (14)$$

rather than coupling (eqn.13).

Although these different reactivities have been investigated for their synthetic potential no explanation has been proposed so far for such subtle dependance on the metal salt nature and quantity. A way to approach the problem is to artificially dissociate the oxidative properties of the metal salt from those relative to specific environment in the coordination shell, i.e., in other words the outer- sphere vis-à-vis the inner-sphere properties. In this regards electrochemical oxidation can be viewed as involving only outer-sphere (the electrode being considered as the oxidant) contributions, except when specific adsorption occurs. This prompted us to investigate the mechanism of oxidative coupling of α-sulfonyl carbanions, under electrochemical conditions (11).

Cyclic voltammetry of the oxidation of benzyl or prenyl-phenyl sulfone anion indicates that the anion is quantitatively oxidized to the radical provided that the time scale is below the microsecond range.

$$\Sigma - \bar{C}H - R \quad - e \rightleftharpoons \Sigma - \dot{C}H - R \qquad (15)$$

In a larger time scale the oxidation still involves a one electron consumption but the radical generated in eqn.15 rapidly evolves.
Peak potential analysis as a function of the scan rate and of the anion concentration establishes that the radical undergoes competitively H atom abstraction from the medium (THF, 0.3M NBu_4BF_4)

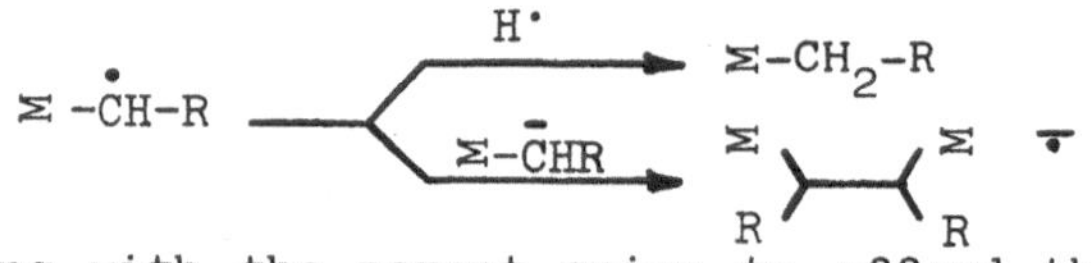

(16a)

(16b)

and coupling with the parent anion to afford the anion radical of the β-disulfone in eqn.16b. The former route is globally equivalent ($-e, +H^{\bullet}$) to a proton transfer to the anion to yield back the starting sulfone. The anion radical obtained by radical/anion coupling was generated independently by electrochemical reduction of the β-disulfone. Preparative scale electrolysis showed that an overall two-electron reduction is observed to afford two benzenesulfinate moieties and

Scheme III :

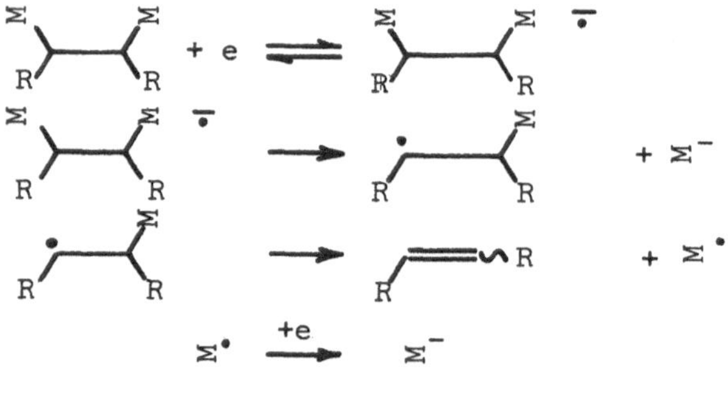

(17)

(18)

(19)

(20)

and the corresponding olefin, as evidenced by the result concerning phenylalkylsulfones. Peak potential analysis of the cathodic wave indicates that the anion radical formed upon the electron transfer in eqn.17, undergoes a rapid cleavage of the carbon-benzene sulfinate bond in eqn.18. The radical thus obtained is known to undergo a rapid elimination of a sulfinate radical (12), which is reduced in eqn.20, to afford the corresponding olefin.

From these independent studies (11) we are thus enclined to propose Scheme IV as a unifying mechanism for the oxidative coupling of α-sulfonyl carbanions. Thus in the presence of a large amount of oxidant, the β-disulfone anion radical is reoxidized to afford the disulfone observed in eqn.13. In the presence of small quantities of the oxidant, i.e. in catalytic conditions, this reaction is too slow to deactivate the disulfone anion radical before it cleaves off the benzene sulfinate anion. The radical thus formed yields then the olefin observed in eqn.12 together with the sulfinate radical which is reduced by the α-sulfonyl carbanion, leading then to a new α-sulfonyl radical which enters again the cycle in Scheme IV.

Scheme IV :

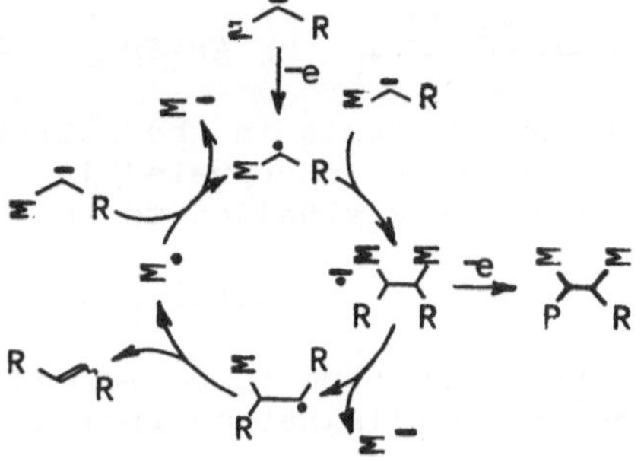

Thus in this description, the dichotomy in eqns.12 and 13, reflects the respective contribution of the electron transfer chain vis-à-vis the termination step. Naturally this interpretation should be transposed with care to the homogeneous systems which involve in addition inner-sphere effects within the transition metal coordination shell.

Finally we want to discuss a possible route for the formation of the α-β-unsaturated sulfone in eqn.14. From our electrochemical data (eqn.16a) the α-sulfonyl radical formed upon the anion oxidation, appears as a rather strong H-atom abstractor. Thus it is conceivable that it may abstract an hydrogen atom from the parent anion (13), as in eqn.21, to afford the rather stable anion radical of the

$$\Sigma\text{-}\bar{C}H\text{-}CHRR' + \Sigma\text{-}\dot{C}H\text{-}CHRR' \longrightarrow \Sigma\text{-}CH_2\text{-}CHRR' + \Sigma\text{-}CH{=}CRR'^{\bullet -} \quad (21)$$

α-β-unsaturated sulfone, which is easily oxidizable to the product in eqn.14. Such a reaction would again act as a termination step for the catalytic radical chain in Scheme IV, being in competition with the radical/anion coupling step in eqn.16b. Yet this must be considered as a tentative explanation since it is seen that the dichotomy in eqns.13 and 14 involves only subtle modification of experimental conditions.

3. BIPHENYL SYNTHESIS CATALYZED BY LOW-LIGATED NICKEL COMPLEXES

Since the pioneering work of Semmelhack and coworkers, synthetic procedures for the preparation of biaryls by the classical Ullman reaction have been supplanted by the use of zerovalent nickel complexes (14). These original methods based upon stoichiometric amounts of nickel were improved by Kumada and coll., using catalytic amounts of

$$2\ ArX \;+\; Zn \xrightarrow{[Ni^{II}]} Ar-Ar \;+\; ZnX_2 \qquad (22)$$

nickel(II) salt and stoichiometric quantities of zinc. Furthermore the presence of excess of reducing metal (Zn, Mn, Mg,...) allows arylchlorides to react (14) although these were reported to react poorly under the Semmelhack's conditions. This enhancement compares with the efficient electrosynthesis of biaryls from arylhalides in the presence of nickel catalysts.

$$2\ ArX \;+\; 2e \xrightarrow{[Ni^{II}]} Ar\text{-}Ar \;+\; 2X^- \qquad (23)$$

The role of the nickel reagents in the activation of the aromatic carbon halogen bond is commonly interpreted by the existence of the rapid oxidative addition of the arylhalide to the in situ generated

$$ArX \;+\; Ni(0)Ln \longrightarrow Ar\text{-}Ni(II)LmX \qquad (24)$$

nickel(0) species. On the other hand there is general agreement on the biaryl formation via reductive elimination from a diarylnickel(III)

$$Ar_2Ni(III)X \longrightarrow Ni(I)X \;+\; Ar\text{-}Ar \qquad (25)$$

species. However the exact sequence of steps leading to such an intermediate from the arylnickel(II) complex in eqn.24, under reductive conditions, is still controversial (14). Elegant mechanistic work by Tsou and Kochi (15) concluded to the involvement of a radical chain process in which paramagnetic Ni(I) and arylnickel(III) are the reactive intermediates. Yet it must be emphasized that the conditions were quite different from those involving a large reductive driving force and catalytic nickel. More recent work by Colon and Kelsey (14), although less documented on kinetic grounds, also retains Ni(I) and Ni(III) species as the key intermediates in biaryl synthesis in conditions akin to those proposed by Kumada and followers. Yet both mechanisms differ significantly in the interpretation of the diarylnickel(III) formation. Kochi and Tsou established, in their conditions, the involvement of an aryl scrambling step as summarized in Scheme V.

<u>Scheme V</u>

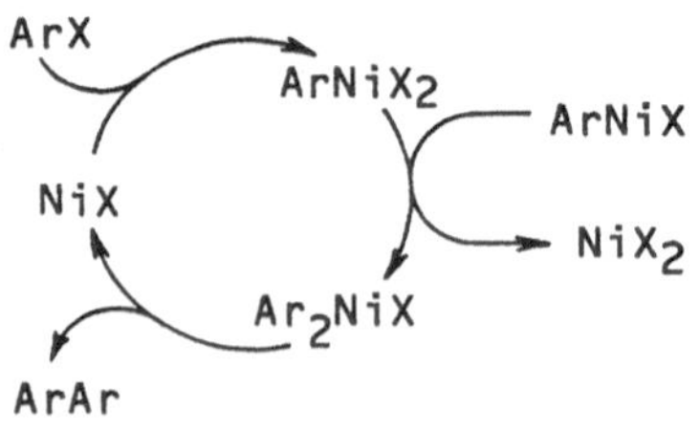

Colon and Kelsey (14) retained an oxidative addition of the organic halide to an arylnickel(I) species as in eqn.25. Other reaction se-

$$ArNi(I) + ArX \longrightarrow Ar_2Ni(III)X \quad (25)$$

quences were also proposed on the basis of electrochemical data but are merely ad-hoc interpretations without firmly established kinetic basis.

The electrochemical reduction of bivalent nickel complexes $Ni(II)Cl_2(dppe)$ to the zerovalent analog involves two successive one-electron transfer (16) as in eqns.26-27. The monovalent nickel electro-

$$Ni(II)Cl_2(dppe) + e \longrightarrow Ni(I)Cl(dppe) + Cl^- \quad (26)$$

$$Ni(I)Cl(dppe) + e \longrightarrow Ni(0)(dppe) + Cl^- \quad (27)$$

generated upon the first electron transfer undergoes a rapid dimerization, with a rate constant of $2.5\ 10^3\ M^{-1}s^{-1}$, to afford dimeric nickel complexes. Yet provided the time scale is short enough, or the reduction performed at the potential of the second wave, a low ligated Ni(0) is obtained quantitatively. The latter undergoes an extremely rapid insertion into aromatic carbon halogen bonds ; the rate constant

$$Ni(0)(dppe) + ArX \longrightarrow Ar\text{-}Ni(II)X(dppe) \quad (28)$$

of the oxidative addition of bromobenzene to Ni(0)(dppe) is evaluated to $10^5\ M^{-1}\ s^{-1}$ by cyclic voltammetry, i.e. ca 50 000 times larger than that corresponding to coordinatively saturated analogs such as $Ni(0)(PEt_3)_4$ and ca 5 000 000 times larger than for the more related

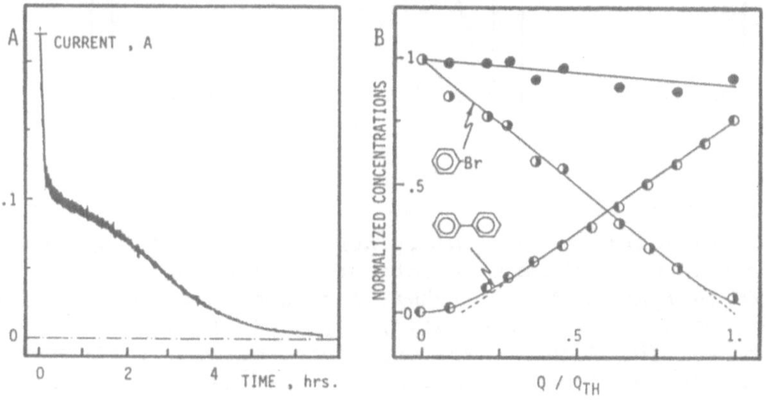

Figure 2 : Biphenyl electrosynthesis from bromobenzene in the presence of 10% $NiCl_2$(dppe) in THF/HMPA (2/1) mixture at -2. V vs SCE, 20°C.(A)Variations of the electrolysis current with time. (B) Solution composition normalized to the initial bromobenzene concentration, as a function of the charge, Q, passed. Q_{TH} : theoretical charge for 1e per PhBr. (●): mass balance including PhBr, PhPh and phenylnickel species.

$Ni(0)(PPh_3)_4$. Yet no evidence of oxidative addition of bromobenzene to the intermediate Ni(I)Cl(dppe) was obtained (17).

On the other hand, although the arylnickel(II) complexe is formed quantitatively upon reduction of the divalent nickel in the presence of bromobenzene, no biphenyl synthesis is observed. Yet if the electrolysis potential is increased cathodically to reach the one electron reduction wave of the arylnickel(II) complexes a nearly quantitative production of biphenyl is obtained as shown in figure 2. Such a result establishes unambiguously the central role of the paramagnetic aryl

$$\text{Ar-Ni(II)X(dppe)} + e \longrightarrow \text{Ar-Ni(I)(dppe)} + X^- \qquad (29)$$

nickel(I) in biphenyl synthesis under reductive conditions. In the presence of an excess of aromatic halide the arylnickel(II)(dppe) reduction wave current magnitude increases which indicates that the arylnickel(I) species obtained in eqn.29 readily reacts with the organic halide to restore finally an arylnickel(II) intermediate (compare figure 3A) (18).

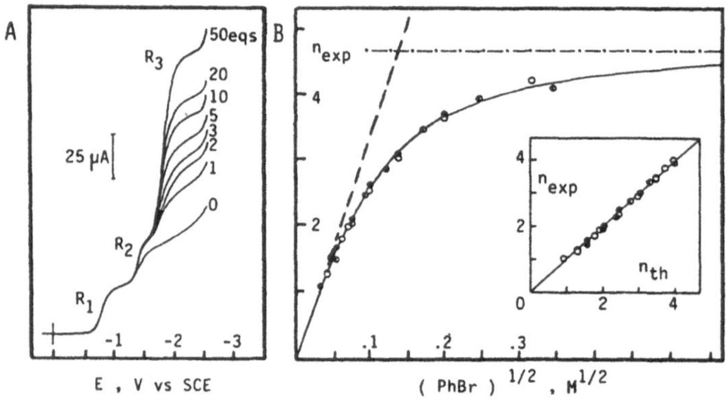

Figure 3 : (A) Voltammetry of $NiCl_2$(dppe) at the Rotating Disk Electrode in the presence of different amounts of bromobenzene. R_1: Ni(II)/Ni(I) ; R_2: Ni(I)/Ni(0); R_3: PhNi(II)X/PhNi(I) reduction waves. The instant turn over number is defined by the ratio, n_{exp}, of the limiting currents of waves R_3 to R_1. (B) Experimental dependance of n_{exp} with PhBr and Ni(II) concentrations. Solid line : theoretical variations (n_{th}) according to Scheme VI. n_{exp} and n_{th} are compared in the insert of (B).

The variations of the overall number of electron consummed at the arylnickel(II) reduction wave, as a function of the bromobenzene and nickel concentrations allows the proposal of the catalytic cycle in Scheme VI. Indeed the non dependance of the instant turn over number

Scheme VI (16)

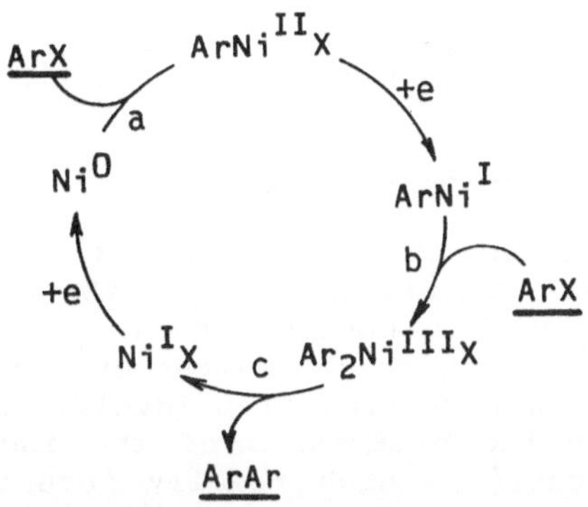

with the nickel concentration rules out the involvement of a bimolecular step between two nickel centered intermediates although such a step has been identified by Tsou and Kochi (compare Scheme V) in the absence of a large reductive driving force (15). Similarly at low bromobenzene concentrations the rate determining step involves bromobenzene and a nickel complex with a molecularity of one for each. At larger concentrations of bromobenzene the turn over tends to be independent of bromobenzene concentration. Both results strongly suggest the involvement of the sequence of steps (b) and (c) in Scheme VI. This is further confirmed by the perfect adequation between the experimental turn over and the theoretical one predicted for the catalytic sequence in Scheme VI, as shown in figure 3B. Thus the pertinent rate constants are determined for the three chemical steps involved in Scheme VI (16).

$$PhBr + Ni(0)(dppe) \xrightarrow{(a)} PhNi(II)Br(dppe)\ ;\ k_a=1.1\ 10^5\ M^{-1}s^{-1} \quad (29)$$

$$PhBr + PhNi(I)(dppe) \xrightarrow{(b)} Ph_2Ni(III)Br(dppe)\ ;\ k_b=960M^{-1}s^{-1} \quad (30)$$

$$Ph_2Ni(III)Br(dppe) \xrightarrow{(c)} PhPh + Ni(I)Br(dppe)\ ;\ k_c=18s^{-1} \quad (31)$$

As explained above the catalytic chain in Scheme VI is initiated by the bivalent nickel reduction into the low ligated Ni(O)(dppe) complex. The chain propagation needs in addition two other electron transfer steps : the one-electron activation of the arylnickel(II) adduct to its nickel(I) analog and the recycling of the Ni(I)X formed after the reductive elimination, into the active Ni(O)(dppe).

The main deactivation route to the catalytic cycle in Scheme VI consists in the rapid (2.5 10^3 $M^{-1}s^{-1}$) dimerization of the nickel(I) species formed in eqn.31. Under electrochemical kinetic conditions because of the millimolar nickel concentration and the time scale of the experiments, this termination step did not interfere significantly. Yet for preparative purposes it may constitute a severe limitation. Indeed the rate at which the nickel(I) is reduced is determinant since it governs the amount of nickel(0) regenerated at each cycle. Thus the lower this rate or the higher the concentration of

$$\mathrm{Ni(I)X(dppe)} \xrightarrow{+e,\,-X^-} \mathrm{Ni(0)(dppe)} \xrightarrow{ArX} \cdots \qquad (32)$$

$$\mathrm{Ni(I)X(dppe)} \xrightarrow{x2} \text{dimer} \qquad (33)$$

nickel, the larger the amount of nickel lost at each cycle. Since on the other hand the overall conversion rate of the organic halide into biphenyl depends on the nickel concentration, it is seen that the conversion rate tends to decay with the number of cycles.

The other one electron reduction step involved in the propagation of the chain consists in the activation of the inert arylnickel(II) complex into the arylnickel(I) which readily inserts into the aryl-halide bond. In the absence of a large cathodic potential this reduction is certainly the rate limiting step in the propagation of the chain, because of the important reduction potential of the arylnickel(II). Thus when the biphenyl synthesis is performed with metal powder as the reducing agent, most of the nickel concentration is stored under the form of the arylnickel(II) intermediate. This results in a "protection" vis-à-vis the deactivation of the cycle since then the Ni(I)X concentration tends to be very small. A correlate is then that the rate of biphenyl formation should be nearly independent of the conversion for a given system, but its value should be extremely dependent on the metal "concentration" as well as on the ease of the arylnickel(II) reduction (19).

4. CONCLUSION

In the above presentation we have tried to illustrate the role of the activation/termination sequences on the overall efficiency of electron transfer chain catalysis. This is a well identified problem in other chain mechanisms such as alcane substitution, polymerizations, combustion,... but tends to be neglected in the discussion of most electron transfer chains. This is certainly due to the difficulty in identifying such steps for efficient chains. In this respect electrochemistry which allows a facile control of the energy and rate flow of the electrons injected into the system proves to be an extremely adequate method.

ACKNOWLEDGEMENTS

The work presented was performed under the auspices of CNRS (UA. 1110, "Activation moléculaire") and Ecole Normale Supérieure. Miss T. El Moustafid is acknowledged for her participation in the α-sulfonyl anion oxidative coupling study, as well as Dr. C. Rolando (CNRS, ENS) for valuable discussions in the early stages of this project. Similarly Pr. D. Astruc (Université of Bordeaux I) is acknowledged for permission of quoting part of our unpublished results on the dithiocarbamate electron transfer induced chelation.

REFERENCES

1. See e.g.(a) J.K. Kochi, in "Organometallic Mechanism and Catalysis", Academic Press, New-York, 1978 ; (b) M. Chanon, Bull. Soc.Chim. Fr.,II, 1982, 197 ; (c) M. Chanon and M.L. Tobe, Angew.Chem. Int. Ed. Engl., 21, 1982, 1 ; (d) J.K. Kochi, J. Organomet. Chem., 300, 1986, 139 ; (e) J.M. Savéant, Acc. Chem. Res., 13, 1980, 323.

2. J.F. Bunnett, Acc. Chem. Res., 11, 1978, 413, and refs therein.

3. N. Kornblum, R.E. Michel, and R.C. Kerber, J. Am. Chem. Soc., 88, 1966, 5662.

4. C. Amatore, J. Pinson, J.M. Savéant and A. Thiébault, J. Am. Chem. Soc., 103, 1981, 6930.

5. R.A. Rossi and R.H. de Rossi, in "Aromatic Substitution by the $S_{RN}1$ Mechanism", ACS Monographs 178, Washington, 1983, and refs therein.

6. C. Amatore, M.A. Oturan, J. Pinson, J.M. Savéant and A. Thiébault, J. Am. Chem. Soc., 107, 1985, 3451.

7. C. Amatore, J. Pinson, J.M. Savéant and A. Thiébault, J. Electroanal. Chem., 107, 1980, 59.

8. (a) C. Amatore, J.N. Verpeaux, A. Madonik, M.H. Desbois and D. Astruc, submitted for publication ; (b) C. Amatore, J.N. Verpeaux and D. Astruc, unpublished results.

9. See ref. 1d for a detailled presentation, and : (a) S.W. Feldberg and L. Jeftic, J. Phys. Chem., 76, 1972, 2439 ; (b) R.D. Rieke, H. Kojima and Ofele, K., J. Am. Chem. Soc., 98, 1976, 6735 ; (c) J. Moraczewski, W.E. Geiger, J. Am. Chem. Soc., 101, 1979, 3407 ; for early examples.

10. See e.g.: (a) M. Julia and J.N. Verpeaux, Tetrahedron Lett., 23, 1982, 2457 ; (b) M. Julia, G. Le Thuillier, C. Rolando and L. Saussine, Tetrahedron Lett., 23, 1982, 2453 ; (c) J.B. Baudin, M. Julia, C. Rolando and J.N. Verpeaux, Tetrahedron Lett., 25, 1984, 3203, and (d) Bull. Soc. Chim. Fr., 1987, 493.

11. C. Amatore, T. El Moustafid and J.N. Verpeaux, manuscript in preparation.

12. T.E. Boothe, J.L. Greene Jr. and P.B. Shevlin, J. Org. Chem., 45, 1980, 794.

13. The proximity of a negative charge, should enhance the lability of the α CH bond as observed e.g. for alcoolates : C. Amatore, J. Badoz-Lambling, C. Bonnel-Huyghes, J. Pinson, J.M. Savéant and

A. Thiébault, J. Am. Chem. Soc., 104, 1982, 1979.

14. See : I. Colon and D.R. Kelsey, J. Org. Chem., 51, 1986, 2627, for a detailled presentation of the reaction and of the pertinent references.

15. T.T. Tsou and J.K. Kochi, J. Am. Chem. Soc., 101, 1979, 7547.

16. C. Amatore and A. Jutand, submitted for publication.

17. Compare with ref.15.

18. This phenomenon is similar to "Redox Catalysis" in terms of kinetic behavior. Compare : (a) R.S. Nicholson and I. Shain, Anal. Chem., 36, 1964, 706 ; (b) C.P. Andrieux, C. Blocman, J.M. Dumas-Bouchiat, F. M'Halla and J.M. Savéant, J. Am. Chem. Soc., 102, 1980, 3806, and refs. therein.

19. All three effects are observed by Colon and Kelsey. Compare figs. 1-4 in ref.14.

MOLECULAR ORBITAL TREATMENT OF ACTIVATION INDUCED BY ELECTRON ATTACHMENT.
APPLICATION TO ELECTRON TRANSFER CATALYSIS

Keith F. PURCELL, Willard Hall, Kansas State Univ., Manhattan(KS)
Michel RAJZMANN and **Michel CHANON**, Faculté Sciences St Jérome, Marseille, 13397 .
Henri CHERMETTE, Chérifa MEHADJI, CPN2 Inst. Phys. Nucl., 43 Bd. 11 Novembre, 69622 Villeurbanne Cédex.

In a first paper dealing with the generalization of electron transfer catalysis, we suggested that the common feature shared by all mechanisms, is the existence, at a critical step, of an activation induced by electron transfer (1). By "activation", we mean a marked enhancement of reactivity for a given transformation; by "induced by electron transfer", we mean that such an activation may result from electron attachment to the substrate as well as from electron removal from the substrate. In that first report, illustrative examples were selected from inorganic and organic chemistry to show that rate enhancements associated with such an activation could reach values as high an 10^{15} (Table 4 in ref.1). Two main types of activation were identified : **associative activation and dissociative activation**. Organometallic examples illustrate these types of activation :

1. 1. Reductive Associative Activation

<u>**Scheme 1**</u> (ref. 2)

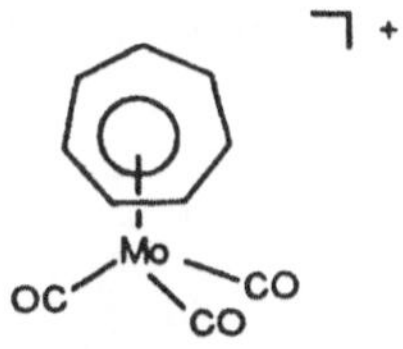

displays no tendency to dimerization, whereas

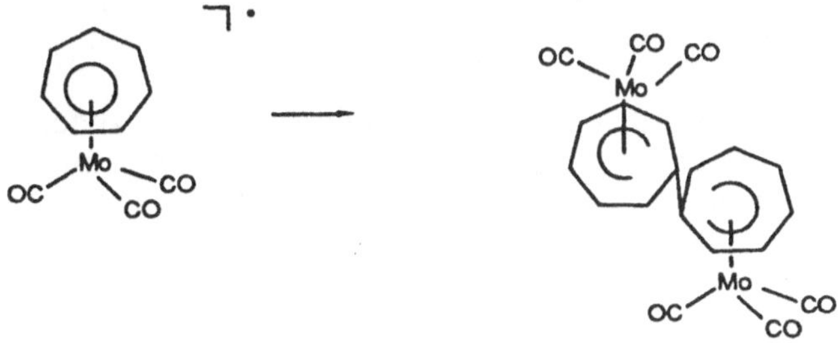

M. Chanon et al. (eds.), Paramagnetic Organometallic Species in Activation / Selectivity, Catalysis, 225–243.

In the same sense Cp_2Rh dimerizes far more easily than does Cp_2Rh^+ (3).

1. 2. Reductive Dissociative Activation

Electrochemical studies (4,5) provide abundant examples of this type of activation. $M(CO)_6$ (M = Cr, Mo, W) is stable whereas $[M(CO)_6]^-$ loses very easily a CO (6,7,8). References 4 and 5 should be consulted for further examples. Reductive dissociative activation has been thoroughly studied on halogeno aromatic substrates (9). The lifetimes of halogeno-aromatic radical anions span more than 10 orders of magnitude. These lifetimes seem to be directly connected with the energy of LUMO to which the electron is added. This is what should be expected (ref 1 p. 11) from a zeroth order approach to reactivity: the more the LUMO is antibonding (i.e. the more the substrate is difficult to reduce), the shorter should be the lifetime in a series of compounds. If this pattern of reactivity applies also to reductive dissociative activation of organometallic substrates, it would be interesting to find correlations for predicting the reduction potential of complexes on the basis of ligand parameters, as it has been done for the oxidation potential $E_{1/2}$ (10,11) :

$$E_{1/2}\,[ML] = Es + \beta\,\rho_L$$

ρ_L is a ligand parameter

β is a measure of site polarizability

Es is the potential when the metal site Ms is occupied by CO, and it reflects approximately the electron richness of this site. Such a correlation would be especially interesting for a VLCEK redox series (12-14), which is a set of compounds with identical composition differing only in the overall number of electrons (n) :

$$ML_m{}^{z},\ ML_m{}^{z-1},\ ML_m{}^{z-2},\ ML_m{}^{z-3}$$

m : number of identical ligands

z : charge of the complex

Of the structural factors governing the stability of a redox series of complexes one is the direct or metal-mediated ligand-ligand interaction, and others are the IP (ionisation potential) and EA (electron affinity) of the central metal and of the the ligands. In a redox series, a certain number of electrons may be successively added to the most oxidized term of the series, but beyond a given threshold, there is usually a strong dissociative reductive activation. For hexacoordinated symmetrical complexes, using the very crude angular ovelap model (15), it was found that, for some configurations of the central metal, one could expect a very marked change in stability on adding an electron (1); these are d^3 (particularly with ligands favoring high spin states in the d^4 configuration), d^6 low spin (organometallics derive from a special class of ligands favouring a low spin configuration). From the rate values of water exchange measured for the first row dipositive transition-metal ions (Fig 1 in ref 16), it would seem that configurations d^2 and d^8 are also good candidates for appropriate ligands. For configurations d^3 and d^6, the marked effect of adding an electron on the stability of the complex corresponds simply to the population of a strongly antibonding σ^* orbital. There is a difference between the d^3 and d^6 configurations: for the d^3 configuration, if one wants to stay with ligands favoring high spin states, an intrinsic limit to reductive dissociative activation appears because after a given threshold of ligand strength, rendering the LUMO more and more antibonding, one switchs to low spin states. In contrast, for d^6 the same parameters which make the antibonding character stronger, as demanded for a good reductive dissociative activation, are consonant with the low spin filling of molecular orbitals.

When the 18 electron rule applies, to predict strong dissociative reductive activation one may simply rely on the number of valence electrons (NVE). This has been done in Stiegman and Tyler's recent review (17), where the structural factors playing a role in the stability of 19e organometallic species are critically discussed.

Two special cases of reductive dissociative activation deserve special comment. The first corresponds to a situation where the activation does not cause a total dissociation of the ligand, but nevertheless labilizes the bond between the metal and the ligand. Cyclopentadienyl cobalt complexes of cyclooctatetraene may present two types of binding :

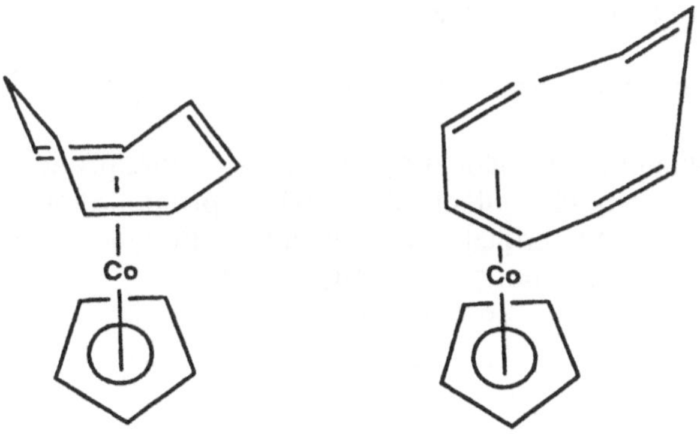

Scheme 2 : a) (1,5 -COT)CoCp b) (1,3 - COT)CoCp

One electron reduction of the (1,5-COT) isomer causes its very rapid transformation to the anion of the 1,3-isomer (18). More examples of such structural changes induced by electron attachment to organometallic substrates may be found in ref 19. Bimetallic organometallic compounds with p-bridging ligands are especially important to study quantitatively for the variation of bond distance and formal bond order with electron transfer (20 - 22). The second special case corresponds to the situation where electron attachment to a substrate provides an indirect labilization of a bond. The behaviour of polynuclear metal carbonyls such as

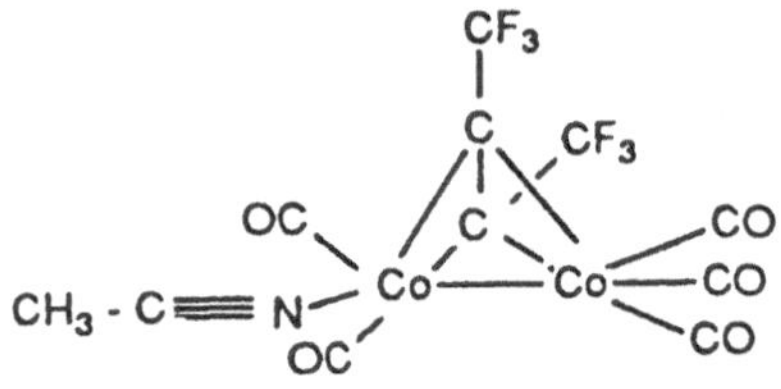

provides such an example. The radical anions of these polynuclear metal carbonyls are much more prone to ligand substitution than their neutral counterparts. But here, in contrast with what is known for the SRN1 substitution, the radical anion does not dissociate into radicals and anions. The added electron occupies an antibonding orbital centered on the metal core, and the increase in ligand lability is an indirect consequence of the metal-metal bond weakening or cleavage (23), which produces a coordinatively unsaturated 17 - electron metal atom. Such an atom is activated towards substitution (17). Other examples of such an indirect activation may be found in ref 24 (see also Kochi's chapter in this book).

Electron deficient organometallics (RHgX, $RTlX_2$) are also prone to dissociative reductive activation (25,26).

1. 3. Oxidative dissociative activation

Oxidative dissociative activation is less widespread than its reductive counterpart. This could reflect the MO representation of bonding, which considers that bonding orbitals display less strongly bonding properties than their antibonding counterparts do antibonding:

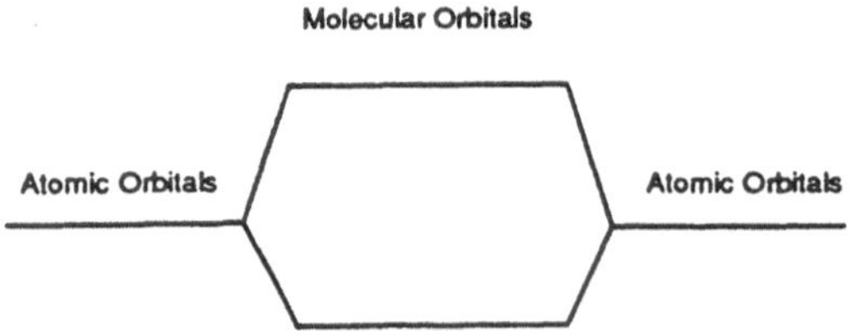

The 18e complex $[Cr(CNR)_7]^{2+}$ dissociates to an octahedral species on one-electron oxidation (27). Particularly clear cut cases of oxidative dissociative activation have been reported by Kochi's group (28) :

$$R_4M + [IrCl_6]^{2-} \dashrightarrow R_4M^+ + [IrCl_6]^{3-} \qquad M = Sn$$
$$R_4M^+ \dashrightarrow R_3M^+ + R^0$$

For alkylmetals derived from the main-group elements (M = Sn, Pb, Hg), the electron is lost from a carbon-metal bonding orbital (HOMO) (28b). The selectivity of cleavage in methylethyltin compounds (cleavage of Sn-Et versus cleavage of Sn-Me) provides insights on the transition state of electron transfer (note 51 in ref 28b). Oxidative dissociative activation has been studied or used by various authors (29,30,31,32).

1.4. Associative Oxidative Activation

For organic substrates, associative activation following oxidation is well known for olefins (1). Some organometallic substrates undergo also this type of activation (33).

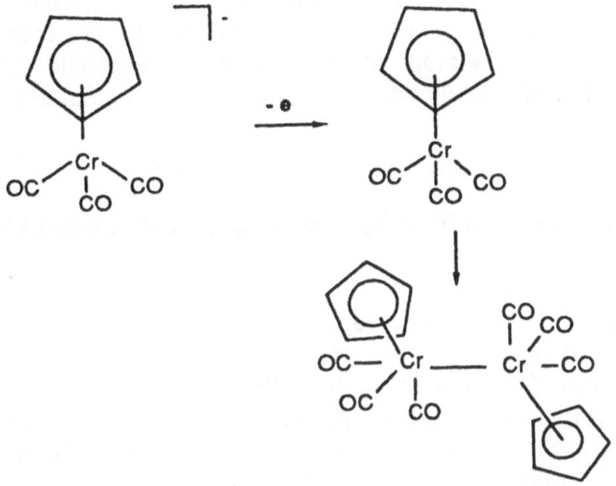

1.5. Disproportionation

In some cases, the addition or removal of an electron activates the substrate toward disproportionation. Vlcek (13) recalled that, in an equilibrium,

$$ML_m^{\,z} \underset{}{\overset{E_1'}{\rightleftharpoons}} ML_m^{\,z-1} \overset{E_2'}{\rightleftharpoons} ML_m^{\,z-2}$$

the comproportionation constant

$$K = \frac{[ML_m{}^{z-1}]^2}{[ML_m{}^{z}]\ [[ML_m{}^{z-2}]}$$

depends upon the difference $E_1^{\circ} - E_2^{\circ}$. For a difference smaller than 180 mV, $ML_m{}^{z-1}$ disproportionates by more than 3% into $ML_m{}^{z}$ and $ML_m{}^{z-2}$. For a difference larger than 300 mV, the disporportionation is less than 0.3%.

At this point, it is important to notice that dramatic activation is mainly expected to occur at the extremes of a redox series. Indeed, all the members of a redox series display a given lifetime, each certainly possesses its own pattern of reactivity different from its oxidized and reduced partners, and the difference may be more or less marked. However, it is at the extremes of the series that the addition or removal of an electron causes drastic changes because these species now have a very fleeting existence. Much more data are needed to identify clearly the types of activation associated with a displacement in a redox series, and these data would probably be of great help in designing new catalytic systems. It is the purpose of this report to show in which way a molecular orbital approach may be used to study reductive dissociative activation and a related electron transfer catalytic scheme. The last part of this report will be devoted to some general aspect of electron transfer catalysis.

2. Molecular Orbital Treatment of Two Typical Cases of Dissociative Reductive Activation

The theoretical treatment of organometallic systems from a perspective of reactivity has been excellently dealt with by Dedieu (34). An in depth treatment of electron transfer induced activation should involve the complete calculation of every term of a redox series, with the inherent difficulties associated with treatment of paramagnetic species. For dramatic activation induced by electron transfer (table 4 ref.1), even crude but consistent treatments using a frontier orbital approach (35) will provide insights, despite their simplicity.

2.1. A straightforward case : $[PtCl_6]^{2-}$

Relativistic MS - Xα calculations were performed on $[PtCl_6]^{2-}$, $[PtCl_6]^{3-}$ and related species (36), to rationalize the spectra of various transients observed during the pulse radiolysis of $[PtCl_6]^{2-}$ (37).

Figure 1 : Total σ electronic isodensity contours, plotted in the zx plane; **A**-$[PtCl_6]^{2-}$(Oh); **B**-$[PtCl_6]^{3-}$(Oh). Contour values of , 2, 3, 4, 5, 6 are equal to 0.006, 0.030, 0.042, 0.090, 0.210 (electron/bohr^3), respectively and are the same for all the total σ electronic isodensity contours.

Figure 1 compares the total σ electronic isodensity in $[PtCl_6]^{2-}$ and $[PtCl_6]^{3-}$ and shows clearly the strong activation towards dissociation, induced by electron attachment to $[PtCl_6]^{2-}$. The prevailing factor in the σ-electronic interactions of $[PtCl_6]^{3-}$ is the population of the σ-antibonding MO 3 eg, which weakens the metal-ligand σ-bonding framework. Formally, the additional electron in 3 eg would be attributed half to the metal and half to the ligands,. The naive picture of electron-transfer activation toward dissociation (population of an antibonding MO) is therefore confirmed by this more elaborate model. Experimentally, and in agreement with such an activation, $[PtCl_6]^{3-}$ loses very rapidly one or two chloride ligands :

$[PtCl_6]^{3-}$ ---------> $[PtCl_5]^{2-} + Cl^-$ (ref 37a)

$[PtCl_6]^{3-}$ ---------> $[PtCl_4]^- + 2\ Cl^-$ (ref 37c)

2.2. A more difficult case : $Fe(CO)_5$

Whereas $Fe(CO)_5$ is inert toward substitution of one CO ligand, the 19-electron complex $[Fe(CO)_5]^-$ dissociates a ligand rapidly to yield a 17 electron complex (7) which then, in effect, dimerizes:

$[Fe(CO)_5]^-$ ---------> $[Fe(CO)_4]^-$ + CO

↓

1/2 $[Fe_2(CO)_8]^{2-}$

Depending upon the conditions, $[Fe(CO)_4]^-$ may have various fates, all described in Connelly-Geiger's review (4) and Krusic has observed up to four paramagnetic species when alkali metals are added to dilute solutions of $Fe(CO)_5$ (7b). It is noteworthy that none of the ESR spectra correspond to the simplest members $[Fe(CO)_5]^-$ and $[Fe(CO)_4]^-$: this suggests that these species are extremely short lived (The radical anions observed in ESR are $[Fe_2CO)_8]^-$, $[Fe_3(CO)_{11}]^-$ and $[Fe_4(CO)_{13}]^-$. As transition metal pentacoordinated complexes have been successfully treated with EHT (38), we decided to adopt this method for the treatment of reductive dissociative activation of $Fe(CO)_5$. The results gathered in **Figures 2a, 2b** and **Table 1** yield an interesting but somewhat unexpected result. The localization of LUMO in $Fe(CO)_5$ is incompatible with the naive picture of an added electron occupying a LUMO strongly antibonding between Fe and C.

Figure 2a : LUMO of $Fe(CO)_5$ as yielded by EHT

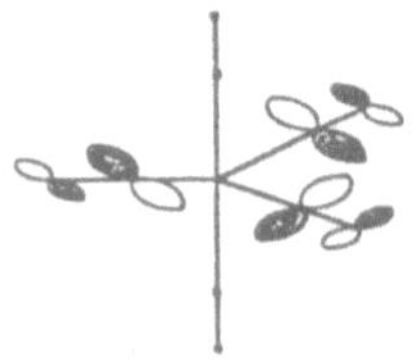

AO contributions

* nothing on Fe
* C_e p_x 27 %
* O_e p_x 7%

Table 1 : EHT Eigenvalues for $Fe(CO)_5$

MO	N° MO	MO Energy (eV)
4 E"	15 - 16	- 8.143
5 A"2	17	- 8.374
7 E'	18 - 19	- 8.686
8 A'2	20 LUMO	- 9.001
6 E'	21 - 22 HOMO	- 12.103
3 E"	23 - 24	- 13.320
4 A"2	25	- 14.263
7 A'	26	- 14.595

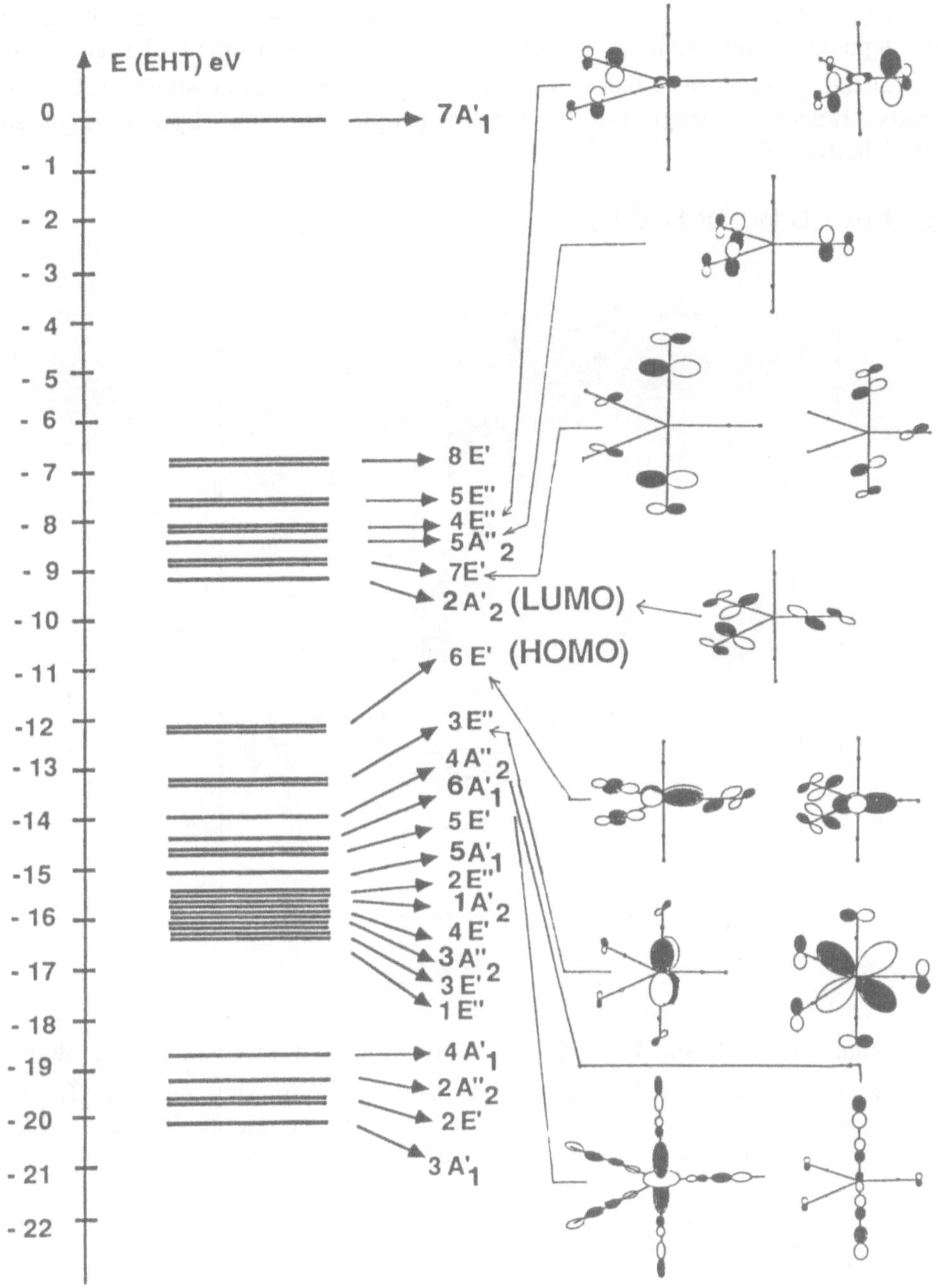

Figure 2b : M.O. diagram of $Fe(CO)_5$ provided by EHT *

* For illustrative specific values see **Table 2**

Indeed the LUMO displayed in **Figure 2a** is localized on the CO ligands and the orientation of the p type orbitals, situated in the plane defined by the 3 CO coplanar ligands is not suitable for direct dissociation of ligands. To verify this ligand-localized nature of LUMO, we have begun Xα calculations (39). The preliminary results support the EHT conclusion that the LUMO is ligand localized (**Figure 3**).

Figure 3 : Xα LUMO for $Fe(CO)_5$

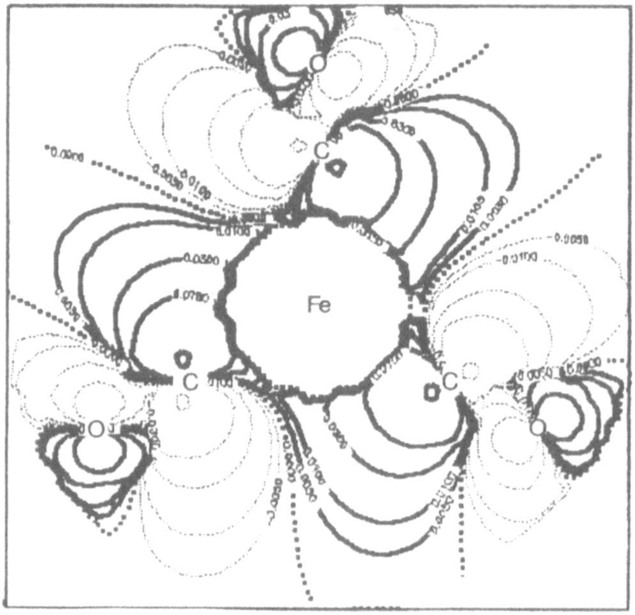

Even though the EHT and Xα LUMO's have different symmetries (a_2' and e', respectively; but notice that the latter is close lying to the former in EHT), they have in common that the LUMO is localized on the equatorial ligands. Thus, the addition of an electron to the $Fe(CO)_5$ LUMO does not allow rationalization of direct ligand labilization. Nevertheless, an EHT calculation shows that the loss of one ligand from the anion is about 30 kcal/mol less endothermic than is the case for the the neutral molecule. There clearly is an activation, but of thermodynamic origin.

The least motion path for dissociation of $[FeCO)_5]^-$ is described in **Figure 4**.

Figure 4 : Vibrational motion leading to $[Fe(CO)_5]^-$ dissociation

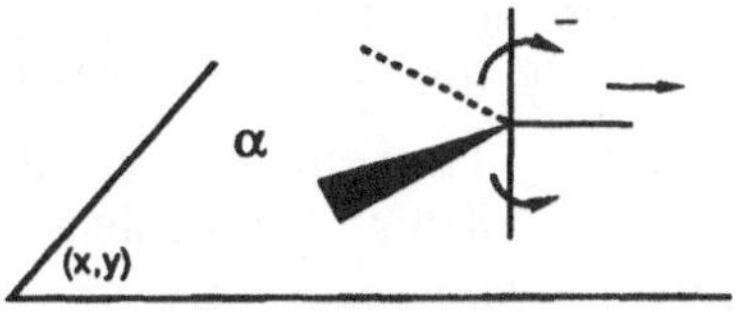

This pathway of dissociation for $[Fe(CO)_5]^-$ leads to the stable form of $[Fe(CO)_4]^-$ which is nearly tetrahedral (**Table 2**).

Table 2: $[Fe(CO)_4]^-$ energies

Total Energy (Kcal)	- 21199.156	- 21206.504	- 21206.641	- 21136.047

3. Molecular Orbital Treatment of on Electron Transfer Catalytic Cycle.

3.1. Various steps involved in the catalytic cycle.

The reaction of $Fe(CO)_5$ with L ($L=PPh_3$, $P(OPh)_3$, $P(OMe)_3$) in the presence of a catalytic amount of iron carbonyl anion, yields $LFe(CO)_4$ (40). One possible catalytic cycle may be written (17) (**Scheme 3**):

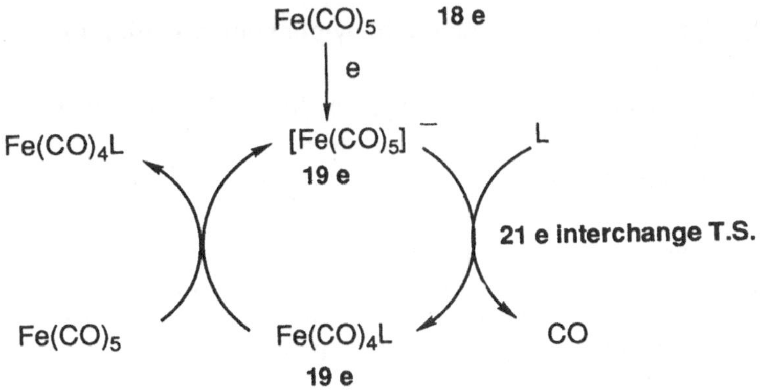

In the ligand substitution section of this cycle one has the counterpart of SRN2 (41,42). According to such a mechanism, the dissociative reductive activation is weak so that a nucleophile is still needed to obtain substitution. Nucleophilic attack does not occur easily on $Fe(CO)_5$ itself, and it does not appear that it will be easier for the anion. Furthermore, we know that formation of $Fe(CO)_4^-$ is rapid. Thus, the more likely mechanism is the reductive dissociation path (SRN1 mechanism (43)) (**Scheme 4**):

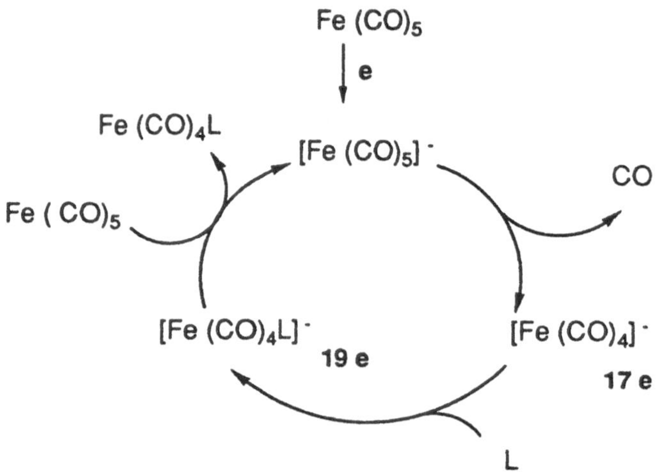

Not surprizingly, we find for organometallic substrates the same dilema always present for the SRN1 mechanism in organic chemistry. The range of halogeno-aromatic radical anions spans more than 9 orders of magnitude (9b) and, nevertheless, all the electron transfer provoked substitutions of halo-aromatics are considered as dissociative (44). Stiegman and Tyler (17) propose that most of the electron transfer catalysed organometallic substitutions involving 19e species should occur via a dissociative mechanism, but we add a word of caution, recalling that 21-electron species such as

Cp_2Mn dppm (dppm=1,2-bis(diphenylphosphino)methane)

are known (45). Probably for this reason, in the electron transfer provoked substitution of $Fe(CO)_5$, they propose as propagation step the interchange mechanism for the reaction between $[Fe(CO)_5]^-$ and L.

At this point there is, as for organic substrates (41-42), not completely compelling evidence for the interchange mechanism of the radical anion. Therefore we will deal here mainly with the dissociative mechanism, which is more amenable to an EHT treatment and that we intuitively favour because even the fastest spectroscopic methods do not allow the direct observation of $[Fe(CO)_5]^-$ (7b). Within such a dissociative mechanism, the frontier orbital approach yields information about

1) the initiation step,
2) the associative reaction between the 17e species and L,
3) the propagation step.

The initiation of this reaction may be effected in different ways. The substitution may be triggered by adding solutions of sodium benzophenone ketyl or $[Fe_x(CO)_y]^{2-}$ to a mixture of Ph_3P and $Fe(CO)_5$. Curiously, the ETH approach suggests that these anions are not strong enough to reduce $Fe(CO)_5$ (the energy of the LUMO of the latter falls at about -9eV while the anion HOMO's fall in the -10 to -9 eV range). This suggests that the rate controlling step may be the initiation step.

The step :

$$\underset{\mathbf{17e}}{[Fe(CO)_4]^-} + L \dashrightarrow \underset{\mathbf{19e}}{[Fe(CO)_4L]^-}$$

like that of analogous organic substrates (46), may be nearly diffusion controlled. As in the absence of L, where the fate of $[Fe(CO)_4]^-$ is to give $[Fe_2(CO)_8]^{2-}$ so rapidly that ESR cannot detect $[Fe(CO)_4]^-$ (7b), in the presence of L, substitution is observed only subsequent to anion formation.

Saveant points out that this type of reaction is the reverse of cleavage of aromatic halide anion radicals, and therefore that the easier ArNu is reduced (i.e. the deeper is the LUMO in ArNu), the faster should be the reverse (addition) reaction. This remark has interesting consequences for the last step of our catalytic cycle. The last step involves an electron transfer from the radical anion $[Fe(CO)_4L]^-$ to $Fe(CO)_5$ to propagate the catalytic cycle. This step is expected to be faster when the gap LUMO of $Fe(CO)_4L$ to LUMO of $Fe(CO)_5$ is the largest possible in the right direction (LUMO $Fe(CO)_5$ deeper). One therefore sees that substitutional effects have conflicting consequences on the dissociation, addition and electron transfer propagation steps. The higher lying the LUMO in $Fe(CO)_5$, the faster the CO dissociation step but the slower the addition and the propagation electron transfer steps. This remark seems to be general in electron transfer catalysis: variations in LUMO energy may display opposite effects on the successive steps so there is a requirement for an overall balance in its energy for an efficient catalytic cycle.

3.2. Innersphere and Outersphere Electron Transfer Catalytic Cycles.

The classical SRN1 mechanism has propagated the belief that most of electron transfer catalysed phenomena involve mainly outer sphere electron transfer. One of the consequences of generalizing the SRN1 mechanism was to propose that, in the spirit of the inorganic chemists who consider two main types of electron transfer mechanism (inner sphere and outer sphere) (15), two sub-classes of electron transfer catalysis are to be expected in general (43d, 46, 47). One may illustrate this proposition with one example taken in ref.17.

The following chain mechanism was proposed for the $Cp_2Mo_2(CO)_6$ catalysed substitution of $CpMo(CO)_3I$ by various L (48). In this chain mechanism, the propagation step may be viewed as an inner sphere electron transfer (i.s.e.t.) with group transfer, and if one considers the propagation step as the most important step in the chain, then one sees an important correspondance between processes classically described as chain reactions and processes called "electron transfer catalysed" (**Scheme 5**):

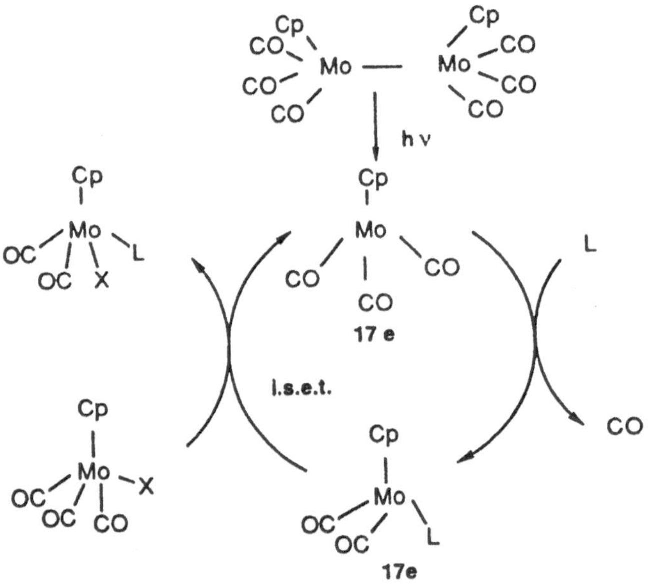

For the same "chain mechanism", Stiegman and Tyler propose an outer sphere electron transfer (**Scheme 6**):

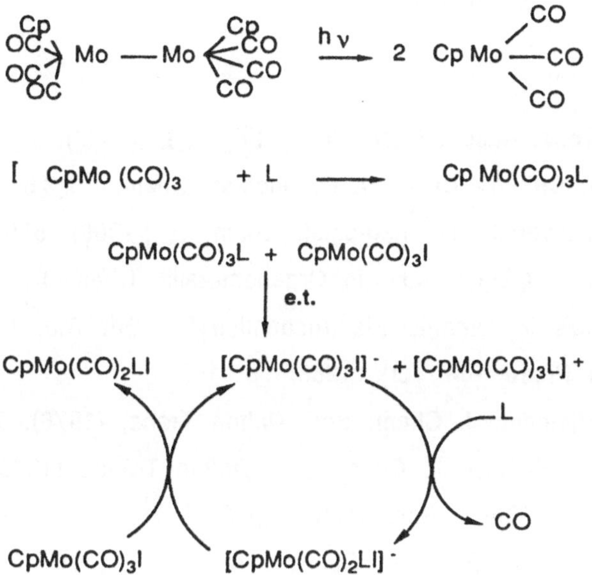

This example recalls for us that several different mechanisms may be found under the label "electron transfer catalysis" (the same holds true for proton transfer catalysis (49)).

In summary, the M.O. approach brings interesting unifying information on the problem of electron transfer activation. The tools have to be sharpened, however, if one wants to address problems such as interchange versus dissociative mechanisms; in terms of a simple frontier orbital approach, there is a need for consistent treatment of the depth of these orbitals and of their nature. One way of reaching consistency is to use different methods (EHT, Xα, Fenske) to treat the molecular targets where electron transfer activation has been clearly shown. Such a consistent treatment would clearly be useful for photochemical, electrochemical and catalytic problems involving paramagnetic organometallic species. Finally, the role of the counterion in electron transfer catalysis has not been addresseed in this work, but will be investigated as our work continues.

References

1. M. Chanon, M.L. Tobe, Angew. Chem. Int. Ed, **21** ,(1982), 1
2. J.A. Armstead, D.J. Cox, R. Davis, J. Organomet. Chem., **236** , (1982) 213
3. E.O. Fisher, H. Wawersik, J. Organomet. Chem, **5** ,(1966), 559
4. N.G. Connelly, W. E. Geiger, Adv. in Organometallic Chemistry, **23**, (1984), 1
5. J.C. Kotz in "Topics in Organic Electrochemistry", Ed. A.J. Fry and W. E. Britton, Plenum Press New York, 1986, p. 142
6. C.J. Pickett , D. Pletcher, J. Chem. Soc. Dalton Trans, (1976), 749
7. a) C.J. Pickett, D. Pletcher, J. Chem. Soc. Dalton Trans., (1975), 879
   b) P.J. Krusic, J. Am. Chem. Soc. ,(1981), **103**, 2129
8. P. Lemoine, M. Gross, C. R. Hebd. Séances Acad. Sci, **280**, (1975), 797
9. a) C.P. Andrieux, J.M. Savéant, K. B. Su, J. Phys. Chem. ,**90** ,(1986) 3815
   b) L.G. Feoktistov in "Organic Electrochemistry", Ed. M. M. Baizer and H. Lund, M. Dekker New York and Basel, 1983 p. 259 and references, cited therein
10. C. J. Pickett, D. Pletcher, J. Organomet. Chem., **102** , (1975) 327
11. J. Chatt, C. T. Kan, G. J. Leigh, C. J. Pickett , D. R. Stanley, J. Chem. Soc. Dalton Trans., (1980), 2032
12. A. A. Vlcek, Rev. Chim. Minérale, **5**, (1968), 297
13. A. A. Vlcek in "Coordination Chemistry 21" , Proceedings of the 21st International Conference on Coordination Chemistry Toulouse, France 7-11 July 1980, Ed. J. P. Laurent , Pergamon Press Oxford, 1981, p. 99
14. A. A. Vlcek, Coordination Chem. Rev., **43**, (1982) ,39
15. K. F. Purcell, J. C. Kotz, "Inorganic Chemistry ", Saunders, Philadelphia, (1977), p. 543
16. G. Linck in "Transition Metals in Homogeneous Catalysis",G. N. Schrauzer Ed. Dekker York 1971 p. 297

17. A. E. Stiegman , D. R. Tyler, Comments on Inorganic Chemistry , Part A **5**, (1986),215

18. J. Moraczewski , W. E. Geiger , J. Am. Chem. Soc. ,**103**, (1981), 4779

19. W. E. Geiger, Progress in Inorganic Chemistry, **33**, (1985), 276

20. F. A. Cotton, Chem. Soc. Reviews, **12**, (1983), 35

21. B. K. Teo, M. B. Hall, R. F. Fenske, L. F. Dahl , Inorg. Chem. ,**14** (1975) , 3103

22. R. E. Ginsborg, R. K. Rothrock, R. G. Finkle, J. P. Collman , L. F. Dahl , J. Am. Chem. Soc. ,**101**, (1979), 6550

23. M. Arewgoda, B. H. Robinson, J. Simpson, J. Am. Chem. Soc., **105**, (1983),1893

24. M. R. Richmond, J. K. Kochi , Inorg. Chem. ,**25**, (1986), 656

25. G. A. Russell, J. Hershberger, K. Owens , J. Organometal. Chem., **225**, (1982), 43

26. H. Kurosawa, H. Okada, M. Sato, T. Hattori , J. Organometal. Chem. , **250** , (1983) 83

27. W. S. Mialki, D. E. Wigley, T. E. Wood, R. A. Walton, Inorg. Chem. , **21**, (1982), 480

28. a) R. J. Klinger, J. K. Kochi, J. Am. Chem.. Soc., **103**, (1981), 5839
b) C. L. Wong, J. K. Kochi ,J. Am. Chem. Soc., **101**, (1979), 5593

29. J. Yoshida, K. Tamao, T. Kakui, A. Kurita, M. Murata, K. Yamada, M. Kumada, Organometallics , **1** ,(1982), 369

30. D. F. Eaton, Pure Appl. Chem. ,**56**, (1982), 1191

31. D. F. Eaton, Adv. Photochem., **13**, (1986), 427

32.a) S. Fukuzumi, S. Kuroda, T. Tanaka, J. Chem. Soc. Perkin Trans. II,(1986), 25
b) S. Fukuzumi, S. Kuroda, T. Tanaka, J. Chem. Soc. Chem. Commun., (1986), 1553

33. T. Madach and H. V. Vahrenkamp, Z. Naturforsch., **33b**, (1978), 1301

34. A. Dedieu, "Topics in Physical Organometallic Chemistry", M. Gielen Ed.

Freund Publishing House Ltd. , London , 1985 p. 1

35. K. Fukui, Angew. Chem. Int. Ed., **21**, (1982), 801

36. a) A. Goursot, H. Chermette, M. Chanon, W. Waltz, Inorg. Chem., **24**, (1985), 1042

    b) A. Goursot, H. Chermette, E. Penigault, M. Chanon, W. L. Waltz, Inorg. Chem. , **23**, (1984), 3618

37. a) G.E. Adams, R.B. Broszkiewicz, B. D. Michael, Trans. Faraday Soc., (1968), 1256

    b) W. L. Waltz, J. Lilie, R. T. Walter, R. J. Woods, Inorg. Chem. ,**19**, (1980), 3284

    c) R. C. Wright, G. S. Laurence, J. Chem. Soc. Chem. Commun., (1972), 132

38. A. R. Rossi, R. Hoffmann, Inorg. Chem., **14**, (1975), 365

39. a) J.C. Slater, Adv. Quantum Chem., **6**, (1972), 1

    b) A. Goursot, H. Chermette, C. Daul, Inorg. Chem., **23**, (1984), 305

40. a) S. B. Butts, D. F. Shriver, J. Organomet. Chem. , **169**, (1979), 191

    b). M. O. Albers, N. J. Coville, T. V. Ashworth, E. Singleton , J. Organomet. Chem., **217** ,(1981), 385

    c) M. O. Albers, N. J. Coville, E. Singleton, J. Organomet. Chem., **232**, (1982), 261

41. a) G. A. Russell , Chemia Stosowana, **26**, (1982), 317

    b) G. A. Russell, B. Mudryk, F. Ros, M. M. Jawdosiuk, Tetrahedron Symp. , **38**, (1982), 1059

    c) G. A. Russell, B. Mudryk, F. Ros, M. Jawdosiuk, J. Am. Chem. Soc., **103** , (1981), 4610

42. M. Julliard, M. Chanon , Chem. Rev. , **83** , (1983), 453. In scheme XII of this paper the activation associated with electron transfer was described as an associative activation because of the associative character of an SN2 like transition state. In the present report, we prefer to describe it as a weak dissociative activation : the C- leaving group bond is weakened

but not enough to undergo an unimolecular homolysis

43. a) J. F. Bunnett, Accts. Chem. Res. , **11**, (1978), 413
   b) N. Kornblum , "The Chemistry of Functional Groups", Supplement F : The Chemistry of Amino, Nitroso and Nitro Compounds and their Derivatives , S.Patai Ed. Interscience, New York, 1982 p. 361
   c) G. Russell, Adv. Phys. Org. Chem. , **23**, (1987), 271
   d) M. Chanon, Accts. Chem. Res., **20**, (1987), 214
44. R.A.Rossi, R.H. Rossi, "Aromatic Substitution by the SRN1 Mechanism", ACS Monograph, 178, A.C.S. Washington ,1978
45. J. P. Blaha, M. S. Wrighton, J. Am. Chem. Soc. ,**107**, (1985), 2694
46. J. M. Saveant , Bull. Soc. Chim., 1988 , Symposium on Organic Reactivity Paris, July 1987
47. M. Chanon , Bull. Soc. Chim. Fr., (1982), 197
48. D. G. Alway, K. W. Barnett, Inorg. Chem. ,**19**, (1980),1533
49. W. P. Jencks , "Catalysis in Chemistry and Enzymology", Mc Graw-Hill, New York, 1969.

Structural Consequences of Electron-Transfer in Dinuclear Iron Polyaromatic Complexes.

Didier Astruc*[‡], Marc Lacoste[‡], Marie-Hélène Desbois[‡]
François Varret[§], Loïc Toupet[#].

[‡]Laboratoire de Chimie Organique et Organométallique, U.A. CNRS n° 35, Université de Bordeaux I, 351, Cours de la Libération 33405 Talence Cédex, France.
[§]Groupe de Physique et Chimie du Solide, U.A. CNRS n° 807, Université du Maine, 72017 Le Mans Cédex, France.
[#]Laboratoire de Physique Cristalline, U.A. CNRS n° 7015, Université de Rennes I, 35042 Rennes Cédex, France.

Abstract

Binuclear complexes of polyaromatics have been synthesized in order to examine their electron transfer (ET) chemistry and the stereoelectronic consequences of ET. The electrochemistry of known complexes bearing two $(Fe^{II}Cp)^+$ units ($Cp = C_5H_5$) shows a single 2-e^- wave for diphenyl in DMF on Hg cathode at − 30°C, two close one-electron waves for dihydrophenanthrene and four one-electron waves with pyrene, triphenylene and phenanthrene. Since reduced states are not stable, $Fe^{II}Cp^*$ analogues were made ($Cp^* = C_5Me_5$). The diphenyl complex now has two one-e^- reductions and the X-ray crystal structures of both the mono- reduced (average valence on the Mössbauer time scale) and of the bi-reduced complexes show that chemical coupling intervenes in the course of the 2nd ET to give the new bi-cyclohexadienylidene ligand. Delocalized mixed valence $Fe^{I}Fe^{II}$ complexes are obtained for all the polyaromatics under study.

Chemical coupling after 2-e^- reduction tolerated in dihydrophenanthrene is inhibited in phenanthrene, triphenylene and pyrene for which bis-reduced complexes are $38e^-$ biradicals, as $[Fe_2(fulvalene)\,C_6R_6]$ (R = H, Me), but with much more spin density on the aromatic ligand than in the latter.

M. Chanon et al. (eds.), Paramagnetic Organometallic Species in Activation / Selectivity, Catalysis, 245–259.

Introduction

Of the many electron transfer (ET) reactions of organometallics, very few structural modifications are known, yet the nature of intermediates is important to the understanding of activation processes. The available information concerns mononuclear species with delocalized ligands such as sandwich complexes for which representatives of 17 e^- and 19 e^- complexes have been known for 35 years (cf discovery of ferricinium and cobaltocene by Wilkinson in 1952)[1]. It has been shown recently that such species[2,3] as well as 20-electron complexes[4] are important intermediates for C-C bond formation.

Binuclear complexes can have an even richer ET chemistry, namely formation of binuclear monoradicals or biradicals. Biradicals, especially, have an extremely rich and versatile ET chemistry leading to odd reactions[4], although we do not detail them here. Rather, we would like to focus on the structural changes accompanying ET process of binuclear complexes and examine the influence of structures on the type of ET (number of ET steps, number of e^- transferred in a given step).

It is common to find transition metal complexes in three oxidation states interrelated by single ET ; then, if two such units are linked, a total of five oxidation states should be accessible, at least by electrochemistry (indeed as many as in the cubane clusters)

$$A^+ \rightleftharpoons A \rightleftharpoons A^- \quad \text{(mononuclear)}$$
$$A^{++} \rightleftharpoons A^+ \rightleftharpoons A \rightleftharpoons A^- \rightleftharpoons A^{--} \quad \text{(binuclear)}$$

On the other hand, one may question the possibility of transferring e^-s two by two or even four at a time in a single steps which would provide redox catalysts for the photosplitting of water and related energy storage problems :

$$A^{++} \underset{}{\overset{2e^-}{\rightleftharpoons}} A \overset{2e^-}{\rightleftharpoons} A^{--} \quad \text{or} \quad A^{++} \overset{4e^-}{\rightleftharpoons} A^{--}$$

Results and Discussion

For the present study, we choose two isomers of bis-sandwich iron structure[5] because of their nice complementarity. The FeCp(arene) unit exists as a stable 18-electron Fe^{II} cation, as a labile but isolable 19-electron Fe^{I} species, or as a transient 20-electron Fe^{0} anion, observed only by cyclic voltammetry (CV)[6]. Two such units can link either through Cp's or through arene ligands (fig. 1). The properties of the central Fe^{I} state in mononuclear complexes are summarized in fig. 2.

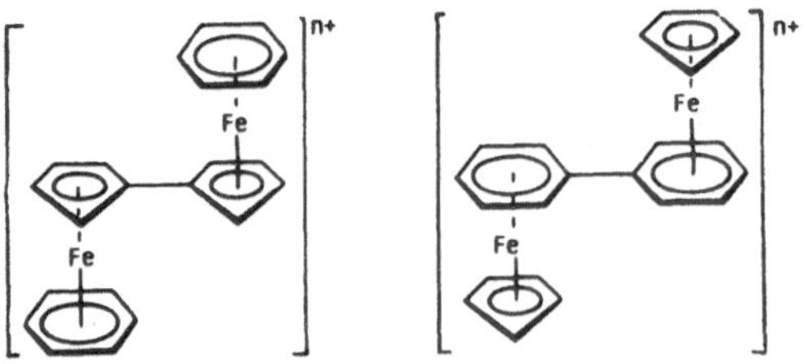

1 2

Figure 1

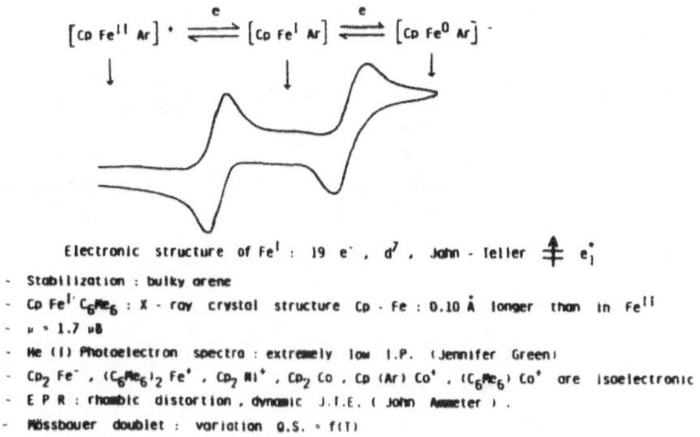
[Cp Fe^II Ar]^+ ⇌ [Cp Fe^I Ar] ⇌ [Cp Fe^0 Ar]^-
e
e
Electronic structure of Fe^I : 19 e^- , d^7 , John - Teller
- Stabilization : bulky arene
- Cp Fe^I C6Me6 : X - ray crystal structure Cp - Fe : 0.10 Å longer than in Fe^II
- μ = 1.7 μB
- He (I) Photoelectron spectra : extremely low I.P. (Jennifer Green)
- Cp2 Fe^- , (C6Me6)2 Fe^+ , Cp2 Ni^+ , Cp2 Co , Cp (Ar) Co^+ , (C6Me6) Co^+ are isoelectronic
- E P R : rhombic distortion , dynamic J.T.E. (John Ammeter) .
- Mössbauer doublet : variation Q.S. = f(T)

<u>Figure 2</u> : Main characteristics of the Fe(I) state in FeCp(arene) complexes.

The cyclic voltammetry of the fulvalene binuclear dication <u>1</u> (36-electron) shows four waves indicating five oxidation states with a high degree of chemical reversibility (- 30°C, permethylated benzene complex)[7]. Only the fourth reduction is slow, with peak potentials typically depending on scan rates[8] (k = 2.6 10^{-3} cm s^{-1}). This CV indicates that the mixed valence 37-e^- species is largely stabilized, with differences between the first and second reduction of ΔE = 480 mV for the C_6Me_6 complex and ΔE = 350 mV for the benzene analogue (fig. 3). Indeed Na/Hg reduction affords first this one-electron reduction products, isolable as purple air sensitive complexes, thermally stable for the C_6Me_6 complex, and, more slowly, the two-electron reduction products, isolable as extremely air sensitive green complexes, also stable only in the C_6Me_6 series (fig. 4).

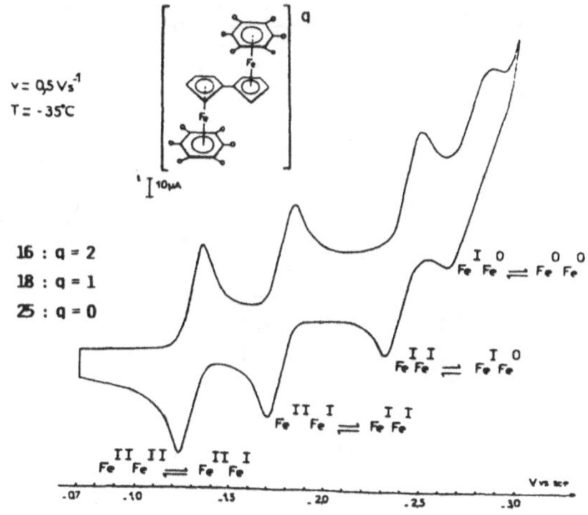

<u>Figure 3</u> : Cyclic voltammogram of $[Fe_2(fulvalene)(C_6Me_6)_2]^{++}$ $(PF_6^-)_2$ (q = 2) in DMF + $n\text{-}Bu_4NBF_4$ on Hg cathode. Only the fourth wave depends on scan rate (slow ET).

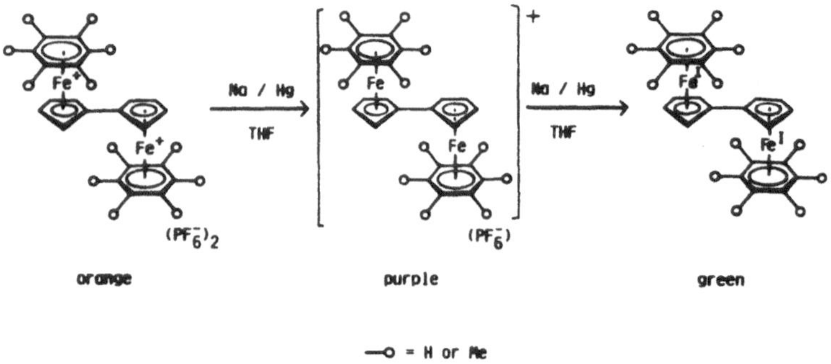

Figure 4

The Mössbauer spectra of the 37-electron complex show only one quadrupole doublet with temperature independant parameters. This indicates the presence of only one type of iron without Jahn-Teller activity (fig. 5). Thus the mixed valency is averaged and the extra (37th) electron is located in a non-degenerate orbital, contrary to the monomer (compare fig. 2 and 6). It is assumed that this

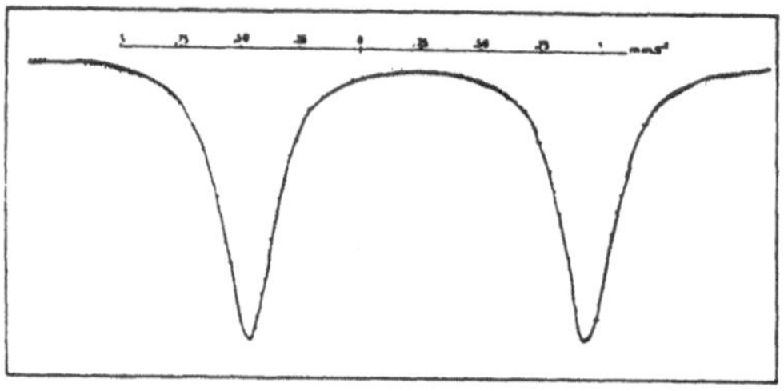

Figure 5 : Mössbauer spectrum of $[Fe_2(fulvalene)(C_6Me_6)_2]^+(PF_6^-)$. The single quadrupole doublet does not depend on temperature, contrary to Jahn-Teller active Fe(I) complexes.

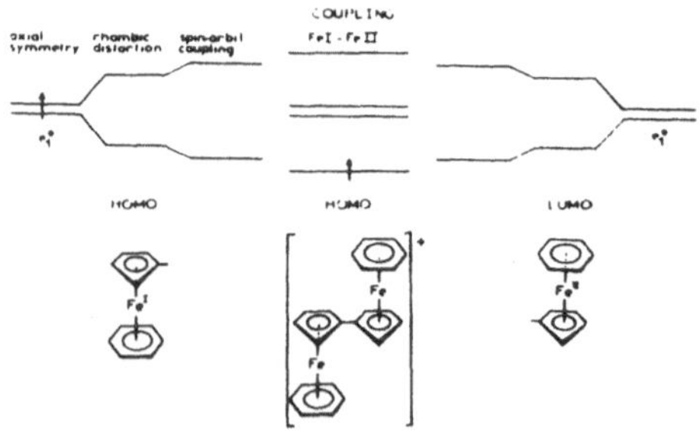

Figure 6 : Molecular orbital diagram (HOMO) of the delocalized mixed valence, 37-electron complexes $[Fe_2(fulvalene)(arene)_2]^+$.

delocalization is allowed by a favorable coplanar geometry of both Cp rings of the fulvalene ligands, the two iron moieties being thus necessarily located at opposite sides of this planar fulvalene ligand.

The situation is drastically different in the 38-electron biradicals which have uv-vis and Mössbauer spectra very analogous to those of the Fe^{I} monomers (temperature dependance of the quadrupole splitting indicating Jahn-Teller activity, e.g. thermal population of the upper Kramer's level). Thus the two Fe^{I} radical units are fairly independent in these 38 electrons $Fe^{I}Fe^{I}$ biradical, except for the dipolar coupling noted from the ESR spectra. This is true only down to 30 K, however, a temperature at which the Mössbauer and magnetic measurements show an antiferromagnetic transition[8]. Thus no chemical coupling is observed at the 38-electron level ; the 3rd reduction is also fast and since the difference in potential between this and the 4th reduction is the same as between the 1^{rst} and the 2nd reduction, one can suggest that no structural rearrangement intervenes in the 3rd reduction. On the other hand the slow 4th reduction might be the sign of such a rearrangement at the 40-electron level. But the very negative potential prevents verifying that the ET does not depend on the electrode material and no conclusion can be made.

This system[7] consisting in well separated single ET's is an excellent reference for the study of the other isomer 2[9]. The latter was supposed to give two waves also corresponding to a stabilization of the $Fe^{I}Fe^{II}$ mixed valence, based on polarography in CH_3CN at 20°C. We indeed also found two CV waves under room temperature conditions (fig. 7), but the low temperature CV of 2 consists in only one two-electron wave (Epc-Epa = 35 mV, chemical and electrochemical reversibility). As

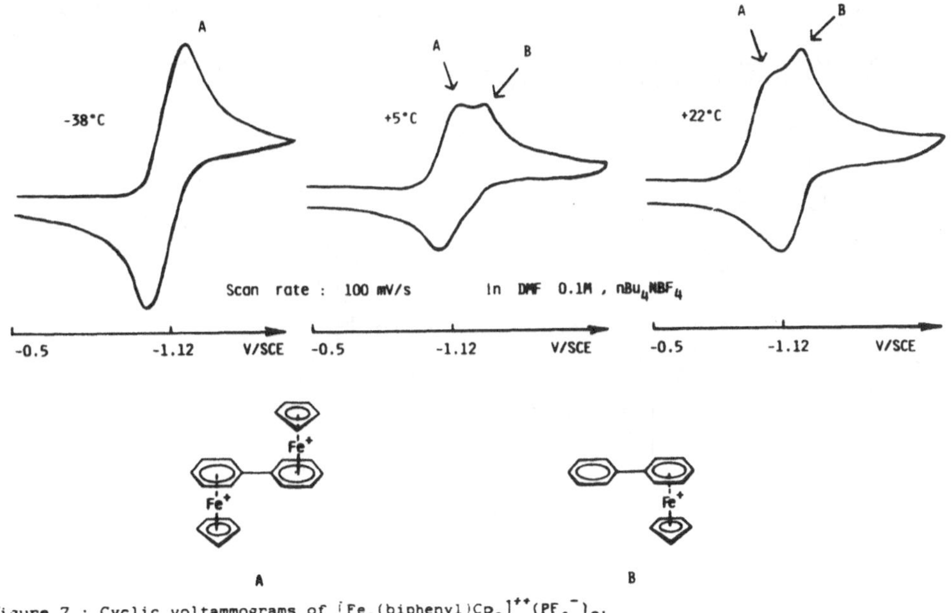

Figure 7 : Cyclic voltammograms of $[Fe_2(biphenyl)Cp_2]^{++}(PF_6^-)_2$, A, at various temperature (cathodic reduction on Hg). Addition of mononuclear complex B also increases the intensity of the second wave.

the temperature is increased, the CV of the monometallic biphenyl complex appears progressively, as can be checked by independent synthesis and CV record or addition to the solution of 1. Thus an E E C E mechanism is responsible for the appearance of the second wave at 20°C, not a mixed valence stabilization[10]. The $LiAlH_4$ reduction of $FeCp(arene)^+$ cations proceeds by ET at low temperature and subsequent H atom transfer with this reagent is found only at higher temperature[11]. Thus the low temperature $LiAlH_4$ reduction of 2 (- 80°C, THF) provides the 2-electron reduction to a deep blue, thermally unstable complex which is ESR silent down to 4.2 K (fig. 8). The fact that the 2nd ET is at least as easy as the first one, contrary to the fulvalene series, and the apparent diamagnetism of the 2-electron reduction product leads to the hypothesis that a stabilization of this 2-electron reduction product occurs by a stereoelectronic rearrangement leading to chemical coupling of the two Fe^I units (fig. 9). A similar rearrangement was

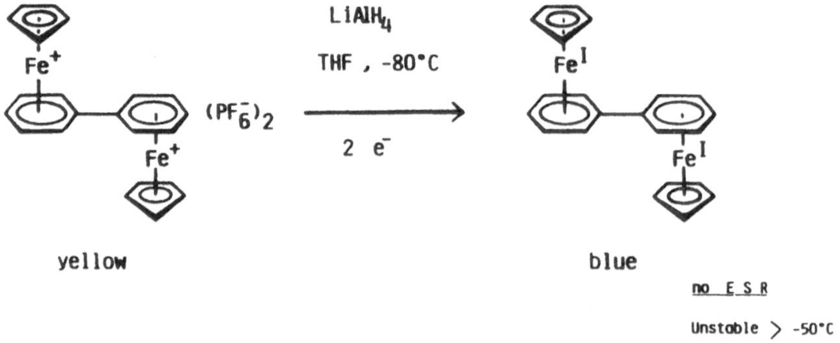

Figure 8 : The expected synthesis of the Fe^IFe^I complex, not taking into account the intramolecular coupling disclosed later.

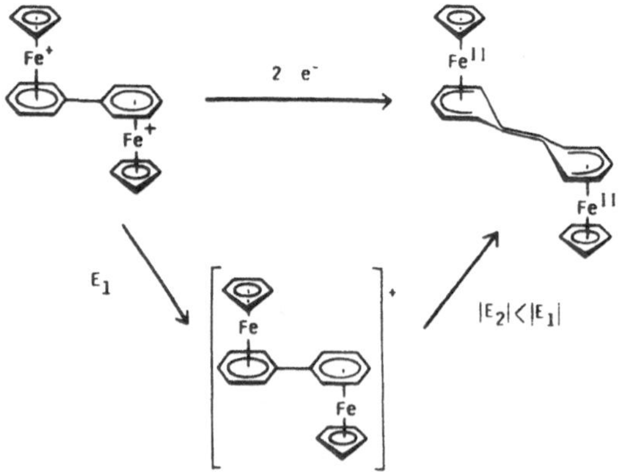

Figure 9 : Tentative mechanism for intramolecular coupling upon reduction.

proposed for the bis $Cr(CO)_3$ complex some years ago[12,13] althought the unstable di-anion could not be characterized by NMR at that time ; the rearrangement was proposed to occur in the first step to give a 35-electron species (fig. 10).

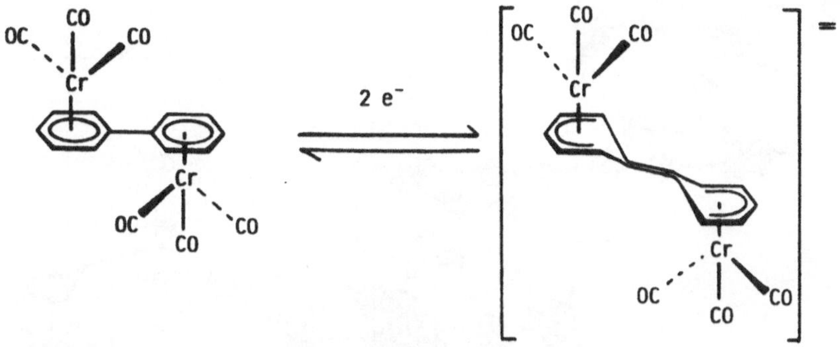

Figure 10 : Analogy between the 18 electron complexes of diphenyl : $Cr(CO)_3$ and $CpFe^+$. Neither 2-e⁻ reduced complex is stable but intramolecular coupling occurs in both cases. For the Cr studies by Rieke, see ref 12.

In order to gain more information on this system, we decided to turn to the C_5Me_5 (Cp^*) series which should provide more stable reduced species[14]. Thus we made the dicationic analogue $(FeCp^*)$biphenyl $]^{++}$ $(PF_6^-)_2$, 3, from $FeCp^*(CO)_2Br$, $AlCl_3$ and biphenyl (neat, 120°, 2 days) and found that its Na/Hg reduction at 20°C (THF, 1 h) also gives a deep blue diamagnetic neutral complex 4 which is now thermally stable at 20°C, althought extremely air sensitive. The ^1H and ^{13}C NMR spectra showed the formation of a new ligand, bis-cyclohexadienylidene with uncoordinated carbons at δ = 103 ppm vs TMS in C_6D_6 (fig. 11). The X-ray crystal structure of 4 shows a double bond with a C-C distance of 1.37 Å, and a folding angle of 25° (fig. 12). This confirms that intramolecular chemical coupling occurs upon addition of two electrons. The CV of 3 (fig. 13), unlike that of 2,

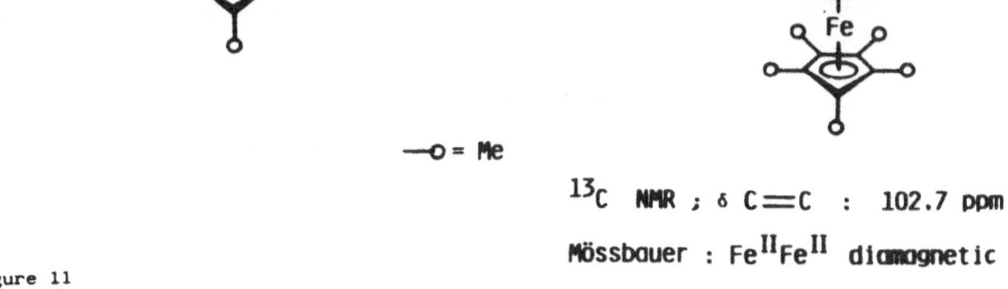

Figure 11

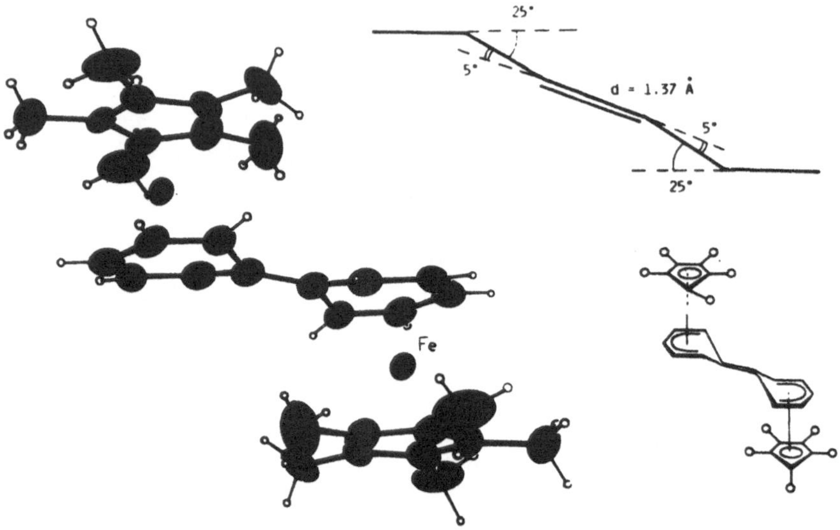

Figure 12 : Ortep view of the X-ray crystal structure of Fe_2(diphenyl)Cp^*_2 showing the distortion of the diphenyl ligand upon intramolecular coupling.

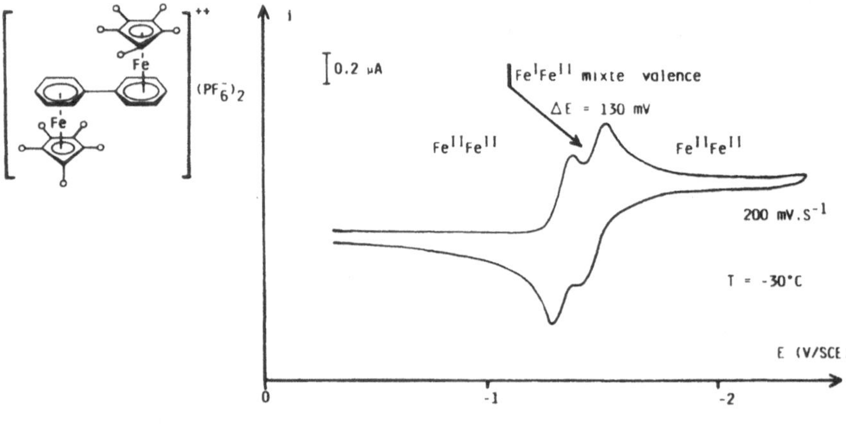

Figure 13 : Cyclic voltammogram of $[Fe_2(diphenyl)Cp^*_2]^{++}{}_2(PF_6^-)_2$. Hg cathode, DMF + n-$Bu_4$ N BF_4.

shows two chemically and electrochemically reversible very close one-electron waves. Thus the coupling energy must be slightly lower than in the parent Cp series, which affords some stabilization of the mixed valence $Fe^{I}Fe^{II}$ species. This separation of the two-electron reduction of 2 into two one-electron waves is of interest with respect to possible informations on the mixed valence species. Indeed the comproportionation constant $K' = 158$ should allow us to make the $Fe^{II}Fe^{I}$ complex from the dicationic and neutral complexes (fig. 14 and 15).

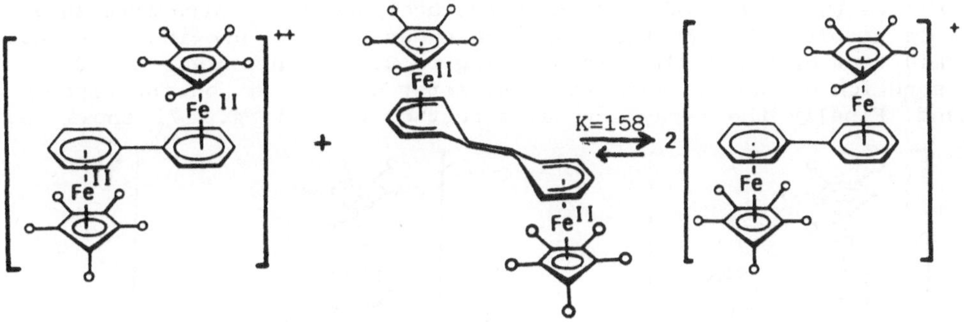

Figure 14

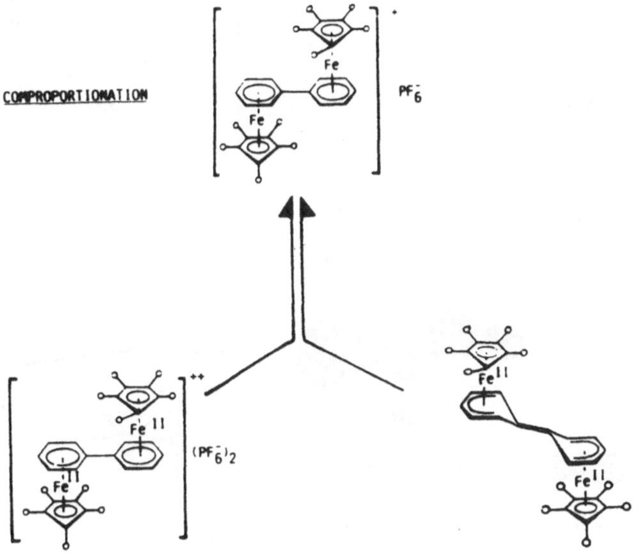

Figure 15

Thus mixing equimolar amounts of the yellow dication 3 and of the blue neutral complex 4 in THF afforded the green, insoluble, mixed valence complex 5. The question concerning 5 is to know wether it is a chemically coupled 35-electron species or a delocalized biphenyl 37-electron species (fig. 16). Its Mössbauer spectra with one quadrupole doublet, indicate only one type of iron without Jahn-Teller activity. In order to insure a maximum conjugation, we reasonned that the aromatic structure of diphenyl should be retained. Since the 35-e^- and 37-e^- structures both represent a departure of one electron with respect to the stable 36-electron structure, the 37-e^- structure should be more stable. The Mössbauer spectra under 6 Tesla provide a contact (Fermi) term of 6-7 Tesla corresponding to 0.25 e^- per ion, i.e. 0.5 e^- per sandwich unit with 50 % metal character and 0.5 e^- on the biphenyl ligand. Finally the X-ray crystal structure of 5 (fig. 17) shows the

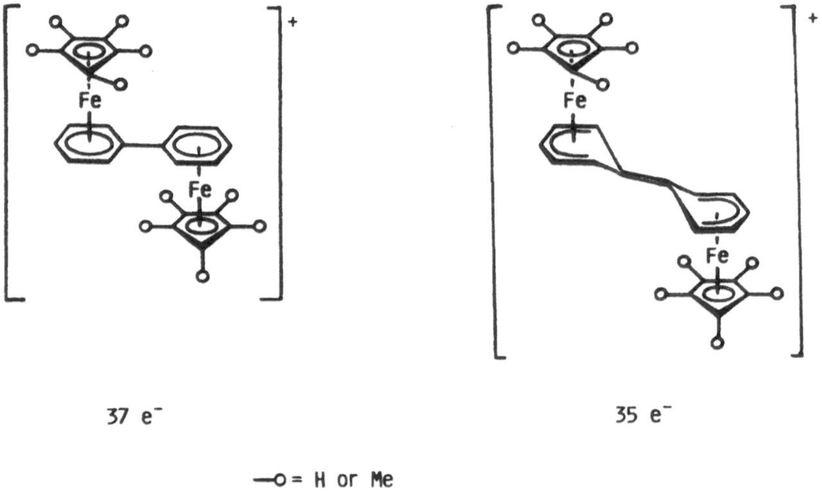

Figure 16 : Two possible structures of the mixed valence monocationic dinuclear Fe complex. In the $Cr(CO)_3$ case (see figure 10), the 35-e^- structure was proposed.

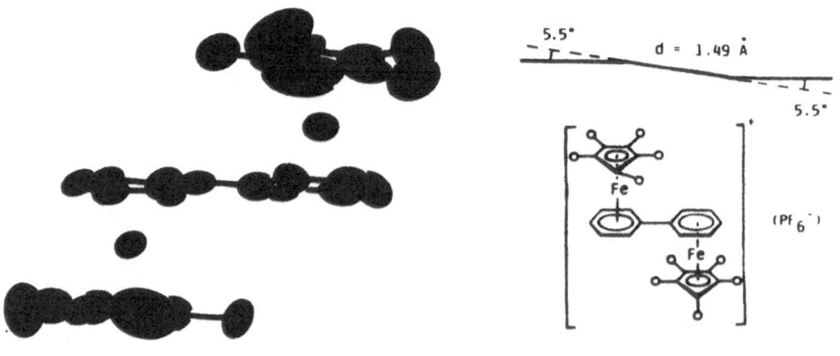

Figure 17 : Ortep view of the X-ray crystal structure of the mixed valence monocation $[Fe_2(biphenyl)Cp^*_2]^+$, PF_6^-. Note the very slight distorsion of the biphenyl ligand (folding : 5°) bearing 0.5 electron.

presence of a diphenyl ligand with a single C–C bond (C–C distance : 1.48 Å) and a small folding angle of 5° consistent with the relatively high spin density on this ligand. It is clear that chemical coupling has not occured in the first ET step and that the average valence species 5 is a 37-electron complex. The interrelated species and their structures are summarized in fig. 18. This structural characterization tells us that the stereoelectronic rearrangement intervenes in the second ET. The comparison with the fulvalene isomer should allow us to estimate the energy of the rearrangement. We know that the two isomers are reduced at the same potential of – 1.2 V vs SCE. The second reduction of 1 needs 350 mV more whereas the second reduction of 2 does not need any additional potential. Thus one may estimate that the

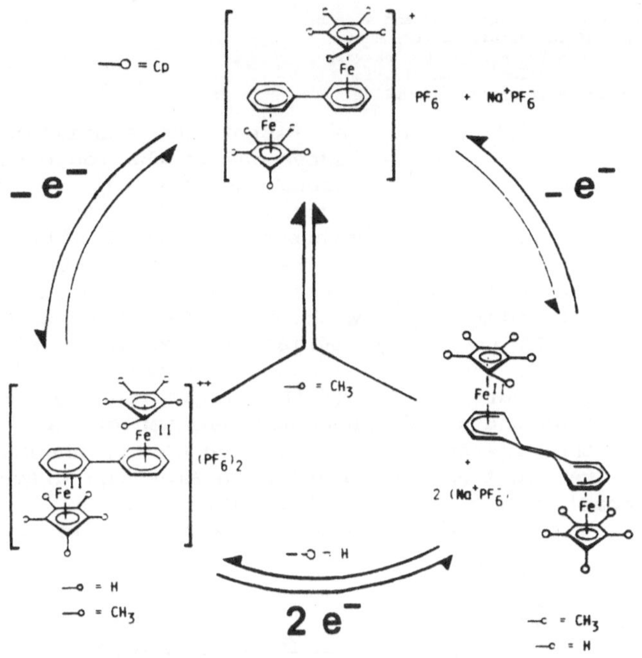

Figure 18 : Electron-transfer reactions in Fe_2 diphenyl complexes. Note the dichromotomy between the Cp and Cp* series.

stabilization gained by coupling in the 2nd ET is of the order of 350 mV, i.e. 7.5 kcal. mol^{-1}. In summary, switching from Cp to Cp* series brings about a change from a single two-electron transfer step[15] to two one-electron transfer steps. Essentially, it affords getting structural information on the average valence species which is not even observable in the CV of the Cp series. Oxidation of 4 with ½ mol O_2 in the presence of 1 eq. $NaPF_6$ rapidly gives 5 and ½ mol H_2O_2 and further reaction with ½ mol O_2 gives 3 (the salt prevents the cage reactivity of $O_2^{-\cdot}$ generated from 4 and O_2 or 5 and O_2).

In order to examine the influence of stereoelectronic constraints on the coupling process, we made a series of polyaromatic binuclear complexes shown in fig. 19. The dihydrophenanthrene series

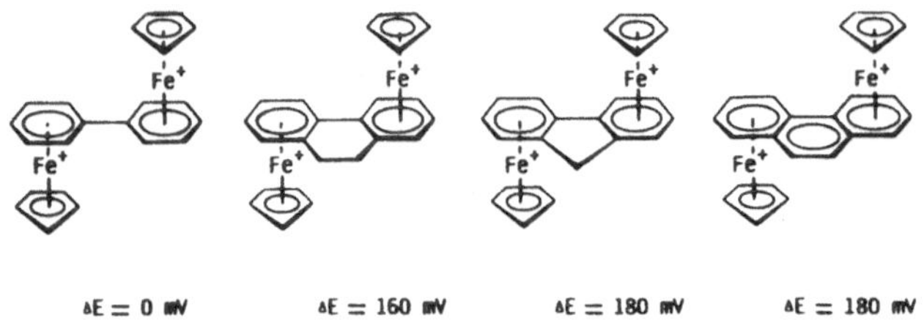

Figure 19 : Differences ΔE between the first and second reduction potentials (E°) of the dications (PF_6 salts) on Hg cathode in DMF + 0.1 M $n\text{-}Bu_4NBF_4$. The ΔE value is characteristic of the mixed valence stabilization and is 0.4 volt in the absence of chemical coupling.

behaves just as the diphenyl series, except the stabilization of the mixed valence species is a little larger due to the loose constraint of the bimethylene bridge. The fluorene series has a much larger difference between the two first reduction potentials($E_2 - E_1$ = 300 mV) as the polyaromatics series (phenanthrene, triphenylene). These series have $E_2 - E_1$ values close to that of the fulvalene series, which tends to indicate that chemical coupling possibly does not occur. We tried to make $[(FeCp^*)_2$ polyaromatic $]^{++}$ with the hope to stabilize reduced species with Cp^* as in the biaryl series and examine their electronic structure. Although the phenanthrene complex is not directly accessible because of extensive hydrogenation of this ligand during the synthesis, we could make it from the dihydrophenanthrene complex as shown in fig. 20 and show that the bis-reduced complex $[(Fe^I Cp^*)_2$, phenanthrene$]$ has an $Fe^I Fe^I$ structure, just as the fulvalene analog (no diamagnetic NMR, EPR spectra characteristic of Fe^I ; this brown complex is only stable up to 0°C). The path of fig. 20 represent a cascade of 4 e^- and 2 H^+

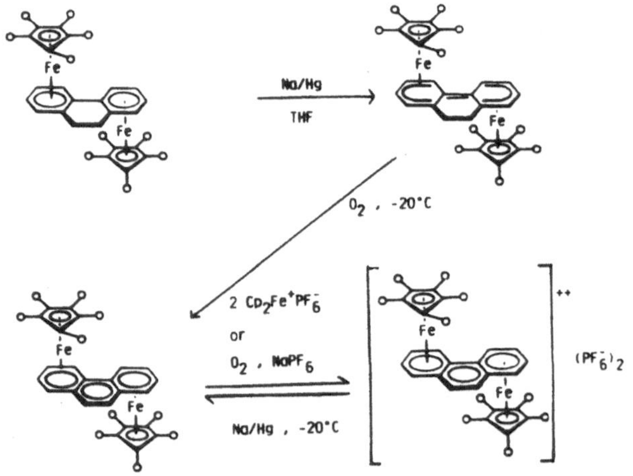

Figure 20

recalling the quinone- hydroquinone system in organic chemistry. Triphenylene also gave a complex [$(FeCp^*)_2$ triphenylene $]^{++}$ which is reduced as the phenanthrene complexes, in four one-electron steps (fig. 21) and the two-electron reduction product also has an Fe^IFe^I structure. The Na/Hg reduction at -15°C gives these thermally unstable, brown neutral complexes which react with 1 eq. of the dication to give

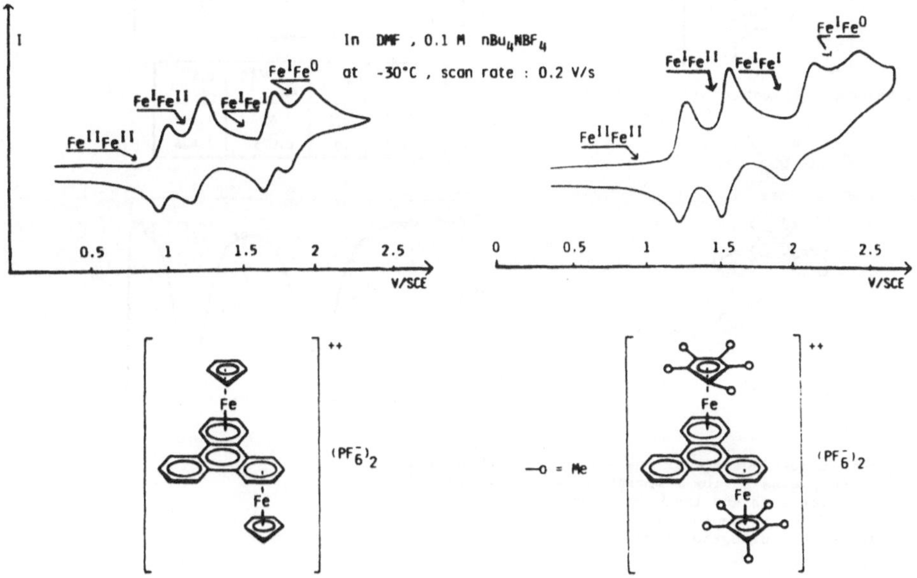

Figure 21 : Cyclic voltammograms of binuclear triphenylene dicationic complexes. Hg cathode, only the fourth wave varies with scan rate (slow ET).

the stable mixed valence Fe^IFe^{II} salts. The Mössbauer spectra (fig. 22) shows the latter are delocalized, giving only one quadrupole doublet as in the bi-aryl series. Addition of the second electron does not influence the Mössbauer parameters, which indicates that the extra electron does not provide additional spin density on iron ; the latter is diluted amoung the polyaromatic ring even thought no chemical coupling occurs. Thus the Fe^IFe^I formalism is far from reflecting the reality in the cases of polyaromatic ligands. This situation parallels what we know of monoiron-polyaromatic complexes.

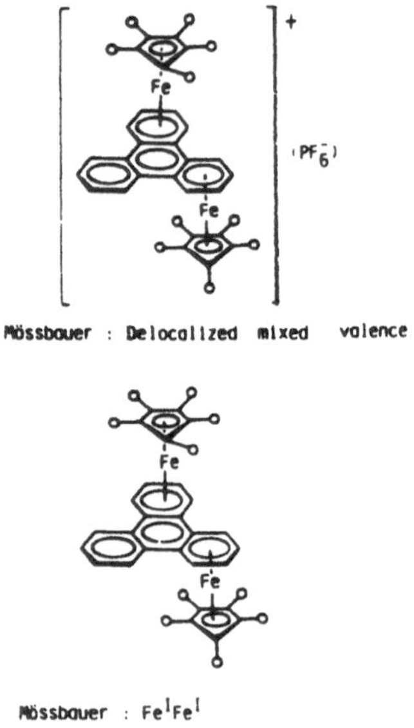

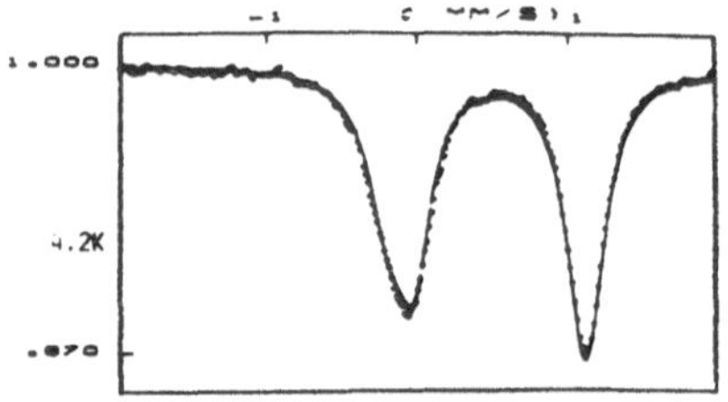

	T °K	I.S.(/Fe) mm.s^{-1}	Q.S. mm.s^{-1}
$Fe^{I}Fe^{II}$	4.2	0.63	1.19
$Fe^{I}Fe^{I}$	10	0.623	1.426

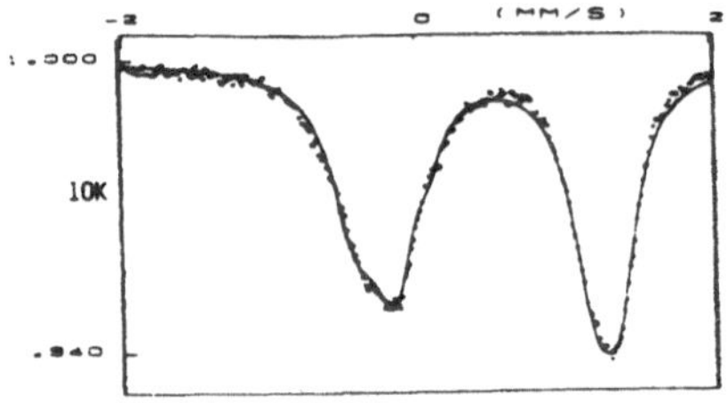

Figure 22 : Mössbauer spectra of the 37- and 38-e^- binuclear triphenylene complexes at low temperature. Only one type of Fe is seen in each case and the 38-e^- sample shows some decomposition. Note the analogies of parameters in spite of the difference in oxidation state.

Conclusion

The main conclusions of this study are :

(i) ligand design controls the number of ET steps and the number of e^-(s) transferred in each step

(ii) electronic transmition is faster than the Mössbauer frequency of $10^8 s^{-1}$, i e it proceeds easily through fulvalene or polyaromatics

(iii) chemical coupling through the biphenyl ligand is structurally characterized for the first time. It occurs essentially during the 2nd ET step of the cathodic reduction as shown by compared X-ray crystal structures of both the one- and two-electron reduction products. Its energy is of the order of 7 5 kcal mol^{-1} and can drop with stereoelectronic constraints, reaching inhibition in the polyaromatic complexes.

(iv) biradicals (38 e^-) are submitted to coupling which may be dipolar antiferromagnetic or chemical.

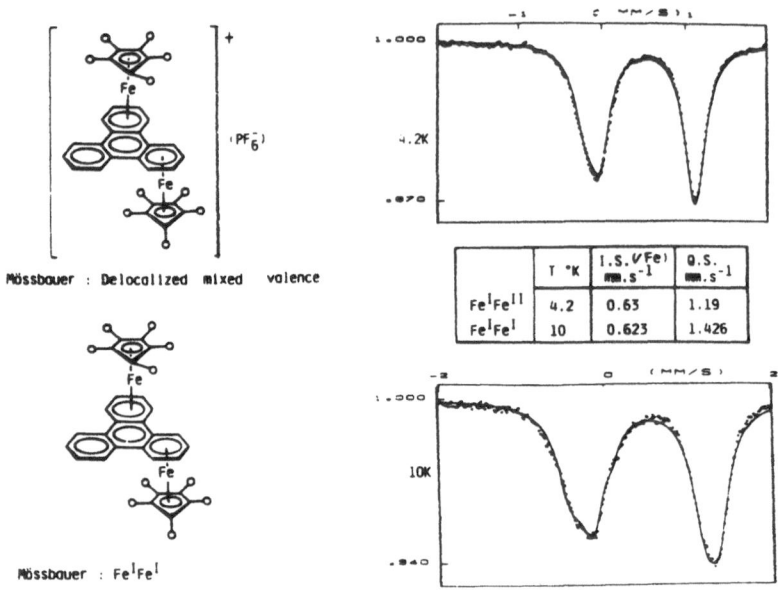

References

1 - G. Wilkinson, J. Organomet. Chem., 1975, 100, 273.
2 - D. Astruc, in "The Chemistry of the Metal-Carbon Bond" S. Patai, F.R. Hartley Ed., Wiley, New-York, 1987, Vol 4, Chapter 7, p. 6625.
3 - **a)** D. Mandon, L. Toupet, D. Astruc, J. Am. Chem. Soc., 1986, 108, 1320 ;
b) A.M. Madonik, D. Astruc, J. Am. Chem. Soc., 1984, 106, 2437.
4 - **a)** M. Lacoste, M.-H. Desbois, D. Astruc, Nouv. Journ. Chem., 1987, 11, 561 ;
b) D. Astruc, Comment. Inorg. Chem., 1987, 6, 61.
5 - W.H. Morrison, E.Y. Ho, D.N. Hendrickson, J. Am. Chem. Soc., 1974, 96, 3603.
6 - Review : N.G. Connelly, W.E. Geiger, Advan. Organomet. Chem., 1984, 23, 1.
7 - M.-H. Desbois, D. Astruc, J. Guillin, J.-P. Mariot, F. Varret, J. Am. Chem. Soc., 1985, 107, 5280.
8 - M.-H. Desbois, D. Astruc, J. Guillin, F. Varret, J.-P. Mariot, A.X. Trautwein, G. Villeneuve, to be submitted.
9 - M. Lacoste, F. Varret, L. Toupet, D. Astruc, J. Am. Chem. Soc., 1987, 109, 6504.
10 - M. Lacoste, L. Toupet, F. Varret, D. Astruc, to be submitted.
11 - **a)** P. Michaud, D. Astruc, J.H. Ammeter, J. Am. Chem. Soc., 1982, 104, 3755 ;
b) P. Michaud, C. Lapinte, D. Astruc, Annals N.Y. Acad. Sci., 1983, 415, 97.
12 - **a)** S.N. Milligan, R.D. Rieke, Organometallics, 1983, 2, 171 ;
b) R.D. Rieke, S.N. Milligan, L.D. Shulte, Organometallics, 1987, 6, 699.
13 - For 2-e^- reduction of η^6-arene to η^4-arene complexes, see :
a) W.E. Geiger, Prog. Inorg. Chem., 1985, 33, 275 ;
b) R.G. Finke, R.H. Volgeli, E.D. Laganis, V. Boekelheide, Organometallics, 1983, 2, 347 ;
c) R.D. Rieke, W.P. Henry, J.S. Arney, Inorg. Chem., 1987, 26, 420.
14 - For complexation of $FeCp^*$ to aromatics, see :
a) D. Astruc, Tetrahedron Report N° 157, Tetrahedron, 1983, 39, 4027 ;
b) D. Astruc, J.-R. Hamon, M. Lacoste, M.-H. Desbois, A. Madonik, E. Roman, "Organomet. Synthesis", 1988, Vol 4 (in press).
15 - For discussions of one step 2-electrons transfers, see :
a) A.G. Sykes, Adv. Inorg. Chem. Radiochem., 1967, 10, 153 ;
b) F. Ammar, J.-M. Savéant, J. Electroanal. Chem., 1973, 47, 115 ;
c) A.J. Bard, Pure Appl. Chem., 1971, 25, 379 ;
d) see also ref 13 and refs cited therein.

ACTIVATION OF THALLIUM-CARBON AND MERCURY-CARBON BONDS WITH ELECTRON DONORS

Hideo Kurosawa and Yoshikane Kawasaki
Department of Applied Chemistry,
Faculty of Engineering,
Osaka University, Suita, Osaka 565
Japan

ABSTRACT. Reactions of organothallium compounds, $RTlX_2$ with electron donors are surveyed. Interaction of the donors with $RTlX_2$ provides the Tl-C bond with the enhanced reactivity compared with that of the original $RTlX_2$ or with the reactivity pattern not found in the original. These include homolysis of the Tl-C bond, enhanced electrophilic attack at the thallium-bound R group, and enhanced nucleophilic substitution at the thallium-bound aryl and vinyl groups. The relevant examples of these trends are described in the reactions employing, as the donors, $Me_2CNO_2^-$, N-benzyl-1,4-dihydronicotinamide, CuX, PR_3, I^-, $NaBH_4$ and hydrazine. Relevance of the reductive activation of $RTlX_2$ as well as related RHgX and $RPbX_3$ in organic synthesis is discussed.

1. Introduction

It is often observed that interaction of electron acceptors or donors with organometallic reagents not only increases the reactivity of the metal-carbon bond but alters its reactivity pattern. The change of the reactivity pattern is exemplified by generation of an alkyl radical intermediate or of a metal-bound alkyl group which is susceptible to nucelophilic attack, via interaction of electron acceptors with organometallics containing alkyl ligands of otherwise carbanionic character [1].

The activation of the metal-carbon bond in electron-deficient organometallics (e.g. RHgX, $RCoX_2$) with electron donors has received increasing attention from a synthetic chemical point of view [2,3]. Organothallium(III) compounds, $RTlX_2$ are among the most electron-deficient organometallic reagents, and are important intermediates in organic synthesis [4]. We wish to summarize here new reactivity patterns of $RTlX_2$ and related RHgX compounds induced by electron-transfer from some electron donors to these organometallics.

M. Chanon et al. (eds.), Paramagnetic Organometallic Species in Activation / Selectivity, Catalysis, 261–274.

2. Outline and Formal Classification of Reactions of $RTlX_2$ with and without Added Donors

Some reactivities exhibited by organothallium compounds, $RTlX_2$ (e.g. nucleophilic substitution) are similar to those of organic halides (e.g. RBr, R_2I^+), but others (e.g. electrophilic substitution) are dissimilar. Since homolysis of the halogen-carbon bond induced by single electron-transfer to the organic halides is a relatively facile process [5], it is expected that the organic radical intermediate is also generated by the electron-transfer from the electron donors to the organothallium compounds. One distinct difference between the two classes of compounds lies in their ability to form coordination complexes (Tl ≫ halogen), which may affect the course of the reactions with the electron donors.

2.1 Nucleophilic Substitution

The most general way of decomposition of $RTlX_2$, where R is an sp^3 hybridized alkyl, is nucleophilic substitution at the thallium-bound carbon atom (eq. 1), a typical example being solvolysis (eq. 2) [4]. Some of the reactions shown in eq. 1 proceed through alkyl radical intermediate (see later). The compounds, $RTlX_2$ where R is an sp^2 hybridized carbon (vinyl, aryl), are not susceptible to solvolysis. However, these do undergo substitution with certain nucleophiles such as $Me_2CNO_2^-$, I^- and R'_3P to give $Me_2(R)CNO_2$, RI and RR'_3P^+, respectively. Mechanistic details of these reactions will be discussed later.

$$RTlX_2 + Nu \longrightarrow R\text{-}Nu^+ + TlX + X^- \qquad (1)$$

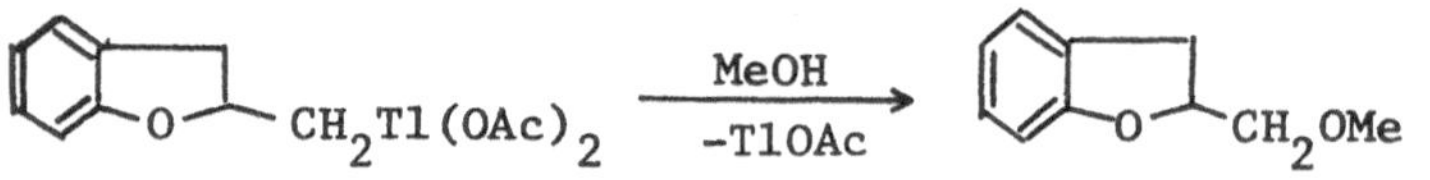

(2)

2.2 Homolytic Cleavage

Thermal homolysis of the Tl-C bond in $RTlX_2$ appears rare. Photoirradiation of $ArTlX_2$ was employed as a mean of generating aryl radical intermediate in synthetically useful transformations (e.g. biaryl and aromatic nitrile synthesis) [4b]. Non-photochemical generation of R radical species is accomplished when $RTlX_2$, especially those containing an sp^3 hybridized alkyl group, are treated with electron donors. A subsequent fate of the alkyl radical generated depends on the nature of the donors used, as described later.

2.3 Electrophilic Substitution

When R group in $RTlX_2$ is aryl, the Tl-R bond is cleaved with electrophiles such as Br_2, I_2 or NO^+ (eq. 3) [4a]. Protonolysis of $ArTl(OOCCF_3)_2$ appears to be enhanced by the action of some reductants, since the reaction of $ArTl(OOCCF_3)_2$ with $NaBH_4$ in EtOD gave ArD very

$$PhTl(OOCCF_3)_2 \;+\; E\text{-}X \longrightarrow Ph\text{-}E \;+\; TlX(OOCCF_3)_2 \qquad (3)$$

E-X= Br-Br, I-I, NO-Cl

selectively [6], whereas $ArTl(OOCCF_3)_2$ is inert to EtOH in the absence of $NaBH_4$ (for more detail, see later).

Transmetallation of R group from $RTlX_2$ to other metal cations (e.g. Hg^{2+}, Pd^{2+}) is another example of the electrophilic cleavage [4a]. Although there is no firm evidence to support occurrence of spontaneous disproportionation of $RTlX_2$ (eq. 4), a kind of homo-transmetallation, action of some electron donors (e.g. hydrazine, $P(OMe)_3$) on $RTlX_2$ induces a ready formation of R_2TlX compounds. Relevance of this reaction to electron transfer activation of Tl-C bond will be discussed later.

$$2RTlX_2 \rightleftharpoons R_2TlX \;+\; TlX_3 \qquad (4)$$

3. Nucleophilic Substitution of $RTlX_2$

With regard to eq. 1, the nature of R in $RTlX_2$ affects both the ease of the reaction and the mechanistic aspect. Some reactions of organomercury (RHgX) and lead ($RPbX_3$) compounds are similar to those of $RTlX_2$, but the three compounds naturally possess the different reactivity trend from each other. In Table I are summarized some results of the relevant reactions of RMX_n (M= Hg, Tl, Pb) which may well illustrate the effects of R and the metal. More details will be described in the following sub-sections according to the reaction type.

3.1 Nucleophilic Substitution Proceeding through Alkyl Radical Intermediate

Detailed mechanistic studies involving stereochemical analysis and a spin trap experiment on reactions of eq. 1 employing $PhCH(OMe)CH_2Tl(OAc)_2$ **1** and Nu= $Me_2CNO_2^-$ [7a,c], H^- (via BNAH → BNA^+ conversion) [7a,d], and Cl^- or Br^- (used in conjunction with CuCl or CuBr) [9] (Scheme 1) revealed existence of alkyl radical intermediate in the main course of the reaction. The yields of the products, **2**-**4** were modest to very high at room temperature in methanol or DMSO (Nu= $Me_2CNO_2^-$), methanol (BNAH) or acetonitrile (X^-/CuX). Although these reactions of **1** proceeded in the dark, analogous reactions of $PhCH(OMe)CH_2HgOAc$ **5** with $Me_2CNO_2^-$ and BNAH required photoirradiation or addition of radical initiator in order for the good yields to be obtained [7a,b,10].

The stereochemical test of the reaction relied on ^1H NMR analysis of a diastereomeric alkylthallium compound **1**-d [11] and organic products derived from **1**-d. Thus, diastereotopically pure **1**-d gave the

Table I. Reaction of $RM(OAc)_n$ with Electron Donors

Compound M	R	Donor[a]	%yield of product[b]	Condition[c]	Ref.
Hg	$PhCH(OMe)CH_2$	A^-	22	DMSO, 65°C	7a
	$PhCH(OMe)CH_2$	A^-	61	DMSO, irrad.	7a
	E-PhCH=CH	A^-	0	DMSO, irrad.	7a
	$PhCH(OMe)CH_2$	BNAH	87	MeOH, irrad.	7b
Tl	$PhCH(OMe)CH_2$	A^-	91	DMSO, r.t.	7c
	E-PhCH=CH	A^-	99	DMSO, r.t.	7c
	p-tolyl	A^-	40	MeOH, 65°C	7c
	$PhCH(OMe)CH_2$	BNAH	65	MeOH, r.t.	7d
	p-tolyl	B^-	0[d]	MeOH, r.t.	7e
Pb	p-tolyl	A^-	75	DMSO, 40°C	8a
	p-tolyl	B^-	32	$CHCl_3$, 40°C	8b

a A: Me_2CNO_2 B: $CH(COMe)_2$ BNAH: N-benzyl-1,4-dihydronicotinamide.

b Product: R-A, R-B or RH (with BNAH).

c Under dark except as indicated.

d Good yield of $(p\text{-tolyl})_2TlOAc$ obtained.

Scheme 1

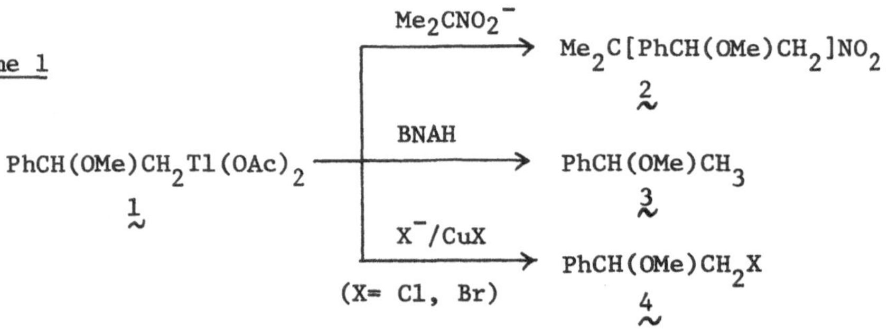

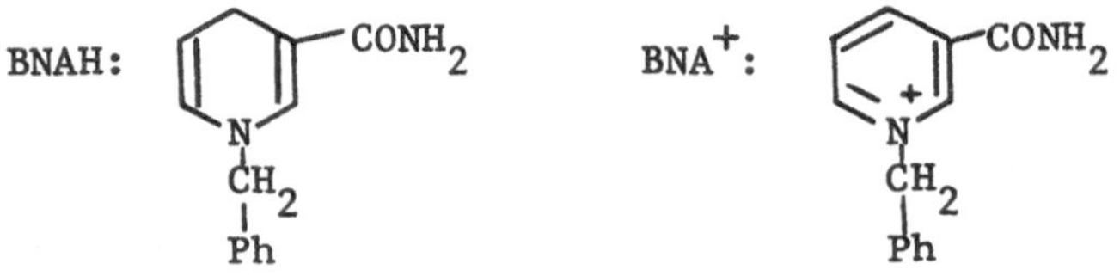

completely epimerized (in the reaction with $Me_2CNO_2^-$) and almost completely epimerized (with Cl^-/CuCl) products 2-d and 4-d (X= Cl), respectively. Incomplete epimerization in the latter case is ascribed to occurrence of the competing concerted pathway (eq. 5), which becomes more dominant at the lower temperatures [9c].

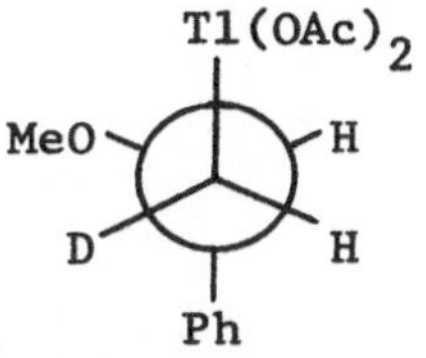

1-d

$PhCH(OMe)CH_2HgOAc$

5

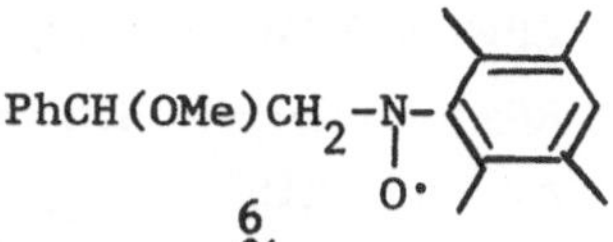

6

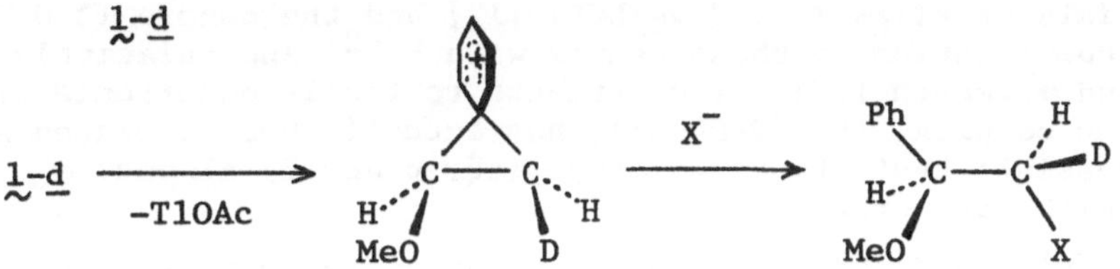

(5)

In all cases of the reactions examined, addition of nitrosodurene resulted in appearance of an ESR signal unambiguously assignable to a radical adduct 6. No such ESR signal could be obtained in the absence of $Me_2CNO_2^-$, BNAH and CuX. The product yields in the reaction carried out under dioxygen decreased considerably from those obtained under nitrogen. Other spin traps (e.g. galvinoxyl, m-dinitrobenzene) were not so effective as dioxygen was in the former two reactions. Addition of nitrosobenzene and o-t-butylcatechol to the reaction mixture of 1 and CuBr reduced the yield of 4 (X= Br).

We propose that the alkyl radical intermediate is generated from homolysis of an alkylthallium(II) species which is formed by single electron reduction of 1 (eq. 6 and 7). Then R· may couple with $Me_2CNO_2\cdot$ (formed in eq. 6) or abstract H from $BNAH^{+\cdot}$ or X from CuX_2, affording $Me_2(R)CNO_2$, RH or RX, respectively. These sequences do not constitute a chain as was often invoked in $S_{RN}1$ mechanism [5a], but it was not possible to determine to what extent the chain mechanism contributes to the overall reaction.

$$RTlX_2 + Nu \longrightarrow RTlX + Nu^{+\cdot} + X^- \quad (6)$$

Nu= $Me_2CNO_2^-$, BNAH, CuX'

$$RTlX \longrightarrow R\cdot + TlX \quad (7)$$

The photo-assisted or initiator-assisted substitution of RHgX with BNAH apparently proceeds via a radical chain mechanism (eq. 8-12, propagation being eq. 10-12) [7b]. The reaction was inhibited by dioxygen, galvinoxyl and m-dinitrobenzene. The $S_{RN}1$ mechanism in the reaction of RHgX with $Me_2CNO_2^-$ has previously been proposed [10].

$$\text{Init.} \longrightarrow R\cdot \quad (8)$$

$$RHgX + BNAH^* \longrightarrow RHg + BNAH^{+\cdot} + X^- \quad (9)$$

$$RHg \longrightarrow R\cdot + Hg \quad (10)$$

$$R\cdot + BNAH \longrightarrow RH + BNA\cdot \quad (11)$$

$$BNA\cdot + RHgX \longrightarrow BNA^+ + RHg + X^- \quad (12)$$

Even though eq. 6 may not be exothermic in view of the estimated redox potentials of $RTlX_2$ (< 0 V vs SCE) [12] and the donors (> 0 V vs SCE), rapid complexation of these donors with RTl^{2+} and relatively weak Tl-C(sp^3) bond strength [13] may contribute to facile occurrence of eq. 6 and 7. The complex, $PhTl(O_2NCMe_2)_2$ has actually been isolated stable. It is also possible that these two steps (eq. 6 and 7) almost synchronize with each other.

The Tl-aryl and Tl-vinyl bonds are expected to be more reluctant to undergo the homolysis than the Tl-alkyl bond. A non-radical substitution of $RTlX_2$ (R= aryl, vinyl) with $Me_2CNO_2^-$ will be described in the next section. Also, generation of the radical intermediate appears to be essential for the reaction with BNAH to proceed smoothly, for this reaction involving $RTl(OAc)_2$ (R= E-PhCH=CH, p-tolyl) afforded not only hydrogenated products (styrene, toluene) but R_2TlOAc in comparable yields. The latter may have been formed by non-radical, electron-transfer activation of the Tl-R bond of $RTlX_2$ (see later). An attempt to couple $Me_2CNO_2^-$ with RHgX (R= aryl, vinyl) via photoirradiation also resulted in the formation of R_2Hg compounds [10].

The thermal electron-transfer from BNAH to RHgX may be still more endothermic, since the oxidizing ability of Hg^{2+} is weaker than Tl^{3+} and the Hg-R bond is stronger than the Tl-R bond [13,14]. However, photo-excited BNAH may well be a sufficiently effective reductant [15] to compel eq. 9; indeed the fluorescence of BNAH was efficiently quenched by **5** (k_q= 2.5 x 10^9 M^{-1} sec^{-1}) [7b]. Moreover, photoirradiation of **5** alone under the condition used in the reaction with BNAH did not result in an efficient generation of the alkyl radical. Finally, a radical, BNA· reacting in one of the propagation steps is known to be a good electron donor [16].

The firm evidence to support occurrence of eq. 11 was obtained from a reaction of 5-hexenylmercury acetate with BNAH (eq. 13) [7b]. Key to this experiment is a facile rearrangement of 5-hexenyl radical to cyclopentylmethyl radical with its rate constant being in the order of 10^5 sec^{-1} [17]. As shown in Scheme 2, the radical rearrangement competes with trap of the 5-hexenyl radical by BNAH. This predicts a linear dependency of the product ratio, 1-hexene vs methylcyclopentane, on the amount of BNAH used; this was indeed the case, and from the slope of

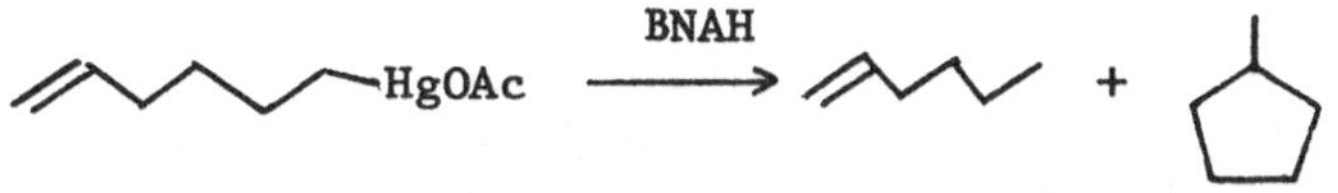

(13)

Scheme 2

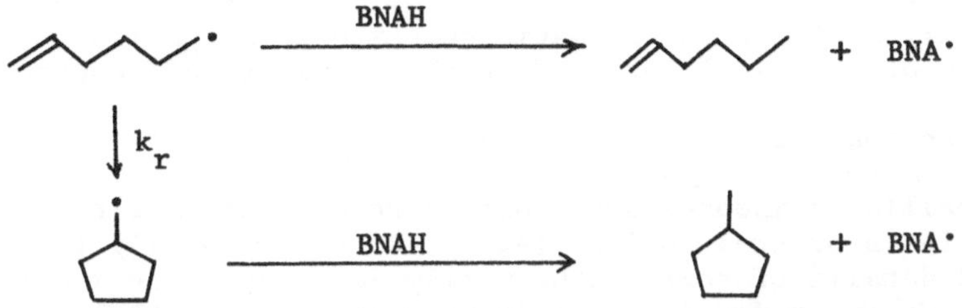

such dependency was calculated the rate constant of the reaction of R· with BNAH as being in the order of 10^5 M^{-1} sec^{-1}.

3.2 Nucleophilic Substitution of Vinylthallium and Arylthallium Compound

i) Reaction with $Me_2CNO_2^-$

Aryl- and vinylthallium compounds, $RTl(OAc)_2$ reacted with $Me_2CNO_2^-$ under dark according to eq. 1 to give the coupling product, $Me_2(R)CNO_2$ in moderate (R= p-tolyl, in MeOH at 65°C) to good yields (R= E- and Z-PhCH=CH, in DMSO at room temperature) [7c]. In the case of the vinyl derivatives, the reaction rate was not enhanced by photoirradiation. More importantly, the stereochemistry of the styryl group was almost retained on going from $RTl(OAc)_2$ to $Me_2(R)CNO_2$. In addition, no spin trap could be detected by adding nitrosodurene to a reaction mixture of the vinylthallium compound and $Me_2CNO_2^-$, nor did dioxygen retard this reaction. These facts may exclude a vinyl radical intermediate. The most probable path is addition of the anion at the α-carbon, followed by elimination of the thallium moiety without rotation of the C-C single bond (eq. 14). It is also possible that single electron-transfer from $Me_2CNO_2^-$ to $PhCH{=}CHTl(OAc)_2$ occurs first, followed by rapid addition of $Me_2CNO_2\cdot$ to the α-carbon of a vinylthallium(II) species and subsequent elimination of Tl(I) species.

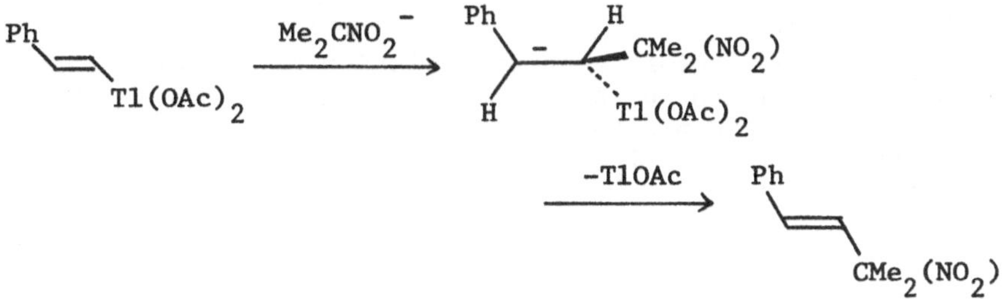

(14)

In contrast to the above result, the product yield in the reaction of the p-tolylthallium compound with $Me_2CNO_2^-$ was slightly raised by photoirradiation. Although a spin adduct was detected upon addition of nitrosodurene to the reaction mixture, dioxygen had very little effect on the product yield. The main course of this reaction would again be a non-radical, nucleophilic aromatic substitution (S_NAr) path. A minor contribution of a radical path, if at all, would be of in-cage character.

ii) Reaction with I^-

Arylthallium compounds are known to react rapidly with I^- ion to give good yields of aryl iodides (eq. 1, R= aryl, Nu= I^-) [4]. No mechanistic details of this synthetically important reaction have been reported. We succeeded in stabilizing a phenyl(iodo)thallium moiety through complexation with a crown ether ligand (see 7) [18]. Without the crown ether addition of I^- to $PhTl^{2+}$ so rapidly results in the formation of phenyl iodide.

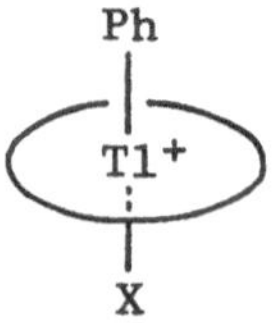

7 X= I
8 X= Cl
9 X= Br
10 X= SPh

(crown= dibenzo-18-crown-6)

Even though the phenyl and the iodo ligands are located distant from each other, photoirradiation of 7 in acetonitrile (pyrex glass tube) at room temperature rapidly afforded a good yield of phenyl iodide together with a small amount of benzene [19]. The yield of phenyl iodide decreased in the photoreaction conducted either under dioxygen or in THF; benzene was the major product in THF. Phenyl radical was trapped as $PhN(O\cdot)C_6HMe_4$ when nitrosodurene and 7 were heated together in MeCN/benzene. These facts suggest generation of phenyl radical upon irradiation.

Significantly, no similar Tl-Ph bond homolysis occurred when the compound containing the Tl-Cl or Tl-Br bond (8, 9) was irradiated, while irradiation of the thiolate complex, 10 did afford Ph_2S and benzene in good yields.

Since it was shown before that dissociation of the organothallium cations from the crown ethers in the complexes of the type 7-10 is extremely slow at room temperature [18], it was thought highly unlikely that the photoirradiation of 7 and 10 begins with decomplexation from the crown ether of the $Ph(X)Tl^+$ moiety. Furthermore, high electron-transfer abilities of I^- and PhS^- ions, compared to Cl^- and Br^- ions, strongly suggest that photolysis of 7 and 10 proceeds via intramolecular electron-transfer from I^- or PhS^- to $PhTl^{2+}$, followed by homolysis of the Tl-Ph bond. A plausible path for the iodide complex is shown

Scheme 3

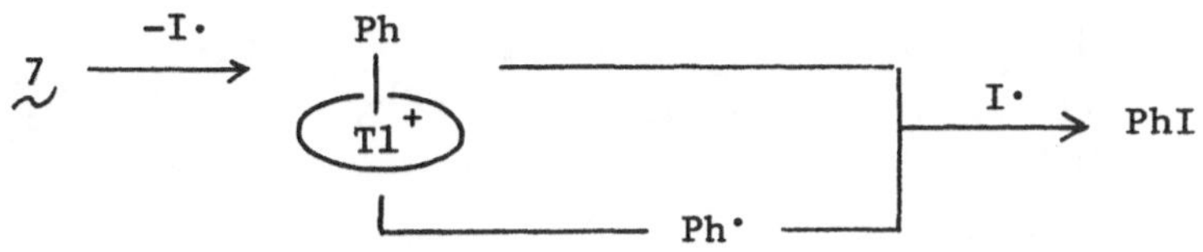

in Scheme 3.

We now refer to the aryl iodide synthesis from $ArTlX_2$ and I^-. It is possible that coordination of I^- to $ArTl^{2+}$ and subsequent electron-transfer make the Tl-Ar bond more or less susceptible to homolysis of unimolecular nature or homolysis assisted by the ipso attack of I·. Owing to availability of two nearby coordination sites for the aryl and the iodo ligands in the absence of the crown ether, occurrence of electron-transfer activation of the Tl-Ar bond to only a very small extent would be enough to compel the coupling of the aryl and the iodo groups.

iii) Reaction with R_3P

It seems of further interest that action of an excess amount of R_3P (R= Ph, CH_2CH_2CN) or $Ph_2(OEt)P$ on another crown ether complex of $PhTl^{2+}$ ion, 11 afforded moderate to good yields of the formal S_NAr products, PhR_3P^+ (R= Ph, CH_2CH_2CN) or $Ph_3(OEt)P^+$ and a Tl(I) compound, together with a small amount of benzene [20]. The yields of the phosphonium ion decreased and that of benzene increased if the molar ratio of R_3P/Tl was unity. The presence of the crown is essential, for it was found that in the absence of the crown the reaction of R_3P with $PhTlX_2$ results in reductive disproportionation of $PhTlX_2$ giving rise to Ph_2TlX. The reaction of 11 with more basic R_3P (R= cyclo-hex, n-Bu) resulted in decomplexation of $PhTl^{2+}$ from the crown, thereby enhancing the disproportionation.

As the disproportionation of $PhTlX_2$ most probably involves the electron-transfer activation of the Tl-Ph bond (see the next section), the formation of the phenylphosphonium ion may also have resulted from similar activation of the Tl-Ph bond (see, e.g. Scheme 4). The role of the crown ether in this reaction would be to prevent the access of two "activated PhTl" moieties (e.g. $PhTl^+$, see the next section), which is required for the disproportionation reaction to proceed.

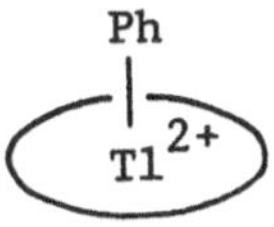

11 (crown= 1,10-dithia-18-crown-6)

Scheme 4

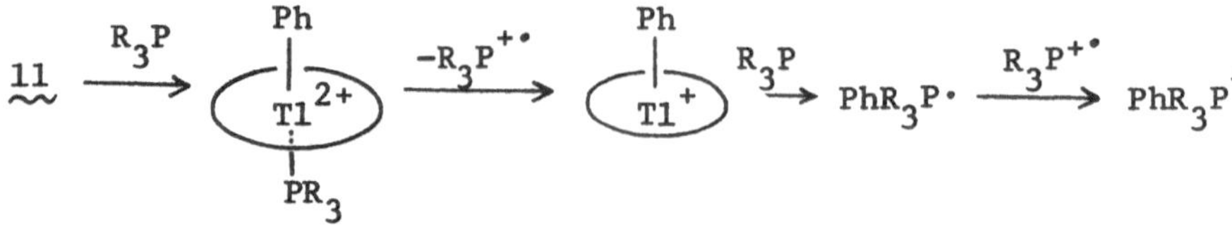

4. Reductant-induced Disproportionation of $RTlX_2$

As decribed in the last section, the reaction of $RTlX_2$ with potential electron donors often leads to the formation of R_2TlX and TlX compounds (eq. 15) [21-23]. There has been a controversy as to whether this reaction proceeds via an initial spontaneous disproportionation (eq. 4) and the subsequent rapid quenching of TlX_3 by the electron donors [21], or via a direct interaction between $RTlX_2$ and the electron donors [22]. Kinetic and stereochemical studies of some of the reactions shown in eq. 15 gave evidence in favor of the direct $RTlX_2$-reductant interaction, as detailed below.

$$2RTl(OAc)_2 \xrightarrow{\text{reductant}} R_2TlOAc + TlOAc \qquad (15)$$

R= Me, $PhCH(OMe)CH_2$, E- and Z-PhCH=CH, Ph, p-tolyl

reductant= hydrazine, ascrobic acid, $P(OMe)_3$, $NaBH_4$

We could not observe occurrence of both the forward and the reverse reactions of eq. 4 for the alkylthallium derivatives under the conditions employed in eq. 15 except that the reductant was absent. For the aryl and vinyl derivatives, the rate of the forward path of eq. 4, which was estimated from the equilibrium constant and the rate of the reverse path of eq. 4, is by far slower than the rate of eq. 15 actually observed. Thus, the forward path of eq. 4 cannot be a rate-determining step of eq. 15.

No stereochemical loss of the styryl group was observed when $(PhCH=CH)_2TlOAc$ was formed from E-PhCH=CHTl$(OAc)_2$ and $P(OMe)_3$ [22]. On the other hand, the reaction of the diastereomer, **1**-d with $P(OMe)_3$, ascorbic acid, and hydrazine gave good yields of dialkylthallium acetate, $[PhCH(OMe)CHD]_2TlOAc$ **12** of which ca. 55-70% of the total alkyl group had undergone the epimerization [22,23]. The stereochemical purity of **1**-d recovered from the incomplete reaction was found to be lower than the original one by some 20%. The compound **12** was found stereochemically stable under the reaction conditions. The compound $[PhCH(OMe)CHD]_2TlOH$, a minor product from reduction of **1**-d with $NaBH_4$, also contained partially epimerized alkyl group. The spin adduct **6** was again detected on addition of nitrosodurene to a mixture of **1** and the reductant. It seems worthy of note that the reaction of $RTl(OAc)_2$ with ascorbic acid under dioxygen afforded moderate to good yields of

ROH where R= $PhCH(OMe)CH_2$, $PhMeC(OMe)CH_2$, $Me_2C(OMe)CH_2$, n-octyl and $MeOOC(CH_2)_{10}$ [23].

We propose the mechanism of eq. 15 (R= alkyl) as involving alkyl radical intermediate which is generated by single electron reduction of $RTlX_2$ and the subsequent homolysis (eq. 6 and 7), followed by combination of the R radical with either RTlX or $RTlX_2$. The latter step requires another reduction. The electrochemical reduction of 1-d did afford 12 containing the partially epimerized alky group [24].

For the reactions of the aryl and vinyl derivatives, it is not certain whether the Tl-R bond homolysis (eq. 7) actually occurs. An active species for the disproportionation can be RTlX, or even RTl, of which R is expected to be more susceptible to the electrophilic attack by different organothallium cationic species. Consistent with this enhanced nucleophilicity of the R group would be selective formation of RD in the reaction of $RTlX_2$ (R= aryl, E- and Z-PhCH=CH) with $P(OMe)_3$ [22] or $NaBH_4$ [6] carried out in MeOD or EtOD. A plausible path to RTl from $RTlX_2$ and $P(OMe)_3$ or H^- is shown in Scheme 5.

<u>Scheme 5</u>

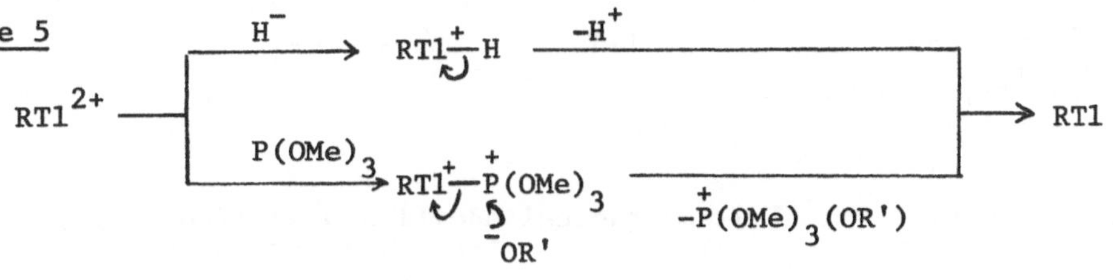

(R'= H, Me, Ac)

As described before, reductive disproportionation of $RTlX_2$ could be suppressed by complexation with crown ethers. Thus, for example, the reaction of 13 with $P(OMe)_3$ and ascrobic acid in methanol gave toluene predominantly [19]. This result may also be taken as evidence to demonstrate the ability of the electron donors to interact directly with RTl^{2+} ion in the course of the reaction of eq. 15.

PhCH(OMe)CHD-Tl-CHDCH(OMe)Ph
            |
           OAc

12

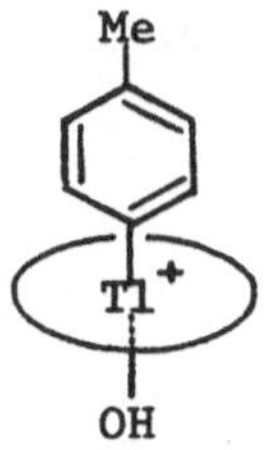

13

(crown= dibenzo-18-crown-6)

5. Relevance of Reductive Activation of Metal-Carbon Bond to Organic Synthesis

One of the most successful applications of the reductive activation of the metal-carbon bond in organic synthesis may be a $NaBH_4$-induced hydrodemercuration of oxymercuration adducts of olefin, which constitutes Markownikov addition of H-OR to olefins (eq. 16) [2b]. This

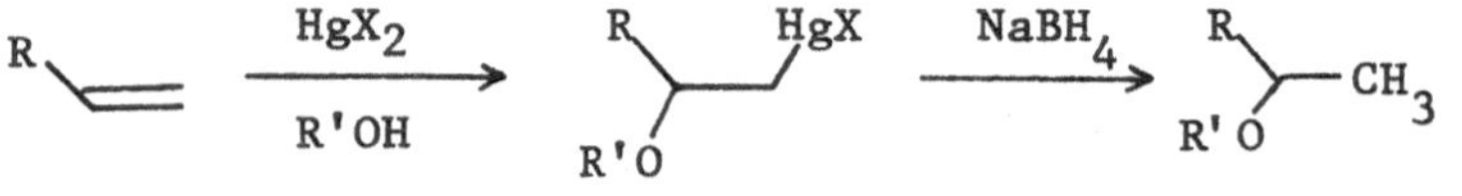

(16)

method is not suited for certain olefins, e.g. allylphenols whose oxymercuration adducts and $NaBH_4$ usually lead to predominant deoxymercuration. In this respect, hydrodethallation (thermal) and hydrodemercuration (irradiation) with BNAH (e.g. eq. 17) may be complementary to the above method. Radical cyclization of ω-alkenylmercury

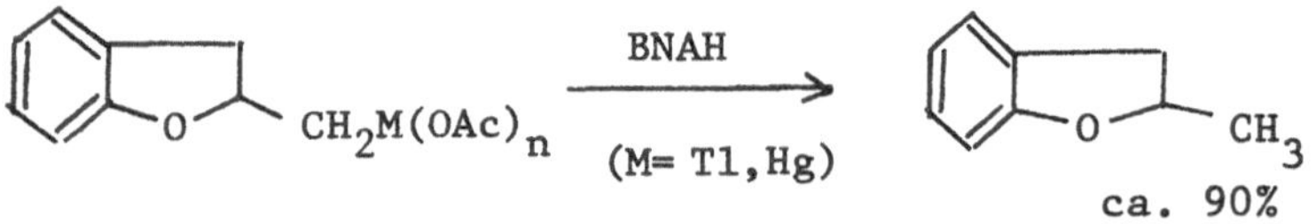

(17)

derivatives (eq. 18) [2a] and radical addition of R (from RHgX) and H to Michael olefins (eq. 19) [2a] are also initiated by the reductive Hg-C bond homolysis by the use of metal hydrides.

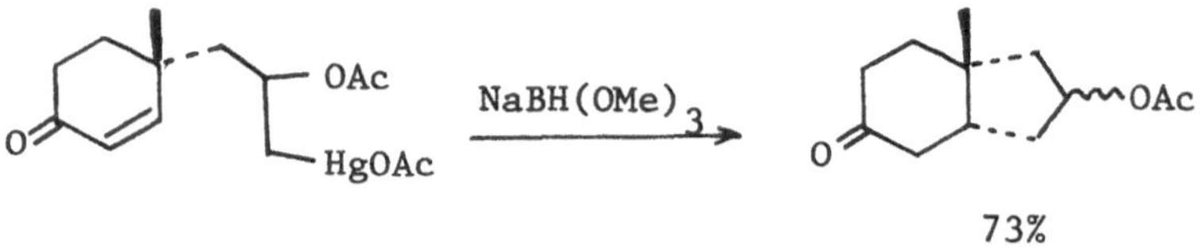

(18)

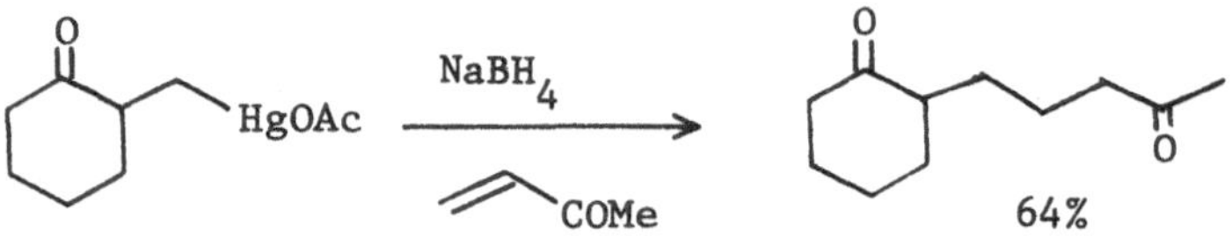

(19)

Eq. 1 is one of the most useful reactions of $RTlX_2$ for organic synthesis [4]. However, since Nu is a potential electron donor, the reductant-induced disproportionation sometimes accompanies or dominates over eq. 1 (e.g. in the reaction using Nu= $CH(COMe)_2^-$, PhS^-, Cl^-/CuCl) [9a,22], which diminishes the synthetic value of eq. 1. Nucleophilic substitution of RHgX, analogous to eq. 1, is more difficult than $RTlX_2$, which in turn is less reactive than $RPbX_3$. Thus, for example, aryl and vinylmercurials did not couple with $Me_2CNO_2^-$ even under photo-irradiation (see Table I) [7a,10], while aryllead compounds coupled with the same carbanion more rapidly than $ArTlX_2$ [8a]. In addition, aryllead compounds readily reacted with β-dicarbonyl anions to accomplish arylation of β-dicarbonyl compounds (Table I) [8b]. An analogous attempt to couple $ArTlX_2$ with $CH(COMe)_2^-$ failed owing to facile disproportionation. In order for the limited availability of aryllead and vinyllead compounds to be overcome, these compounds were generated from R_2Hg or RBu_3Sn and $Pb(OAc)_4$ in situ, followed by successful coupling with the carbanions [25]. These results intensify the synthetic value of the reductive activation of organometallic reagents.

References

1. J. K. Kochi, "Organometallic Mechanisms and Catalysis," Academic Press, New York, 1978, Chapt. 16.
2. a) B. Giese, Angew. Chem., Int. Ed. Eng., 24, 553 (1985) and references therein.
   b) W. Carruthers, "Comprehensive Organometallic Chemistry," ed. by G. Wilkinson, F. G. A. Stone, E. W. Abel, Pergamon Press, Oxford, 1982, Chapt. 49, p. 671.
3. R. Scheffold, G. Rytz, L. Walder, R. Orlinski and Z. Chilmonczyk, Pure Appl. Chem., 55, 1791 (1983).
4. a) H. Kurosawa, Ref. 2b, Chapt. 8.
   b) A. McKillop and E. C. Taylor, Ref. 2b, Chapt. 47.
5. a) N. Kornblum, Angew. Chem., Int. Ed. Eng., 14, 734 (1975).
   b) C. P. Andrieux, I. Gallardo, J. M. Saveant and K. B. Su, J. Am. Chem. Soc., 108, 638 (1986) and references therein.
6. R. B. Herbert, Tetrahedron Lett., 1973, 1375.
7. a) H. Kurosawa, H. Okada, M. Sato and T. Hattori, J. Organomet. Chem., 250, 83 (1983).
   b) H. Kurosawa, H. Okada and T. Hattori, Tetrahedron Lett., 22, 4495 (1981).
   c) H. Kurosawa, M. Sato and H. Okada, ibid., 23, 2965 (1982)
   d) H. Kurosawa, H. Okada and M. Yasuda, ibid., 21, 959 (1980).
   e) H. Kurosawa and M. Sato, unpublished results.
8. a) R. P. Kozyrod and J. T. Pinhey, Tetrahedron Lett., 22, 783 (1981)
   b) J. T. Pinhey and B. A. Rowe, Aust. J. Chem., 32, 1561 (1979).
9. a) S. Uemura, K. Zushi, A. Tabata, A. Toshimitsu and M. Okano, Bull. Chem. Soc. Jpn., 47, 920 (1974).
   b) S. Uemura, A. Toshimitsu, M. Okano, T. Kawamura, T. Yonezawa and K. Ichikawa, J. Chem. Soc., Chem. Comm., 1978, 65.

c) J. E. Bäckvall, M. U. Ahmad, S. Uemura, A. Toshimitsu and T. Kawamura, Tetrahedron Lett., 21, 2283 (1980).

10. G. A. Russell, J. Herschberger and K. Owens, J. Organomet. Chem., 225, 43 (1982).

11. H. Kurosawa, R. Kitano and T. Sasaki, J. Chem. Soc., Dalton Trans., 1978, 234.

12. a) S. Faleschini, G. Pilloni and L. Doretti, J. Electroanal. Chem., 23, 261 (1969).
b) H. Kurosawa and M. Sato, unpublished results.

13. S. J. Price, J. P. Richard, R. C. Rumfeldt and M. G. Jacko, Can. J. Chem., 51, 1397 (1973).

14. H. A. Skinner, Adv. Organomet. Chem., 2, 49 (1964).

15. F. M. Martens, J. W. Verhoeven, R. A. Gase, U. K. Pandit and T. T. de Boer, Tetrahedron, 34, 443 (1978).

16. E. M. Kosower, "Free Radicals in Biology," ed. by W. A. Pryor, Academic Press, New York, vol. 2, 1976, p. 1.

17. D. Lal, D. Griller, S. Husband and K. U. Ingold, J. Am. Chem. Soc., 96, 6355 (1974).

18. Y. Kawasaki, W. Yokota and N. Enomoto, Chem. Lett., 1982, 941.

19. H. Kurosawa, N. Okuda and Y. Kawasaki, J. Organomet. Chem., 255, 153 (1983).

20. Y. Kawasaki, N. Enomoto and J. Tomioka, unpublished results.

21. a) U. Pohl and F. Huber, Z. Naturforsch., B, 33, 1188 (1978).
b) F. Huber, U. Schmidt and H. Kirchmann, "Organometallics and Organometalloids. Occurrence and Fate in the Environment," ACS Symposium Series, ACS, Washington, D. C., 1978, p. 65.

22. H. Kurosawa and M. Sato, Organometallics, 1, 440 (1982).

23. H. Kurosawa and M. Yasuda, J. Chem. Soc., Chem. Comm., 1978, 716.

24. J. Nogami, private communication.

25. M. G. Moloney and J. T. Pinhey, J. Chem. Soc., Chem. Comm., 1984, 965.

ELECTROCHEMISTRY OF NEW MOLYBDENOCENE DIHYDROCARBYLS

M.J. Calhorda, A.R. Dias*, M.H. Garcia, A.M. Martins,
C.C. Romão
Centro de Química Estrutural, Instituto Superior Técnico
1096 LISBOA CODEX
Portugal

ABSTRACT. Study of the electrochemical behaviour of the complexes $|Mo(\eta^5\text{-}C_5H_5)_2R\ R'|$ ($R=R'=CH_3$, C_2H_5, ${}^nC_4H_9$; $R=CH_3$, $R'={}^nC_4H_9$), by cyclic voltammetry in acetonitrile and dichloromethane show that the reversible 1-electron oxidation of $|Mo(\eta^5\text{-}C_5H_5)_2(CH_3)_2|$ (-270 mV) occurs at lower potential than $|Mo(\eta^5\text{-}C_5H_5)_2\ R\ R'|$ ($R=R'=C_2H_5$, ${}^nC_4H_9$) (ca -210 mV) lying the oxidation potential of $|Mo(\eta^5C_5H_5)_2CH_3({}^nC_4H_9)|$ in the middle of that range (-240 mV). These results are explained on the basis of the α-H/metal antibonding interaction in the HOMO orbital, as shown by calculations of the extended Hückel type, for the molibdenocene dimethyl complex.

INTRODUCTION

Cooper and co-workers have demonstrated that hydrogen atom abstraction from paramagnetic tungstenocene dihydrocarbyl cations is an α-selective process (1) and on the basis of the results of an electrochemical investigation of the oxidation of the neutral tungstenocene dihydrocarbyls, they suggest the occurrence of a three-center, three-electron agostic interaction in the cationic complexes (2). Those results prompted us to synthetise for the first time some of the related molybdenocene compounds and to study their properties (3). In the present communication we report a detailed study of the electrochemistry of $|MoCp_2RR'|$ ($Cp=\eta^5\text{-}C_5H_5$; $R=R'=CH_3$, C_2H_5, ${}^nC_4H_9$; $R=CH_3$, $R'={}^nC_4H_9$) and discuss the observed trends in the oxidation potentials in terms of the nature of the molecular orbital involved.

MATERIALS AND METHODS

Compounds $MoCp_2RR'$ were prepared and purified as described in the literature ($R=R'=CH_3$) (4), $R=R'=C_2H_5$, ${}^nC_4H_9$ (3) $R=CH_3$, $R'=C_4H_9$ (8). The electrochemistry instrumention consisted of a Princeton Applied Research Model 173 potentiometer, Model 175 voltage programmer, Model 179 digital coulometer and an Omnigraphic 2000 X-T recorder

M. Chanon et al. (eds.), Paramagnetic Organometallic Species in Activation / Selectivity, Catalysis, 275–281.

of Houston Instruments.

Potentials were refered to a calomel electrode (SCE) containing a saturated solution of potassium chloride checked relative to a $1.0x10^{-3}$M solution of ferrocene in acetonitrile containing 0.10 M $LiClO_4$ for which the ferrocinium/ferrocene potential was in agreement with the literature (5).

The working electrode was made of a 2-mm piece of Pt wire for cyclic voltammograms, while the working electrode for the coulometry experiments was Pt gauze.

The secondary electrode was a Pt wire coil. Cyclic voltametry experiments were performed in a PAR polarographic cell, at room temperature, being used solutions 1 mM in solute and 0.1 M in supporting electrolyte, tetrabutylammonium hexafluorophosphate. Coulometry experiments were done in a three-compartment cell, where the working electrode compartment was separated from the compartment for auxiliary and reference electrodes by medium porosity glass frits.

The solvents, CH_3CN and CH_2Cl_2 were reagent grade materials, were dried over CaH_2 and P_2O_5 and were distilled just before use, under argon atmosphere.

Solutions were degassed with dry nitrogen before each experiment and a nitrogen atmosphere was mantained over the solution during the experiments.

The calculations were of the extended Hückel type (6) with weighted Hij's (7). The basis set for the metal atom consisted of (n-1)d, ns and np orbitals. The s and p orbitals described by simple Slater-type wave functions and the d orbitals as contracted linear combinations of two Slater-type wave functions.

RESULTS

The cyclic voltammogram of the complexes $MoCp_2RR'_2$ (1 $R=R'=CH_3$, 2 $R=R'=C_2H_5$, 3 $R=R'={}^nC_4H_9$, 4 $R=CH_3$, $R'={}^nC_4H_9$) in acetonitrile and dichloromethane solutions between the solvent limits (ca -1.5 and 1.6 V) exhibited a chemically quasi-reversible oxidation around -215 mV at scan between 20 and 1000 mV/s (see Table)

However, when the solvent was CH_2Cl_2 a second irreversible wave was found at 1.05 V for $MoCp_2Et_2$, 1.2 V for $MoCp_2{}^nBut_2$, 1.10 V for $MoCp_2Me_2$ and 1.05 for $MoCp_2Me^nBu$.

Figures 1 and 2 show the cyclic voltammograms of the complex $MoCp_2Et_2$ which exemplify the kind of behaviour of the complexes $MoCp_2RR'$ in CH_3CN and CH_2Cl_2.

TABLE

Electrochemical Data (a)

COMPLEX	ΔEp(mV)		(Ea + Ec)/2(mV)		ia/ic	
	CH_3CN	CH_2Cl_2	CH_3CN	CH_2Cl_2	CH_3CN	CH_2Cl_2
1	90	90	-275	-260	0.98	1.00
2	90	135	-215	-207	1.03	0.98
3	80	90	-220	-215	0.97	1.00
4	80	80	-270	-240	0.98	0.98

[a] Experimental conditions: cyclic voltametry was performed at room temperature on solutions 1.0×10^{-3} M in substrate and 0.10 M on $|^nBu_4N|PF_6$ as supporting electrolyte at scan rate of 100 mV s^{-1}. The auxiliary and working electrodes were respectively a Pt wire coil and a 2 mm Pt sheet, and the reference electrode was a saturated calomel calibrated relative to a 1.0×10^{-3} M solution of ferrocene in acetonitrile containing 0.10 M $LiClO_4$.

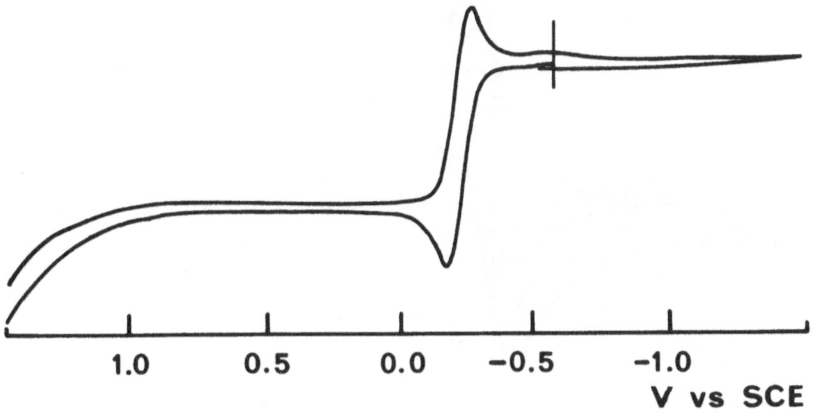

Figure 1. Cyclic voltammogram of $MoCp_2Et_2$ in CH_3CN, 10^{-1} M $^nBu_4NPF_6$. Starting potential -0.6 V. Sweep rate 100 mV/s.

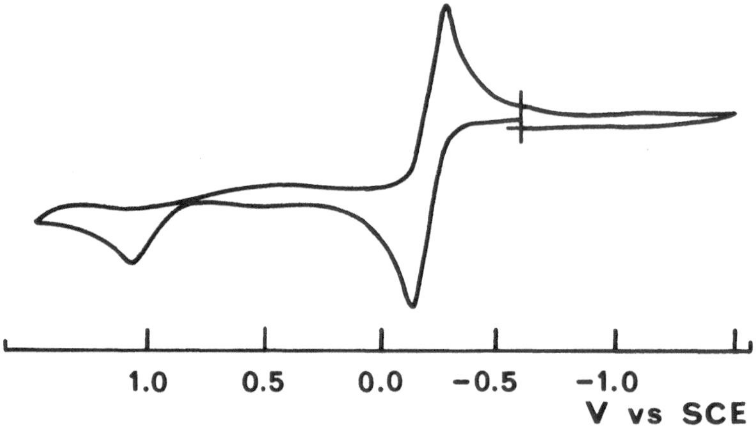

Figure 2. Cyclic voltammogram of $MoCp_2Et_2$ in CH_2Cl_2 $10^{-1}M$ $^{n}Bu_4NPF_6$. Starting potential -0.6 V. Sweep rate 100 mV/s.

A plot of i_a against the square root of the scan speed was linear in the range of those potentials, as expected for a diffusion-controled electrode process.

The average of the anodic and cathodic peak potentials was independent of the scan speed and the separation of these waves ΔE_p was ∿80 mV (see table).

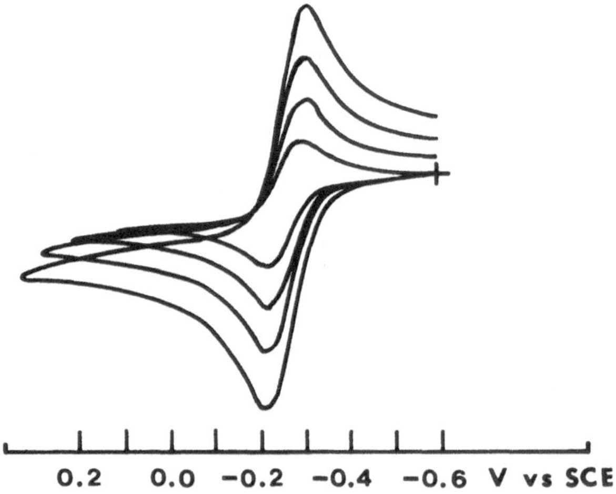

Figure 3. Cyclic voltammograms of $MoCp_2{}^{n}Bu_2$ in CH_3CN 10^{-1} M $^{n}Bu_4NPF_6$. Starting potential -0.6 V. Sweep rate were 20,50, 100 and 200 mV/s.

The example showed in Fig. 3 for the cyclic voltammograms of $MoCp_2{}^nBu_2$, in CH_3CN at scan speeds 20, 50, 100 and 200 mV/s tipifies the CV oxidation of the compounds $MoCp_2RR'$.

The oxidation of the bulk acetonitrile solutions at potentials 200 mV more positive than the corresponding CV oxidation by controlled potential coulometry, of the compounds $MoCp_2RR'$ showed to involve the transfer of an average of 0.98 electrons.

DISCUSSION

As shown in the table, the oxidation potential for the complexes $|MoCp_2RR'|$ are in the reverse order from what should be expected on the basis of the inductive effects of the alkyl groups. These results are similar to those obtained by Cooper for the tungstenocene analogues. Substitution of C_2H_5 for CH_3 in MCp_2R_2 originates a difference in the oxidation potentials of 65 mV for M=W and 63 mV for M=Mo.

The oxidation of MCp_2X_2 complexes has been interpreted by the removal of one electron from the filled "Alcock" orbital located on the mirror plane between the cyclopentadienyl ligands and perpendicular to the plane bisecting the XMX angle. The interaction of the resulting half-filled orbital with an α-C-H bond located on the same plane would originate a stabilizing three-center three electron agostic interaction, invoked by Cooper to interpret the trend found for the oxidation potentials of the WCp_2RR' derivatives.

In order to test Cooper's proposal (3) we carried out extended Hückel type calculations (6) for $MoCp_2Me_2$. The HOMO, localized on the above mentioned plane, Fig. 4, has a major contribution from the

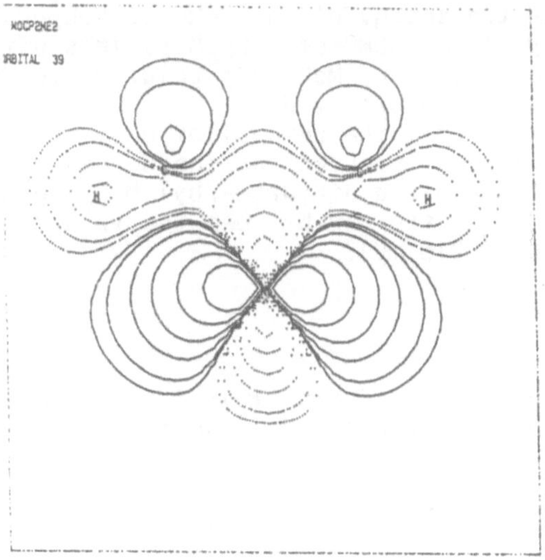

Figure 4. Molecular wave function contour of the HOMO orbital for $MoCp_2(CH_3)_2$, showing two nodals pTanes between Mo and H.

metal but is also delocalized over the α-C-H bonds that lie on the same plane in the more stable conformation, e.g. Ha and Hb in Fig. 5.

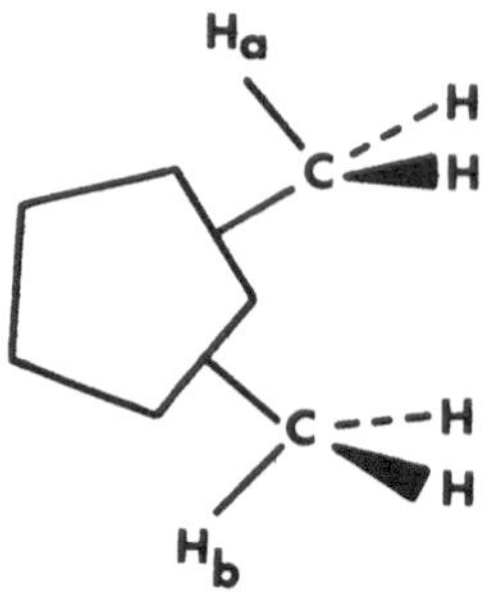

Figure 5. More stable geometric conformation for the complex $MoCp_2(CH_3)_2$

From Fig. 4 it can be seen that there is an anti-bonding interaction between the α-C-H bond and the metal. In the oxidized complex, $MoCp_2Me_2^+$ this repulsive interaction will be smaller since only one electron is left in this HOMO orbital.

Stereochemical arguments show that when a longer alkyl chain is substituted for methyl, the stable conformation no longer has α-C-H bonds on the plane of the "Alcock" orbital as it has been confirmed in the crystal structure of $MoCp_2{}^nBu_2$ (8). It may therefore be expected that the removal of this α-C-H to metal destabilizing (anti-bonding) interaction will lower the energy of the HOMO of $MoCp_2R_2$ (R=Et, ^{n}Bu) relative to the energy of the HOMO of $MoCp_2Me_2$. This effect will result in the easier oxidation of $MoCp_2Me_2$ relative to $MoCp_2R_2$ (R=Et, ^{n}Bu) as it was experimentally observed.

As could be expected from the removal of only one α-C-H to metal interaction, and was already observed for the W analogues (2) the oxidation potential of the mixed dihydrocarbyl $MoCp_2Me^nBu$ (-240 mV) lies close to the middle of the gap between the values for $MoCp_2Me_2$ (-270 mV) and $MoCp_2{}^nBu_2$ (-215 mV).

REFERENCES

1) (a) J.C. Hayes; N.J. Cooper, *J.Am. Chem. Soc.* 1981, **103**, 4648
(b) J.C. Hayes; G.D.N. Pearson; N.J. Cooper, *J. Am. Chem. Soc.*, 1981, **103**, 4648.

2) M.F. Asaro; R.S. Cooper; N.J. Cooper; *J. Am. Chem. Soc.* 1986, **108**, 5187.

3) A.R. Dias; M.H. Garcia; A.M. Martins; C.I. Pinheiro; C.C. Romão; L.F. Veiros. *J. Organomet. Chem.* 1987, **327**, C59.

4) F.W.S. Benfield, M.L.H. Green; *J. Chem. Soc. Dalton*, 1974, 1324.

5) I.V. Nelson; R.T. Iwamoto. *Anal. Chem.* 1963, **35**, 867.

6) R. Hoffmann. *J. Chem. Phys.*, 1963, **39**, 1397. R. Hoffmann; W.N. Lipscomb. *J. Chem. Phys.*, 1962, **36**, 2197.

7) J.H. Ammeter; H.B. Bürgi; J.C. Thibeault; R. Hoffmann, *J. Am. Chem. Soc.*, 1978, **100**, 3686.

8) M.J. Calhorda, M.A.A.F. Carrondo, A.R. Dias, M.H. Garcia, C.C. Romão; J.A.M. Simões, *manuscript in preparation*.

ELECTRON TRANSFER INDUCED METAL ADDITION AND LIGAND EXCHANGE IN ORGANOMETALLIC ANION RADICAL COMPLEXES

Wolfgang Kaim, Barbara Olbrich-Deussner, Renate Gross, Sylvia Ernst, Stephan Kohlmann and Christian Bessenbacher
Institut für Anorganische Chemie der Universität
Niederurseler Hang
D-6000 Frankfurt am Main 50
West Germany

ABSTRACT. Transition metal complexes containing anion radical ligands display a strong tendency towards full coordinative saturation at the singly reduced ligand and towards substitutional activation of co-ligands at the metal center. Examples involving metal carbonyls show how both of these reactivities can be employed either separately or in a combined fashion for electron transfer catalyzed substitution processes and for the construction of new polynuclear complexes with unusual properties.

INTRODUCTION

The essential role of odd electron intermediates in many catalytic processes [1] is due to their increased electron transfer reactivity [2] and substitutional lability [3]. For instance, complexes with 17 or 19 valence electrons at the metal center are generally far more reactive than their 18 electron counterparts, and they can thus become important intermediates in electron transfer catalysis (ETC) [3,4].

Alternatively, the electron added or removed in single electron transfer reactions of organometallics may be localized on a ligand. In particular, complexes of singly reduced, i.e. anion radical ligands [5,6] enjoy a special stabilization because of the strongly increased basicity of such ligands. Most startling effects have thus been observed for chelate ligands of the 2,2'-bipyridine type which receive an increase of pK_{BH^+} from 4.5 to approximately 24 on going from the neutral to the singly reduced form [7]. It will be demonstrated here that metal-to-ligand charge transfer and the formation of anion radical complexes can lead to enhanced coordination activity of the reduced ligand as well as to substitutional activation of co-ligands and that some of these reactions can proceed via ETC.

1. POLYNUCLEATION FACILITATED BY METAL-TO-LIGAND CHARGE TRANSFER

4,4'-Bipyrimidine, bpm (1), is a rather weakly basic (pK_{BH^+} = 1.5)

M. Chanon et al. (eds.), Paramagnetic Organometallic Species in Activation / Selectivity, Catalysis, 283–294.

ambidentate ligand [8] which can bind e.g. two $W(CO)_5$ fragments at the peripheral $N^1,N^{1'}$ centers (compound 2) before converting thermally to the more stable chelate complex 1 (1) [9]. Spectroscopic studies of deeply coloured tetracarbonylmetal species such as 1 have established efficient back bonding in these complexes, i.e., a considerable amount of electron density is transferred to a low-lying π^* level of the ligand from the filled metal d orbital of appropriate symmetry [8,9].

A chemically significant consequence of this back donation is that the bpm ligand becomes sufficiently basic to bind yet another two neutral metal carbonyl fragments $W(CO)_5$ to form the first trinuclear carbonylmetal complex 3 of a bidiazine ligand:

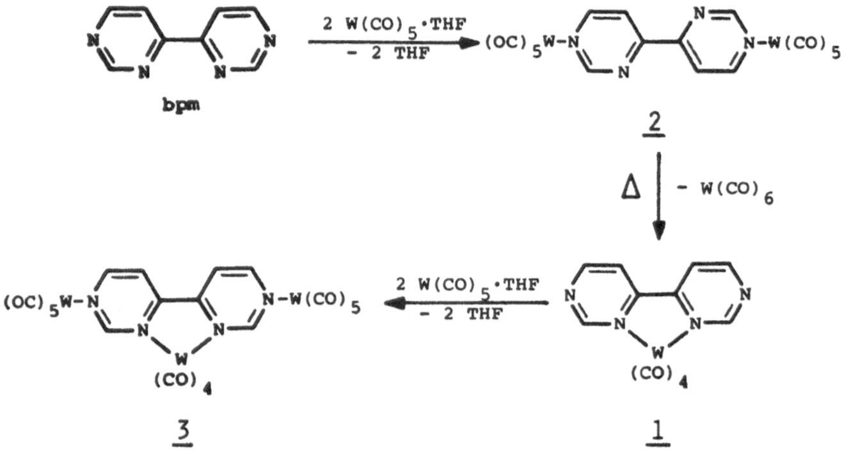

(1)

The little soluble tetradecacarbonyltritungsten complex 3 could be characterized by vibrational and electronic spectroscopy, by cyclic voltammetry and by EPR studies of the anion radical form [10]. As would be expected, the CO stretching frequencies comprise typical values for penta- and cis-tetracarbonyls. Electrochemistry reveals the very facile one electron uptake by the complex in DMF (E_{red} = -0.43 V vs. SCE), yielding a stable anion radical complex with a well balanced spin distribution (g = 2.0031)

Irreversible oxidation waves corresponding to electron removal from the tetracarbonyltungsten ($\sim$+0.8 V vs. SCE) and $W(CO)_5$ groups ($\sim$+1.1 V) are also observed, yet despite the obvious narrowing of the redox potential (and HOMO/LUMO) difference, the energy of the metal-to-ligand charge transfer (MLCT) absorption maximum (17 240 cm^{-1} in THF) is higher than that of the mononuclear complex 1 (16 450 cm^{-1}) [9,10].

The reason for this initially puzzling effect lies in the larger width of the charge transfer absorption bands for complexes with penta-carbonylmetal fragments (band halfwidths $\Delta\nu_{1/2}$ = 6000–8000 cm^{-1}). The structural flexibility of "freely" rotating and vibrating $W(CO)_5$ groups results in considerable geometrical change between the ground and MLCT excited state so that higher vibrationally excited states are significantly populated by "vertical" optical excitation and a broader absorption band with hypsochromically shifted maximum results. Much more pronounced restrictions pertain to rotation and vibration in a rigid chelate complex such as 1 which displays a narrower MLCT absorption band ($\Delta\nu_{1/2}$ = 3800 cm^{-1}) and, consequently, a bathochromically shifted absorption maximum. In accordance with this interpretation, the difference between the transition energy at the absorption maximum (in eV) and the electrochemical potential difference (in V) is twice as large for 2 and 3 than for 1 [10].

Summarizing, effective back donation alone can already activate further, supposedly weakly basic coordination centers in a polydentate π acceptor ligand, even in the absence of complete intramolecular electron transfer. An increasing number of such examples has been observed and reported recently [11-13], involving other bidiazine ligands (i.e. isomers of bpm) [12], 3,6-bis(2'-pyridyl)-1,2,4,5-tetrazine (bptz) [11], or pteridine (pte) [13], the parent compound of the biochemically important pterins and folic acid derivatives [14]. In fact, it has sometimes been difficult to obtain and isolate partially coordinated complexes of such polyfunctional π acceptor ligands because of their pronounced tendency to have most available coordination sites saturated even with neutral metal fragments [11] - a phenomenon which is not common in the chemistry of ordinary, "innocent" ligands.

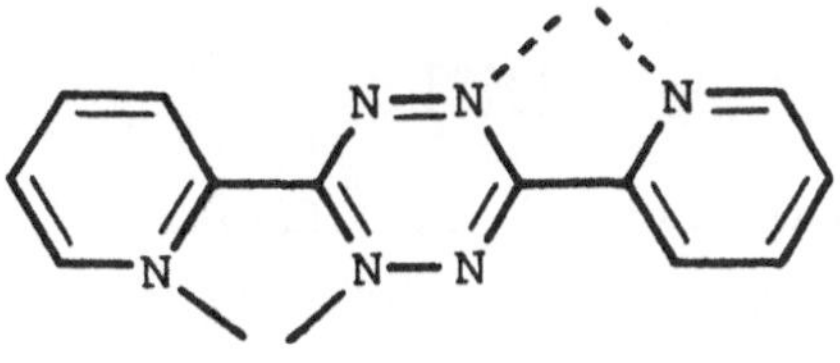
bptz

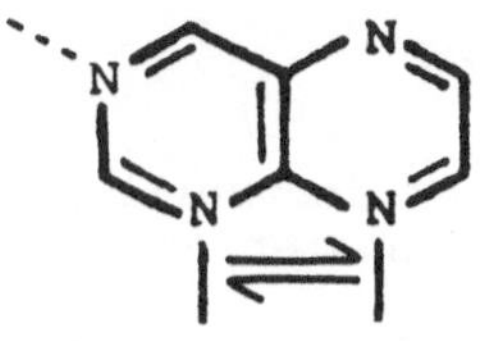
pte

(2)

2. POLYNUCLEATION BROUGHT ABOUT BY METAL-TO-LIGAND ELECTRON TRANSFER

Complete electron transfer occurs [15] between the widely used acceptors TCNE [16] and TCNQ [17] and the organometallic fragments $(C_5R_5)(CO)_2Mn$ [18]. Relatively stable σ adducts are formed in this "single electron oxidative addition" [19] (or "substitution", considering the use of the THF solvate of the metal fragment).

$$\text{TCNX} + (C_5R_5)(CO)_2Mn^{I}(\text{THF}) \xrightarrow[-\text{THF}]{\text{SET}} (C_5R_5)(CO)_2Mn^{II}(\eta^1\text{-TCNX}^{\cdot-}) \qquad (3)$$

TCNE: 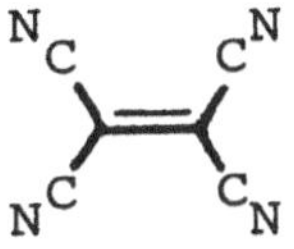

TCNQ: 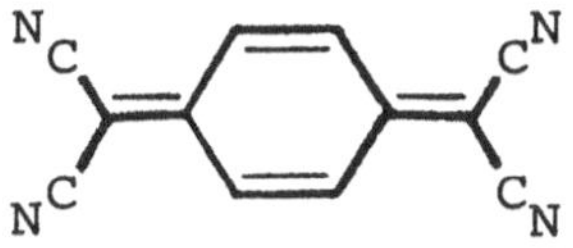

The characterization of the lowest electronic transitions in these complexes as ligand-to-metal charge transfer (LMCT) after electron transfer [27,28] and not as MLCT, as before a complete electron transfer in the ground state [21,28], is most clearly illustrated by the different effects of destabilization of metal levels through methyl group substitution at the cyclopentadienyl rings. Whereas this introduction of donor groups to the metal fragment leads to the expected bathochromic shift in an MLCT situation [21], the reverse holds e.g. for the TCNE complex series: Here, destabilization of the metal levels affects the LUMO more than the HOMO so that a high energy shift of the (ligand-to-metal) charge transfer band accompanies donor substitution at the metal fragment (Scheme 1) [15,28].

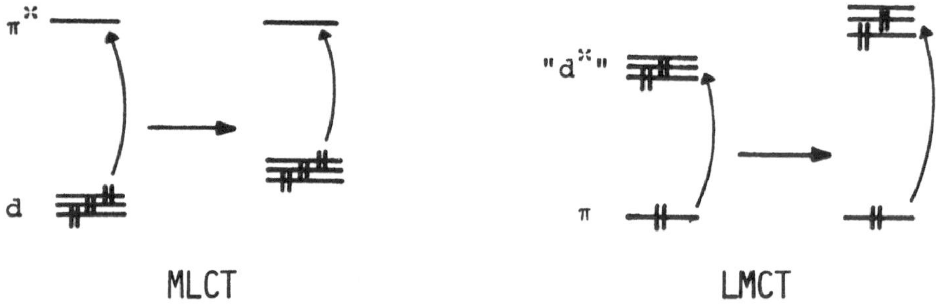

Scheme 1. Effect of destabilization of metal orbitals on different kinds of charge transfer transitions.

The most intriguing and chemically most significant consequence of the single electron shift situation (3) is the increase in basicity at the three remaining nitrile coordination centers in mononuclear complexes of $TCNE^{\bullet-}$ and $TCNQ^{\bullet-}$. Using this concept [6], di-, tri- and the first tetranuclear complexes (Scheme 2) of these commonly used acceptors could be obtained as stable compounds with a unique electronic structure and spectroscopy [15].

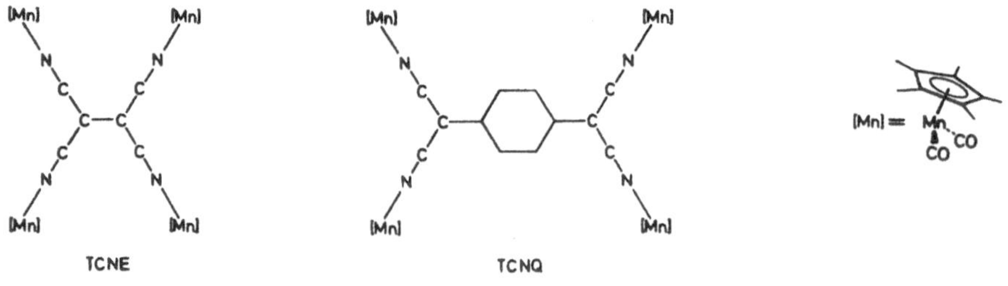

Scheme 2. Scale drawings of tetranuclear TCNE and TCNQ complexes.

Arguments for the formulation of these compounds as electron exchange products (3) containing anion radicals $TCNX^{\bullet -}$ and low-spin Mn(II) (S = 1/2) are summarized in the following:

Facile thermal population of a triplet state close to the singlet ground state of $(C_5Me_5)(CO)_2Mn(TCNE)$ has been established by measurements of the magnetism in the solid state [20]. While the TCNE complexes show only little broadened ^1H- and ^{13}C-NMR signals in toluene or THF solution, the TCNQ derivative tends to dissociate in more polar solvents such as THF or acetonitrile, yielding the anion radical $TCNQ^{\bullet -}$ and a low-spin manganese(II) complex, as determined by EPR [15].

Electron transfer can be expected between TCNE or TCNQ and the $(C_5R_5)(CO)_2Mn(THF)$ solvates because the reduction potentials of the acceptors are larger than the oxidation potentials of the metal fragments [15]. Most characteristically, the 1:1 complexes formed are reduced at more negative potentials than the free TCNX ligands, a relation which is just the opposite of that found in "normal" complexes [6], i.e. in the absence of complete ligand-to-metal electron transfer; coordination of an electrophilic metal center should generally facilitate reduction of the ligand. However, after electron exchange, it is predominantly the metal which gets reduced at rather negative values due to the strong ligand field (basicity! [7]) exerted by a radical anion [6].

The high rate of product formation, leading to complete conversion within a few seconds, is indicated by the spontaneous colour change from red to green upon mixing and stands in stark contrast to the sluggish substitution of $(C_5R_5)(CO)_2Mn(THF)$ by N-donors which often requires days for completion [14,21]. This vastly enhanced rate is compatible with an electron transfer catalysis (ETC) mechanism [4], in which paramagnetic intermediates propagate a fast chain reaction for substitution of THF, as reported by Kochi et al. for electrode catalyzed reactions [22]. However, here it is not necessary to employ an anode since TCNE and TCNQ produce the electron "holes" necessary for this "autocatalytic" variant of ETC [4].

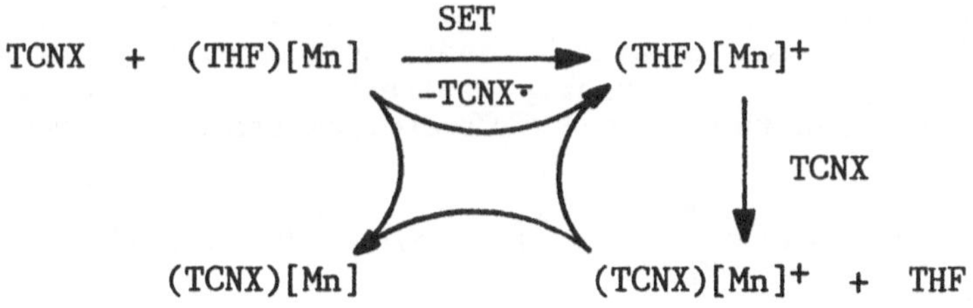

(4)

The oxidation states of TCNE and of TCNQ in their complexes can be conveniently estimated from vibrational spectroscopy because C=C and C≡N stretching frequencies are very characteristic for neutral compounds, anion radicals and dianions, respectively [23–26]; a similar relationship exists between carbonyl stretching frequencies and the oxidation state of manganese in $(C_5R_5)(CO)_2Mn$ complexes [27]. All the measured data [15] point to a situation (3) in which one electron has been transferred from the metal fragment to the acceptor ligand.

The species are simply obtained by coordination of up to three more uncharged $(C_5R_5)(CO)_2Mn$ fragments to the 1:1 complex (5), they may be described either as mixed valence metal oligomers bridged by anion radical ligands (5:A) or, perhaps more appropriately, as delocalized multimetal/π-ligand species (5:B), a view which is supported by the successful application of Hückel MO calculations [29] to the extremely intense electronic transitions in the near infrared region [15,29].

$$(\eta^1\text{-TCNX}^{\bar{\cdot}})[\text{Mn}^+] \;+\; n\,(\text{THF})[\text{Mn}] \xrightarrow[-\,n\,\text{THF}]{} [\text{Mn}]_n(\eta^{n+1}\text{-TCNX}^{\bar{\cdot}})[\text{Mn}^+] \qquad (5\text{:A})$$

$$\updownarrow$$

$$(\eta^{n+1}\text{-TCNX}^{x-})[\text{Mn}^{(x/4)+}]_{n+1} \qquad (5\text{:B})$$

The success in obtaining unexpected compounds of high nuclearity as a result of metal-to-ligand electron transfer suggests that other unprecedented polynuclear species may be obtained via coupled electron transfer/coordination reactions.

3. ACTIVATION OF CO-LIGANDS: CREATION OF OPEN COORDINATION SITES

Activation of co-ligands by electron transfer to a redox-active ligand can involve a variety of processes, stoichiometric and catalytic reactions. It is obvious, for instance, that single electron reduction of a complex $LMX_n(Hal)$, Hal = halide, with a reversibly reducible ligand L can lead to expulsion of the good leaving group Hal^- in a second reaction step:

$$LMX_n(Hal) \;+\; e^- \rightleftharpoons [(L^{\bar{\cdot}})MX_n(Hal)]^- \longrightarrow (\;)LMX_n \;+\; Hal^- \qquad (6)$$

The thus generated open coordination site at the metal may be occupied by a solvent molecule or a species to be activated, e.g. in a catalytic cycle. Such halide loss of anion radical complexes has been observed e.g. for ruthenium(II) [30] and rhenium(I) complexes [31,32] of reduced 2,2'-bipyridine (bpy) and related ligands. Complexes such as fac-$(bpy^{\bar{\cdot}})Re(CO)_3$ (Hal) are particularly interesting because these anion radical complexes and the related MLCT excited states (7) of the neutral form

$$(bpy)Re^{I}(CO)_3(Hal) \xrightarrow[\text{MLCT}]{h\nu} {}^*[(bpy^{\bar{\cdot}})Re^{II}(CO)_3(Hal)] \qquad (7)$$

were found to be dissociative also with respect to loss of CO (cf. Chapter 4) under certain circumstances, thereby making these compounds photoactive <u>and</u> catalytic centers for the photoreductive activation of CO_2 (8) [31]. The parallel between the anion radical ground state and the MLCT excited state of a complex (both containing a singly reduced ligand!) is becoming increasingly important for efforts to connect photo- and electron transfer chemistry, i.e. for systems designed for light/redox energy conversion [1,4,28,31].

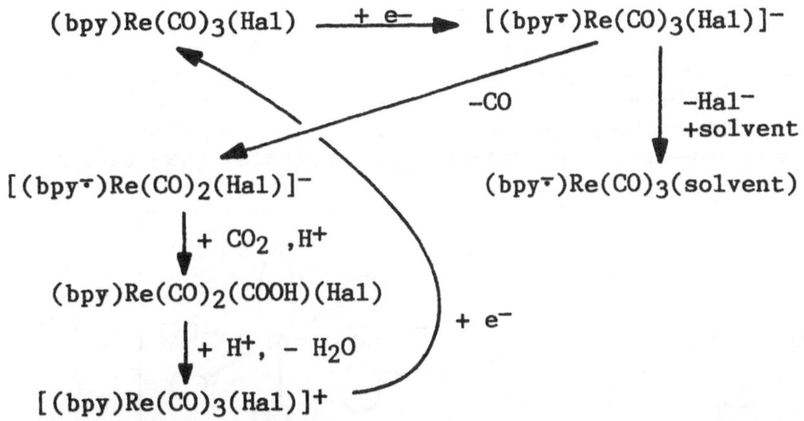

(8)

This remarkable ambiguity between activation of CO or halide in anion radical complexes of $Re(CO)_3(Hal)$ is currently under further investigation [33], EPR spectroscopic monitoring of the reduction of such complexes has shown a stronger leaving tendency for the bromide than for the chloride ion, as supported by a pronounced σ^*_{Re-Br}/π^* hyperconjugative spin delocalization from the heterocyclic anion radical to cis-bromide (fac-isomer, 9: Hal=Br) [32,33]. Notwithstanding these results, there is now a considerable number of radical anion complexes which exhibit exclusively CO activation.

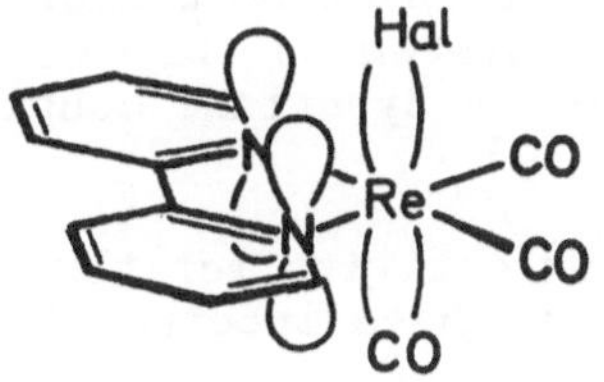

(9)

4. ELECTRON TRANSFER-CATALYZED CARBONYL SUBSTITUTION BASED LIGAND REDUCTION

Facilitated CO substitution e.g. by phosphanes in carbonylmetal complexes of anion radicals has been reported for a variety of metal/ligand combinations [6]. These include chelate and non-chelate complexes of group 6 [34-37] and group 7 transition metals [38-42] with nitrogen- and oxygen-containing ligands. Monitoring of this relatively rapid exchange in the absence of the two common activating effects, heat and light [43-46], could be conveniently carried out by EPR since the introduction of ^{31}P or ^{75}As nuclei gives rise to profound changes in the hyperfine pattern. In only a few instances could the hyperfine interaction of the unpaired electron with the ^{13}C nuclei of labilized carbonyl groups be detected in the precursor complexes [47,48], yet these data have proved to be crucial to the understanding of the selectivity of the activation process [35-37,40-42].

Hyperfine coupling was only detected for the CO substituents cis to the anion radical ligand in chelate [48] and non-chelate complexes [47] (10), indicating that an overlap between the spin-bearing π system and the metal-carbon σ bond is necessary for that activation. This "hyperconjugative" interaction involves the singly occupied MO (SOMO) of the π ligand and low-lying unoccupied σ^*_{M-C} orbitals [49] (10, X=CO)

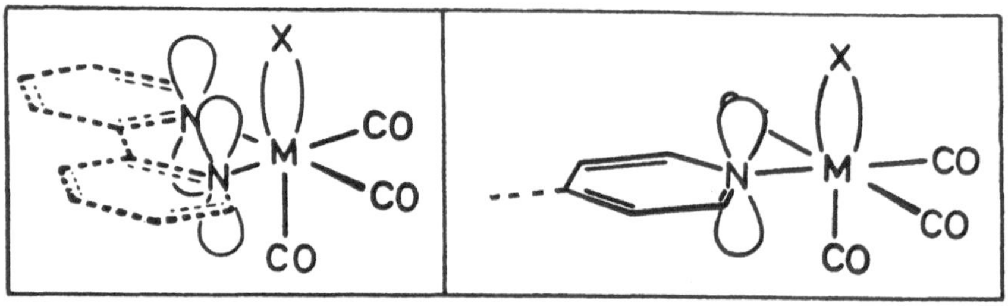

(10)

Apparently, the small delocalization of spin (and negative charge) to the metal fragment to give an "18+δ" electron species is sufficient to induce labilization of X = CO, however, in most instances there is no further substitution after single ligand exchange at each metal center [35-37,39]. EPR data suggest that the hyperconjugatively delocalized spin is largely confined to the metal/phosphorus σ bond (X = PR_3) after single substitution because of the low lying σ^* level of such a bond; accordingly, a second activation of CO cannot take place. The effective transfer of spin from the π system to the σ_{M-P} bond in a cis-configuration is evident from large ^{31}P hyperfine coupling in the substituted anion radical product [32-34].

In an effort to use this selectivity for the synthesis of "normal", i.e. non-radical products, we have incorporated this ligand centered electron transfer activation of CO in an electron transfer catalytic cycle [6,37]. Equation 11 shows the general scheme:

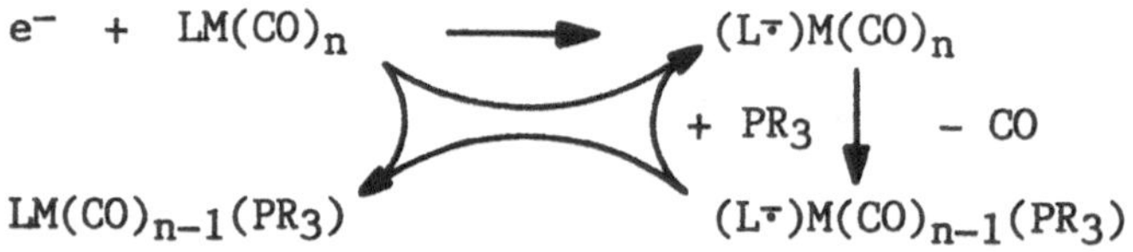

(11)

A chemically or electrochemically generated carbonylmetal complex of an anion radical ligand $L^{\cdot-}$ undergoes CO substitution, as indicated above, in the rate determining step. The ETC cycle is closed if the substituted anion radical complex can reduce the starting material in a homogenous electron transfer step, a requirement which is fulfilled for several systems [37,50,51,52]. A less than stoichiometric amount of electrons, introduced either cathodically [51] or chemically [37,50], can thus effect substitution of CO by PR_3 under mild conditions, especially when the usual routes fail because starting materials or products are thermolabile or light-sensitive. The chemical method, employing e.g. potassium metal in THF solution, has been used in the synthesis of several new carbonylmetal complexes with mixed donor and acceptor ligands, such as tricarbonylphosphane-molybdenum(0) and -tungsten(0) complexes of the bidiazines and tetracarbonylphosphane-tungsten(0) complexes of electron deficient pyridines [37,50]:

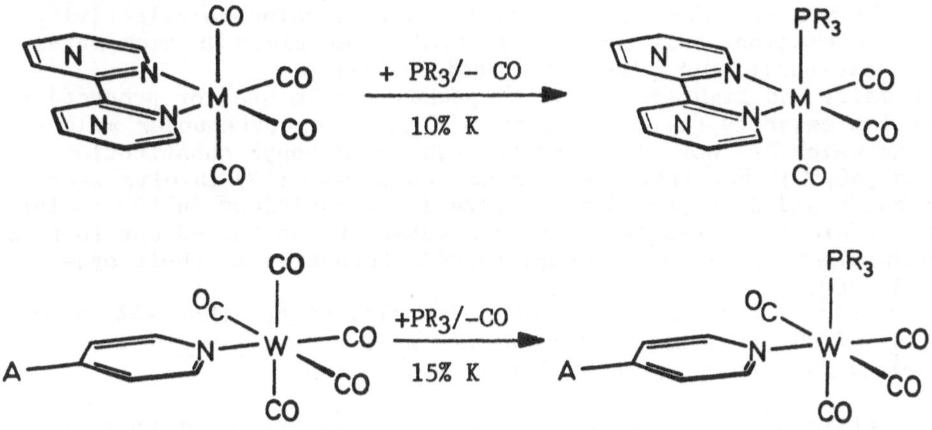

(12)

A = COR, COOR, CN

Advantages and disadvantages of this ligand-centered electron transfer catalytic CO substitution are summarized in the following:

The substitution (11,12) is fairly slow in comparison to metal centered ETC processes [1,3,4], requiring from 1 hr to some days for completion. This is due to the partial character of electron transfer to the metal center via hyperconjugation (10), yet the substitution is significantly enhanced relative to the very slow daylight-induced reaction [53].

The amount of catalyst (electrons!) needed is approximately 10-20 mol%, as evident from electrochemistry [51] and preparative studies [37,50]. When employing alkali metals as source of electrons, some of that amount is also used for internal drying of the reaction apparatus.

The alkaline metals and the reduced species derived thereof are not recovered because of their deactivation in side reactions, their sensitivity and polar character.

A reversibly reducible ligand is required for this kind of ETC reaction, while neither the metal nor the ligands introduced must be easily reduced under the reaction conditions. This requirement favours application of the reaction to low-valent metal species, unfortunately, it has been difficult to use aryl phosphanes for substitution because of their facile reductive cleavage by alkaline metals [54].

The use of such metals, on the other hand, permits convenient application of this method in the hands of an organometallic chemist, neither photochemical nor electrochemical set-ups are needed. Also, work-up is not impaired by the presence of electron transferring mediators such as benzophenone ketyl [46,55] or by alkali metal activators such as crown ethers or cryptands.

The redox potential requirements for the catalytic cycle are well defined and can be easily checked e.g. by cyclovoltammetry.

Both essential anion radical intermediates are usually persistent and can be characterized by various spectroscopies, in particular high resolution EPR.

For the reasons discussed above, there is a valuable selectivity in product formation, only one substitution takes place at each metal center in cis-position to the anion radical ligand.

Thermally and light-sensitive compounds may be used or produced in the reaction sequence (11) because the ETC process works under mild conditions which are not often encountered in carbonyl substitution reactions [46,56]. For instance, the new complexes (12) involve weakly basic ligands and have photodissociative d-d transitions in the visible so that the electron transfer catalyzed substitution turned out to be a welcome straightforward and uncomplicated alternative to their synthesis [37,50].

Conversion to the products (12) is usually higher than 75%, isolated yields of the sensitive compounds are generally above 50% even after column chromatographic purification [37,50].

Most attractive for the synthetic chemist is the opportunity to obtain valuable new products by this procedure. On the one hand, the simultaneous coordination of a donor such as PR_3 and an electron-deficient pyridine or a-diimine acceptor ligand to a carbonylmetal center yields complexes with small frontier orbital differences, characterized e.g. by long-wavelength charge transfer transitions [37].

However, the fact that anion radical complexes are the essential, relatively long-lived intermediates in process (11), can have yet another intriguing consequence, as illustrated by example (13): Weakly coordinating polydentate ligands such as 4-cyanopyridine do not show detectable double coordination of $W(CO)_5$ fragments, however, after ET catalyzed substitution such double coordination of substituted metal fragments is observed in a stable compound (13) [50].

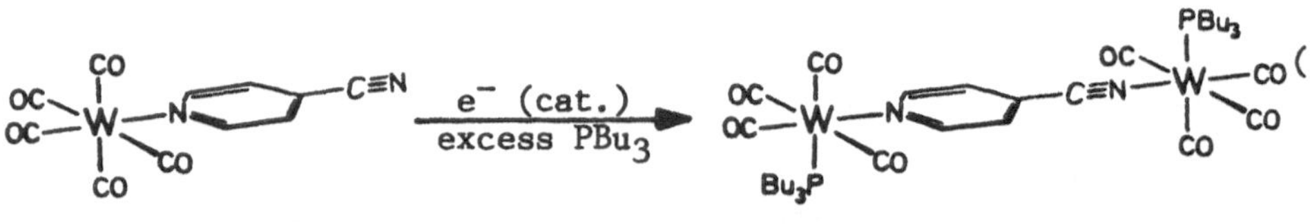

(13)

Obviously, both the labilization effect of co-ligands and the tendency towards additional metal coordination can combine to yield polynuclear substituted complexes in one-pot syntheses.

ACKNOWLEDGEMENT

We thank Deutsche Forschungsgemeinschaft (DFG) for support of work on electron transfer induced reactions (Grant Ka 618/1-2). Further support by Stiftung Volkswagenwerk, Flughafen Frankfurt Main/AG, and Hermann Willkomm-Foundation is also gratefully acknowledged. W.K. thanks the Karl Winnacker-Foundation of Hoechst AG for a most generous fellowship.

REFERENCES

[1] J. K. Kochi, Organometallic Mechanisms and Catalysis, Academic Press, New York, 1978.
[2] T. L. Brown, Ann. N. Y. Acad. Sci. **80** (1980) 333; A. E. Stiegman and D. R. Tyler, Comments Inorg. Chem. 5 (1986) 215; D. J. Kuchynka, C. Amatore and J. K. Kochi, Inorg. Chem. **25** (1986) 4087; P. Rushman and T. L. Brown, J. Am. Chem. Soc. **109** (1987) 3632, and literature cited.
[3] J. K. Kochi, J. Organomet. Chem. **300** (1986) 139.
[4] M. Chanon and M. L. Tobe, Angew. Chem. **94** (1982) 27; Angew. Chem. Int. Ed. Engl. **21** (1982) 1. M. Chanon, Acc. Chem. Res. **20** (1987) 214.
[5] W. Kaim, Acc. Chem. Res. **18** (1985) 160.
[6] W. Kaim, Coord. Chem. Rev. **76** (1987) 187.
[7] C. V. Krishnan, C. Creutz, H. A. Schwarz and N. Sutin, J. Am. Chem. Soc. **105** (1983) 5617.
[8] S. Ernst and W. Kaim, Angew. Chem. **97** (1985) 431; Angew. Chem. Int. Ed. Engl. **24** (1985) 430.
[9] S. Ernst and W. Kaim, J. Am. Chem. Soc. **108** (1986) 3578.
[10] S. Ernst and W. Kaim, Inorg. Chim. Acta, in print.
[11] W. Kaim and S. Kohlmann, Inorg. Chem. **26** (1987) 68.
[12] E. S. Dodsworth, A. B. P. Lever, G. Eryavec and R. J. Crutchley, Inorg. Chem. **24** (1985) 1906. H. E. Toma and A. B. P. Lever, Ibid. **25** (1986) 176.
[13] C. Bessenbacher and W. Kaim, unpublished results.
[14] Cf. literature in W. Kaim, Rev. Chem. Intermed. **8** (1987) 247.
[15] R. Gross and W. Kaim, Angew. Chem. **99** (1987) 257; Angew. Chem. Int. Ed. Engl. **26** (1987) 251.
[16] A. Fatiadi, Synthesis **1986**, 249.
[17] Cf. H. Endres in Extended Linear Chain Compounds, Vol. 3, Ed. J. Miller, Plenum Press, New York, 1983, pp 263-317.
[18] K. G. Caulton, Coord. Chem. Rev. **38** (1981) 1.
[19] Cf. J. P. Collman and L. S. Hegedus, Principles and Applications of Organotransition Metal Chemistry, University Science Books, Mill Valley (California), 1980, p 229.
[20] R. Gross, J. Jordanov, W. Kaim and E. Roth, unpublished results.
[21] R. Gross and W. Kaim, J. Organomet. Chem. **292** (1985) C21 and in print.
[22] J. W. Hershberger, R. J. Klingler and J. K. Kochi, J. Am. Chem. Soc. **105** (1983) 61.

[23] M. F. Rettig and R. M. Wing, Inorg. Chem. **8** (1969) 2685.
[24] W. Beck, R. Schlodder and K. H. Lechler, J. Organomet. Chem. **54** (1973) 303.
[25] D. A. Dixon and J. S. Miller, J. Am. Chem. Soc. **109** (1987) 3656
[26] M. S. Khatkale and J. P. Devlin, J. Chem. Phys. **70** (1979) 1851.
[27] R. Gross and W. Kaim, J. Chem. Soc., Faraday Trans. 1 and Inorg. Chem., in print.
[28] W. Kaim, S. Ernst and S. Kohlmann, Chem. Unserer Zeit **21** (1987) 50.
[29] R. Gross, Ph.D. Thesis, Universität Frankfurt, 1987.
[30] B. P. Sullivan, D. Conrad and T. J. Meyer, Inorg. Chem. **24** (1985) 3640.
[31] J. Hawecker, J. M. Lehn and R. Ziessel, Helv. Chim. Acta **69** (1986) 1990.
[32] W. Kaim and S. Kohlmann, Chem. Phys. Lett., in print.
[33] S. Kohlmann and W. Kaim, unpublished work.
[34] D. Weir and J. K. S. Wan, J. Organomet. Chem. **220** (1981) 323.
[35] W. Kaim, J. Organomet. Chem. **262** (1984) 171.
[36] W. Kaim, Inorg. Chem. **23** (1984) 3365.
[37] B. Olbrich-Deussner and W. Kaim, J. Organomet. Chem., in print.
[38] A. Alberti and C. M. Camaggi, J. Organomet. Chem. **181** (1979) 139.
[39] K. A. M. Creber and J. K. S. Wan, Chem. Phys. Lett. **81** (1981) 453.
[40] K. A. M. Creber, T. I. Ho, M. C. Depew, D. Weir and J. K. S. Wan, Can. J. Chem. **60** (1982) 1504.
[41] K. A. M. Creber and J. K. S. Wan, Transition Met. Chem. (Weinheim) **8** (1983) 253.
[42] K. A. M. Creber and J. K. S. Wan, Can. J. Chem. **61** (1983) 1017.
[43] L. H. Langford and H. B. Gray, Ligand Substitution Processes, Benjamin, New York, 1965.
[44] F. Basolo and R. G. Pearson, Mechanisms of Inorganic Reactions, 2nd Ed., Wiley, New York, 1967.
[45] G. L. Geoffroy and M. S. Wrighton, Organometallic Photochemistry, Academic Press, New York, 1979.
[46] M. O. Albers and N. J. Coville, Coord. Chem. Rev. **53** (1984) 227.
[47] W. Kaim, Inorg. Chim. Acta **53** (1981) L151 and Chem. Ber. **115** (1982) 910.
[48] A. Alberti, M. C. Depew, A. Hudson, W. G. Gimpsey and J. K. S. Wan, J. Organomet. Chem. **280** (1985) C21.
[49] J. H. Moore and J. A. Tossell, J. Am. Chem. Soc. **103** (1981) 6632.
[50] B. Olbrich-Deussner, Diploma Thesis, Universität Frankfurt, 1987.
[51] D. Miholova and A. A. Vlček, J. Organomet. Chem. **279** (1985) 317.
[52] Cf. also observations by M. W. Kokkes, W. G. J. de Lange, D. J. Stufkens and A. Oskam, J. Organomet. Chem. **294** (1985) 59.
[53] Cf. N. Leventis and P. J. Wagner, J. Am. Chem. Soc. **107** (1985) 5807.
[54] W. Kaim, P. Hänel and H. Bock, Z. Naturforsch. **37b** (1982) 1382.
[55] M. I. Bruce, T. W. Hambley, B. K. Nicholson and M. R. Snow, J. Organomet. Chem. **235** (1982) 83.
[56] R. J. Angelici and J. R. Graham, J. Am. Chem. Soc. **87** (1965) 5590.

PARAMAGNETIC SPECIES IN PHOTOINDUCED OR PHOTOCATALYTIC PROCESSES INVOLVING SOME COBALT, TITANIUM AND IRON ORGANOMETALLIC COMPLEXES

C. GIANNOTTI
Institut de Chimie des Substances Naturelles
C.N.R.S., Gif-sur-Yvette, 91190 - FRANCE

Abstract

In this lecture, photochemically induced and catalysed reactions with cobalt and titanium-carbon bond and also iron arene compounds are discussed. Now, compounds having cobalt-carbon σ bond are used as source of alkyl free radicals or Co^{II} species. Alkyl free radicals rearrange in various ways in a more stable compounds. In the case of complexes having titanium - carbon σ bond, the photocatalytic polymerisation was believed to go through radical mechanism, but in the case of fluorinated derivatives, there is no evidence of formation of any radical intermediates and the nature of the species inducing the polymerisation is unclear.

The photoinduced polymerisation of epoxide by iron-arene compounds is probably going through formation of a paramagnetic intermediate having a crown ether as a ligand.

1. INTRODUCTION

The aim of this lecture is to bring into focus the area of photocatalysis of paramagnetic species. If the photochemistry of coordination and organometallics compounds has been discussed recently in a number of books and reviews only few papers deal with photocatalysis, the most noteworthy are those of Hennig et al (1), Juliard and Chanon (2,3).

Photochemical reactions of coordination and organometallic compounds have particular interest because there is the possibility of generating a diverse number of electronically excited states which are characterized by different kinds of electron density distribution.

The primary physical processes which occur immediately after photoexcitation and which eventually cause the rupture of the metal-carbon σ bond are very complicated and are only known for a few simple complexes. Depending on their structure a multitude of excited states must often be considered. We can distinguish ligand field (LF), intra-ligand (IT) and charge transfer (CT) excited states . CT processes are divided into CT from metal-to-ligand (CTML) or from ligand-to-metal (CTLM), from the complex to the solvent (CTTS) and ion

M. Chanon et al. (eds.), Paramagnetic Organometallic Species in Activation / Selectivity, Catalysis, 295–309.

pairs charge transfer (IPCT). These excited states are often superimposed and are rarely separable and hence occur together, especially when the phtoexcitation is not conducted monochromatically.

2 . GENERAL CONSIDERATIONS

2.1 Ligand field excited states

The photoexcitation of d orbitals of transition metal complex can cause important consequence as far as the reactivity towards ligands substitution or isomerization is concerned. Such a consequence can be easily visualised, in the case of strong field octahedral complexes. A photoexcitation following spin allowed transition corresponds to the promotion of an electron from a t 2g orbital, which is directed between the ligand to an eg orbital, which is directed to the ligand can cause (figure 1) :

a - an increase in the metal ligand repulsion which can conduct to the detachment of a ligand or rearrangement of the molecule towards a more stable structure.

b - a decrease of electron density in some directions between the ligands which will facilitate a nucleophilic attack on the central metal by solvent molecules or other ligands present in the solution.

In consequence the type of the reaction occuring after such an excitation is a photosubstitution reaction without any change of the degree of the oxidation of the initial complex, (eqn.1)

$$(M^{z+}\ L_6)^{z+} \quad + X \xrightarrow{h\ v} (M^{z+}L_5\ X)^{z+} \quad + L \qquad (1)$$

2.2 Charge transfer excited states

Charge transfer excited states arise from transition between MO's principaly localised on the metal and MO's principally localised on the ligands. These transitions cause a radical redistribution of the electronic charge between the central metal and ligands and thus a change in their oxidation state. Two types of charge transfer excited states can be obtained if the electron transition is from the ligands to the metal (LMCT) (eqn 2) or from the metal to the ligands (MLCT), (eqn 3).

$$(M^{z+}Ln)^{z+} \xrightarrow[L \rightarrow M]{h\ v} {}^{*}(M^{(z-1)+}\ Ln^{+})^{z+} \longrightarrow \underset{\text{products of L}}{M^{(z-1)+} + (n-1)L = \text{oxidation}} \qquad (2)$$

$$(M^{z+}Ln)^{z+} \xrightarrow[M \rightarrow L]{hv} {}^{*}(M^{(z+1)+}Ln^{-})^{z+} \longrightarrow \underset{\text{products of L}}{M^{(z+1)+} + (n-1)L + \text{reduction}} \qquad (3)$$

The charge transfer excited states will give intramolecular redox reaction coming from the reduction of the metal and oxidation of a ligand or vice-versa. These processes depend on a number of factors such as : the stability of the oxidation states of

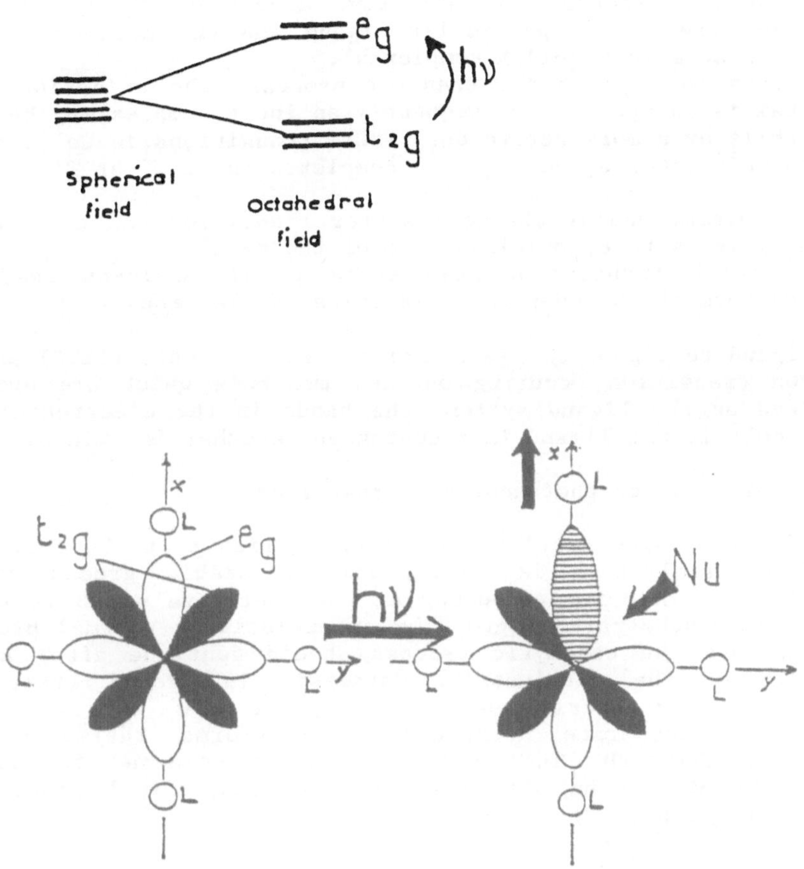

figure 1

metal and the ligands, the effective amount of charge transfer is localised between one particular ligand and the metal (X→M charge transfer states in M(n-1) X complexes).

Because of the charge transfer process, the oxidation state of the metal is changed and consequently an inert complex may be changed into labile or a more active one. LMCT. transitions in Co^{III} complexes lead to Co^{II} species and for Ti^{IV} complexes to Ti^{III} or Ti^{II}.

An intermolecular charge transfer transition can also occur and generally leads to an oxidized form of the metal.

A charge transfer excited state to the solvent leads to an oxidized form of the complex and a solvated electron.

Ligand to ligand charge transfer excited states (ILTC) arise from electron transition occuring between two MO's which are principally localised on the ligand system, the bands in the electronic spectrum appear only if one ligand is reducing and another is oxidizing.

2.3 Photoinduced or photogenerated reactions.

A photoinduced catalytic reaction implies that the photochemical generated catalyst B is coming from a stable ground state and catalytically inactive precursor A. The photocatalyst B is then able to convert a substrate C into D by an exclusively thermal process and in such case the catalytic process should continue after an inital formation of the catalyst B. Catalysis in ground state require coordinative interaction between catalyst and substrate C. The catalyst - substrate complex BC transforms thermally into a catalyst-product BD. The product could be released to generate D (Scheme 1) or may be displaced by C to give final product D and regerate BC (Scheme 2)

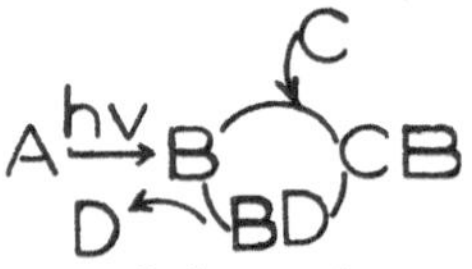

Scheme 1

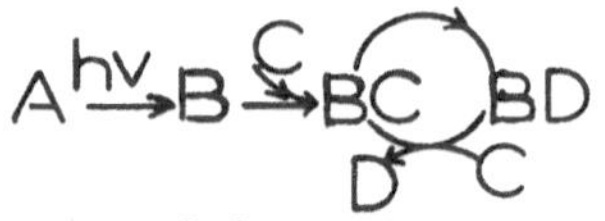

Scheme 2

The catalyst B could be a coordinatively unsaturated complex species, a photoexpulsed free ligand or any kind of highly reactive species (long lived excited state or exciplex, hot ground state) formed in the course of the photoexcitation reaction. Unfortunately , quite often the catalyst after its formation undergoes side reactions which consume the catalyst and explain why the overall quantum yield does not exceed one. The light can induce the formation of an initiator of a thermal chain reaction and may promote transformation of a great number of molecules of substrate in a thermal chain reaction such as polymerisation.

2.4 - Photoassisted or photoactivated reactions

In a such reaction a catalytically inactive precursor A is transformed into a photocatalyst B (Could be A in one of its excited state ^1S or ^3T), (eqn 4). This excited state will have a structure different from that of the ground state because of its peculiar electronic distribution structure and energy content, it may bring about the transformation of the substrate C into the product D, which B itself is converted back to the precursor A (eqn 5). Thus A is not consumed in this process, but a new photon is required to start the next cycle.

$$A \xrightarrow{h\nu} B(A^*) \qquad (4)$$
$$B+C \longrightarrow D+A \qquad (5)$$

2.5 Sensitized photoreactions

In such reactions a sensitizer A, which is inactive in its electronic ground state is absorbing the incident light to give an electronically excited state molecule A^* (generally ^1S, ^2S state) which transfer its excitation energy to a substrate B leading to an excited substrate B^* (eqn 6) which can react directly to give the final compound C, (scheme 3). A^* can give an exciplex with B producing the final product C, (Scheme 4)

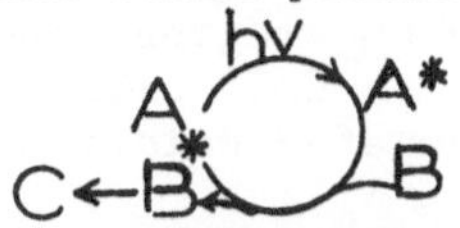

Scheme 3

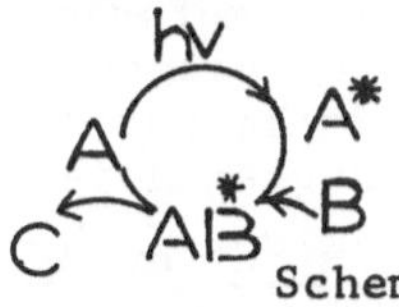

Scheme 4

The sensitization may also proceed as an electron tranfer process giving B^- or B^+ (eqns 7,8).

$$A^* + B \longrightarrow A + B^* \qquad (6)$$
$$A^* + B \longrightarrow A^- + B^+ \qquad (7)$$
$$A^* + B \longrightarrow A^+ + B^- \qquad (8)$$

The excited state A^* can react with one of these species and give the final compound C, (eqn 9). The reaction can be catalytic if

$$A^* + (B^+, B^-) \longrightarrow C \qquad (9)$$

a compound D red or D ox is able to reduce B^+ or oxidize B^- to give back B, (eqns 10,11).

$$B^+ + D\ red \longrightarrow B + D\ ox \qquad (10)$$
$$B^- + D\ ox \longrightarrow B + D\ red \qquad (11)$$

3 . Photoinduced reactions containing cobalt-carbon σ bond

We have been interested in alkyl transition-metal complexes because they are considered key intermediates in many industrially catalytic processes : low pressure polymerisation of olefins in the Ziegler-Natta process, olefin isomerization, hydroformylation, ethylene oxidation in the Wacker-Hoechst process and also in the Fischer-Tropsch hydrocarbon synthesis (4). Nature presents an important representative of this class of compounds, it is a protective factor against pernicious anemia: the vitamin B12 (5). This compound contains a cobalt-carbon σ bond and it is sensitive to light.

Exposure of an aqueous solution of adenosylcobalamin (figure 2) to light causes a loss of coenzyme activity and a new absorption spectrum appears (6). The first step of this photolytic reaction is supposed to be a homolytic cleavage of the cobalt-carbon σ bond to Co^{II} alamin and a 5' deoxyadenosyl radical. Several laboratories have studied the photolysis of many corrinoids derivatives and the results are very well described by Hogemkamp (7). Cobalt II containing corrinoids shows typical well resolved ESR spectra; however in the case of adenosylcobalamin a poorly resolved spectra is obtained , this signal increases several folds in intensity when the preparation is warmed up in the dark and refrozen (8). Love et al (8) suggest that at low temperature the cobalt-carbon σ bond is cleaved to an intermediate not detectable by ESR spectroscopy. Endicott et al (9,10) using a picosecond flash photolysis technique studied the methylcobalamin and the adenosylcobalamin and were unable to detect any long lived intermediate. For methylcobalamin and adenosylcobalamin they found a rapid substrate bleaching (⩽ 8ps) in the 565 nm region and corresponding increased 474 mn absorbance attribuable to Co^{II} alamin. The following sequence is suggested (eqns 12,14). In aerated solution

$$\overset{CH_3}{\underset{B}{Co}} \xrightarrow{h\nu} \underset{B}{Co^{II}} \; (B\ 12r) + {}^{\bullet}CH_3 \qquad (12)$$

$$\underset{B}{Co^{II}} \xrightarrow{k_2} \overset{CH_3}{\underset{B}{Co}} \qquad k_2 = 1.5x\ 10^9\ M^{-1}s^{-1} \qquad (13)$$

$$2{}^{\bullet}CH_3 \xrightarrow{k_3} C_2H_6 \qquad k_3 = 10^{10}\ M^{-1}\ s^{-1} \qquad (14)$$

the initial B_{12r} absorbance decays rapidly and apparently completely in what appears to be a second order process. This decay is attribuated to a combination of two reactions (eqns, 15,16).

$${}^{\bullet}CH_3 + O_2 \longrightarrow {}^{\bullet}O_2-CH_3 \qquad (15)$$

$$\overset{OH_2}{\underset{B}{(Co)}} + {}^{\bullet}O_2CH_3 \xrightarrow[k_4]{h\nu} \underset{B}{Co}{-}O{-}O{-}CH_3 \quad k_4 = 2.4x109\ M^{-1}\ s^{-1} \qquad (16)$$

Similar flash photolysis experiments on methyl (pyridinato) cabaloxime: CH_3-Co $(dmgH)_2-$ py have been performed (10,11). The formation of Co^{II} $(dmgH)_2$-py occured during the flash pulse and decayed rapidly in perfect accord with Endicott et al. results (10): $k=5x10^7\ M^{-1}\ s^{-1}$. When ${}^{\bullet}CH_3$ is trapped with the nitrosodurene, the absorption corresponding to Co^{II} $(dmgH)_2$– py is stabilised (11). A large number of ESR study have been carried out and characteristics of homolytic cleavage of Co^{III} -R σ bond in Co^{II} and ${}^{\bullet}R$, and also its recombination process and evidence of R and Co^{II} have been reported (12-22). Kemp et al. (23) reported that during the photolysis of alkylcobaloximes in acidic medium the aquation of the axial ligands of

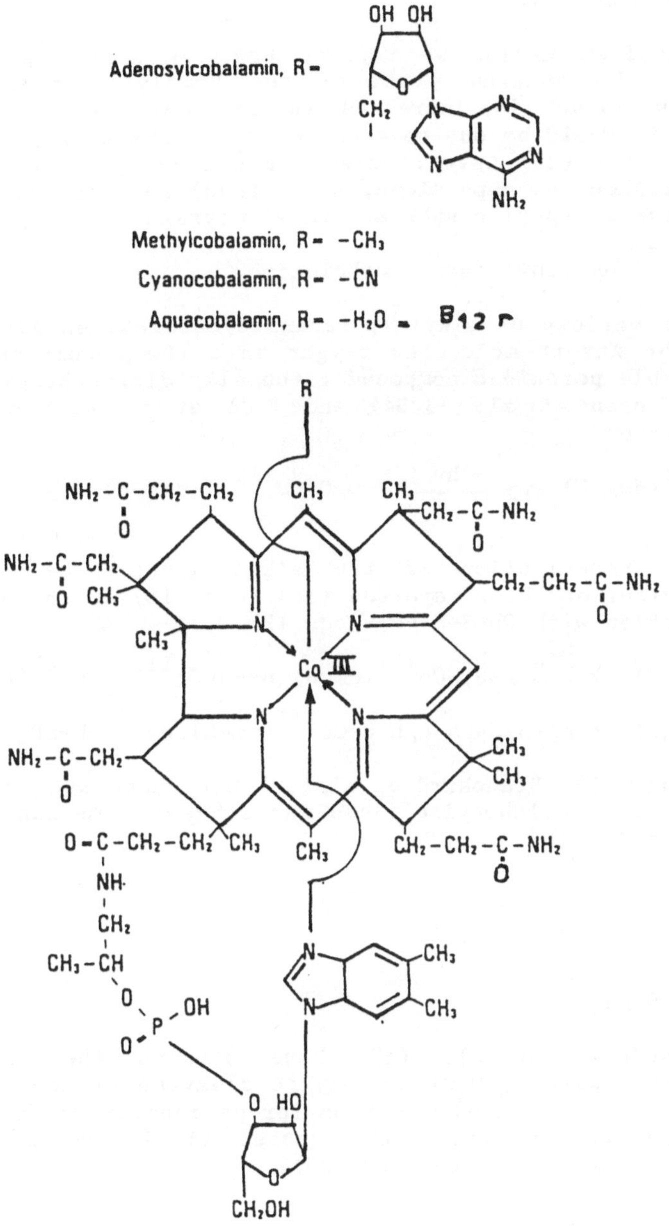

figure 2 Structure of cobalamins.

the cobaloxime occurs.

In organised media, we report that for alkyl(aquo)cobaloximes (figure 3), the quantum yield of the homolytic cleavage of the cobalt-carbon σ bond is increased to the one obtained in organic solution, this could be due to a cooperative effect (24).

Kemp et al. (25) have studied the quenching of organic triplet states (benzophenone, naphtalene, β-carotene) and also of a variety of excited inorganic species such as uranyl nitrate and

$(Cr\ (bipy)_3)^{3+}$ by ethyl (aquo) cobaloxime.

A large variety of alkyl (pyridinato)cobaloximes irradiated with visible light insert molecular oxygen into the cobalt-carbon σ bond and give stable peroxidic compound : the alkyldioxycobaloxime (26-32), (eqn 17). Jensen et al. (33,34) and Tada et al. (35) observed that oxygen insertion

$$R\text{-}Co^{III}(dmg\ H)_2\text{-}py \xrightarrow[O_2]{hv} R\text{-}O\text{-}O\ Co^{III}\ (dmg\ H)_2\ py \qquad (17)$$

occurs with racemisation of the alkyl group. The insertion of elemental sulfur has been reported (36) (eqn, 18) and also photo-induced reaction with Ph-S-S-Ph, (eqn 19).

$$R\text{-}Co^{III}(dmg\ H)_2py \xrightarrow{nS_8} R\text{-}S_4\text{-}Co^{III}(dmgH)_2py\text{+}pyCo^{III}S_4\text{-}Co^{III}py\text{+}RS_4R \quad (18)$$

$$R\text{—}Co^{III}(dmgH)_2 + ph\text{-}S\text{-}S\text{-}ph \rightarrow ph\text{-}S\text{-}Co^{III}(dmgH)_2py + ph\text{-}SR \qquad (19)$$

Johnson et al. (37), Branchard et al. (38) have used such a reaction to prepare various alkylphenylsulfides and suggest a mechanism radically induced (eqns 20-22).

$$RCo(dmgH)_2\ py \rightarrow R^{\bullet} + Co^{II}(dmgH)_2\ py \qquad (20)$$

$$R + ph\text{-}S\text{-}S\text{-}ph \rightarrow RSph + phS^{\bullet} \qquad (21)$$

$$R + phS^{\bullet} \rightarrow R\text{-}S\text{-}ph \qquad (22)$$

Johnson, Gaudemer et al. (39) have studied the reactivity of alkylcobaloxime with R_3SO_3Cl. Homolytic cleavage of the cobalt-carbon σ bond of 2 - acetyl - 2 methoxycarbonylpropylcobaloxime and 2-benzoyl-2 phenylpropylcobaloxime gives the enones which are formed by 1,2 migration of acyl group (40), (eqn 23).

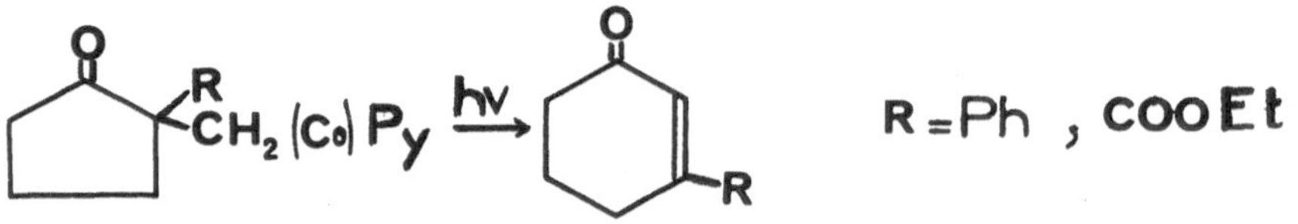

(23)

Under irradiation by visible light, in pyridine solution, substituted alkylcobaloximes undergo rearrangement to more stable substituted alkyl or alkenyl cobaloximes (41). Photolysis of 2-aryl-2 ethoxy 2-phenylethylcobaloxime followed by hydrolysis gave two kinds of substituted 1,2-diphenylethanones arising via phenyl or substituted phenyl migration (42). Yurkevich et al. (43,44) have shown that photolysis of adenosylcobalamin in presence of SH compounds give 5'-deoxyadenosine. On photolysis of MeCbl (5,6-dimethylbenzimidazolyl), methylcobalamide) with ethylene gycol and in presence of SH-Compounds led to conversion of ethylene glycol into acetaldehyde (43.44) and photolysis of methyl (aquo) cobaloxime with 1,2 ethanediol or 2-aminoethanol gives also acetaldehyde (45,46).

Photolysis of R-Co^{III}Salen, R=C1-4 alkyl, Salen=bis (salicylaldehydediethylenediamine) (47) in alcohol mixture led to the evolution of hydrogen, catalytic properties of the Co^{II} produced in this system have been examined (figure 4).

A homogeneous catalytic system for efficient photochemical generation of hydrogen and water reduction under visible light irradiation has been described (48), it uses $Co^{II}(dmgH)_2py$, $Ru(bipy)_3^{2+}$ and a tertiary amine respectively as catalyst, photosensitiser and electron donor. Same type of reaction is obtained when one use alkyl $Co^{III}(dmgH)_2py$ as catalyst (49) (figure 3).

In monolayers Co^{II} tetraphenylporphyrine derivatives give cobalt superoxide complex $TPPCo^{III}OO^{\bullet-}$ trapped between the layers (50,51) . But in micelles and in microheterogeneous systems the reversible fixation of oxygen is observed (52,53).

The effect of various solvents have been studied on the photoinitiated decomposition of 1,2,3,4-tetrahydro-1-naphthylhydroperoxide with various cobalt 2,4-pentanedionate (54).

Cobalt-carbonyl is of interest because $[Co(CO)_3L_2]$ $[Co(CO)_4]$ has been shown to be the photoactive catalyst precursor in the photochemical hydroformylation of olefins in methanol solution (55). Irradiation of the ionic complex $[Co(CO)_3L_2]$ $[Co(CO)_4]$ itself $L=Bu_3P$ leads to dissociative loss of either CO or L to produce a coordinatively unsaturated species (55,56).

Work mainly by Osborn et al. (57) and Brown et al. (58,59) on cobalt-carbonyl complexes has established the importance of the radical chain pathway in the substitution reactions of metal carbonyl complexes.

In a recent review Coville et al. (60) reported photo-induced radical catalysis of several metal carbonyl complexes of Re, Mn, W, Mo and Fe; we also notice that metal-carbonyl catalysed polymerisation of propylene oxide and isomerization of 1-butene with photolysed $W(CO)_6$ on porous vycor glass were investigated (61).

4. Photoinduced catalysis of alkyl complexes of titanium, zirconium and Hafnium.

The photoactivation of alkyl complexes of titanium, zirconium and Hafnium are very important because they are involved in catalysis of

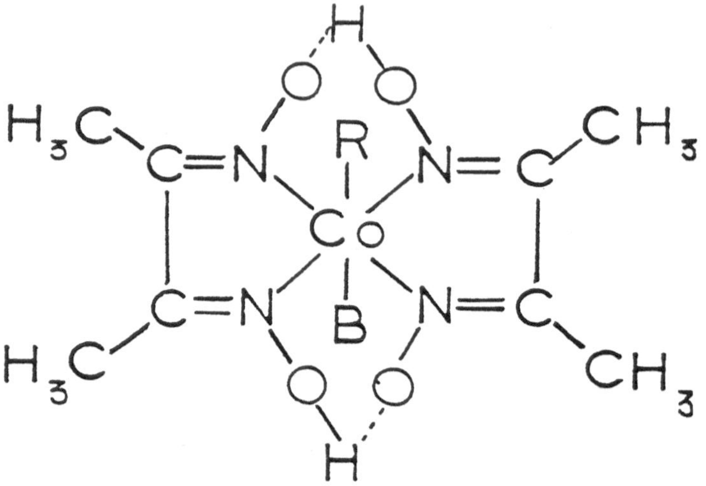

ALKYLCOBALOXIME

figure 3

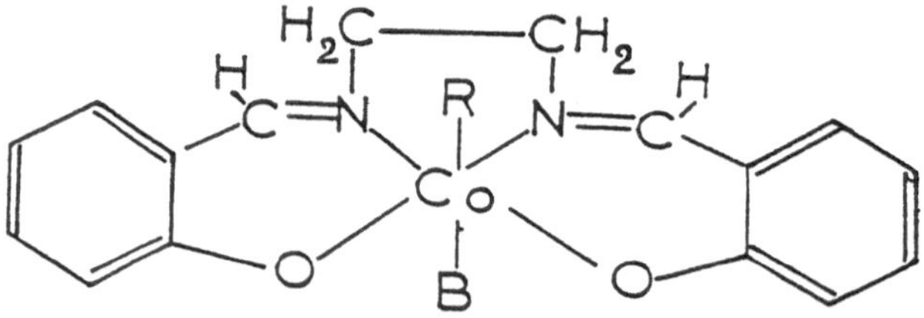

ALKYLSALEN

figure 4

methyl-metacrylate (62)
In the dark, the alkyl complexes $M=(CH_2Ph)_4$ (63,64) (M=Ti,Zr) and Np_4Ti(65) catalyse the polymerisation of methyl methacrylate, styrene or other olefins eitheir slowly or not at all; however upon irradiation with UV light they develop considerable catalytic activity, Ti^{III} or Zr^{III} species as well as alkyl radicals have been proposed as the initiators or catalysts for the onset of polymerisation. Titanocene bis aryl compounds have been known for a number of years (66,67) and their photolysis involve alkyl free radical and Ti^{III} (eqn 24) species and titanocene derivatives (68-70).

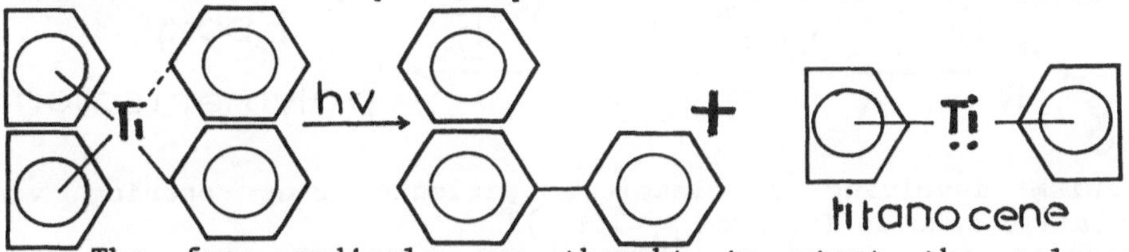

(24)

The free radicals are thought to start the polymerisation reaction. A number of titanium compounds have been reported to serve as photoinitiators (71,72) but all the compounds used so far were either oxidatively or thermally unstable. Riediker et al (73,74) found when a fluorinated aryl ligand is introduced into the titanocene moiety, thermal and oxidative stability is reached and a very efficient curing rate is preserved (73,74). From diphenyl titanocene in toluene solution biphenyl and benzene were detected as the major products. In the case of fluorinated compound the photolysis is found to be totally different and the pentafluorinated substituted cyclopentadiene is isolated as the main organic product, in presence of a radical scarenger no organic radical could be trapped . From these results it is unclear that in the case of fluorinated compounds, which species initiate the polymerisation.
The polymerisation of $Ti(OR)_4$, R= Bu, Me_2CHCH_2, Me_2CH in different solvents such as tetrahydrofuranne, benzene, ethanol and under fluorescent lamp give polyalkyl titanates). The catalytic activities of the polymers for the photolysis of 1/1 mixture H_2O/Me_2CHOH in the presence of Pt as co-catalyst have been measured by the quantum efficiency of the formed hydrogen. The size of the alkyl groups affect strongly the light absorption wavelength and the generation paramagnetic species during the photolysis. The potential gradient created by the addition of metal ions of Pt, Mn or V is probably responsible for the catalytic activities (75). The complex formed by direct reaction between polymerised n-butyl orthotitanate and methanol in benzene solution catalyses the photocatalytic decomposition of H_2O (76).

5. Iron-arene photoinitiators

The light-induced homogeneous catalytic reactions of olefins mediated by metal carbonyl compounds and their derivatives has been reviewed by Moggi et al (77) photocatalytic olefin reactions have been investigated and reviewed by Salomon (78,79) and in the case of copper

(I) photocatalysis by Sykora (80) and Kutul and Grutsch (81). In recent years several types of organometallic photoiniators have been described (82,83) Most of these photoinitiators have certain drawbacks which have so far withheld then from achieving a major break-through. In many cases there is a lack in thermal stability. New organometallic iron arene based photoinitiators which are supposed overcome most of these shortcomings have been described (84,5)
Meier et al (84,85) found that ($C_5R_5Fe^+$ Arene) PF_6 photoinitiate the epoxy-polymerisation in two steps: i) irradiation at low temperature (a bleaching process) and ii) thermal activation (50-100°C) (84).

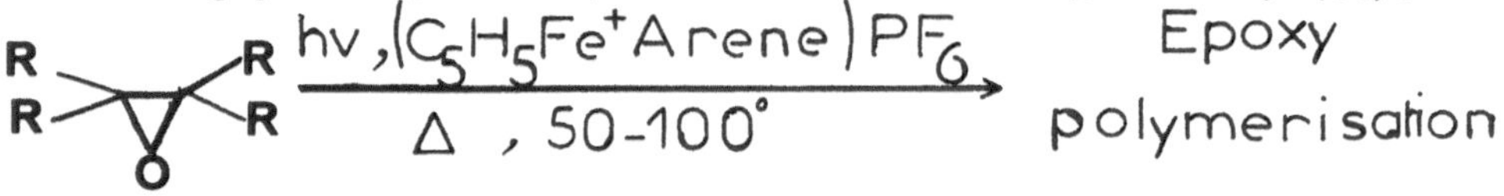

A mechanism involving a classical cationic ring opening via organometallic species (C_5R_5Fe 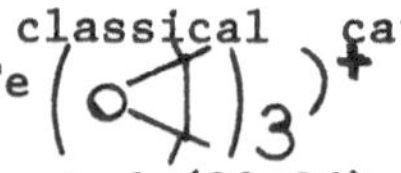 $)^+$

as intermediate has been suggested (82-86). The ring opening start, in the ligand sphere of the ($CpFe^+$) uncoordinated species. But, after our photochemical and ESR study of the same (C_5R_5Fe Arene) PF_6 complexes we have shown that in CH_2Cl_2 as solvent free radicals are generated from the excited state $3E_1$ of $(C_5H_5Fe^+ \text{ Arene})^*$ giving fast electron transfer process,(87).

$$(Cp\ Fe^+\ \text{Arene})^* \longrightarrow \begin{cases} \xrightarrow{D} Fe^{I}\,(S)_m + D^+ + {}^{\bullet}Cp \\ \xrightarrow{A} Fe^{III}(S)m + A^{-\bullet} + {}^{\bullet}Cp \end{cases}$$

(A = Electron acceptor, D= Electron donor)

So we suggest that the classical cationic ring-opening initiation, previously suggested is not so simple (87). ESR and photochemical study of this new photoiniation of epoxypolymerisation are underway.

ACKNOWLEDGMENTS :

It is with pleasure that I acknowledge the following co-workers : G. Merle, P. Maillard, A.M. Ducourant, F. Ricchiero, B. Septe, C. Fontaine, Q. Barcello, C. Merienne, M. Barrera, H. Hernandez and collaborative contributions from Professors D. Benlian, A. Gaudemer, E. Samuel, D. Lerner, K.N.V. Duong, J.R. Bolton and E. Roman. The work described has been supported by several grants from C.N.R.S.

REFERENCES

1 - H. Hennig and D. Rehoreck, Coordination Chemistry Reviews 61 (1985) 1.
2 - M. Juliard and M. Chanon, Chem. Rev., 83 (1983) 425.
3 - M. Chanon, Bull. Soc. France, 7-8 (1982) II-198.
4 - H.G. Alt, Angew. Chem. Int. Ed. Engl., 23 (1984) 766.
5 - K. Folkers, B_{12} Ed. D. Dolphin, John Wiley and Sons (1982), Vol.1 p.1.
6 - H. Weissback, J.N. Ladd, B.E. Volcani, R.D. Smyth and H. Barber J. Biol.Chem. , 235 (1960) 1462.
7 - H.P.C. Hogenkamp, B_{12} Ed. D. Dolphin, John Wiley (1982) Vol 1 p. 295.
8 - D.J. Lowe, K.N. Joblin and D.J. Cardin , Biochem. Biophys. Acta , 539 (1978) 398.
9 - J.F. Endicott and T.L. Netzel, J.Am.Chem.Soc., 101 (1979), 4000.
10- J.F. Endicott and G.J. Ferraudi , J.Am.Chem.Soc., 99 (1977) 243.
11- D. Lerner, R. Bonneau and C. Giannotti, J.Photochem., 11 (1979) 73.
12- D.R. Eaton and S.R. Suart, J.Phys.Chem., 72 (1968) 400.
13 - C. Giannotti and J.R. Bolton, J.Organometal.Chem , 80 (1974) 379.
14 - C. Giannotti, G. Merle, C. Fontaine and J.R. Bolton, J.Organometal.Chem. , 91 (1975) 357.
15 - C. Giannotti, G. Merle and J.R. Bolton, J.Organometal.Chem. , 99 (1975) 154.
16 - K.N. Joblin, A. W. Johnson, M.F. Lappert and B.K. Nicholson J.Chem.Soc.Chem.Commun. (1975) 441.
17 - V.D. Ghanekar and R.E. Coffman, J.Organometal.Chem., 198 (1980) C45.
18 - V.D. Ghanekar, R.J. Lin and R.E. Coffman and R.L. Blakley, Biochem.Biophys.Res.Commun., 101 (1981) 215.
19 - P. Maillard and C. Giannotti, Can.J.Chem., 60 (1982) 1402.
20 - D.N.R. Rao and M.C.R. Symons, J.Chem.Soc.Chem.Commun. (1982) 954.
21 - D.N.R. Rao and M.C.R. Symons , J.Chem.Soc., Perk.Trans.II, (1983) 187.
22 - D.N.R. Rao and M.C.R. Symons, J.Chem.Soc. Farad. Trans I, (1984) 423.
23 - B.T Golding, T.S. Kemp, P.J. Sellers and E. Mocchi, J.Chem.Soc. Dalton, (1977) 1266.
24 - D.A. Lerner F. Ricchiero and C. Giannotti J. Phys.Chem., 84 (1980) 3007.
25 - H.Y. Al-Saigh and T. Kemp, J.Chem.Soc.Perk.Trans.II., (1983) 615.
26 - C. Giannotti, A. Gaudemer, C. Fontaine, Tetrahedron.Let.,3209 (1970).
27 - K.N.V. Duong, C. Fontaine, C. Giannotti and A. Gaudemer Tetrahedron Let. , 1187 (1971)
28 - C. Fontaine, K.N.V. Duong, C. Merienne, A. Gaudemer and C. Giannotti J.Organometal.Chem., 38 (1972) 167.
29 - C. Giannotti, B. Septe and Benlian J.Organometal.Chem., 39 (1972) 381.
30 - C.Giannotti, C. Fontaine and A. Gaudemer J.Organometal.Chem., 39

(1972) 381.
31 - C. Giannotti and B. Septe. J.Organometal.Chem., 52 (1973) C45
32 - C. Merienne , C. Giannotti and C. Gaudemer J.Organometal.Chem. 71 (1974) 107.
33 - F.R. Jensen and R.C. Kiskis, J.Organometal.Chem, 49 (1973) C46.
34 - F.R. Jensen and R.C. Kiskis, J.Amer.Chem.Soc., 97 (1975) 5825.
35 - H. Shinozaki and M.Tada, Chem.and Ind., (1975) 179.
36 - C. Giannotti and G.Merle, J.Organometal.Chem. , 113 (1976) 45.
37 - J. Deniau, K.N.V. Duong; A. Gaudemer, P. Bougeard and M.D. Johnson. J.Chem.Soc. Perkin II, (1981) 393.
38 - B.P. Branchaud, M.S. Meir and M.N. Malekzadeh, J.Org.Chem., 52 (1987) 212.
39 - A.E. Crease, B. Gupta, M.D. Johnson, E. Bialkouska, K.N.V. Duong and A. Gaudemer, J.Chem.Soc.Perkin I ,(1979) 2611.
40 - M. Okabe, T. Osawa and M. Tada, Tetrahedron Let., 20 (1981) 1899.
41 - P. Bougeard, C.J. Cooksey, M..D. Johnson, M.J. Lewin, S. Mitchell and P.A. Owens. J.Organometal.Chem., 288 (1985) 349.
42 - M. Tada, S. Akinaga and M. Okabe, BullChem.Soc.Jpn. , 55 (1982) 3939.
42 - M. Tada, K. Inoue and M. Okabe, Bull.Chem.Soc. Jpn., 56 (1983) 1420.
43 - I.P. Rudakova, E.G. Chauser and A.M. Yurkevich, Bio.Organic.Chem., 1 (1975) 616.
44 - I.P. Rudakova, T.E. Ershova, A.B. Belikov and A.M. Yurkevich J.Chem.Soc.Chem.Commun., (1978) 592.
45 - B.T. Golding, T.J. Kemp, E. Nocchi and W.P. Watson. Angew.Chem.Int.Edit., 14 (1975) 813.
46 - A.J. Hartshorn, A.W. Johnson, S.M. Kennedy, M.F. Lappert and J.J. Macquitty. J.Chem.Soc.Chem.Commun., (1978) 643;
47 - G.A. Shagisultanova and A. Maslov, Zh.Fiz.Khim., 59 (1985) 1305.
48 - J. Hawecker, J.M. Lehn and R. Ziessel, Nouveau J.Chim., 7 (1983) 271.
49 - C. Giannotti, Unpublished results.
50 - A. Barraud, A. Ruaudel-Teixier, M. Vandevyver, P. Maillard and C. Giannotti, J.Colloid.Interf.Science, 85 (1982) 571.
51 - C. Lecomte, C. Baudin, F. Berleur, A. Ruandel-Teixier, A. Barraud and M. Momenteau. Thin.Sol.Films, 133 (1985) 103.
52 - D.A. Lerner, F.R. Ricchiero, P. Maillard and C. Giannotti? J.Photochem., 18 (1982) 193.
53 - D.A. Lerner, M. Barcello, P. Maillard and C. Giannotti, Submitted for publication.
54 - E. Macova, S. Polakova, P. Lederer, React.Kinet.Catol.Lett. 32 (1986) 97.
55 - M.F. Mirback, M.J. Mirback and R.W. Wegman, Organometallico., 3 (1983) 900.
56 - M. Absi-Halabi, J.D. Atwood, N.P. Fabius and T.L. Brown. J.Amer.Chem.Soc., 102 (1980) 6248.
57 - J.A. Labinger, J.A. Osborn and N.J. Coville, Inorg.Chem., 19 (1980) 3326.
58 - T.L. Brown, Ann.N.Y.Acad.Sci., 330 (1980) 80.
59 - R.W. Wegman and T.L. Brown, Organometallic., 1 (1982) 47.

60 - M.O. Albers and N. Coville, Cordin, Chem.Rev., 53 (1984)227.
61 - D.J. Peretti, M.S. Paquette, R.L. Yates, H.D. Gafney, Nato.Asi.Ser. SerB, (1984) 105.
62 - H. Alt, Angew.Chem.Int.Ed., 23 (1984) 766.
63 - U. Zucchini, E. Albizzati and U. Giannini, J.organometal.Chem., 26 (1971) 357.
64 - D.G. Ballard and P.W. Van.Lienden, Makromol.Chem., 154 (1972) 177.
65 - J.C.W. Chien, J.C. Wa and M.D. Rausch, J.Am.Chem.Soc., 103 (1981) 1180.
66 - L. Summers, R. Uloth and A. Holmes, J.Am.Chem.Soc., 77 (1955) 3604.
67 - H. Alt and M.D. Raush, J.Am.Chem.Soc., 96, 5936 (1974)
68 - E. Samuel and C. Giannotti, J. Organometal. Chem., 113 (1976) C17.
69 - E.Samuel, P. Maillard and C. Giannotti, J.Organometal.Chem., 142 (1977), 289.
71 - U. Zucchini, E. Albizati, U. Giannini J.Organiomet.Chem., 26 (1971) 357.
72 - K/ Kaeriyama, Y. Shimura, J.Polym.Sci., 10 (1972) 2833.
73 - A. Roloff, K. Meier and M. Riediker, Pure and Applied Chem., 58 (1986) 1267.
74 - M. Riediker, M. Roth, N. Bühler and J. Berger, Eur.Pat.Appl. n°0122223 U.S.Patent 45990287.
75 - S. Tadao, Y. Michoko, U.Hisashi, Nippon Kagaku Kaishi (1985) 605.
76 - T. Shinoda, K. Ohkawa, J. Photochem.31 (1985) 57.
77 - L. Moggi, A. Jurris, D. Sandrini and M.F. Manfrin, Rev.Chem.Intermed., 4 (1981) 171.
78 - R.G. Salomon, Adv.Chem.Ser., 168 (1978) 174.
79 - R.G. Salomon, Tetrahedron, 39 (1983) 485.
80 - J. Sykora, Chem. Listy, 76 (1982) 1947.
81 - C. Kutal and P.A. Grutsch, Adv.Chem.Ser., 173 (1979) 325.
82 - D.L.S. Brown, J.A. Connor, B. Dobinson, B.P. Stark, Angew.Macromol.Chem., 50 (1976) 9.
83 - H. Curtis, E. Irwing, B.F.G. Johnson, Chem.Britain.(1986) 327.
84 - K. Meier, N.Bühler, H. Zweigfel, G. Berner, F. Lohse, Eur.Pat.Appl. n°094915 (1984)
85 - H. Zweifel, K. Meier, Polym.Preprints, 26 (1985) 347.
86 - K. Meier, Technical paper, AFP-SME , 1, (1985).
87 - E. Roman, M. Barrera, S. Hernandez and C. Giannotti, NATO Ser. this book. (1987).

MONOELECTRONIC PROCESSES IN IRON COMPLEXES AND CONTROLLED REACTIONS BY ANCILLARY PR_3 LIGANDS

Daniel TOUCHARD, Jean-Luc FILLAUT, Hubert LE BOZEC, Claude MOINET and Pierre H. DIXNEUF.
Laboratoires de Chimie, Campus de Beaulieu, Université de Rennes, 35042 Rennes-Cedex (France).

ABSTRACT. Carbon disulfide is markedly activated, by coordination to a metal center, towards alkylhalides and electrophilic alkynes. The resulting metalladithioester or metal-carbene complexes are activated by monoelectronic processes, but the orientation of the reaction is strongly controlled by the nature of ancillary PR_3 ligands. $Fe(\eta^2CS_2R)(CO)_2(PR_3)_2^+$ cations are reduced by borohydride either by hydride transfer with electron donating PR_3 groups or by electron transfer with electron poor PR_3 ligands. Their reduction with sodium-amalgam or activated magnesium gives either C-C bond coupling with weak basic and labile phosphine ligands (PPh_3) or desulfurization when basic phosphines (PMe_3, PMe_2Ph, PBu_3) are used. These reactions show that the nature of the PR_3 ancillary ligand and that of the reducing reagent can be chosen to produce selectively new dithioformate, tetrathiooxalate or thiocarbonyl iron derivatives. Electron-transfer-induced ligand substitution has been used for the selective substitution of phosphines by more electron donating phosphines even in the presence of carbonyl ligands. Electrochemical or chemical oxidation of iron(o)-carbene $Fe(=CY_2)(CO)_2(PR_3)_2$ complexes is a one-electron oxidation and is markedly dependent on the nature of PR_3 groups. PR_3 ligands can be selected either for the reversibility of the oxidation process, the dimerization of the 17 electron intermediate or the dimerization of the carbene ligand, with carbon-carbon bond coupling and formation of tetrathiafulvalene derivatives.

Introduction

Heteroallenes constitute an useful class of reagents for the functionalization of substrates. Among them, carbon dioxide and carbon disulfide, the most stable representative examples allow easy carbon-heteroatom bond formation for the direct synthesis of carbonates, xanthates, carbamates or dithiocarbamates. The insertion of a heteroallene unit into a variety of metal-carbon bonds represents a classical route to carboxylic acid, ester or dithioester derivatives and the easiest method to form carbon-carbon bonds involving heteroallenes. The high stability of heteroallenes has attracted interest for the search of new activation processes, particularly in view of the discovery of novel carbon-carbon bond forming reactions. These processes have implied either activation at a metal center (1,7) or photo- and electro-

M. Chanon et al. (eds.), Paramagnetic Organometallic Species in Activation / Selectivity, Catalysis, 311–325.

chemical reduction (8-11). Examples of C-C bond coupling at an electron-rich metal center, between a heteroallene and an unsaturated substrate such as alkyne (12), diene (13) or olefin (14,15), have been described, and some of them have led to catalytic reactions (4,16-18). Electrochemical reductions of carbon dioxide can produce oxalate (8) or be favored by the presence of transition metal complexes (9). Electrochemical reductions combined with activation at a metal center have appeared to constitute a powerful tool for the functionalization of unsaturated substrates into carboxylic acids from CO_2 (19-22).

Metal centers have been shown to promote the dimerization of heteroallenes, but generally with head-to-tail coupling and formation of heteroatom-carbon bonds (23,24). Only a few processes involving head-to-head, oxidative C-C coupling of heteroallenes have been discovered (10,25-30). The first example was shown by electrochemical studies of a bis (dithiooxalato)copper(II) complex which showed the reversible C-C bond coupling between two SCO ligands, sulfur-coordinated to a copper center (10). C-C bond formation involving carbodiimide was observed by reaction with $Cp_2Ti(CO)_2$ (25). C-C coupling of carbon disulfide was achieved by reaction with $Fe_3(CO)_{12}$ (26) and $Ni_2(C_5H_5)_2(CO)_2$ (27). The treatment of (triphos) $MCl(CS_2)$ (28) and (triphos)$MCl(CSe_2)$ (29) derivatives of rhodium and iridium, with Lewis acids or even alcohols underwent to a C-C bond coupling affording the dimeric (triphos)$M(X_2C$-$CX_2)M$(triphos)$^{2+}$ cations. This dimerization appears to result from promoted homolytic cleavage of metal carbon bond (30).

These studies showed that most of C-C couplings of heteroallene moieties involved reduction at or by a metal center and monoelectronic processes. These observations have motivated investigations for the discovery of new ways of C-C bond formation, promoted by monoelectronic processes, from carbon disulfide derivatives at a metal center. Carbon disulfide iron complexes $Fe(\eta^2\text{-}CS_2)(CO)_2(PR_3)_2$ (31,32) have given evidence for the nucleophilic activation of carbon disulfide towards alkylhalides (33) and electrophilic alkynes (7) and allowed the access to metalladithioester $Fe(\eta^2\text{-}CS_2R)(CO)_2(PR_3)_2{}^+$ cations and iron-carbene $Fe(CS_2C_2R_2)(CO)_2(PR_3)_2$ complexes respectively. The readily access to these carbon disulfide-iron derivatives, containing a variety of phosphorus ligands, allowed the investigation for new processes of C-C bond formation depending on the nature of the ancillary PR_3 ligands. We report herein that reduction of $Fe(\eta^2\text{-}CS_2R)(CO)_2(PR_3)_2{}^+$ cation by borohydride or sodium-amalgam is very versatile and markedly depends on that nature of the PR_3 ligands. It selectively allows hydride or electron transfer and the formation of novel dithioformate, tetrathiooxalate resulting from C-C bond coupling, or thiocarbonyl iron derivatives. The electrochemical or chemical oxidation of iron(o)-carbene complexes is also controlled by the nature of the ancillary PR_3 ligands and appears to be an unprecedented route for the dimerization of carbene ligands, by C-C bond coupling, and the formation of tetrathiafulvalene derivatives.

Experimental Section

Syntheses and materials. All reactions were carried out under a nitrogen atmosphere using Schlenk techniques. Solvents were dried by reflux over appropriate drying agents. Tetrahydrofuran and diethylether were distilled over sodium benzophenone ketyl; pentane, hexane and acetonitrile over calcium hydride; dichloromethane over phosphorus pentoxide first and then over calcium hydride.

Preparation of starting iron complexes
All starting complexes are derivatives of $Fe(\eta^2\text{-}CS_2)(CO)_2(PPh_3)_2$ **1a**, readily available in one-step from $Fe(CO)_5$ (32). $Fe(\eta^2\text{-}CS_2)(CO)_2(PR_3)_2$ complexes **1** were obtained by phosphine substitution reaction of **1a** (31). Cationic derivatives $Fe(\eta^2\text{-}CS_2R)(CO)_2(PR_3)_2^+\ X^-$ **2** (33) were prepared by reaction of an excess of alkylhalide RX with the corresponding complexes **1**, in the presence of $NaPF_6$ for **2a** and **2d**.

Electrochemical Studies. Cyclic voltammetric measurements were obtained with the use of a conventionnal three electrode system. The working electrode was a platinum electrode. A Tacussel UAP-4 or PAR Model 362 potentiostat, and a Kipp and Zonen BD90 X-Y recorder were used for cyclic voltammetric experiments. All the measurements were carried out at 25°C under a nitrogen atmosphere with a solution of the complex (1.5 - 3. 10^{-3}M) in 0.1M Bu_4NBF_4 or Bu_4NPF_6/CH_3CN or CH_2Cl_2. All the potentials are referred to a saturated calomel electrode.

Controlled potential electrolyses were performed with a Tacussel PRT-20-2 potentiostat and a Tacussel IG-5 integrator using a cell (34) equipped with a vitreous carbon electrode (Ø : 4 cm) as the working electrode, a platinum electrode (Ø : 4 cm) as the auxiliary electrode and a saturated calomel electrode (SCE). Rotating-disc experiments were performed in the same cell by using a platinum RDE as the working electrode. The electrolyses were carried out at 25°C under a nitrogen atmosphere in a CH_2Cl_2 or CH_3CN solution with $LiClO_4$ as electrolyte.

Results and Discussion :

1. Reduction of $Fe(\eta^2\text{-}CS_2R)$ Cations.

Access to tetrathiooxalate iron complexes.

A simple route to low valent group 8 and 9 thiocarbonylmetal complexes have been previously elaborated using the reduction of $L_nM(\eta^2\text{-}CS_2R)^+$ cations by sodium borohydride. $Os(CS)(CO)_2(PPh_3)_2$ (35) was thus obtained from $Os(\eta^2\text{-}CS_2Me)(CO)_2(PPh_3)_2^+$ and via the isolable intermediate $Os(H)(\eta^1\text{-}CS_2Me)(CO)_2(PPh_3)_2$, resulting from hydride transfer from Na BH_4. The treatment of the salt $[N(CH_2CH_2PPh_2)_3Co(\eta^2\text{-}CS_2Me)]\ BPh_4$ with $NaBH_4$ led directly to $[N(CH_2CH_2PPh_2)_3Co(CS)]\ BPh_4$ (36) without observation of a hydride-cobalt intermediate. By contrast, we showed that the cation $Fe(\eta^2\text{-}CS_2Me)(CO)_2L_2^+$ **2a**, containing the basic phosphines L=PMe_3 (or PMe_2Ph), on reaction with $NaBH_4$ led to complex **3a** (37) (Scheme 1). In that case we could observed, using NMR at low temperature (38) , the fixation of the hydride to the iron atom followed by 1,2 migration to the carbon atom, affording the stable dithioformate-iron(o) complex **3a**. This indicated that no thiocarbonyliron(o) derivatives could be obtained by classical reduction with borohydride. Attempts to extend this reaction to cations **2b** and **2c**, containing the weakly basic phosphine L=PPh3, led us to give evidence for a new carbon-carbon bond forming reaction (39).

Scheme 1

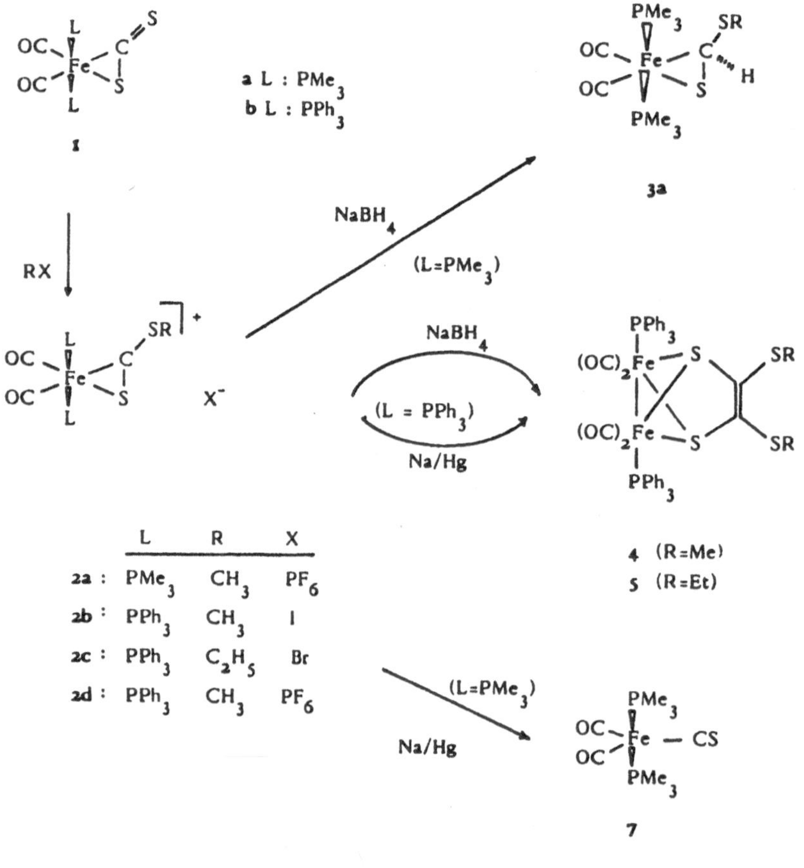

	L	R	X
2a :	PMe_3	CH_3	PF_6
2b :	PPh_3	CH_3	I
2c :	PPh_3	C_2H_5	Br
2d :	PPh_3	CH_3	PF_6

Complex **2b** was treated with two equivalents of $NaBH_4$ in THF and at room temperature. After 3 h complex **2b** was completely transformed into a purple air-sensitive derivative **4** which was purified by chromatography and isolated in 64 % yield. Similarly, complex **2c** led to 55% of purified complex **5**. (Scheme 1). The derivative **4** was obtained in 62-65 % yields either by reaction of **2b** with two equivalents of $(Ph_3PNPPh_3)BH_4$ or by that of $Fe(\eta^2\text{-}CS_2Me)(CO)_2(PPh_3)_2^+ PF_6^-$ **2d** with two equivalents of $NaBH_4$, showing that the nature of the counter cation or anion had no effect on the reaction.

The structure of complex **4** was established by an X-ray diffraction study and showed that a tetrathiooxalate ligand is bridging two $Fe(CO)_2(PPh_3)$ units linked by a Fe-Fe bond (39). The complexes **4** and **5** gave similar four terminal carbonyl absorption bands in their infrared spectra. Their ^1H NMR spectra indicated the presence of one SR group per PPh_3 ligand. Consequently, the formation of complexes **4** and **5** formally results from C-C and Fe-Fe bond couplings. This requires one-electron transfer from borohydride to the cation **2b** and the loss of one PPh_3 group. This observation suggested that a similar transformation **2b** → **4** could be performed by the direct one-electron reduction of **2b**. Complex **2b** was reacted with an excess of 1 % sodium-amalgam in THF at room temperature and was completely transformed after a 20 min time. Column chromatography allowed the successive isolation of traces of the blue complex identified to $Fe(\eta^2\text{-}S_2C_2(SMe)_2)(CO)_2(PPh_3)$ (39), a degradation product of **4**, and of 40 % of the purple complex **4** . The sodium-amalgam reduction of **2c** led to the isolation of 35 % of **5**

The striking difference in the behavior toward borohydride of **2a**, which captured one hydride leading to **3a**, (37) and of **2b**, which gave one electron transfer and formation of **4** and **5**, suggested that the reaction might be controlled by their reduction potentials. Complexes **2** were then studied using cyclic voltammetry in acetonitrile with $(Bu_4N)PF_6$ as electrolyte (Table I). All complexes **2** showed one **irreversible** reduction wave at negative potential at a scan rate of 100 mV. s^{-1}. A reversible wave was observed only for **2a** but at a higher scan rate. The reduction potential depends strongly on the nature of both the phosphine and R groups. The easiest reduction occurs with **2b-2d** containing the weak electron donating PPh_3 groups and leading to complexes **4** and **5** with borohydride. The large difference between the potential of borohydride (for the $\frac{1}{2}$ H_2 / H^- system E° is - 2.25 Vsce) and that of **2b-2d** (E^c ca -0.75 Vsce) might be responsible for electron transfer. This observation is opposed to hydride addition occuring with **2a** and **2f** containing basic phosphines, which are reduced at more negative potentials (-0.94 ; -1.04 Vsce). The lability of the PR_3 ligands appears also critical in determining the reactivity pattern. In the Na-Hg reduction of $Fe(\eta^2\text{-}CS_2Me)(CO)_2(PMe_2R)_2^+$ cations **2a** and **2f** the elimination of PMe_2R groups was not observed (vide supra). Thus the lability of the $Fe\text{-}PPh_3$ bond in the reduced species of **2b-2d** determined their transformation into **4** and **5**.

We have shown previously (40) that two fragment addition of a proton and the uncoordinated sulfur atom of complex **1b** to an α,β-unsaturated ketone afforded complexes of type **2d**. Pulegone and complex **1b**, in the presence of HBF_4, led to the isolation of 80 % of the red salt **2e** (Scheme 2). The treatment of complex **2e** with $NaBH_4$ led to deprotonation and the release of **1b** and pulegone. The reduction of **2e** into complex **6** would give access to iron complex of tetrathiooxalate containing two chiral ketone groups.

Scheme 2

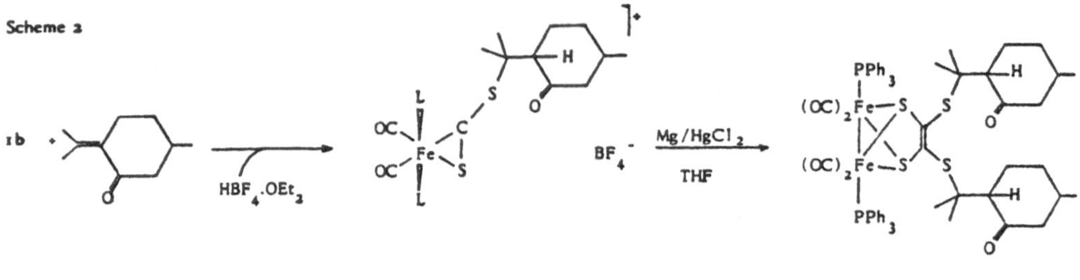

L = PPh_3 **2e**   **6**

Cyclic voltametry of **2e** in acetonitrile indicated that the reduction potential (E^c = -0.98 Vsce) was much more negative than those of complexes **2b-2d** containing the same PPh_3 groups (Table I). This observed shift may be due to a distorsion of the structure of **2e**, as compared to that of **2b**, resulting from interaction of the hindered pulegone fragment with the PPh_3 groups. To produce complex **6**, and avoid deprotonation and/or reduction of the ketone carbonyl, a milder reducing agent than Na/Hg had to be found. Magnesium ($E°$ (Mg^{2+}/Mg) = -2.62 vs $E°$ (Na^+/Na) = -2.96 Vsce) was the most efficient. Complex **2e** was reacted with magnesium, activated by treatment with mercury dichloride. After 24 h at room temperature complex **6** was isolated in 35% yield (Scheme 2).

Table I. Reduction potentials and infra-red absorptions of complexes $Fe(\eta^2\text{-}CS_2R)(CO)_2L_2{}^+X^-$

	L	R	X^-	$E^c(V_{SCE})$ (a)	νCO (b)
2b	PPh_3	Me	I^-	-0.74	1978-2050
2e	PPh_3	Me	PF_6^-	-0.74	1978-2050
2c	PPh_3	Et	Br^-	-0.76	1975-2050
2g	$P(OMe)_3$	Me	PF_6^-	-0.81	2020-2075
2f	PMe_2Ph	Me	PF_6^-	-0.94	1970-2040
2a	PMe_3	Me	PF_6^-	-1.04	1965-2035
2e	PPh_3	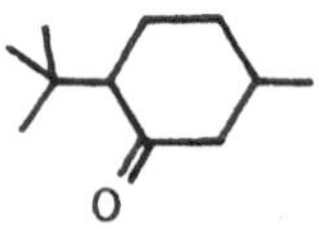	BF_4^-	-0.98	1980-2045

(a) determined by cyclic voltammetry with a $1.5 \cdot 10^{-3}$ M of complex in 0.1 M of Bu_4NPF_6 in CH_3CN, (200 $mV.s^{-1}$). (b) cm^{-1}, (nujol mull).

Access to thiocarbonyliron(0) complexes and ligand substitutions.

The electron-withdrawing thiocarbonyl ligand has been used for the stabilization of low-valent metal moieties (41). Four methods have been described for the access to low-valent thiocarbonyl-metal complexes: the desulfurization of (η^2-CS_2) metal complexes by phosphines (41,42), the reaction of metalcarbonyl anions with thiophosgene (43-44), the addition of NaSH to dihalocarbene-metal complexes (45) and the addition of borohydride to $M(\eta^2\text{-}CS_2R)^+$ cations (M = Os, Co, Ir), via $M(H)(\eta^1\text{-}CS_2R)$ intermediate (35,46). Although thiocarbonyliron(II) complexes have been described (41,47) no general route has been found for the access to thiocarbonyliron(o) derivatives (44,48). The latter method for instance can not be used for their preparation, for cations **2a** and **2b** react quite differently with borohydride whatever are the ancillary ligands affording **3a** and **4** respectively (Scheme 1). We have now found a simple and novel route to thiocarbonyliron(o) derivatives, but containing basic ligands, by reduction of $Fe(\eta^2\text{-}CS_2R)^+$ cations (49).

The reduction of complex **2a** with an excess of 1 % sodium-amalgam, under conditions similar to that of the transformation **2b** → **4**, after 4 h at room temperature and purification by chromatography, led instead to 35 % of the yellow thiocarbonyliron(o) complex **7** (Schemes 1,3). Under similar conditions complexes **2f** and **2h** led to the isolation of small amounts of unstable thiocarbonyliron(o) complexes **8** (8 %) and **9** (9%) (Scheme 3). Complex **7** was also obtained by reduction of **2a** with magnesium-amalgam $Mg/HgCl_2$, in 16 % yield. This low yield may be due to the in situ formation of the adduct $(PMe_3)_2(OC)_2(SC)Fe \rightarrow HgCl_2$ which was obtained and characterized by addition of $HgCl_2$ to an ether solution of **7** (49).

Scheme 3

$$[Fe(\eta^2\text{-}CS_2Me)(CO)_2L_2]^+PF_6^- \xrightarrow[\text{THF}]{\text{Na/Hg}} Fe(CS)(CO)_2L_2$$

2a	L = PMe_3	**7**
2f	L = PMe_2Ph	**8**
2h	L = PBu_3	**9**

The moderate to low yields of FeCS complexes **7-9**, led us to look for a more direct way of access, starting from the precursor **2b** containing PPh_3 groups. We were facing two problems: (i) **2b** by reduction leads to dimerization into **4** and (ii) although PPh_3 ligand substitution of $Fe(\eta^2\text{-}CS_2)(CO)_2(PPh_3)_2$ complex **1b** by PMe_3 takes place easily leading to high yield of **1a** (31), the same type of substitution, involving the cationic complexes $Fe(\eta^2\text{-}CS_2Me)(CO)_2(PPh_3)_2^+$ **2b-d** did not occur even on heating.

Electrochemical studies of metal complexes (50) have shown that selective ligand substitution reactions, essentially carbonyl substitution reactions, could be induced by electrode electron transfer. These studies have allowed to perform selective carbonyl substitution reactions using chemical redox reagents (51). To be efficient, the process involving an initial reduction has to fulfill three conditions according to Scheme 4 (52). (i) the final product (B) is more difficult to reduce than the initial derivative (A) (e.g. $E^c_1 > E^a_2$) although this condition can be overcome (53), (ii) the intermediate ($A^{-\cdot}$) is more reactive than its precursor (A) to give a fast $A^{\cdot-} \rightarrow B^{\cdot-}$ reaction and (iii) the redox system ($B/B^{\cdot-}$) is reversible to allow the formation of (B), by oxidation of ($B^{\cdot-}$) by (A).

Scheme 4

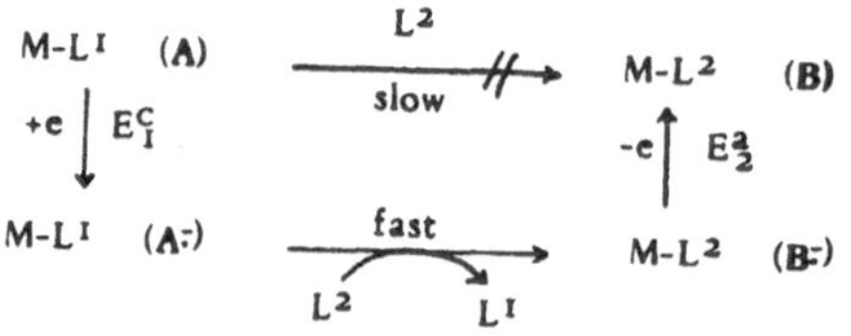

We can now show that this process can be used for the transformation **2b**→**2a**, by selective phosphine substitution, a process which is not possible thermally (Scheme 5). The cathodic potentials for **2b** and **2a** are respectively E^c_1 = -0.74 Vsce > E^c_2 = -1.04 Vsce (Table I). The system **2a**$^+$/**2a** appears to be reversible by cyclic voltammetry. Between 0 and -1.10 Vsce the anodic peak is observed at E^a_1 = -0.94 Vsce, but only for a sweep rate above 100 $mV.s^{-1}$. The electron-transfer-induced ligand substitution of **2b** by PMe_3 was then possible between - 0.74 and -0.94 Vsce.

Scheme 5

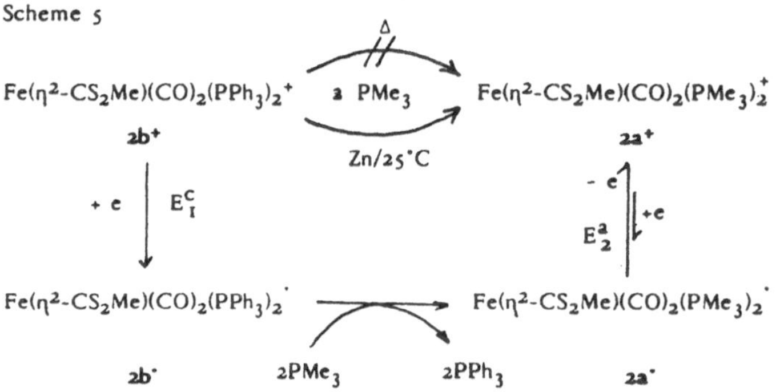

However, the electron promoted reaction of **2b** with PMe_3 could also lead to carbonyl substitution under the same conditions. The solution of 5.10-4 M of **2b** and 0.1 M of Bu_4NPF_6 in 15 mL of acetonitrile was studied by cyclic voltammetry : (i) a first scan showed the presence of **2b** (E^c_1= -0.74 Vsce), (ii) an excess of PMe_3 was added to the solution and (iii) after homogeneization at room temperature a second scan was immediately recorded; it showed essentially the presence of **2a** (E^c_2=- 1.04 Vsce; E^a_2= -0.94 Vsce) without a significant peak at -0.74 Vsce indicative of the presence of **2b**. These observations were consistent with the electrode electron-transfer-induced substitution of two PPh_3 groups by two PMe_3 ligands at room temperature.

We have turned this observation into a preparative transformation of **2b** into **2a** using zinc as reducing reagent. Zinc powder was added to a solution of complex **2d**, or **2b** with $NaPF_6$, in acetonitrile or dichloromethane and an excess of trimethylphosphine was immediately added. After 6h at 25°C, complex **2d** had disappeared and **2a** was isolated in 70 % yield (Scheme 5). This transformation shows an example of electron promoted selective substitution of two phosphine ligands by two more basic phosphines (38,54).

The above studies have been used to perform the direct transformation of **2b** into thiocarbonyliron(O) derivatives. It corresponds to the coupling of two reactions in a one-pot **2b** → **2a** and **2a** → **7**, both resulting from electron promoted reactions. When a solution of **2b** with 3 equivalents of PMe_3 was added to 1 % sodium-amalgam a slow transformation occurs at room temperature and after 5 h complex **7** was isolated in 52 % (Scheme 6). Similarly when PMe_2Ph or PBu_3 were used complex **8** (12 %) and **9** (14 %) were obtained. The low yield in **8** and **9** are due to their high instability as compared to that of **7**. This method was used for the access to a thiocarbonyliron(O) complex **10**, for which its formal precursor $Fe(\eta^2\text{-}CS_2)(CO)_2(dppe)$ was not possible to produce. Complex **10** (12%) was thus obtained by reaction of **2b** and 1.5 equivalent of Bis(diphenylphosphino)ethane (dppe) with sodium-amalgam (38,54).

Scheme 6

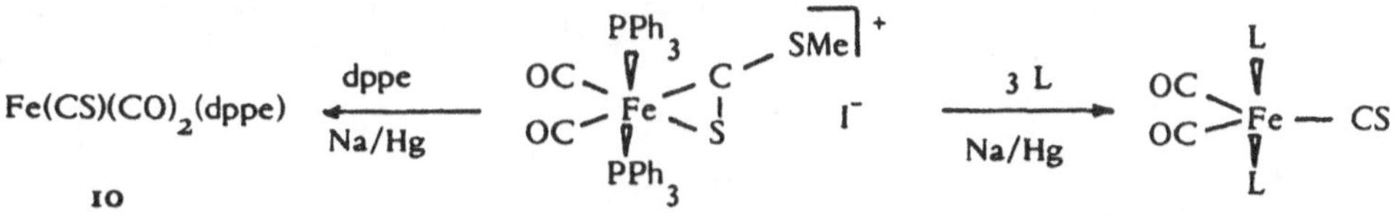

10

7,8,9

7 ($L=PMe_3$); **8** ($L=PMe_2Ph$); **9** ($L=PBu_3$); **10** (dppe : $Ph_2PCH_2CH_2PPh_2$)

The formation of **7** by reduction of **2b** occurs without any evidence for the formation of **4**, the normal product of reduction of **2b** under the same conditions except the presence of PMe_3. This indicates that the substitution of PPh_3 groups by PMe_3 in the reduced species of **2b** is fast as compared to the dimerization of the $Fe(\eta^2CS_2Me)(CO)_2(PPh_3)$ radical. The lability of the Fe-PPh_3 bond in the reduced species of **2b** is likely to control its evolution into **4** whereas the presence of non labile PMe_3 groups in the reduced species of **2a** may favor its evolution into **7**.

2. Oxidation of metal-carbene complexes.

Access to tetrathiafulvalene derivatives

Metal-carbene complexes are useful in synthesis and catalysis for the formation of carbon-carbon bonds. The coupling of the carbene ligand with unsaturated substrates such as olefins and alkynes in metathesis or carbon-heteroatom double bonds, usually requires the coordinative unsaturation of the metal site (55). Dimerization of the carbene ligand into an olefin, especially involving d^6 metal(o)-carbene complexes, has already been achieved by thermolysis and the reaction is initiated by the dissociation of a metal-carbonyl bond (56,57). Moreover, it has been established that the thermolytic reaction does not involve the free carbene and proceeds by a bimetallic intermediate (58).

Electron-rich cyclic carbene-iron(o) complexes 11 are readily obtained by cycloaddition of electrophilic alkynes to the activated carbon disulfide ligand of complexes $Fe(\eta^2-CS_2)(CO_2)(PR_3)_2$ 1 (7,59) (Scheme 7). Thermolysis of complex 11b also led to the dimerization of the carbene ligand giving tetrathiafulvalene derivatives via unsaturated metal-carbene intermediate (60). TTF 12 was also obtained in 83% yield by treatment of complex 11b with an excess of iodine (60). The electrochemical study of the latter oxidation reaction has led to give evidence for a new dimerization of carbene ligand by electron transfer catalysis depending on the nature of ancillary PR_3 ligands.

Scheme 7

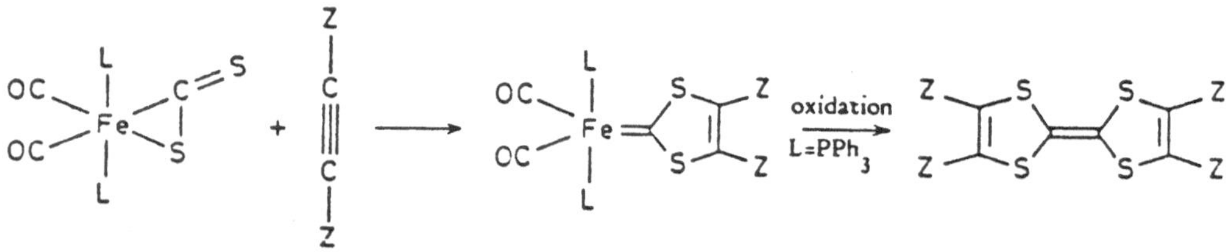

1a-e 11a-e 12

Z = CO_2Me

a ($L=PMe_3$) ; b ($L=PPh_3$) ; c ($L=PMePh_2$) ; d ($L=PMe_2Ph$) ; e ($L=PPh_3$ and PMe_3)

The cyclic voltammetry (CV) of complexes 11a-e, which were made by adding an excess of dimethylacetylene dicarboxylate to the corresponding $(\eta^2-CS_2)Fe$ complexes 1a-e (Scheme 7) was carried out at a stationary platinum electrode (Table II). All complexes 11 show a first anodic wave corresponding to a one electron oxidation between - 0.1 and - 0.3 V/SCE, as indicated by stoechiometric reaction with ferrocenium ion. The chemical stability of the anodically generated species 11^+, as shown by the ip^a/ip^c ratio (Table II), depends on the nature of the solvent and of the phosphine ligands. For example, with the complex 11b ($L = PPh_3$) the reversibility is observed in dichloromethane whereas it disappears in acetonitrile. This observation is thought to be due to the coordinating properties of CH_3CN which can displace one triphenylphosphine ligand (61).

The nature of L markedly affects the easiness of oxidation of complexes 11 : E_p^a is lowered as the electron donor capability of L is increased and follows the sequence $PPh_3 < PMePh_2 < PMe_2Ph < PMe_3$. Moreover the chemical stability of 11^+ is also strongly dependent on the nature of L : $11c^+$ ($L = PMePh_2$) appears to be more stable than $11d^+$ ($L = PMe_2Ph$) and $11a^+$ ($L = PMe_3$) which contain more basic phosphines and $11b^+$ which contain the more hindered and labile PPh_3 ligands.

Table II . Cyclic Voltammetric Data for Complexes 11 (a)

			oxidation			reduction	
Complex	L/L	Solvent	E_p^a(V)	E_p^c(V)	ip(c)/ip(a)	E_p^c1(V)	E_p^c2(V)
11b	PPh_3/PPh_3	CH_3CN	-0.19	-	0	-0.50	-0.70
"	" "	CH_2Cl_2	-0.11	-0.20	0.8		-0.65
"	" "	CH_3COCH_3	-0.20	-0.29	1		
11c	$PMePh_2/PMePh_2$	CH_3CN	-0.23	-0.32	0.8	-0.50	-0.70
"	" "	CH_2Cl_2	-0.12	-0.27	1		
11d	PMe_2Ph/PMe_2Ph	CH_3CN	-0.29	-	0	-0.54	-0.74
"	" "	CH_2Cl_2	-0.15	-0.26	0.3		-0.73
11a	PMe_3/PMe_3	CH_3CN	-0.30	-	0	-0.59	-0.78
11e	PPh_3/PMe_3	CH_3CN	-0.24	-0.32	0.3		

(a) performed with a 3.10^{-3}M of complex in 0.1 M solution of Bu_4NBF_4 or Bu_4NPF_6 at 0.1 $V.s^{-1}$. Potentials are vs SCE.

The cyclic voltammograms of **11b** and of **11d** are distinguished by their behavior when recorded between - 1 and + 1 V : For **11b** only new intense waves are observed at more anodic potentials, which are identified to the oxidation waves of the tetrathiafulvalene (TTF) **12** resulting from the dimerization of the carbene ligand, and to the oxidation of the free triphenylphosphine.

The controlled potential electrolysis of 1 mmol of **11b** in acetonitrile, at a large vitreous carbone anode, at 0 V corresponding to the first anodic wave, involves nearly 0.1 F/mol^{-1} and leads to TTF **12** in approximatively 60 % yield. This study reveals the electrocatalytic nature of the TTF formation. To confirm the electrocatalysis, the electrolysis was stopped after that 0.01 $F.mol^{-1}$ was passed through the solution and the reaction was then monitored by following the changes at a rotating disc electrode (rde): the first oxidation wave rapidly disappeared with the concommitant increase of a new wave, at more positive potential, corresponding to the TTF formation.

We have previously reported that addition of a stoechiometric amount of iodine to complex **11b** gave rise to TTF **12** (60). In view of the preceding electrochemical results we have investigated the oxidation of complex **11b** with a catalytic amount of iodine. The reaction was followed by voltammetry at rde. At room temperature **11b** disappeared in a first-order reaction ($t_{\frac{1}{2}}$ 0.5 h) and simultaneously **12** was formed. This experiment corroborates the electrocatalytic process and shows that the reaction can be initiated chemically by an oxidant, such as iodine or the ferrocenium cation.

Cyclic voltammetry of **11b** in CH_2Cl_2 gives a quasi irreversible wave with $E\frac{1}{2}$= -0.12 V whereas $Fe(CO)_3(PPh_3)_2$ reversibly oxidizes at $E\frac{1}{2}$ = + 0.34 V (62). This shows the much better donor ability of the carbene ligand vs the CO ligand. The chemical instability of the oxidized species 11^+ contrasts with the stability of paramagnetic carbene iron(I) salts. $Fe(CO)_2L_2L^{Me}$ (L=PR_3or $P(OR)_3$; L^{Me} = $\overline{CN(Me)CH_2CH_2N}$-Me) (63). $Fe(CO)_2(PPh_3)_2(L^{Me})$ oxidizes reversibly at -0.5 V and gives a stable green paramagnetic cation whereas oxidation of **11b** leads to the dimerization of the carbene ligand. Moreover attempts to isolate **11d**$^+$ (L=PMe_2Ph) by addition of one equivalent of $AgCF_3SO_3$ (or $Cp_2Fe^+PF_6^-$) to a dichloromethane solution of **11d** results in the formation of new **diamagnetic** salt which corresponds, as suggested by elemental analysis and spectroscopic studies, to the dimeric form (**11d-11d**)$^{2+}$ by Fe-Fe bond formation (61).

The electrochemical studies also show the crucial role of triphenylphosphine in the electrocatalytic dimerization of the carbene ligand. They can be related to the synthesis of TTF **12** by thermolysis of **11b** which was dependent on the lability of PPh_3 ligand and did not result from dimerization of a free carbene but suggested a binuclear complex as intermediate (60).

From these observations it is possible to propose a mechanism which involves an electrochemical step followed by several chemical steps (Scheme 8). The first step is a one electron oxidation leading to a paramagnetic 17-electron species. Then two possible pathways can be considered:

Scheme 8 (L=PPh_3)

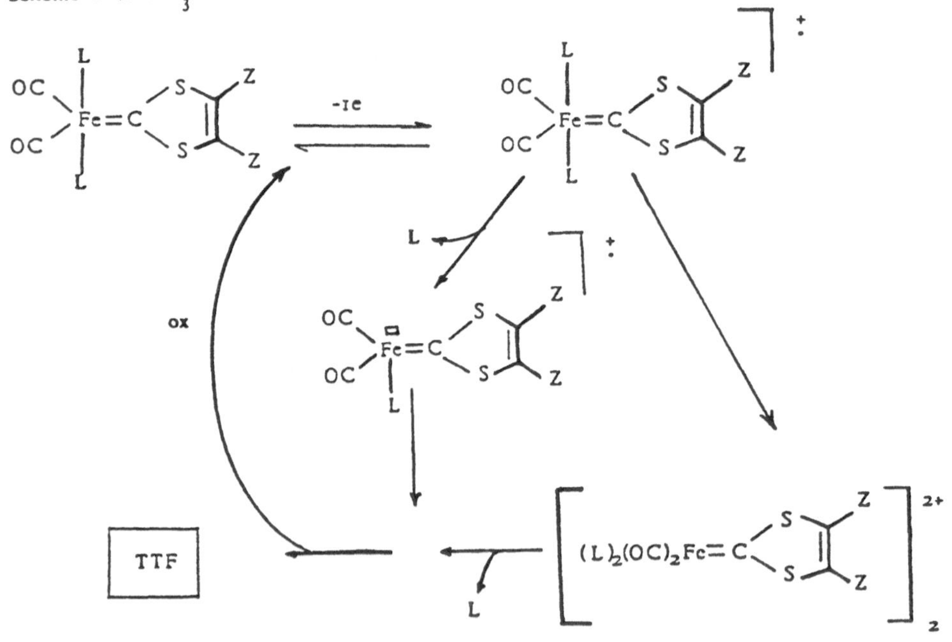

- As the formation of a binuclear complex seems necessary to explain the dimerization of the carbene ligand, the next step could be the dimerization of the 17-electron complex, leading to a diamagnetic species analogous to $(\mathbf{11d}\text{-}\mathbf{11d})^{2+}$ and followed by the loss of PPh_3 and carbene transfer to give TTF.

- Another route could involve the initial loss of PPh_3, since the lability of 17 electron cation radicals is well known and has intensively been used in ligand exchange reactions (50,51). The intermediate could then dimerize to afford TTF **12**.

Finally, to explain the electrocatalysis, one of the intermediates formed in the chemical evolution of the 17-electron complex has to be an oxidative species for the starting complex **11b**.

The one electron oxidation of complex **11b** leads to the formation of tetrathiafulvalene **12** by carbon-carbon coupling of two carbene ligands, with a high turnover number above 100. This process occurs either electrochemically or chemically with the use of an appropriate oxidant. As the precursor **1b** can be made in only one step from iron pentacarbonyl, this reaction appears to be an efficient and very simple method to prepare TTF **12** and could be generalized for the preparation of other tetrathiafulvalene derivatives bearing at least one electron withdrawing substituent.

Finally it is worthy to note that if many ligand substitution reactions are known to occur by electron-transfer catalysis, this carbene reaction represents the first example of ligand dimerization promoted by a one electron oxidation.

References

1 P.H. Dixneuf, and R.D. Adams, Report of the International Seminar on the Activation of Carbon Dioxide and Related Heteroallenes on Metal Centers, Rennes, France, **1981**.

2 (a) J.A. Ibers, *Chem. Soc. Rev.* **1982**, 11, 57. (b) H. Werner, *Coord. Chem. Rev.* **1982**, *43*, 165 and references therein.

3 C. Bianchini, C. Mealli, A. Meli, M. Sabat, in "*stereochemistry of organometallic and Inorganic Compounds*", Eds. I. Bernal, Elsevier science Publishers B.V. Amsterdam **1986**, 146.

4 "organic and bioorganic chemistry of carbon dioxide". S. Inoue and N. Yamazaki eds., Kodansha Ltd., Tokyo; Wiley, New-York, **1981**.

5 D.J. Darensbourg and R.A. Kudaroski, *Adv. Organomet. Chem.* **1983**, *22*, 129.

6 'Carbon Dioxide as a source of Carbon'. M. Aresta and G. Forti, Eds., D. Reidel, Dordrecht **1987**, NATO ASI series vol. 206.

7 H. Le Bozec, A. Gorgues and P.H. Dixneuf, *J. Am. Chem. Soc.* **1978**, *100*, 3946.

8 (a) J.C. Gressin, D. Michelet, L. Nadjo and J-M. Savéant, *Nouv. J. Chim.*, **1979**, *3*, 545. (b) C. Amatore and J-M. Savéant, *J. Am. Chem. Soc.* **1981**, *103*, 5021.

9 B.J. Fisher, and R.J. Eisenberg, *J. Am. Chem. Soc.* **1980**, *102*, 7361.

10 D. Coucouvanis, *J. Am. Chem. Soc.* **1971**, *93*, 1786.

11 R. Ziessel, *Nouv. J. Chim.* **1983**, *7*, 613.
12 H. Hoberg, D. Schaefer, G. Burkhart, *J. Organomet. Chem.* **1982**, *228*, C21.
13 H. Hoberg, K. Jenni, C. Krüger, E. Raabe, *Angew. Chem. Int. Ed. Engl.* **1986**, *25*.
14 R. Alvarez, E. Carmona, D.J. Cole-Hamilton, A. Galindo, E. Guitterrez-Puebla, A. Monge, M.L. Poveda and C. Ruiz, *J. Am. Chem. Soc.* **1985**, *107*, 5529.
15 (a) H. Hoberg, Y. Peres and A. Milchereit, *J. Organomet. Chem.* **1986**, *307*, C41. (b) H. Hoberg, K. Jenni, K. Angermund and C. Krüger, *Angew. Chem. Int. Ed. Engl.* **1987**, *26*, 153.
16 H. Hoberg, E. Hermandez, *J. Chem. Soc., Chem. Commun.* **1986**, *544*.
17 Y. Inoue, Y. Itoh and H. Hashimoto, *Chem. Lett.* **1978**, *633*.
18 A. Musco, C. Perego and V. Tartiari, *Inorg. Chim. Acta*, **1978**, *28*, L 147.
19 G. Silvestri '*Carbon dioxide as a source of carbon*' M. Aresta and G. Forti, Eds., D. Reidel, Dordrecht, Nato ASI series, **1987**, *Vol. 206*, 339.
20 I.B.M. Tkatchenko and D.A. Ballivet-Tkatchenko, Fr. Pat. **1984**, *2*, 542, 764.
21 J.F. Fauvarque, C. Chevrot, A. Jutand, M. François, J. Perichon, *J. Organomet. Chem.* **1984**, *264*, 273.
22 S. Torii, H. Janaka, T. Hamatani, K. Morisaki, A. Jutand, F. Peluger, and J.F. Fauvarque, *Chem. Lett.* **1986**, 169.
23 (a) H. Werner, O. Kolb, R. Feser, U. Schubert, *J. Organomet. Chem.* **1980**, *191*, 283. (b) M. Cowie, S.K. Dwight, *J. Organomet. Chem.* **1981**, *214*, 233. (c) D. H. M. W. Thewissen, *J. Organomet. Chem.* **1980**, *188*, 211.
24 T. Herskovitz, L.J. Guggenberger, *J. Am. Chem. Soc.* **1976**, 1815.
25 M. Pasquali, C. Floriani, A. Chiesi-Villa, C. Guastini, *Inorg. Chem.* **1981**, *20*, 349.
26 P.V. Broadhurst, B.F.G. Johnson, J. Lewis, and P.R. Raithby *J. Chem. Soc., Chem. Commun* **1982**, 140.
27 J.J. Maj, A.D. Rae and L.F. Dahl *J. Am. Chem. Soc.* **1982**, *104*, 4278.
28 C. Bianchini, C. Mealli, A. Meli, and M. Sabat, *J. Chem. Soc., Chem. Commun.* **1984**, 1647.
29 C. Bianchini, C. Mealli, A. Meli, and M. Sabat, *Inorg. Chem.* **1984**, *23*, 4125.
30 C. Bianchini, C. Mealli, A. Meli, M. Sabat, and P. Zanello, *J. Am. Chem. Soc.* **1987**, *109*, 185.
31 H. Le Bozec, P.H. Dixneuf, A.J. Carty, and N.J. Taylor, *Inorg. Chem.*, **1978**, *17*, 2568.
32 H. Le Bozec, J. Fournier, A. Samb and P.H. Dixneuf *Organomet. Synth.* **1986**, vol. *3*, 297.
33 D. Touchard, H. Le Bozec, P.H. Dixneuf, A.J. Carthy, and N.J. Taylor, *Inorg. Chem.*, **1981**, *20*, 1811.
34 G. Jacob and C. Moinet, *Bull. Soc. Chim. Fr*, **1983**, I, 291.
35 T.J. Collins, W.R. Roper, K.G. Town, *J. Organomet. Chem.*, **1976**, *121*, C41
36 C. Bianchini, D. Masi, C. Mealli, A. Meli, M. Sabat, and G. Scapacci, *J. Organomet. Chem* **1984**, *273*, 91.
37 D. Touchard, P.H. Dixneuf, R.D. Adams, and B. E. Segmuller, *Organometallics*, **1984**, *3*, 640.
38 J-L. Fillaut, these de doctorat, Université de Rennes, **1987**.
39 (a) D. Touchard, P.H. Dixneuf, R.D. Adams, B.E. Segmuller, *Organometallics*, **1984**, *3*, 640.
40 D. Plusquellec, and P.H. Dixneuf, *Organometallics*, **1982**, 1, 1401.
41 I. S. Butler, *Acc. Chem. Res.*, **1977**, *10*, 359.
42 M.C. Baird, G. Hartwell, and G. Wilkinson, *J. Chem. Soc.*, A, **1967**, 2037.
43 B. D. Dombek and R.J. Angelici, *Inorg. Chem*, **1976**, *15*, 1089.
44 W. Petz, *J. Organomet. Chem.*, **1978**, *146*, C23.
45 G. R. Clark, K. Marsden, W.R. Roper and L.J. Right, *J. Am. Chem. Soc*, **1980**, *102*, 1206.
46 C. Bianchini, A. Meli and G. Scapacci, *Organometallics*, **1983**, *2*, 1934.
47 L. Busetto, V. Belluco and R.J. Angelici, *J. Organomet. Chem.*, **1969**, *18*, 213.

48 P. Conway, A.R. Manning and F.R. Stevens, *J. Organomet. Chem*, **1980**, *186*, C64.
49 D. Touchard, C. Le Lay, J-L. Fillaut and P.H. Dixneuf, *J. Chem. Soc., Chem. Commun.*, **1986**, *1*, 37.
50 (a) G. J. Bezems, P.H. Rieger, and S. Visco, *J. Chem. Soc., Chem. Commun.*, **1981**, 265. (b) A. Darchen, C. Mahé, and H. Patin, *J. Chem. Soc., Chem. Commun.*, **1982**, 243. (c) C.M. Arewgoda, P.H. Rieger, B.H. Robinson, and J. Simpson, *J. Am. Chem. Soc.*, **1982**, *104*, 5633. (d) B.A. Narayanan, C. Amatore and J.K. Kochi, *J. Chem. Soc., Chem. Commun.* **1983**, 397. (e) J.W. Hershberger, R.J. Klingler and J.K. Kochi, *J. Am. Chem. Soc.*, **1983**, *105*, 61. (f) J.W. Hershberger, C. Amatore and J.K. Kochi, *J. Organomet. Chem.*, **1983**, *250*, 345. (g) J. Rimmelin, P. Lemoine, M. Gross, A.A. Bahsaun and J.A. Osborn, *Nouv. J. Chim.*, **1985**, *9*, 181.
51 M.I. Bruce, J.G. Matisons and B.K. Nicholson, *J. Organomet. Chem.*, **1983**, *247*, 321,
52 J.M. Savéant, *Acc. Chem. Res.*, **1980**, *13*, 323.
53 C. Amatore, personnal communication.
54 D. Touchard, J-L Fillaut and P.H. Dixneuf, unpublished results.
55 K.H. Dötz, H. Fischer, P. Hofman, F.R. Kreisse, U. Schubbert and K. Weiss '*Transition Metal Carbene Complexes*", Verlag Chemie, Weinheim, 1983.
56 (a) E.O. Fischer and K.H. Dotz, *J. Organomet. Chem.* **1972**, *36*, C4. (b) E.O. Fischer and D. Plabst. *Chem. Ber.* **1974**, *107*, 3326.
57 C.P. Casey T.J, Burkhardt, C.A. Bunnel and J. C. Calabrese *J. Am. Chem. Soc.* **1977**, *99*, 2127.
58 C.P. Casey and R.L. Anderson *J. Chem. Soc. Chem. Commun.* **1975**, 895.
59 H. Le Bozec, A. Gorgues and P.H. Dixneuf. *Inorg. Chem.* **1981**, *20*, 2486.
60 H. Le Bozec and P.H. Dixneuf, *J. Chem. Soc. Chem. Commun.* **1983**, 1462.
61 H. Le Bozec, C. Moinet and P.H. Dixneuf unpublished results.
62 N.G. Connelly and K.R. Somers, *J. Organomet. Chem.* **1976**, *113*, C39.
63 M.F. Lappert ; J.J. Mac Quitty and P.L. Pye, *J. Chem. Soc. Chem. Commun.* **1977**, 411.

PHOTOLYTIC GENERATION OF RADICALS FROM [η-$C_5R_5Fe^+$ARENE]* COMPLEXES. IMPLICATIONS FOR ELECTRON TRANSFER CATALYSIS AND RADICAL REACTIONS.

Enrique Roman,[1] Mauricio Barrera,[1] Sergio Hernandez,[1] and Charles Giannotti.*[2]*
*[*1] Catholic University of Chile. Faculty of Chemistry. P.O. Box 6177. Santiago-Chile. [*2] Institut de Substances Naturelles, CNRS, F-91190, Gif-sur-Yvette. France.*

ABSTRACT. Visible irradiation of [η-$C_5R_5Fe^+$Arene] in polar solvents leads to generation of $(C_5R_5)^{\bullet}$ radicals, (R=H, CH_3), which are identified as their spin-trapped adducts. Photophysical studies show that the excited triplet state, $a\ ^3E_1$, of these organometallic complexes can be efficiently quenched by classical redox quenchers such as methylviologen or dimethylaniline. The excited state of [η-$C_5R_5Fe^+$Arene]* complexes can also be quenched by hexamethyl (Dewar) benzene, HMDB. In this case two decay types have been determined: (i) a photochemical path occuring in the excited state surface, where simultaneous regioisomerization of HMDB and photoligand exchange take place, affording the hexamethylbenzene derivative [η-$C_5R_5Fe^+$HMB], and (ii) a photoinduced radical chain mechanism (detected on the e.s.r. time scale) for HMDB isomerization, caused by an ET photoprimary process between HMDB (an electron donor molecule) and the [η-$C_5R_5Fe^+$Arene]* complex. Electrocatalytic studies suggest that in the decay of [η-$C_5R_5Fe^+$Arene]* complexes, two kinds of transient species (17 or 19 electron) are photogenerated depending on the redox-potentials of the quenchers.

INTRODUCTION

The chemistry of cationic metallocenic iron Π-complexes of the form [n-$C_5R_5Fe^+$Arene] has been developed extensively since 1975. Several reviews have appeared in recent years.[1] Interesting

M. Chanon et al. (eds.), Paramagnetic Organometallic Species in Activation / Selectivity, Catalysis, 327–343.

aspects of these complexes include (i) reversible ET between cationic 18 electron and neutral 19 electron complexes in mononuclear series (eq.1),[2] (ii) multi-step reversible ET's that occur in binuclear series of the $\{[Cp^*Fe]_2(\mu\text{-}\eta\text{-}\eta\text{-polyarene})\}^{2+}$ complexes[3] (eq.2), ($Cp^*=C_5Me_5$).

$$[CpFe^{+}Arene] \underset{e}{\overset{e}{\rightleftharpoons}} [CpFe^{\bullet}Arene] \qquad (ref.2) \qquad (1)$$

$$\underset{36e}{[Fe(II),Fe(II)]^{2+}} \underset{-e}{\overset{e}{\rightleftharpoons}} \underset{37e}{[Fe(I),\ Fe(II)]^{+}} \underset{-e}{\overset{e}{\rightleftharpoons}} \underset{38e}{[Fe(I)Fe(I)]} \underset{-e}{\overset{e}{\rightleftharpoons}} \underset{39e}{[FeI,Fe(o)]^{-1}}$$

$$(ref.3) \qquad (2)$$

An electrocatalytic ligand exchange (coulombic efficiency 0.03) has been reported for the mononuclear series in which intermediate species of 19 electron $[Cp\ Fe^{I}S_3]$, (S = solvent or the entering ligand) and the 17 electron $[CpFe^{I}(\eta^4\text{-Arene})]$ have been suggested[4] (eq. 3).

$$\underset{18e}{[CpFe^{+}Arene]} \underset{-e}{\overset{e}{\rightleftharpoons}} \underset{19e}{[CpFe^{\bullet}Arene]} \rightleftharpoons \underset{17e}{[CpFe^{\bullet}\ \eta^4\ Arene]} \longrightarrow \underset{19e}{[CpFe^{\bullet}\ S_3]} \qquad (3)$$

A redox catalyst, in the zwitterionic form $[\eta\text{-}C_5H_4COO^{-}Fe^{+}HMB]$ was synthesized[5] to carry out a rapid and efficient catalysis of NO_3^- ion to NH_3 reduction in homogeneous basic media,[6] ($k \approx 10^2 Mol^{-1}\ l\ s^{-1}$). The intermediate anionic species $[\eta\text{-}C_5H_4COO^{-}Fe^{\bullet}HMB]$ in 19 electron configuration was probed by e.s.r. spectroscopy.[5] Under photochemical conditions $[\eta\text{-}C_5R_5Fe^{+}Arene]$ undergoes many photoligand exchanges. A cationic, unstable 18 electron species $[CpFe^{+}S_3]$ has been claimed by several authors[7], (S=solvent or a two electron L ligand, or a η^6-ligand as cyclooctatetraene and cycloheptatriene), (eq.4).

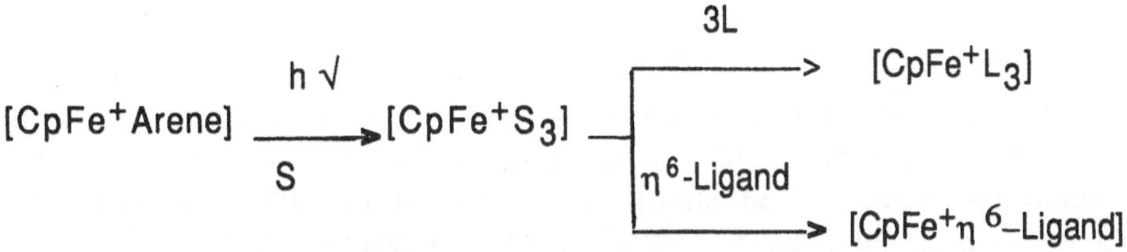

(4)

In this last field, no photochemical and photophysical study of [η–$C_5R_5Fe^+$Arene] complexes has been carried out to determine the nature of the intermediates or transient species involved. Recently, [8] we reported the luminescence of [η–$C_5R_5Fe^+$Arene], this being only the second metallocene to show an emission in the visible spectral region, (λ_{em}= 530 nm ; λ_{exc}= 440nm; 293 K). The first emissive metallocene complex, (in solid state), ruthenocene, $(C_5H_5)_2Ru$, was reported by Wrighton, [9] at 4.2-77K.

In this work we describe the use of the emission phenomena of these complexes [η–$C_5R_5Fe^+$Arene], coupled with e.s.r and n.m.r. spectroscopic studies. The emission phenomena allow an examination of the phototransient species of short lifetime, generated by an ET process that occurs on the excited surface belonging to *a* $^3E_1$ excited state of these complexes. An application of a photoelectron catalysis as the regioisomerization of hexamethyl(Dewar)benzene, is described. Therefore, this work can be classified in the general area of ET catalysis, an area developed by Chanon,[10] Turro,[11] Kochi,[12] Astruc,[13] and other investigators by which critical views on the multiple aspects of ET reactions and ET catalysis have been put forward.

EXPERIMENTAL

[$CpFe^+$Arene]X, (X=Cl, PF_6, BF_4) were prepared by standard procedures[14] and were purified by crystallization from acetone and by alumina chromatography. Spectroscopic grade solvents were used in e.s.r. and photochemical reactions. The e.s.r. spin trapping determinations were as described in previous publications.[15] Hexamethyl(Dewar)benzene, (Aldrich), was distilled under reduced pressure and stored in the dark at 253K.

Methylviologen dichloride, $MVCl_2$, (Aldrich), was crystallized 3 times from methanol and dried at 343K under vacuum for 24h.

Dimethylaniline, DMA, (Aldrich), was distilled before use. Luminescence spectra, emission quantum yield and fluorescence-quenching experiments were described in a previous report. [8b]

RESULTS AND DISCUSSION

1. Visible irradiation of [η-$C_5R_5Fe^+$Arene]PF_6 generates radicals, (R=H, CH_3).

The ligand field irradiation (λ_{irrad}= 441 nm) corresponding to the ^1A ---> $a\ ^1E_1$ electronic transition, (λ_{max}= 450nm; ε = 69 dm l^{-1} mol^{-1}) of the [η-$C_5R_5Fe^+$Arene], (arene: naphthalene, benzene, toluene, *P*-xylene, durene, pentamethylbenzene, hexamethylbenzene, and hexaethylbenzene), in dichloromethane, in the presence of a spin-trapping reagent such as nitrosodurene, ArNO, shows two kinds of signals in its e.s.r. spectrum. The first, **1**, observed during the irradiation time, corresponds to a quadruplet with relative intensities of 1:2:2:1, (Fig.1), with coupling constant shown in Table1. When irradiation is stopped, the signal decays to a six lines signal. This sextuplet **2** is invariant in the dark but further irradiation leads to its disappearance and replacement by the quadruplet **1** observed during initial irradiation (Fig 1). Radical formation is solvent independent, the same radicals **1** and **2** being observed in CH_3CN, acetone and CH_2Cl_2. Visible photolysis of [η-$C_5R_5Fe^+$Arene] complexes with different arene ligands gives similar e.s.r. spectra with similar $\mathbf{a}_N$, $\mathbf{a}_{\beta H}$ e.s.r. parameters (see Table 1). When the cyclopentadienyl ring is permethylated as in [η-C_5Me_5Fe+Arene] complexes, (arene = *p*-xylene, benzene), and when the photolysis is carried out in the cavity of the e.s.r. spectrometer in the presence of nitrosodurene, a new signal corresponding to a triplet signal (1:1:1) is observed. This signal **3** is stable both during and after irradiation. It is represented in Fig 1, corresponding to the adduct **3**, ArNO$^{\cdot}C_5Me_5$. The $\mathbf{a}_N$ and $\mathbf{a}_{\beta H}$ values are in good agreement with those observed during the UV-photolysis of C_5Me_5H in spin trapping experiments. [17] The radical **2**, $ArNO^{\bullet}C_5H_5$ has already been mentioned in other literature reports concerning the UV-photolysis of $(\eta\text{-}C_5H_5)_2MX_2$, (M=Zr, Hf, Th, U), [18] and $(C_5H_5)_2Hg$. [19] The other spin trapped radical **1** corresponds exactly to the multiplet ascribed elsewhere[19] corresponding to the (aminoxyl) radical adduct $ArN(O^{\bullet})H$. It suggests that a H-transfer occurs simultaneously with the generation of $C_5H_5^{\bullet}$ radicals. This reduction of spin trap is not observed in the case of the irradiation of the pentamethylated derivative $[\eta\text{-}C_5Me_5Fe^{+}Arene]$.

Table 1. Spin adducts obtained from $[C_5R_5Fe^{+}Arene]PF_6$ visible irradiations in the presence of nitrosodurene. [a]

Spin adducts	a_N(G)	$a_{\beta H}$(G)	signal
1^b, H-NOAr	13.5	4.5	quadruplet (1:2:2:1)
2^c, C_5H_5-NOAr	14.0	14.0	sextuplet (1:1:1:1:1:1)
3^d, C_5Me_5-NOAr	14.2	—	triplet (1:1:1)
4^e, $[C_6Me_6\text{-}NOAr]^{+\bullet}$	13.0	—	triplet (1:1:1)

a: 1 mM of $[C_5R_5Fe^{+}Arene]PF_6$ and 5 mM of nitrosodurene, (ArNO), (Arene : naphthalene, benzene, toluene, *p*-xylene, durene, PMB, HMB, and C_6Et_6); λ_{irrad} = 441 nm (Oriel Filter); argon atmospher; **b**: the aminoxyl radical adduct is observed only under irradiation; in CD_2Cl_2 no change of this signal is observed; **c**: observed during and after irradiation. **d, e**: stable signals during and after irradiation.

Fig. 1. e.s.r. spectra on photolysis of $[C_5R_5FeArene]PF_6$ complexes in CH_2Cl_2 solution in the presence of nitrosodurene.

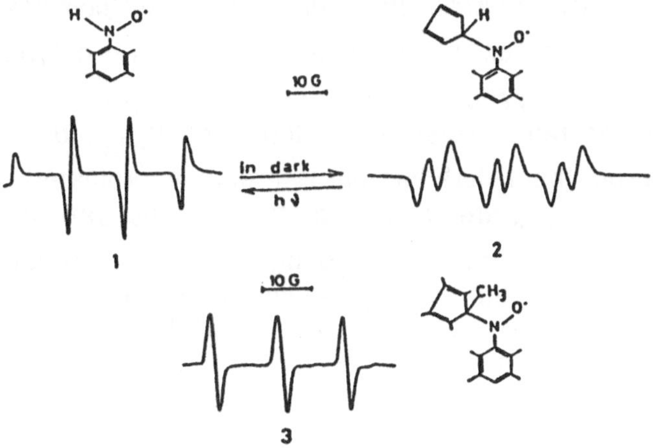

2. Methylviologen, MV^{2+}, irreversibly quenches the excited states of $[CpFe^{+}Arene]^{*}$ (arene=p-xylene, p-dimethoxybenzene, hexamethylbenzene and hexaethylbenzene).

Irradiation of water or methanol solutions of $[CpFe^{+}Arene]X$, (X = Cl, BF_4), ($3x10^{-3}M$) and $MVCl_2$ ($1.5x10^{-3}M$) at 380-400 nm under argon atmosphere at room temperature induces an electron transfer from the excited state of the iron (II)-complexe to MV^{2+} giving $MV^{+\bullet}$ radical which persists for some days in this system (eq.5). $MV^{+\bullet}$ radical was observed in e.s.r. and by its visible spectrum.

$$[\eta-C_5R_5Fe^{+}Arene]^{*} + MV^{2+} \xrightarrow{h\nu} MV^{+\bullet} + FeS_n{}^{2+} + CpR \qquad (5)$$

S: solvent
R: H, Cp
λ_{irr} : 400nm

The generation of $MV^{+\bullet}$ is a linear function of the irradiation time; $[MV^{+\bullet}]$ was measured at 605 nm by using its molar absorption ($1.37X10^{4}$ dm l^{-1}), (Fig.2). After the initial induction period which is atributed to the scavenging of residual dioxygen by $MV^{+\bullet}$, the absorption of the radical $MV^{+\bullet}$ increases linearly with time. From a knowledge of the slope of the plot a value of k_{obs} for the formation of $MV^{+\bullet}$ can be calculated for each $[CpFe^{+}Arene]BF_4$ electron donor complex, (Table 2). Quantum yields of formation of $MV^{+\bullet}$, ($\Phi_{MV^{+\bullet}}$), were determinated as a function of the arene ligands coordinated to the $[CpFe^{+}]$ moiety. The photoelectron transfer rate, expressed in terms of k_{obs}, does not show a clear dependence on the nature of the π-arene coordinated ligand. The lowest and highest k_{obs} are observed for the hexaethylbenzene and *p*-dimethoxybenzene derivatives respectively. This results can be explained in terms of (i) the varying π-electron density present in the arene ligands and (ii) a steric hindrance caused by the six ethyl groups of the π-hexaethylbenzene derivative, which prevents effective collision between MV^{2+} and the $[CpFe^{+}C_6Et_6]^{*}$ complexe and then the formation of the encounter excited complexe. This argument is

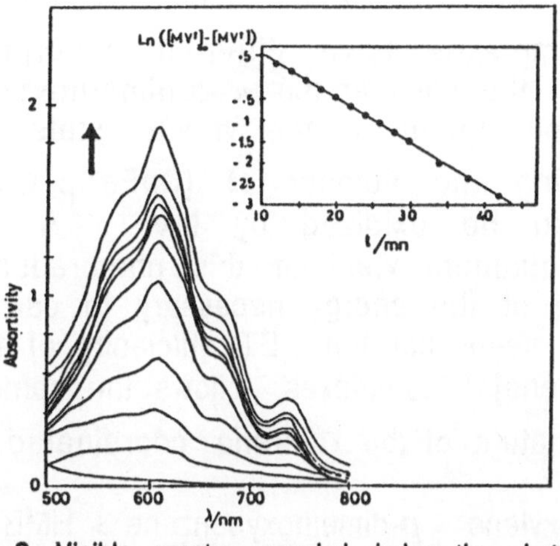

Fig. 2. Visible spectra recorded during the photolysis of $[CpFe^{+}p\text{-xylene}]BF_4$, (2 mM), and $MVCl_2$ (1.5 mM) in CH_3OH solution under argon atmosphere at room temperature.

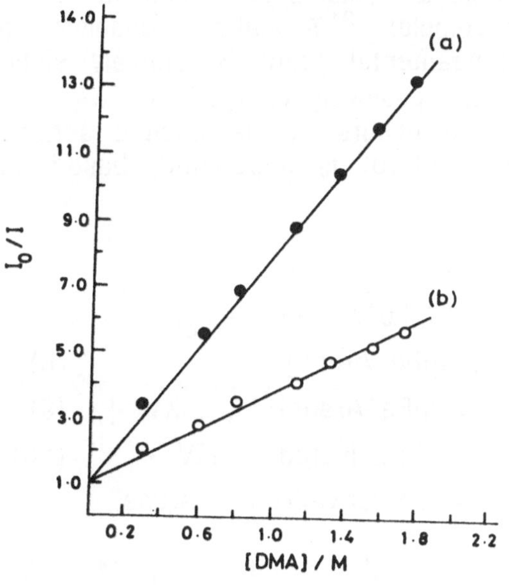

Fig. 3. Stern-Volmer plots for the quenching of the emission of (*a*) $[CpFe^{+}, HMB]PF_6$, (•), $k_Q\tau = 7M^{-1}$, and (*b*) $[CpFe^{+}p\text{-xylene}]PF_6$, (O), $k_Q\tau = 2.4\ M^{-1}$ both complexes in dichloromethane using dimethylaniline, DMA (mol l^{-1}) as quencher. Excitation wavelength: 370 nm; room temperature.

weakened by the fact that at the monochromatic λ_{irr}= 390nm, the [CpFe$^+$Arene] complexes have different absorptions. The greatest absorption occurs in the case of the π-p-dimethoxybenzene derivative, **i.e.** the excited population of the $a\ ^3E_1$ state is greatest in this derivative, increasing the number of [CpFe$^+$p-dimethoxybenzene]* molecules that can be oxidized by MV^{2+}. A clearer comparative parameter is the quantum yield of this photoreduction of MV^{2+}. This gives an estimation of the energy necessary to carry out the ET. From table **2**, it can be seen that the ET efficiency of the photoreduction of MV^{2+} by [CpFe$^+$Arene] * complexes, follows the same order as $\Phi_{MV^{+\bullet}}$, with regard to the nature of the π-arene coordinated ligand:

p-xylene > *p*-dimethoxybenzene > HMB > C_6Et_6

Stern-Volmer relation, [20] linear plots of $(\Phi_{MV^{+\bullet}})^{-1}$ **v/s** $[MV^{+\bullet}]^{-1}$ for this photoreduction of MV^{2+} by [CpFe$^+$Arene]* complexes, are deduced from the results shown in Table **2.** This is in agreement with a dynamic mechanism via an encounter complex. [21] A static mechanism via a charge transfer complex in the fundamental state is unlikely since both reagent MV^{2+} and [CpFe$^+$Arene] are positively charged and we have already shown that this type of reaction occurs in the excited surface state.[8b] A mechanism described in eqs. (6-10) is suggested, based on photokinetic results.[20]

[CpFe$^+$Arene]+ $h\nu$	⟶	[CpFe$^+$Arene]*	(6)
[CpFe$^+$Arene]*	⟶	[CpFe$^+$Arene] + $h\nu_{Fl}$	(7)
[CpFe$^+$Arene]*	⟶	[CpFe$^+$Arene]	(8)
[CpFe$^+$Arene]* + MV^{2+}	⟶	{[CpFe$^+$Arene]$^{2+}$......$MV^{+\bullet}$}	(9)
{[CpFe$^+$Arene]$^{2+}$......$MV^{+\bullet}$}	⟶	Fe(II)-solvated + $MV^{+\bullet}$ + Cp-derivatives + Arene	(10)

This result probes that the photochemical reduction of MV^{2+} by [CpFe$^+$Arene]* complexes is an efficient and fast process. In this case, the organometallic iron (II)-cations studied, have a sacrificial electron donor behaviour in the excited surface. They can eventually compete with the classical sacrificial electron donors such as EDTA and TEOA which are employed in sensitized photoreduction of water and in solar energy conversion studies. [22]

Table 2. Rate constants and quantum yields for MV^{2+} photoreduction by $[CpFe^{+}Arene]^{*}$ complexes. [a]

Arene	k_Q $mol^{-1}min^{-1}$	$\Phi_{MV^{+\cdot}}$
p-xylene	0.285	0.299
p-dimethoxybenzene	0.311	0.243
HMB	0.209	0.115
C_6Et_6	0.121	0.052

[a] $[CpFe^{+}$ Arene$]BF_4$ in methanol, ($2x10^{-3}$M), and $MVCl_2$ ($1.38x10^{-2}$M); 20°C, λ_{irrad} = 410 nm (Oriel filter).

3. [CpFe+Arene]* is quenched by an electron donor such as dimethylaniline, DMA.

Quenching experiments for the $[CpFe^{+}Arene]PF_6$ fluorescence by DMA in dichloromethane were carried out. No shape change was observed in the fluorescence spectra but their intensity changed upon addition of DMA. The Stern-Volmer relationship was obtained between the ratio of the fluorescence intensity in the absence and presence of the quencher, Io/I, and [DMA] concentration, as expressed by the equation (11).

$$Io/I = 1 + k_Q \tau [DMA] \qquad (11)$$

where k_Q is the quenching rate constant and τ is the fluorescence lifetime of $[CpFe^{+}Arene]^{*}$, (Fig 3). The fact that the plots are linear up to Io/I values as high as 13 indicates that the luminescence observed is being emitted from a single excited state, $^1E_1$. Furthermore, the $k_Q \tau$ = 2.4 M^{-1}, for $[CpFe^{+}$*p*-xylene], and $k_Q \tau$ = 7.0 M^{-1} for $[CpFe^{+}HMB]$, are in accordance with their Φ value of 0.9 x 10^{-3} and 4.6 x 10^{-3} respectively. The singlet lifetimes in these cases have already been determined by our group, [8b] being τ > 7 nsec. These short lifetimes are also supported by the fact that the observed luminescence is nearly independent of oxygen. We suggest that a 19 electron excited complex $[CpFe^{\cdot}Arene]^{*}$ is the result of an ET from DMA to the $[CpFe^{+}Arene]^{*}$ complex in the excited surface; the ET being a primary photoprocess, (eq. 12), ($k_T = 3 \times 10^8\ M^{-1} \times sec^{-1}$).

$$[CpFe^{+}Arene]^{*} + DMA \longrightarrow \underset{19e}{[CpFe\cdot\ Arene]^{*}} + DMA^{+\cdot} \qquad (12)$$

4. Photocatalytic regioisomerization of hexamethyl (Dewar) benzene by [CpFe⁺Arene]* complexes.

We have recently shown [8a] that HMDB under mild conditions, (273 K and solar irradiation), can be photoactivated in the presence of catalytical amounts of $[CpFe^{+}Arene]$ complexes, (eq. 13).

(13)

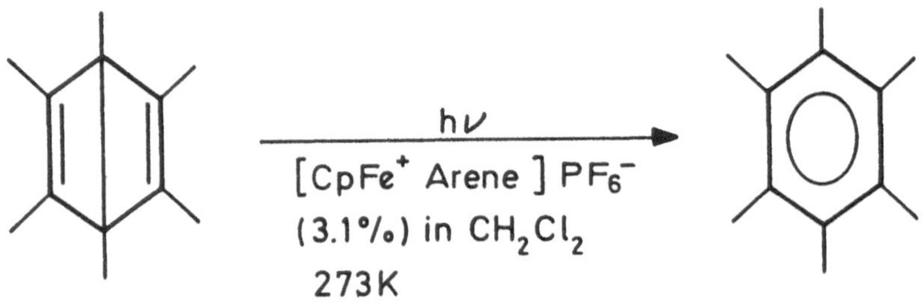

This photo-assisted regioisomerization of HMDB in hexamethylbenzene (HMB) was followed by e.s.r. and n.m.r. spectroscopy. Firstly, in e.s.r., visible irradiation (LF, λ_{irr}= 450 nm) of HMDB (0.16M) in dichloromethane in the presence of catalytical amounts of $[CpFe^{+}benzene]PF_6$, (0.006 M) and in the presence of the air stable free radical 2,2,6,6-tetramethyl piperidyl N-oxyl, (TEMPO), was carried out. A total inhibition of the regioisomerization of HMDB was observed. The TEMPO-signal (triplet a_N =13.0G) decayed in 10 min during time irradiation (Fig. 4). In the same way nitrosodurene inhibits this photoisomerization, an e.s.r. spectrum showing the radical adduct **1** and a new spin-trapped adduct **4**, (triplet 1:1:1), is observed; this adduct probably being a classical cyclohexadienyl cation derivative, $[ArNO^{\bullet}-HMB]^{+}$. When a CD_2Cl_2 solution of HMDB (0.16 M) and $[CpFe^{+}benzene]PF_6$ (0.006M) in the presence of TEMPO is irradiated (λ_{irr}= 410 nm), (Fig. 5),

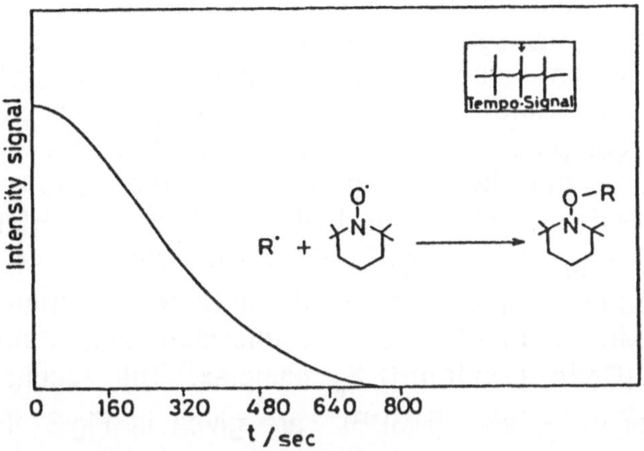

Fig. 4. Total and rapid e.s.r. signal decay from the TEMPO radical, during photolysis of HMDB (0.16 M) and catalytical amounts of $[CpFe^{+}benzene]PF_6$ (0.006 M) in CH_2Cl_2 solution. (R˙ : $HMDB^{+}$, HMB^{+}, Cp˙).

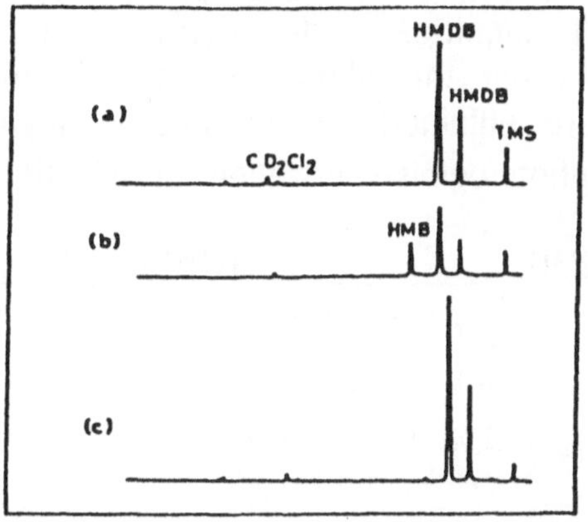

Fig. 5. ^{1}H-n.m.r. monitoring during photolysis of HMDB (0.2 M), and catalytical amounts of $[CpFe^{+}benzene]PF_6$, (0.006 M), (*a*) before irradiation; (*b*) after 20 min of irradiation, and (*c*) after 1h of irradiation in the presence of the radical inhibitor TEMPO (0.006 M).

and the solution is monitored by ^{1}H-n.m.r spectroscopy, inhibition of photoregioisomerization is again observed. The n.m.r and e.s.r. results suggest that a chain reaction is involved in the photocatalytic regioisomerization of HMDB induced by the excited state of the $[CpFe^{+}Arene]^{*}$ complexes. Photophysical studies on this system, previously carried out by our group, [8b] have shown that the $[CpFe^{+}Arene]^{*}$ emission is photoquenched by addition of HMDB or HMB. The emission yield Φ_{Fl} of a given $[CpFe^{+}Arene]^{*}$ complex is sensitive to the presence of other arenes. A detailed study was carried out on the effect of added HMB or HMDB upon the emission yield obtained for the irradiation of the $[CpFe^{+}p\text{-xylene}]PF_6$ complex. The results, plotted as Φ_{Fl}^{-1} v/s $[HMB]^{-1}$ or Φ^{-1}_{Fl} v/s $[HMDB]^{-1}$ are given in Fig.6. The emission at $[HMB]_\infty$ or $[HMDB]_\infty$ is 8 x 10^{-4}, a value very close to that for the $[CpFe^{+}HMB]$ complex directly irradiated. Furthermore, the increase in emission takes place without noticeable changes in the spectral distribution. The fact that at very high HMB concentrations the observed yields approach those obtained in direct irradiation of $[CpFe^{+}HMB]$, indicates that ligand exchange is another photoefficient reaction that takes places without significant deactivation of the excited mixed metallocene complex. From the slope and the intercept of the plot in Fig.6, values of $k_Q \tau$ are obtained for the photocatalytical ligand exchange (eq. 14) and for the photoregioisomerization of HMDB (eq. 15).

$$[CpFe^{+}p\text{-xylene}]^{*} + HMB \longrightarrow [CpFe^{+}HMB]^{*} + p\text{-xylene} \qquad (14)$$

$$k_{14}\tau = 1.3\ M^{-1}$$
$$\tau < 7\ nsec$$
$$k_{14} > 2 \times 10^{8}\ M^{-1}\ sec^{-1}$$

$$[CpFe^{+}p\text{-xylene}]^{*} + HMDB \longrightarrow [CpFe^{+}HMB]^{*} + p\text{-xylene} \qquad (15)$$

$$k_{15}\tau = 0.4\ M^{-1}$$
$$\tau < 7\ nsec$$
$$k_{15} > 6 \times 10^{7}\ M^{-1}\ sec^{-1}$$

The latter reaction (eq. 15), although highly efficient, is thus considerably slower than the reaction in eq. 14, which involves exchange without regioisomerization. This results provides, moreover, a photochemical path for the valence isomerization of HMDB, probably via an exciplex species represented by the following scheme:

$$[CpFe^{+}Arene]^{*} + HMDB \longrightarrow \underset{Ex_{I}}{[CpFe^{+}Arene \ldots\ldots HMDB]^{*}} \qquad (16)$$

$$E_{xI} \longrightarrow \underset{Ex_{II}}{[CpFe^{+}Arene \ldots\ldots HMB]^{*}} \qquad (17)$$

$$E_{x\,II} \longrightarrow \underset{Ex_{III}}{[CpFe^{+}HMB \ldots\ldots Arene]^{*}} \qquad (18)$$

$$E_{x\,III} \longrightarrow [CpFe^{+}HMB]^{*} + Arene \qquad (19)$$

where process (16) represents the reversible formation of a $[CpFe^{+}Arene \ldots HMDB]^{*}$ exciplex, and process (19) corresponds to the dissociation of an $[CpFe^{+}Arene \ldots HMB]^{*}$ exciplex. Formation of HMDB and HMB exciplexes have been reported both from excited singlet and triplet states. [23] Regioisomerization of HMDB exciplexes has also been reported. [24] In general, very high yields of adiabatic isomerizations in excited singlet complexes, formed between excited electron acceptors and HMDB have been reported. [25] Thus the ocurrence of reaction (17) provides a photochemical path for the valence isomerization of HMDB. However, the e.s.r. and n.m.r. measurements for inhibition of the radical isomerization of HMDB by radical trapping reagents suggest a radical path for this process. Moreover, quantum yields for this photochemical regioisomerization show considerably larger values than the unity, (Table 3), this being indicative of a chain process. It is clear that a primary photoprocess of ET is necessary for the radical initiation (eq. 20).

$$[CpFe^{+}Arene]^{*} + HMDB \xrightarrow{ET} \underset{19\ e}{[CpFe^{\bullet}Arene]^{*}} + HMDB^{+} \qquad (20)$$

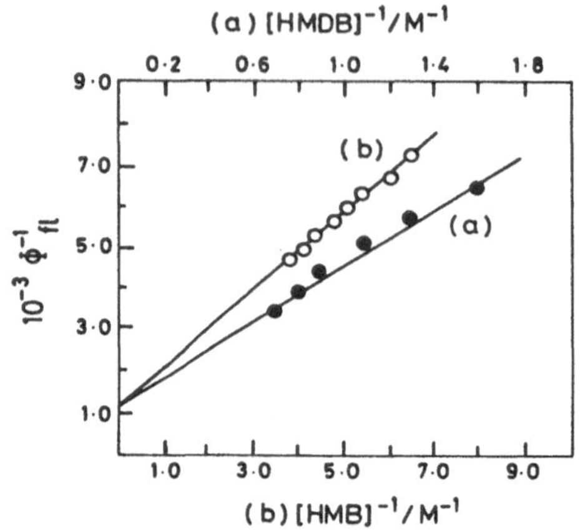

Fig. 6. The dependence of fluorescence quantum yield on addition of: (*a*) hexamethyl(Dewar)benzene, HMDB, (•) or (*b*) hexamethylbenzene, HMB, (O), into a dichloromethane solution of [CpFe$^+$*p*-xylene]PF_6, (1 x 10^{-2}M.) $\Phi_{El\infty}$ =8 x 10^{-4} determined from the same intercept.

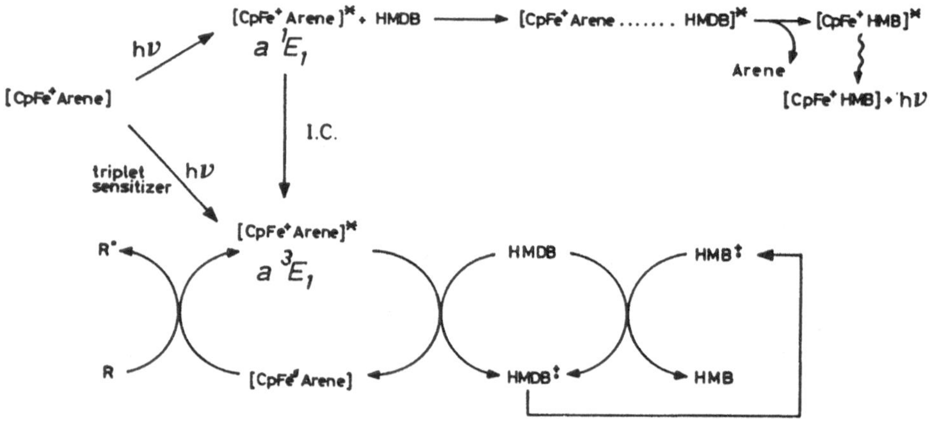

<u>Scheme I</u> : (I.C. intersystem crossing; R = Cp$^\cdot$, H^+, $HMB^{+\cdot}$, $HMDN^{+\cdot}$).

Table 3. Quantum yields for photoinduced regioisomerization of HMDB by [CpFe⁺Arene]* complexes. [a]

Arene	Φiso
Benzene	58
p-xylene	22
HMB	-1.3[b]

[a] λ_{irrad} = 441 nm ; 273K ; 0.16 M of HMDB in CH_2Cl_2 solution in the presence of 0.006 M of $[CpFeArene]PF_6$ complexes during 5 min of irradiation. [b] Estimated at long time of irradiation.

In order to obtain more precise information about the excited state of the [CpFe⁺Arene] complexes implicated in all of these photoelectron transfer studies, we carried out a study via the triplet excited state $a\,^3E_1$ of the [CpFe⁺Arene] complexes. This state corresponds to a very low intensity absorption,[16] (λ_{max} = 650 nm, ε = 1.5 ; $^1A \longrightarrow a^3E_1{}^*$), and it can be achieved by a triplet sensitizer such as anthracene, (E_T = 14.700 cm^{-1} of anthracene, very near of the E_T = 15.400 cm^{-1} corresponding to the $^1A \longrightarrow a\,^3E_1{}^*$ electron transition). So by irradiation of [CpFe⁺Arene] complexes in dichloromethane in the presence of nitrosodurene , new radicals are observed. The nature of these radicals will discussed in an other paper.[26] If the photoregioisomerization of HMDB, assisted by [CpFe⁺Arene] complexes, is carried out in the presence of the triplet sensitizer, anthracene, a more efficient photocatalysis is observed. This efficiency can be demonstrated by measurements of the half lifetime of HMDB in this photochemical systems by n.m.r. (Fig. 8). The triplet sensitized reactions is also inhibited by TEMPO. In the experiments with anthracene as a sensitizer, a dramatic acceleration of the kinetic photoelectron transfer via the excited electronic $a\,^3E_1{}^*$ state of the [CpFe⁺ Arene] complexes has been observed. The results can be explained by the Scheme I.

CONCLUSIONS

The study of photochemical ET reactions promoted by an excited organometallic species as [CpFe^{+}Arene]* show that these species have an electron acceptor or electron donor behaviour. 17-electron and 19-electron organometallic transients can be obtained by ET good electron acceptor such as MV^{2+} (A/A$^{\cdot}$= -0.45 Volts, SCE), [11] and from good electron donors such as DMA, (D^{+}/D = + 0.81 Volts, SCE),[11] respectively. The photochemical reactions described in this study could throw light on many photoinduced reactions in the fields of organic and organometallic chemistry.

Acknowledgements

We thank the DIUC-Catholic University of Chile for sponsoring this work and Professor E. Lissi (University of Santiago of Chile) for helpful collaboration, and CNRS for a fellowship to M.B. at ICSN Gif-sur-Yvette, France.

REFERENCES

1 (a) R.G. Sutherland, M. Igbal, A. Piorko, *J . Organomet. Chem,* 309 (1986) 307; (b) D. Astruc Tetrahedron Report n° 157, *Tetrahedron*, 39 (1983) 4027 ; (c) D. Astruc, *Comments Inorg. Chem.*, 6 (1987), 61 ; (d) D. Astruc , *Acc. Chem. Res,* 19 (1986) 377.

2 (a)D. Astruc, E. Roman, J.R. Hamon, P. Batail, *J. Am. Chem. Soc.* 101 (1976) 2240; (b) D. Astruc, J.R. Hamon, E. Roman *ibid*, 103 (1981) 7502; (c) C. Moinet, E. Roman, D. Astruc, *J. Electroanal. Chem.* 241 (1981) 121.

3 (a) M. Lacoste, *Thèse de Doctorat de l'Université de Bordeaux I* (1987) (b) D. Astruc *et al.*, *J. Am. Chem. Soc.* (in press).

4 A. Darchen, *J. Chem. Soc. Chem. Commun.* (1983) 768

5 (a) E. Roman, R. Dabard, C. Moinet, D. Astruc, *Tetrahedron Lett.*, 16 (1979) 1433; (b) V. Guerchais, E. Roman, D. Astruc *Organometallics* , 5 (1986) 2505. (c) V. Guerchais, D. Astruc, *J. Chem. Soc, Chem. Commun,* (1984), 881.

6 A. Buhet, A. Darchen, C. Moinet, *J. Chem. Soc. Chem. Commun.* (1979), 447.

7 (a) T.P. Gill, K.R. Mann, *Inorg. Chem.,* 22 (1983) 1986; (b) A.M. Mc Nair, J.L. Schrenk, K.R. Mann, *Inorg. Chem,* 23(1984)2633; (c) D. Catheline, D. Astruc, *Organometallics*, (d) E. Roman, M. Gonzales, manuscript in preparation.

8 (a) E. Roman, S. Hernandez, M. Barrera, *J.C.S. Chem Commun*, (1984) 1067; (b) E. Roman, S. Hernandez, M. Barrera, E. Lissi *J.C.S. Perkin II*,

(in press).
9 (a) A. J. Lees, *Chem. Rev.* 87 (1987), 711 ; (b) M. Wrighton, L. Pdungsap; D.L. Morse, *J. Phys. Chem.*, 79 (1975) 66.
10 (a) M. Chanon *Chem. Rev,* 83 (1983) 425; (b) M. Chanon *Bull. Soc. Chim. Fr.*, (1985) 209.
11 G.J. Kavarnos, N. J. Turro, *Chem. Rev.,* 86 (1986) 401
12 J.K. Kochi "*Organometallic Mechanisms and Catalysis*", Academic Press, New York, (1978).
13 D. Astruc, *Angew. Chem. Int. Ed. Engl.,* 27 (1988) 643.
14 A. Nesmeyanov, N. Vol'kenau, I. Boleslova, *Dokl. Akad. Nauk SSRR* 149 (1963) 615.
15 (a) P. Maillard and C. Giannotti, *Can. J. Chem,* 60 (1982) 1402, and references there in ; (b) J. Delaive, T.K. Foreman, C. Giannotti and P. Whitten, *J. Amer. Chem. Soc. 102,* (1980), 5627.
16 (a) Y.S. Sohn, D. Hendrickson, H. Gray, *J. Am. Chem. Soc.* 93 (1971) 3603; (b) K. D. Warren, *Inorg. Chem.* 13 (1974) 1243; (c) J.C. Green, M.R. Kelly, M.P. Payne, E.A. Seddon, D. Astruc, J.R. Hamon, P. Michaud, *Organometallics*, 2 (1983) 211.
17 A.G. Davies, J. Lusztyk, *J.C.S. Perkin II,* (1981) 692
18 (a) E. Samuel, P. Maillard, C. Giannotti, *J. Organometal. Chem.* 142 (1977) 289 ; (d) Z. Tsai, C.H. Brubaker Jr. *ibid*, 166 (1979) 199 ; (c) P.B. Brindley, A.G. Davies, J. A. Hawari, *ibid,* 250 (1983) 247.
19 (a) P.J. Barker, A.G. Davies, M.W. TSe , J.C.S., *Perkin Trans. II,* (1980) 941; (b) P.J. Barker, S.R. Stobart and P.R. West. *ibid*, (1986), 127.
20 E. Roman, E. Lissi, G. Ferraudi *et. al.*, photonanosecond study in $[CpFe^{+}Arene]^{*}$ complexes are under study. Manuscript in preparation.
21 (a) B.T. Ahn ; D.R. Mc Millin; *Inorg. Chem.* 20 (1981) 1427. (b) *ibid,* 17 (1978), 2253; (c) R.E. Gamache, D.R. Mc Millin, *J. Am. Chem. Soc.,* 107, (1985), 1141.
22 (a) K. Kalyanasundaram, *Coord. Chem. Rev.* 159 (1982) 244; (b) K. Kalyanasundaram, J. Kiwi, M. Gratzel, *Helv. Chimica Acta* , 61(1978), 2720.
23 N.C. Peacok, G.B. Schuster, *J. Am. Chem. Soc.,* 105 (1983) 3632.
24 G. Jones II, S.H. Chaing, *J. Am. Chem. Soc.,* 101 (1979) 7421
25 G. Jones II, W.G. Becker, *J. Am. Chem. Soc.* 105 (1983) 1276.
26 E. Roman, Ch. Giannotti, M. Barrera, S. Hernandez, manuscript in preparation

PHOTOINDUCED RADICAL CHAIN BROMINATION OF DECARBONYLDIMANGANESE

Nancy M. J. Brodie and Anthony J. Poë
Erindale College and the Chemistry Department,
University of Toronto, Mississauga, Ontario,
Canada L5L 1C6

ABSTRACT. The reaction of bromine with $Mn_2(CO)_{10}$ in tetrachloro methane proceeds quite rapidly and cleanly on exposure to light to form $Mn(CO)_5Br$ as the only IR detectable product. The kinetics of this reaction have been studied under a wide variety of conditions, reaction being initiated by photodissociation of bromine with light of wavelength ca. 435nm. It is concluded that the bromine atoms initiate a chain reaction by attacking the $Mn_2(CO)_{10}$ and forming $Mn(CO)_5Br$ and an $Mn(CO)_5$ radical. This in turn reacts with bromine to generate $Mn(CO)_5Br$ and another bromine atom to continue the chain. The absence of appropriate effects on varying the intensity of absorbed light shows that chain termination does not involve radical recombination. It is probably effected by reaction of bromine atoms and $Mn(CO)_5$ radicals with the solvent or, more effectively, with $CBrCl_3$ when added in small amounts. Reaction in perfluorodecalin proceeds by a similar path but with higher quantum yields. Only bromine atoms, and not $Mn(CO)_5$ radicals, appear to be involved in the termination step.

1. INTRODUCTION

Kinetic studies of reactions of the simple metal-metal bonded carbonyl [1] $Mn_2(CO)_{10}$ with halogens were initiated over 20 years ago [2] in an attempt to see whether the rate determining step was homolysis of the Mn-Mn bond to form $Mn(CO)_5$ radicals which would then be rapidly scavenged by the halogen to form $Mn(CO)_5X$. Were it to operate, this mechanism would enable one to obtain activation enthalpies for homolysis of the Mn-Mn bond and hence a rather precise measure of its "kinetic strength". Since recombination of the $Mn(CO)_5$ radicals was expected, and has since been shown [3], to be a low energy process the kinetic strength would be only a slight overestimate of the thermodynamic enthalpy change in the homolysis equilibrium. The only attempt to

M. Chanon et al. (eds.), Paramagnetic Organometallic Species in Activation / Selectivity, Catalysis, 345–356.

measure this enthalpy by thermochemical means had led [4] to a value of $\Delta H° = 142$ kJ mol^{-1} with an uncertainty of at least 54 kJ mol^{-1} and the kinetic method would have led to a much more precise value than that.

Reaction with iodine was found to proceed in part by a path independent of $[I_2]$ and the first order rate constants coincided with limiting values found for reaction with oxygen as a radical scavenger [5]. This, together with other more complex features of the kinetics [6], led to the belief that rate determining homolysis was in fact occurring and that the kinetic strength of the Mn-Mn bond was 154 ± 2 kJ mol^{-1}. Similar, but less extensive, kinetic results for reactions of $Re_2(CO)_{10}$ [7], $MnRe(CO)_{10}$ [6] and $Tc_2(CO)_{10}$ [8] were also thought to indicate rate determining homolysis of the metal-metal bonds in these carbonyls. However, crucial studies of the implied scrambling reaction between $Mn_2(CO)_{10}$ and $Re_2(CO)_{10}$ [9], or between different isotopically labelled forms of $Re_2(CO)_{10}$ [10] or $Mn_2(CO)_{10}$ [11], showed that scrambling did not occur as it should if homolysis were occurring, and homolysis cannot, therefore, be the rate determining step in reactions of these carbonyls. It has, nevertheless, been shown to be involved in reactions of a number of octacarbonyldimanganese complexes that contain a series of P-donor substitutents in the axial positions [12,13]. Evidence included observation of relevant scrambling reactions. The activation enthalpies for homolysis showed [14] that the strength of the Mn-Mn bond is greatly affected by the size of the axial substitutents as measured by their Tolman cone angles, θ. When θ is less than ca. 125° the Mn-Mn bond is sufficiently strong that a mechanism other than homolysis is favoured. A steric effect has also been shown to retard the rates of recombination and other reactions of $Mn(CO)_4L$ and $Mn(CO)_3L_2$ radicals [15-17].

Studies of reactions of $Re_2(CO)_8\{P(C_6H_{11})_3\}_2$ demonstrate that even the very large $P(C_6H_{11})_3$ substituents ($\theta = 170°$) are not capable of weakening the Re-Re bond sufficiently for $Re(CO)_4\{P(C_6H_{11})_3\}$ radicals to be formed by thermal homolysis, and ligand dissociative rate determining steps are preferred [18]. This contrasts with $Mn_2(CO)_8\{(P(C_6H_{11})_3\}_2$ for which the activation enthalpy for homolysis is only 99 kJ mol^{-1} [12,19] compared with > 154 kJ mol^{-1} for $Mn_2(CO)_{10}$ [20].

Studies of the kinetics of $Mn_2(CO)_8(PPh_3)_2$ [13,21] have shown that another reaction path is available apart from spontaneous homolysis. The data led to the proposal of a rather unusual process in which reversible isomerization of the complex led to a reactive form that was capable of undergoing fragmentation to radicals when attacked by a variety of reactants. Studies of the substitution reaction with $P(OPh)_3$ [21] led to the firm conclusion that the $Mn(CO)_4(PPh_3)$ radicals formed by homolysis underwent a rapid bimolecular substitution reaction with the $P(OPh)_3$. Such associative substitution reactions of metal centered radicals were not then believed to be likely, dissociative processes being favoured [22]. However, it was later demonstrated by product analysis of the competition reactions shown in eq.(1)-(3) (L = PPh_3 or $P\text{-}n\text{-}Bu_3$) that reaction (2) proceeds via an associative path [23], the $Re(CO)_5$ radicals being

$$Re(CO)_5 \xrightarrow{CCl_4} Re(CO)_5Cl \quad (1)$$

$$Re(CO)_5 + L \longrightarrow Re(CO)_4L + CO \quad (2)$$

$$Re(CO)_4L \xrightarrow{CCl_4} Re(CO)_4LCl \quad (3)$$

generated by photolysis of $Re_2(CO)_{10}$. A similar approach to reactions of $Mn(CO)_5$ showed that it too undergoes associative substitution [24] as do other 17-electron metal centered radicals such as $Mn(CO)_3L_2$ [25], $V(CO)_6$ [26], and $CpW(CO)_3$ [27]. It has always been our view [21] that the ability to form 2-center 3-electron bonds of order ca. 0.5 between the incoming nucleophile and the metal would be sufficient to favour associative rather than dissociative substitution.

The studies of the reactions of $Mn_2(CO)_8(PPh_3)_2$ also led [13], via some rate constant ratios, to a series of rate constants for bimolecular reactions of the $Mn(CO)_4(PPh_3)$ radical. These varied very little with the nature of the reactant, substitution reactions with $P(OPh)_3$, $P(OMe)_3$, and CO all proceeding with second order rate constants of (6-8) x 10^2 $M^{-1}s^{-1}$ at 49.9°C while reactions with $C_2H_4Cl_2$ and $C_{16}H_{33}I$ have rate constants of ca. 3 x 10^2 and 3.6 x 10^3 $M^{-1}s^{-1}$, respectively, at that temperature. These values remain to be confirmed by absolute rate measurements on radicals generated by flash photolysis but they do suggest a very low degree of substitutional selectivity for this 5-coordinate 17-electron metal centered radical. An approximate value of 53 kJ mol^{-1} was found for $\Delta H_2^{\ddagger}$ for substitution by $P(OPh)_3$, a value comparable to those for associative substitution of $P(OPh)_3$ into the 18-electron carbonyls $Co(CO)_3(NO)$ [28] and $Ru_3(CO)_{12}$ [29] for which other facile means of bond making are available [28,30]. The reasonably substantial difference in rates of what is stoichiometrically Cl and I transfer to $Mn(CO)_4(PPh_3)$ from $C_2H_4Cl_2$ and $C_{16}H_{11}I$ is in accord with an important contribution from C-X bond breaking in the reaction but the intimate mechanism may not be as simple as atom transfer.

Reverting to the thermal reaction of halogens with $Mn_2(CO)_{10}$, it was found [2] that the reaction of Br_2 proceeded fairly cleanly in CCl_4 to form $Mn(CO)_5Br$ at rates much faster than those of I_2 and much slower than those of

Cl_2. The kinetics of the reaction with Br_2 showed all the features of a radical chain reaction, being strongly retarded by hydroquinone, and accelerated by I_2, benzoyl peroxide, and light. Reaction in cyclohexane was accompanied by formation of large amounts of HBr. The kinetics of reactions with Br_2 alone were very irreproducible but those in the presence of added benzoyl peroxide were quite well behaved. A complex rate equation was obtained in which the observed rate was first order in $[Mn_2(CO)_{10}]$, $[Mn(CO)_5Br]$, and [benzoyl peroxide], and was of 3/2 order in Br_2. No mechanism corresponding to this behaviour has yet been suggested.

The rates when the reacting solutions of Br_2 and $Mn_2(CO)_{10}$ were exposed to room light were also quite reproducible and a photoinduced radical chain reaction appeared to be occurring. Radical chain reactions involving organometallic complexes have been receiving increasing attention recently (particularly with respect to their application in synthetic chemistry) along with related electron transfer reactions [31-39]. Rather few detailed kinetic studies have been made on such reactions. Since $Mn_2(CO)_{10}$ is the archetype of unbridged metal-metal bonded carbonyls [1] and still the subject of considerable interest [40], we were led recently to investigate the photokinetics of its reaction with bromine since this seemed likely [2] to provide reproducible and significant data. The results of these studies are reported here.

2. EXPERIMENTAL

The complex $Mn_2(CO)_{10}$ (Strem Chemicals) and bromine (BDH, Analar) were used as received. Spectroscopic grade tetrachloro methane, bromo-trichloro methane, and perfluorodecalin were obtained from BDH, Aldrich, and PCR Research Chemicals, respectively, and dried over molecular sieves.

Reactant solutions were prepared freshly each day and manipulated under red light. Reactions were carried out at ambient temperatures of ca. 24°C in 10mm path length silica cells irradiated with light from a projector lamp and passed through a Corning CS-5-774 narrow band pass filter (422-450nm with a maximum transmittance at ca. 435nm). Incident light intensities were varied by passing the light beam through neutral density filters (Kodak Wratten, No. 96). Rates were determined by monitoring the decrease in absorbance of the band at 343nm due to the $Mn_2(CO)_{10}$. Actinometry and calculation of quantum yields were carried out exactly as described elsewhere [41].

3. RESULTS

The photoreaction of Br_2 with $Mn_2(CO)_{10}$ in CCl_4 leads only to $Mn(CO)_5Br$ as evidenced by th IR spectra. Quantum yields in CCl_4 were found to depend significantly on the purity of the solvent, values of Φ(obsd) increasing with increasing purity. The following results were therefore all obtained by using spectroscopic grade CCl_4.

3.1. Concentration effects

At constant $[Mn_2(CO)_{10}]$ the values of Φ(obsd) increase rapidly with $[Br_2]$ and reach a limiting value Φ(lim)(1) when $[Br_2] > 10^{-3}$M. The values of Φ(lim)(1) depend linearly on $[Mn_2(CO)_{10}]$. An average of 36 measurements of $\Phi(\text{obsd})/[Mn_2(CO)_{10}] = \Phi(\text{lim})(1)/[Mn_2(CO)_{10}]$ for $[Br_2] > 10^{-3}$M and $10^5[Mn_2(CO)_{10}] = 1.85$ to 4.18M is $(1.74 \pm 0.04) \times 10^5 M^{-1}$ with a probable error of ± 13% in the determination of each value of Φ(lim)(1).

At constant $[Br_2]$, values of Φ(obsd) increase rather slowly with $[Mn_2(CO)_{10}]$ to a limiting value Φ(lim)(2). Inverse plots of $1/\Phi$(obsd) vs $1/[Mn_2(CO)_{10}]$ are linear at each bromine concentration with intercepts and gradients as shown in Table I. The gradients decrease quickly with increasing $[Br_2]$ and reach a lower limit when $[Br_2] \geq 10^{-3}$M, the unweighted average of which is $(3.5 \pm 0.7) \times 10^{-6}$M. The intercepts are a measure of Φ(lim)(2) but are not very precisely defined by the data. However, if we assume that the gradients are, in fact, constant at 3.5×10^{-6}M when $[Br_2] > 10^{-3}$M then values of Φ(lim)(2) can be calculated from each pair of $1/\Phi$(obsd) and $1/[Mn_2(CO)_{10}]$ values by application of this gradient. The scatter of the values obtained in this way for Φ(lim)(2) at any given bromine concentration is rather large but, if the highest and the lowest values are omitted, a weighted linear least squares analysis of the dependence of Φ(lim)(2) on $[Br_2]$ leads to eq.(4) with a probable error of ± 19% for

$$\Phi(\text{lim})(2) = (7.5 \pm 0.5) + (7.7 \pm 1.2) \times 10^2[Br_2] \quad (4)$$

each estimation of Φ(lim)(2).

3.2. Intensity effects

The effect of varying the intensity, Ia, of the light absorbed by the reaction mixture was studied with $[Br_2] >$ ca. 1.5×10^{-3}M where $\Phi(\text{obsd}) = \Phi(\text{lim})(1)$ is linearly dependent on $[Mn_2(CO)_{10}]$. The gradients of the linear plots varied randomly between a minimum of $1.02 \times 10^5 M^{-1}$ to a maximum of $1.82 \times 10^5 M^{-1}$ over a range of 10^{-7}Ia from 0.3 to 2.33 E $l^{-1}s^{-1}$ and a range of $10^3[Br_2]$ from 1.5 to 12M. The average of 11 values of the gradient, each corresponding to a different value of Ia, was $(1.50 \pm 0.08) \times 10^5 M^{-1}$ with a probable error of ± 17% for each gradient. Incorporated in these runs were some in which oxygen was bubbled through the solutions before photolysis but no effect due to this was observed.

3.3. Effects of added $CBrCl_3$

The effect of added $CBrCl_3$ was studied with $[Br_2] = 1.7 \times 10^{-3}$M. Quantum yields were always decreased by the presence of $CBrCl_3$ and, at each $[CBrCl_3]$, there was a linear dependence of $1/\Phi$(obsd) on $1/[Mn_2(CO)_{10}]$ as in the absence of $CBrCl_3$. The intercepts, $\underline{a}$, and gradients, $\underline{b}$, of these inverse plots both increased with $[CBrCl_3]$ as shown in Table II. The gradients increase linearly with $[CBrCl_3]$ according to eq.(5) while the intercepts rise towards a limiting value, Φ(lim)(3), according to the relationship in eq.(6), provided

$$\underline{b} = (0.56 \pm 0.01) \times 10^{-5} + (5.1 \pm 0.8) \times 10^{-5}[CBrCl_3] \qquad (5)$$

$$\underline{a} = \underline{c}\Phi(lim)(3)[CBrCl_3]/(1 + \underline{c}[CBrCl_3]) \qquad (6)$$

$[CBrCl_3] >$ ca. 0.005M. An inverse plot of $1/\underline{a}$ vs $1/[CBrCl_3]$ is found to be linear according to eq.(7) and the probable errors in the estimates in $\underline{a}$ and $\underline{b}$ are ± 12% and ± 17%,

$$1/\underline{a} = (0.27 \pm 0.06) + (1.2 \pm 0.2) \times 10^{-2}/[CBrCl_3] \qquad (7)$$

respectively.

3.4. Photolysis in perfluorodecalin

Reactions were carried out in perfluorodecalin with $[Br_2] >$ ca. 2×10^{-3}M and the rates found to be considerably faster than those in CCl_4. A set of 22 values of Φ(obsd) at $[Br_2] = (3.0 \pm 0.4) \times 10^{-3}$M showed a linear dependence on $[Mn_2(CO)_{10}]$ over a concentration range $10^5[Mn_2(CO)_{10}] = 0.84$ to 8.22M. The average value of $\Phi(obsd)/[Mn_2(CO)_{10}]$ was $(8.8 \pm 0.6) \times 10^5 M^{-1}$ with a probable error of ± 30% for each measurement of Φ(obsd). Values of $\Phi(obsd)/[Mn_2(CO)_{10}]$ varied randomly with $10^3[Br_2]$ over the range 2-12M, with an average of 30 values being $(10.9 \pm 0.7) \times 10^5 M^{-1}$ and a probable error of ± 34% for each estimation of Φ(obsd)/[$Mn_2(CO)_{10}$].

No dependence of Φ(obsd)/[$Mn_2(CO)_{10}$] at [Br_2] = 2 x x 10^{-3}M was observed over a range of 10^7Ia from 0.76 to 6.6 E $l^{-1}s^{-1}$, the average of 8 values being $(9.2 \pm 0.6) \times 10^5 M^{-1}$ with a probable error of ± 16%.

4. DISCUSSION

The photolytic reaction of Br_2 with $Mn_2(CO)_{10}$ in CCl_4 leads to $Mn(CO)_5Br$ as the only observable carbonyl product so that the reaction can be written stoichiometrically as in eq.(8). The variation in quantum yields with purity of the

$$Mn_2(CO)_{10} + Br_2 \xrightarrow{h\nu} 2Mn(CO)_5Br \qquad (8)$$

CCl_4 solvent, and the fact that quantum yields considerably in excess of unity were usually obtained, show conclusively that a chain reaction is occurring. Since essentially all the light is absorbed by the Br_2 rather than the $Mn_2(CO)_{10}$, chain initiation must be photohomolysis of the Br_2 molecules to form Br atoms. The absence of intensity dependence of Φ(obsd) suggests that chain termination does not involve radical recombination and we propose the following mechanism for the reaction. Reactions (10) and (11) are the chain propagating steps, and X in reactions (12) and (13) is

$$Br_2 \xrightarrow[\Phi_1]{h\nu} 2Br \qquad (9)$$

$$Br + Mn_2(CO)_{10} \xrightarrow{k_2} Mn(CO)_5Br + Mn(CO)_5 \qquad (10)$$

$$Br_2 + Mn(CO)_5 \xrightarrow{k_3} Mn(CO)_5Br + Br \qquad (11)$$

$$Br + X \xrightarrow{k_4} \text{chain termination} \qquad (12)$$

$$Mn(CO)_5 + X \xrightarrow{k_5} \text{chain termination} \qquad (13)$$

probably the solvent in the absence of any deliberately added chain terminator. The rate equation for this mechanism is shown in alternative forms in eq.(14) and (15).

$$\Phi = \frac{2\Phi_1(k_3[Br_2] + k_4[X])k_2[Mn_2(CO)_{10}]}{k_4[X](k_3[Br_2] + k_4[X]) + k_2k_5[X][Mn_2(CO)_{10}]} \quad (14)$$

$$1/\Phi = k_5[X]/2\Phi_1(k_3[Br_2] + k_4[X]) + $$

$$+ k_4[X]/2\Phi_1k_2[Mn_2(CO)_{10}] \quad (15)$$

The data are all qualitatively and quantitatively in excellent agreement with these equations. The increase in Φ(obsd) with increasing $[Br_2]$ or $[Mn_2(CO)_{10}]$ is as expected from eq.(14), as is the absence of any dependence of Φ(obsd) on Ia.

The linearity of the plots of $1/\Phi$(obsd) vs $1/[Mn_2(CO)_{10}]$ is in good quantitative accord with eq.(15) and the lack of dependence of the gradients on $[Br_2]$ (see Table I), at least when $[Br_2] > 10^{-3}$M, is also as predicted. The inverse of the intercepts also shows the required linear dependence on $[Br_2]$, with a finite intercept at $[Br_2] = 0$. When the data for the plots of $1/\Phi$(obsd) vs $1/[Mn_2(CO)_{10}]$ were obtained from a single run the internal consistency of the data was excellent with probable errors of below ± 10% for each determination of Φ(obsd) (Table I). When values from separately performed runs are considered the probable error rises to ca. ± 20% but this is still very good for a radical chain process with each value of Φ(obsd) being estimated from two quite similar absorbance measurements [41].

The effect of added $CBrCl_3$ is evidently to provide additional bimolecular termination processes for removal of Br atoms and $Mn(CO)_5$ radicals. These can be taken to be governed by effective second order rate constants k_4' and k_5' respectively. In this case the inverse form of the rate equation is as in eq.(16). When $[CBrCl_3]$ is large enough for $k_5'[CBrCl_3]/k_5[X] >> 1$ the intercept,

$$1/\Phi = \frac{k_5[X] + k_5'[CBrCl_3]}{2\Phi_1(k_3[Br_2] + k_4[X] + k_4'[CBrCl_3])} + $$

$$+ \frac{k_4[X] + k_4'[CBrCl_3]}{2\Phi_1k_2} \quad \frac{1}{[Mn_2(CO)_{10}]}$$

$$\ldots\ldots\ldots\ldots\ldots(16)$$

<u>a</u>, of the $1/\Phi$(obsd) vs $1/[Mn_2(CO)_{10}]$ plots will rise to a limiting value so that a plot of 1/<u>a</u> vs $1/[CBrCl_3]$ will be linear as shown by eq.(7). The gradient, <u>b</u>, of the $1/\Phi$(obsd) vs $1/[Mn_2(CO)_{10}]$ plots should increase linearly with $[CBrCl_3]$ in agreement with eq.(5).

Since $CBrCl_3$ is evidently an effective chain terminator it seems reasonable to suppose that the chain terminator X is actually the CCl_4 solvent. The chain termination seems unlikely, however, to involve simple Cl or Br transfer from the CCl_4 or $CBrCl_3$, respectively, since the common CCl_3 product would itself be expected to react with Br_2 to provide $CBrCl_3$ and a Br atom to initiate another chain. A more complex series of reactions must be involved before chain termination is final.

In spite of this ambiguity in the precise nature of the termination step the primary participation of Cl atom transfer from the CCl_4 solvent does suggest that termination would be much less efficient in a solvent such as perfluorodecalin and the much higher quantum yields in this solvent bear out this expectation. Since the quantum yields in perfluorodecalin increase linearly with $[Mn_2(CO)_{10}]$ it follows from eq.(14) that termination by reaction (13) must be relatively unimportant.

The quantitative kinetic data given above in eq.(4)-(7) and elsewhere can be related to the rate equation so as to lead to rate constant ratios shown in Table III. These show that reaction of bromine atoms with $Mn_2(CO)_{10}$ is ca. 10^3 times faster than reaction of Br_2 with $Mn(CO)_5$ radicals. Bromine atoms react ca. 4 times more rapidly with the CCl_4 solvent than do $Mn(CO)_5$ radicals whereas they react about twice as slowly as $Mn(CO)_5$ radicals with $CBrCl_3$.

Although there is a much greater effective concentration of CCl_4 in the pure CCl_4 solvent, bromine atoms and $Mn(CO)_5$ radicals react with the CCl_4 solvent 2 or 3 orders of magnitude more slowly than with 1M $CBrCl_3$. This is probably simply due to the greater strength of C-Cl bonds compared with C-Br bonds. On the other hand, bromine atoms react with CCl_4 solvent only 6 times more rapidly than with the perfluorodecalin solvent which is surprising in view of the greater strength of C-F bonds compared with C-Cl bonds. However, the quantitative analysis of the rate data agrees extremely well with the proposed mechanism and leads to values for rate constant ratios for the various steps involved that are generally quite reasonable.

TABLE I

Intercepts, a, and gradients, b, of linear plots of 1/Φ(obsd) vs 1/[$Mn_2(CO)_{10}$]

$10^4[Br_2]$, M	No. of points	10^2a	10^7b, M	σ[Φ(obsd)], %
3.86	4	14 ± 6	161 ± 22	3.4
5.60	5	23 ± 5	85 ± 15	4.3
7.60	12	15 ± 4	49.5 ± 7.3	9.6
11.4	9	10 ± 3	39.7 ± 5.1	7.2
16.1	3	8.5 ± 3.0	26.2 ± 2.7	6.6
25.5	11	12.0 ± 4.0	28.4 ± 7.6	7.4
55.0	11	8.0 ± 1.4	35.0 ± 1.5	4.4
60.0	7	5.6 ± 2.4	42.3 ± 2.2	5.7
121	8	3.3 ± 3.0	39.5 ± 6.0	6.1

TABLE II

Intercepts, a, and gradients, b, of linear plots of 1/Φ(obsd) vs 1/[$Mn_2(CO)_{10}$] in CCl_4 in the presence of $CBrCl_3$ [Br_2] = 1.7×10^{-3}M

[$CBrCl_3$], M	No. of points	10^2a, M	10^7b, M	σ[Φ(obsd)], %
0	11	9.5 ± 4.8	51 ± 6	13.8
0.002	8	21 ± 3	51 ± 3	5.6
0.005	10	34 ± 2	67 ± 3	3.9
0.016	10	80 ± 10	68 ± 17	16.2
0.081	4	137 ± 6	116 ± 18	1.1
0.407	4	216 ± 11	243 ± 32	1.4

TABLE III

Rate constant ratios derived from the kinetics

$k_2/k_3 = (8.4 \pm 2.4) \times 10^2$ $k_4/k_5 = 3.8 \pm 0.3$

$k_4'/k_5' = 0.44 \pm 0.03$ $k_4[X]/k_4' = (1.3 \pm 0.3) \times 10^{-2}$M

$k_5[X]/k_5' = (1.5 \pm 0.3) \times 10^{-3}$M

$k_4[X]$ (in CCl_4)/k_4 (in perfluorodecalin) PFD = ca. 6[a]

[a]Assuming k_2 is approximately the same in CCl_4 and PFD.

REFERENCES

(1) Dahl, L. F.; Ishishi, E.; Rundle, R. E. J.Chem. Phys. 1957, 26, 1750.
(2) Hopgood, D. Ph.D. Thesis, University of London, 1966.
(3) Hughey, J. L.; Anderson, C. P.; Meyer, T. J. J. Organomet.Chem. 1977, 125, C49. Waltz, W. L.; Hackelberg, D.; Dorfman, L. M.; Wojcicki, A. J.Am. Chem.Soc. 1978, 100, 7259. Wegman, R. W.; Olsen, R. J.; Gard, D. R.; Faulkner, L. R.; Brown, T. L. J.Am.Chem.Soc. 1981, 103, 6089.
(4) Cotton, F. A.; Monchamp, R. R. J.Chem.Soc. 1960, 533.
(5) Haines, L. I. B.; Hopgood, D.; Poë, A. J. J.Chem. Soc.A. 1968, 421.
(6) Fawcett, J. P.; Poë, A. J.; Sharma, K. R. J.Am. Chem.Soc. 1976, 98, 1401.
(7) Fawcett, J. P.; Poë, A. J.; Sharma, K. R. J.Chem. Soc., Dalton Trans. 1979, 1886.
(8) Fawcett, J. P.; Poë, A. J. J.Chem.Soc., Dalton Trans. 1976, 2039.
(9) Schmidt, S. P.; Trogler, W. C.; Basolo, F. Inorg. Chem. 1982, 21, 1098.
(10) Stolzenberg, A. M.; Muetterties, E. L. J.Am. Chem.Soc. 1983, 105, 822.
(11) Coville, N. J.; Stolzenberg, A. M.; Muetterties, E. L. J.Am.Chem.Soc. 1983, 105, 2499.
(12) Poë, A. J.; Sekhar, C. J.Chem.Soc., Chem. Commun. 1983, 566.
(13) Poë, A. J.; Sekhar, C. J.Am.Chem.Soc. 1985, 107, 4874.
(14) Poë, A. J. Chem.in Brit. 1983, 19, 997.
(15) Walker, H. W.; Herrick, R. S.; Olsen, R. J.; Brown, T. L. Inorg.Chem. 1984, 23, 3748.
(16) Herrick, R. S.; Herrinton, T. R.; Walker, H. W.; Brown, T. L. Organometallics 1985, 4, 42.
(17) McCullen, S. B.; Brown, T. L. J.Am.Chem.Soc. 1982, 104, 749.
(18) Poë, A. J.; Sampson, C. N.; Sekhar, C. Inorg. Chem. 1987, 26, 1057.
(19) Jackson, R. A.; Poë, A. J. Inorg.Chem. 1978, 17, 997.
(20) Marcomini, A.; Poë, A. J. J.Chem.Soc.,Dalton Trans. 1984, 95.
(21) Fawcett, J. P.; Jackson, R. A.; Poë, A. J. J. Chem.Soc.,Chem.Commun. 1975, 733. Fawcett, J. P.; Jackson, R. A.; Poë, A. J. J.Chem.Soc., Dalton Trans. 1978, 789.
(22) Kidd, D. R.; Brown, T. L. J.Am.Chem.Soc. 1978, 100, 4095.

(23) Fox, A.; Malito, J.; Poë, A. J. J.Chem.Soc., Chem.Commun. 1981, 105.
(24) Herrinton, T. R.; Brown, T. L. J.Am.Chem.Soc. 1985, 107, 5700.
(25) McCullen, S. B.; Walker, H. W.; Brown, T. L. J. Am.Chem.Soc. 1982, 104, 4007.
(26) Shi, Q.-Z.; Richmond, T. G.; Trogler, W. C.; Basolo, F. J.Am.Chem.Soc. 1982, 104, 4032.
(27) Turaki, N. N.; Higgins, J. M. Organometallics 1986, 5, 1703.
(28) Thorsteinson, E. M.; Basolo, F. J.Am.Chem.Soc. 1966, 88, 3929.
(29) Poë, A. J.; Twigg, M. V. J.Chem.Soc., Dalton Trans. 1974, 1860.
(30) Poë, A. J. Pure and App. Chem. to be published, Brodie, N. M. J.; Chen, L.; Poë, A. J. Intl.J. Chem.Kin., to be published.
(31) Bruce, M. I., Coord.Chem.Rev. 1987, 76, 1.
(32) Samsel, E. G.; Kochi, J. K. J.Am.Chem.Soc. 1986, 108, 4790. Narayanan, B. A.; Amatore, C.; Kochi, J. K. Organometallics 1986, 5, 926.
(33) Goldman, A. S.; Tyler, D. R. J.Am.Chem.Soc. 1986, 108, 89.
(34) Downard, A. J.; Robinson, B. H.; Simpson, J. Organometallics 1986, 5, 1140.
(35) Hill, R. H.; Puddephatt, R. J. J.Am.Chem.Soc. 1985, 107, 1218.
(36) Stiegman, A. E.; Tyler, D. R. J.Am.Chem.Soc. 1985, 107, 967.
(37) Absi-Halabi; Atwood, J. D.; Forbus, N. P.; Brown, T. L. J.Am.Chem.Soc. 1980, 102, 6248. Hoffman, N. W.; Brown, T. L. Inorg.Chem. 1978, 17, 613. Byers, B. H.; Brown, T. L. J.Am.Chem.Soc. 1977, 99, 2527.
(38) Fabian, B. D.; Labiinger, J. A. J.Am.Chem.Soc. 1979, 101, 2239.
(39) Kochi, J. K. Organometallic Mechanisms and Catalysis, Academic Press, New York, 1978.
(40) Meyer, T. J.; Caspar, J. V. Chem.Rev. 1985, 85, 187.
(41) Poë, A. J.; Sekhar, C. V. J.Am.Chem.Soc. 1986, 108, 3673.

CHEMISTRY AND REACTIVITY OF METAL CLUSTER CARBONYL RADICAL ANIONS

Brian H. Robinson and Jim Simpson,
Department of Chemistry,
University of Otago,
P.O. Box 56,
Dunedin,
NEW ZEALAND.

ABSTRACT. Metal carbonyl clusters generally undergo one-electron reduction to reactive radical anions. Reactivity is dependent on the nuclearity of the cluster and whether capping or bridging groups 'clamp' the metal-metal bonds. Because nucleophilic substitution on the radical anion is fast even the most unstable anions can participate in efficient electron transfer chain catalysis. Illustrative examples using C_2Co_2, CM_3 and M_3 clusters are given and the influence of the nucleophile on the chain length and possible mechanisms highlighted. Selective substitution, isomerization and metal-metal bond formation are attractive features of ETC reactions. Synthetic and catalytic possibilities using electrogeneration of reactive cluster fragments are discussed.

1. REDOX CHEMISTRY OF METAL CLUSTER CARBONYLS.

The application of metal carbonyl clusters to electron transfer reactions arose from our work on carbyne-capped clusters, particularly $RCCo_3(CO)_9$, 1. It was evident from the chemistry of these molecules that the delocalized bonding in the CCo_3 core provided a 'push-pull' mechanism for the transfer of charge via the capping group.[1,2] Theoretical and spectroscopic data from other work[3] suggested that most

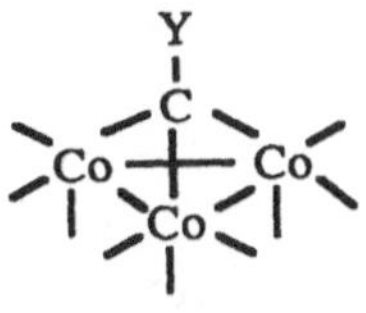

1

Y = Ph, a
Me, b
H, c

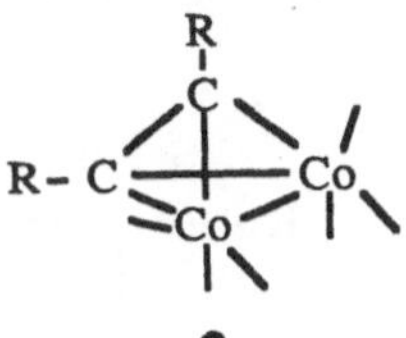

2

R = CF_3, a
Ph, b
Bu, c

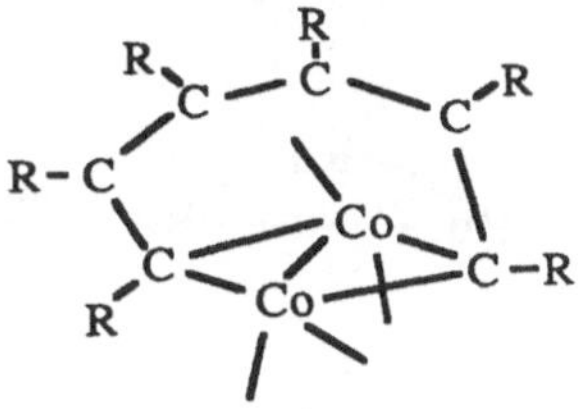

3

R = CF_3

M. Chanon et al. (eds.), Paramagnetic Organometallic Species in Activation / Selectivity, Catalysis, 357–374.

metal carbonyl clusters had metal-centred delocalized cores so we embarked on a systematic survey of the redox properties of carbonyl clusters,[4] the only previous work in this area being that of Dessy and co-workers.[5]

Because of the electron acceptor properties of the CO ligands and the low energy, metal-centred LUMO, the majority of metal carbonyl clusters undergo facile one-electron reduction to radical anions.[6] Detailed spectroscopic studies[7-10] on C_2M_2, CM_3, and M_4 radical anions have confirmed the a_2^* metal-centred LUMO. Subsequent electron transfer behaviour and stability of the radical anions is dependent on the structural type, metal and temperature, Scheme 1.

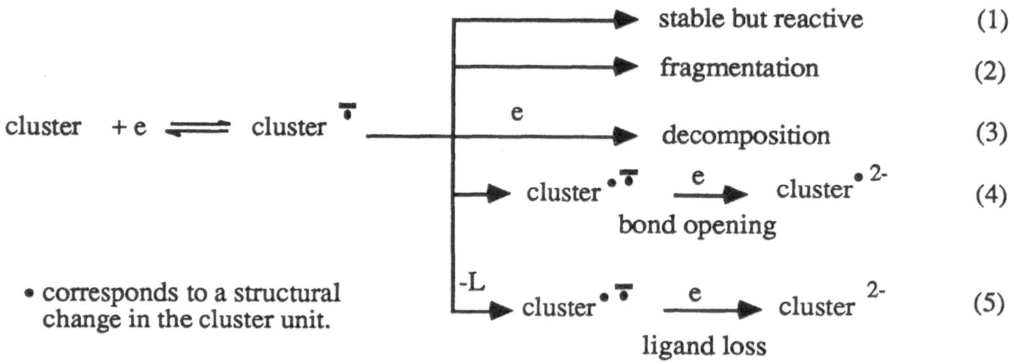

Scheme 1

Maximum stability of the radical anions occurs with the capped metal clusters (eg. 1, 2, P_2Fe_3, PCo_3) or where the metal-metal bond is 'clamped' by a suitable bridging group as in 3. Stability decreases with heterometal cores[11] or if the CO groups are replaced by weaker π-acceptors (eg. H^-, PR_3).[12] Unless a structural change occurs, the addition of a second electron leads to decomposition. path 3 of Scheme 1; when conproportionation is fast a relatively high concentration of radical anion may be maintained at the electrode surface.[13] Non-capped,

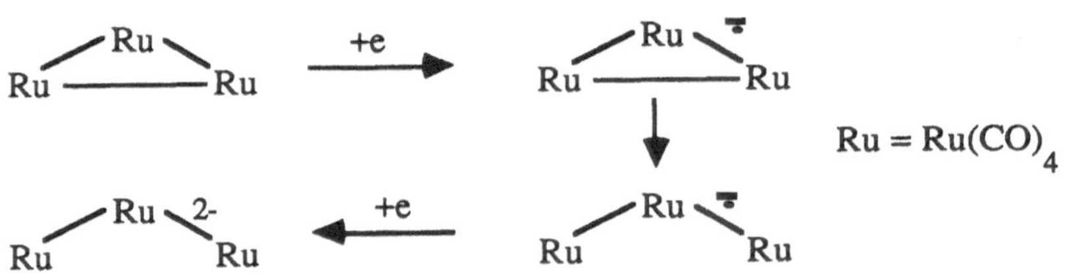

Scheme 2

first row radical anions tend to fragment. path 2, upon reduction at ambient temperature although the lifetime at low temperatures is often sufficient to enable their characterization ($Fe_3(CO)_{12}$[14], $H_3Mn_3(CO)_{12}$[15]). The redox chemistry of second row radical anions or those of higher nuclearity usually follow path 4, but the electron transfer chemistry is often so complex that few systems have been unravelled.[16,17] $Ru_3(CO)_{12}$ undergoes an apparent 2e reduction at ambient temperatures but this is due to an ECE mechanism involving Ru-Ru bond cleavage,[16] Scheme 2.

The importance of ligand dissociation, path 5, in the redox chemistry of the substituted clusters is related to the degree of steric congestion around the cluster core. Thus multi-electron reduction arising from an ECE process is common for derivatives of CCo_3 clusters, Scheme 3, and ligand lability on reduction is dependent on the cone-angle of the ligand.[18,19] In contrast, the less crowded C_2Co_2 derivatives have the expected one-electron i/E curves.[20,33]

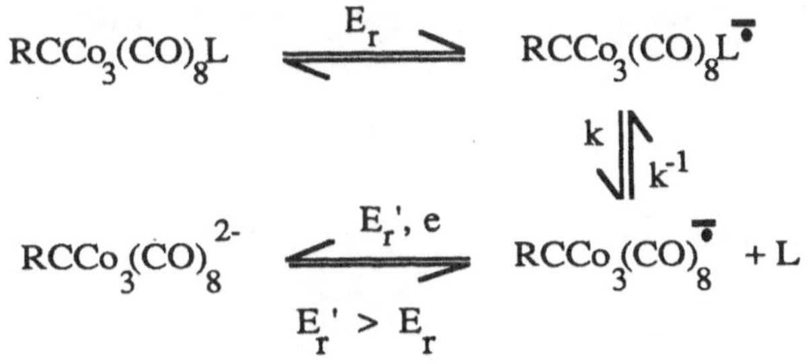

Scheme 3

Utilization of these cluster radical anions in electron transfer processes depends on the rate of intermolecular electron transfer, the rates of the alternative reaction pathways 2 → 5 and the rate of nucleophilic substitution on the radical anion. Electrochemical parameters as well as qualitative data suggest that intermolecular electron transfer is fast given the geometry of metal carbonyl clusters. We believe that the CO groups are central to the transfer mechanism and direct evidence of this is seen in a study of cluster-derivatized electrodes.[21] Even where the radical anion is extremely unstable, the rate of nucleophilic substitution is often competitive and synthetically useful reactions can be devised. $Ru_3(CO)_{12}$ is a case in point; although $k < 10^{-6}$ s for the ECE step[16] the rate of ligand substitution is so efficient that $Ru_3(CO)_{12}$ is one of best the substrates for ETC reactions (vide infra). Other factors such as ion pairing and solvent (THF has a ubiquitous place in this chemistry) also influence yields and product composition.[22] Careful choice of cluster should therefore give access to the radical anions as reactive intermediates or allow incorporation of clusters in conducting materials. In this article we discuss the former role, but recent work has shown that the latter is an exciting possibility.[23]

2. ELECTRON TRANSFER CHAIN CATALYSED NUCLEOPHILIC SUBSTITUTION.

ETC catalysis of nucleophilic substitution was first recognised in our laboratories during electrochemical studies of the complexes 2 in the presence of phosphorus(III) donors. Cyclic voltammetry of 2a in the presence of a Lewis base, Figure 1, shows the reduction of the parent molecule together with reduction waves at more negative potentials which could be assigned to the reduction of the substituted derivatives $(CF_3)_2C_2Co_2(CO)_{6-n}L_n$, [n = 1, L = PPh_3; n = 1,2, L = $P(OMe)_3$].[20] On repeat scans the diffusion current due to 2a decreased and ultimately disappeared, while waves due to substituted products increased.

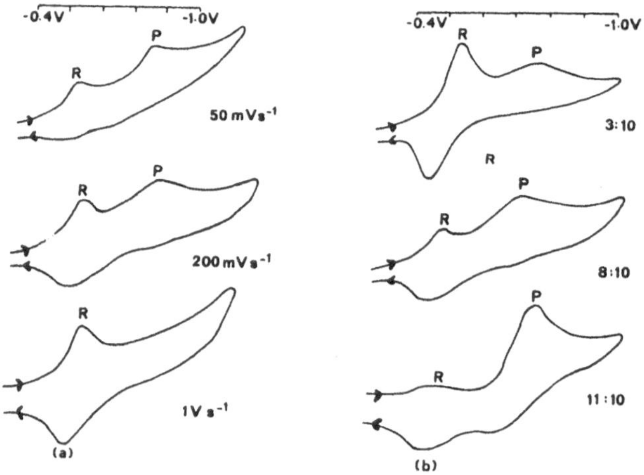

Figure 1. Cyclic voltammograms of 2a in acetone, 0.88 x 10^{-3} M in acetone (0.1 M TEAP) at 293 K. (a) Varying the sweep rate with a Ph_3P:cluster concentration 1:2. (b) Varying the Ph_3P:cluster concentration at 200 mV s^{-1} **R** = reactant; **P** =product.

Such behaviour is characteristic of an ETC process.[24] Scheme 4. Replacement of CO by an alternative nucleophile with poorer π-acceptor capabilities increases the charge density on the metal atoms causing a negative shift in the one-electron reduction potentials of the substituted products. Thus a kinetically fast and exogenic electron transfer occurs from the carbonyl substrate, regenerating the parent radical anion and perpetuating the catalytic cycle.[25]

The ability to form radical anions of reasonable stability, a criterion for ETC catalysis, is a common feature of the carbonyl substrates. Reduction to the radical anion is brought about either by controlled potential electrolysis at the reduction potential of the parent carbonyl or by using a one electron reducing agent such as sodium naphthalenide or sodium benzophenone ketyl (BPK). The latter reagent has proved particularly useful for general synthetic purposes. It has a characteristic purple colour in THF solution, is self-indicating and can be syringed readily into a reaction vessel under anaerobic conditions.

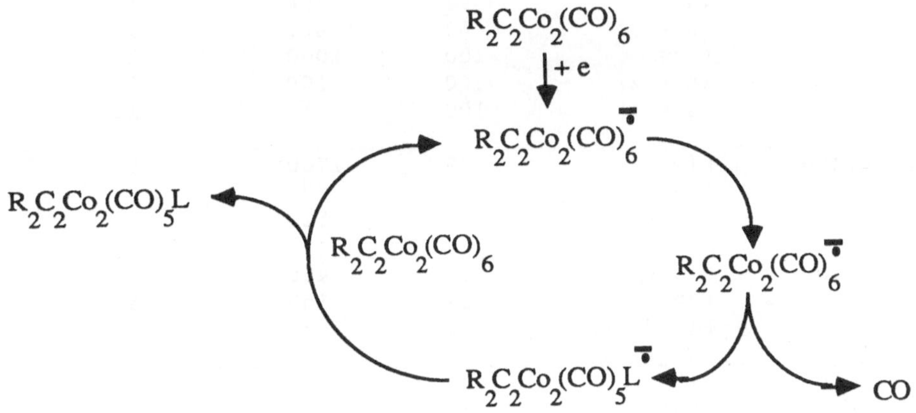

Scheme 4

Subsequent investigations revealed that ETC nucleophilic substitution is a general phenomenon for capped polynuclear cobalt carbonyl systems. Substituted products could be isolated in high yield from reactions that were complete in minutes at ambient temperatures. The method has since been successfully applied by in our laboratories[15,18,26-34] and by others[9,35-39] to the synthesis of a wide variety of derivatives of polynuclear complexes including derivatives, such as those of acetonitrile, not readily synthesised by classical substitution reactions. Tables 1 and 2 illustrate the variation in efficiency and the scope of nucleophilic activity with some of the substrates we have used. In the majority of cases, the ETC route contrasts sharply with the classical thermal substitution of cluster carbonyls, which may require long reaction times, generally gives poorer yields of complex product mixtures, and often leads to fragmentation of the cluster substrate. It is however usually possible to form more highly substituted derivatives by thermal than by electrochemical initiation.

TABLE 1

Synthesis of Cluster Derivatives by Electrolysis[a]

Cluster	Ligand	Yield[b]	Current[c] Efficiency	Time/min
$(CF_3)_2C_2Co_2(CO)_6$	CH_3CN	70	960	5
	$P(C_6H_{11})_3$	83	1520	1
	PPh_3	85	2100	1
	(n = 2)	60	512	1
	$P(OMe)_3$	100	3100	1
	(n = 2)	94	864	1
	(n = 3)	64	460	1
	ttas	100	>1000	<1
	(n = 2)	100	>100	1
	(n = 3)	100	>1	15
$(CF_3)C_2HCo_2(CO)_6$	$P(OMe)_3$	79	1700	5
	PPh_3	65	870	5
	$P(C_6H_{11})_3$	49	520	5
$t\text{-}BuC_2HCo_2(CO)_6$	$P(OMe)_3$	74	456	5
	PPh_3	17	159	5
	$P(C_6H_{11})_3$	2	33	5
$PhC_2HCo_2(CO)_6$	$P(OMe)_3$	36	220	5
	PPh_3	15	61	5
	$P(C_6H_{11})_3$	4	31	5
$Ph_2C_2Co_2(CO)_6$	$P(OMe)_3$	20		5
	PPh_3	8		5
	$P(C_6H_{11})_3$	<5		5
$(CF_3)_6C_6Co_2(CO)_4$	MeCN	70		5
	(n = 1)	20		5
	$P(OMe)_3$	60		3
	(n = 2)	40		3
$(CF_3)_3H_3C_6Co_2(CO)_4$	MeCN		2.5	10

[a] Data from references 20, 25, 32, 47
[b] For mono-substitution unless stated otherwise.
[c] Moles of product per Faraday passed.

TABLE 2

Electron-Induced Reactions with $PhCCo_3(CO)_9$[a]

nucleophile	product (yield%)
PPh_3	$PhCCo_3(CO)_8PPh_3$ (90)
$P(OMe)_3$	$PhCCo_3(CO)_{9-n}[P(OMe)_3]_n$ (100) (b)
$P(OPh)_3$	$PhCCo_3(CO)_{9-n}[P(OPh)_3]_n$ (100) (b)
$P(C_6H_{11})_3$	$PhCCo_3(CO)_8P(C_6H_{11})_3$ (65)
P^nBu_3	$PhCCo_3(CO)_8PBu_3$ (85)
dppm	$PhCCo_3(CO)_7$dppm (70) (c)
dppe	$[PhCCo_3(CO)_8]_2$dppe (80)
	$PhCCo_3(CO)_7$dppe
tpme	$PhCCo_3(CO)_{9-n}$tpme (70) (c)(j)
ttas	ir evidence for chelation (d)
diars	$PhCCo_3(CO)_7$diars(<20) (d) (e)
Chiraphos	$PhCCo_3(CO)_7$(chiraphos) (88) (h)
CH_3CN	$PhCCo_3(CO)_8(CNCH_3)$ (35)
$CpFeC_5H_4PPh_2$	$PhCCo_3(CO)_8(PPh_2C_5H_4FeCp)$ (95)
$(C_5H_4PPh_2)_2Fe$	$PhCCo_3(CO)_7[(C_5H_4PPh_2)_2Fe]$ (85)
$CpFe_5C_5H_3(NMe_2)(PPh_2)$	$PhCCo_3(CO)_7[CpFeC_5H_3(NMe_2)PPh_2]$(90) (g)
Ph_3C_4P	$PhCCo_3(CO)_8(Ph_3C_4P)$ (55) (i)
$CpFeC_4(Me_2)(H)_2P$	$PhCCo_3(CO)_8[CpFeC_4(Me_2)(H)_2P]$(30) (i)
$MoCp(CO)_3^-$	$PhCCo_2Mo(CO)_8Cp$ (90)
$WCp(CO)_3^-$	$PhCCo_2W(CO)_8Cp$ (5)
$Hg[CpW(CO)_3]_2$	$PhCCo_2W(CO)_8Cp$ (35)
$Hg[CpFe(CO)_2]_2$	$PhCCo_2Fe(CO)_7Cp$ (40)
norb	$PhCCo_3(CO)_7$norb (20) (e)
COT	$PhCCo_3(CO)_7$COT (10) (e)
C_2H_4	$PhCCo_3(CO)_8(C_2H_4)$(10) (e)
$Co(CO)_4^-$	$Ph_2C_2Co_4(CO)_{10}$(30)
^{13}CO	23% isotopic substitution

a) Data taken from references 18, 27, 29, 50, 61, 62
b) Yield decreases n=1>>n=2>>n=3 . c) Isomers formed. d) Product unstable
e) electron induced,not catalysed f) Interacting redox centres
g) coordination via P/N h) Chiral i) Phosphole
j) stoichiometry solvent dependent

2.1. Influence of Nucleophiles.

Product distribution in ETC substitution reactions is largely determined by the nature of the entering nucleophile and the reaction stoichiometry. Fine-tuning of the product distribution is often more easily achieved by electrochemical rather than BPK initiation. A wide range of 'soft' nucleophiles have been used in our laboratories including phosphines, acetonitrile, ferrocenylaminophosphines metal carbonyl anions, ^{13}CO, alkenes and alkynes, Table 2. Well defined trends are found with these ligands. The efficiency of substitution decreases with the number of CO groups replaced and, depending on the thermodynamic stability of the products, this often allows selective control of stoichiometry which cannot be achieved by thermal methods. This is due to the increasing charge density at the 17e substitution centre as CO is replaced by a better donor. Detailed electrochemical studies using the well-behaved 2a show that the initial rate-determining step is dissociative and ligand independent whereas the overall rate of substitution follows the order Bu_3P > Ph_3P > $(C_6H_{11})_3P$ > $(PhO)_3P$ > $(MeO)_3P$ and dppm > Ph_3P, an order dependent on both steric and electronic factors. This is confirmed by competition studies.[40] The trends are not straightforward with 1a (when the cluster is congested) as the lability of the substituted radical anion is an important factor especially in competition studies although the rate-determining step is still dissociative.[18] Solvent is also important; THF gives rates some five orders of magnitude faster than CH_2Cl_2. For higher substitution the order is $(PhO)_3P$ > Ph_3P reflecting the greater stabilization of the 17e transtion state. Other intriguing observations remain to be explained. For example, catalysis of $Ru_3(CO)_{12}$ reactions only occurs within a narrow concentration range of BPK ($>10^{-4}<10^{-2}$ mol dm^{-3}); PPN encourages ($Ru_3(CO)_{12}$)[41] or discourages ($PhCCo_3(CO)_9$)[22] electro-catalysis.

2.2. Mechanism of the ETC Substitution Reactions.

The importance of radical anion formation in these electron transfer processes is unequivocally demonstrated by adding a single crystal of the radical anion [Na(2,2,2,kryptand)$^+$][$PhCCo_3(CO)_9^{-\cdot}$] to a THF solution of neutral 1a in the presence of a slight molar excess of PPh_3. Analysis of the solution after 1 minute reveals quantitative conversion to the derivative $PhCCo_3(CO)_8PPh_3$.[22] The successful operation of the ETC process is also crucially dependent on the longevity of the radical anion. Simple alkyl and aryl acetylene

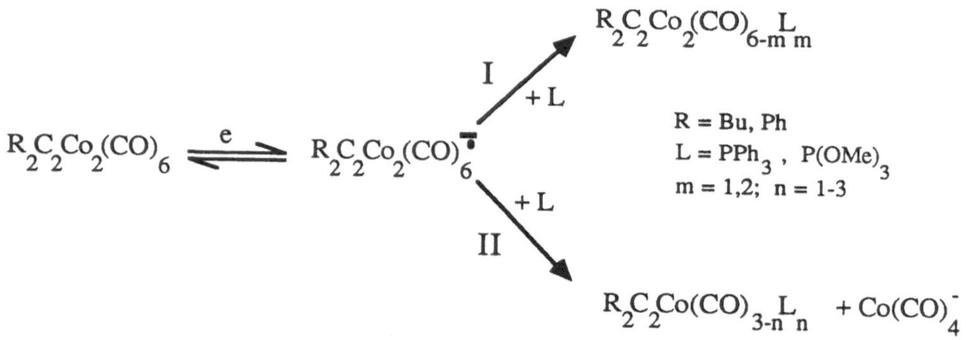

Scheme 5

complexes 2b, 2c show two competing reaction pathways when reduced in the presence of phosphorus(III) ligands at ambient temperatures.[20] In addition to the straightforward ETC substitution of the dicobalt acetylene moiety, **I**, fragmentation of the tetrahedral unit results from Co-Co bond cleavage to produce $Co(CO)_4^-$, the ubiquitous product of reductive degradation of cobalt carbonyl clusters, together with the unusual monocobalt radical species $R_2C_2Co(CO)_{3-n}L_n$ which have been characterised by esr spectroscopy.[42] The relative importance of the fragmentation pathway **II** decreases with temperature in accordance with the observed stability of the radical anion precursor.

The intimate mechanism of the substitution step of the ETC cycle remains arcane. Simulation of the electrochemical behaviour of the substrate $PhCCo_3(CO)_9$ in CH_2Cl_2 leads[30] to the steady state rate law

$$rate = k_1[PhCCo_3(CO)_9^{-\cdot}]/[1 + (k_{-1}[CO]/k_2[L])]$$
$$L = PPh_3,\ dppm,\ dppe$$

and for PPh_3 the value of $k_1 = 3.2\ s^{-1}$ at 293 K shows that a rate enhancement of at least 5 orders of magnitude is obtained in the ETC reaction when compared with the thermal substitution process ($k_{obsd} = 3.6 \times 10^{-5}\ s^{-1}$ at 308 K) which was shown to proceed via a normal CO dissociative mechanism.[43] The i-E responses in these systems show that the current due to the reduction of $PhCCo_3(CO)_9$ becomes independent of ligand concentration at high concentrations, indicating a rate determining dissociative step, implicit in the enhanced reactivity of the radical anions towards nucleophiles. Two dissociative pathways have been proposed for the formation of an activated complex in the substitution process, Scheme 6. First dissociative loss of CO,[36,37] mirroring the classic dissociative mechanism for thermal CO substitution. Second a coordinatively unsaturated 17-electron centre generated by homolytic cleavage of a Co-Co bond in the radical anion.[20,25] The steady state rate laws do not allow an experimental

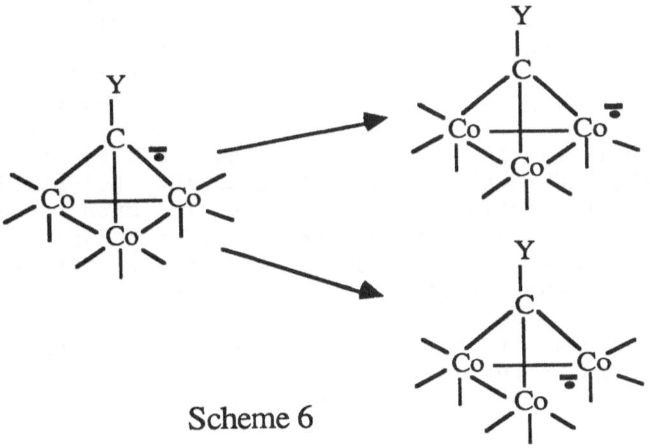

Scheme 6

distinction between these two mechanistic possibilities but the evidence leads us to the conclusion that Co-C bond cleavage is a viable path to a substitution-labile, 17-electron metal centre, particularly for cobalt complexes in which the Co-Co bond is "clamped" by bridging ligands.

The inherent lability of metal-metal bonds is widely acknowledged and facile M-M cleavage forms the basis of a number of synthetic strategies in organometallic chemistry.[44] Furthermore, radical anion formation requires the occupancy of orbitals that are Co-Co antibonding in character, with concomitant weakening of the metal-metal interaction. This is clearly evidenced by Co-Co bond extension in the paramagnetic species $ECo_3(CO)_9$ (E = S,Se).[45] For the carbon capped or bridged moieties, a decrease in ν(CO) on radical anion formation signals an increase in the M-CO bond order at the expense of the M-M interaction. Fragmentation of the dicobalt acetylene radical anions when the C_2Co_2 core is electron rich (vide supra), formation of the butterfly complex $Ph_2C_2Co_4(CO)_{10}$ in the exhaustive electrolysis of $PhCCo_3(CO)_9$, and the facile extrusion of $Co(CO)_4^-$ from the radical anion in ETC isolobal replacement reactions (vide infra) all point to a remarkable tendency to Co-Co bond rupture in these reduced species. The ability of the bridging ligand to stabilise a bond-opened intermediate is common to these polynuclear cobalt compounds. Ohst and Kochi[9] provided incontravertable esr evidence for the production of a coordinatively unsaturated 17-electron Fe centre on reduction of the bicapped phosphinidene cluster $Fe_3(CO)_9(PPh)_2$, by the cleavage of an Fe-P bond to one of the phosphinidene capping groups. Here again, the bond opened intermediate is stabilised by the presence of a second capping group, which maintains the structural integrity of the Fe_3P_2 unit during the substitution process.

2.3. Directed Synthesis by ETC Substitution Reactions.

The reactions described so far have utilised the ETC substitution method as a synthetic tool for the facile, high yield preparation of compounds that could be produced by alternative substitution reactions. A major additional benefit of the ETC reaction is that it provides routes to novel compounds that were hitherto unavailable by the classical reactions of organometallic chemistry. Access to such innovative chemistry is provided by the ability to tailor the conditions of the ETC reaction, by control of the potential at which radical anions are formed. This results in much greater product selectivity than is possible when the metal carbonyl substrate is activated by thermal or photochemical methods. A striking example is provided by the substitution of the CCo_3 cluster with a phenyltricarbonylchromium moiety as the apical substituent. ETC activation of the cluster substrate in the presence of phosphine leads exclusively to substitution of Co bound carbonyls. The corresponding thermal reaction gives a complex mixture of products resulting from replacement of both Co and Cr bound carbonyl groups.[15]

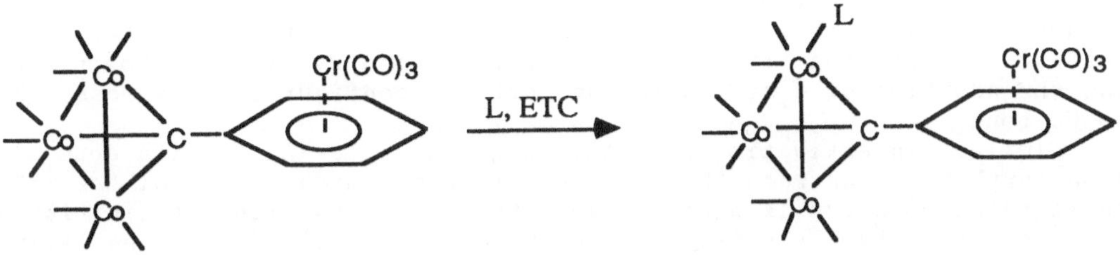

2.3.1. Regioselectivity in Reactions with Polydentate Ligands.

The reaction of 2a with the tridentate arsine $\{o\text{-}C_6H_4[AsMe_2]\}_2AsMe$, ttas, nicely illustrates the ability to control the extent of substitution in the catalytic process,[28] Scheme 7. The sole product of the thermal reaction between 2a and ttas was the purple, tri-susbstituted derivative 6. ETC substitution initiated by BPK gave the bis-substitued complex 5. However, using specific electrochemical initiation at increasingly cathodic potentials it was possible to selectively form each of the three possible substitution products in high yield. While the first and second substitution processes are clearly catalytic, with extremely high current efficiencies, electrochemical conversion of 5 to 6 requires 1 F mol^{-1} and may be

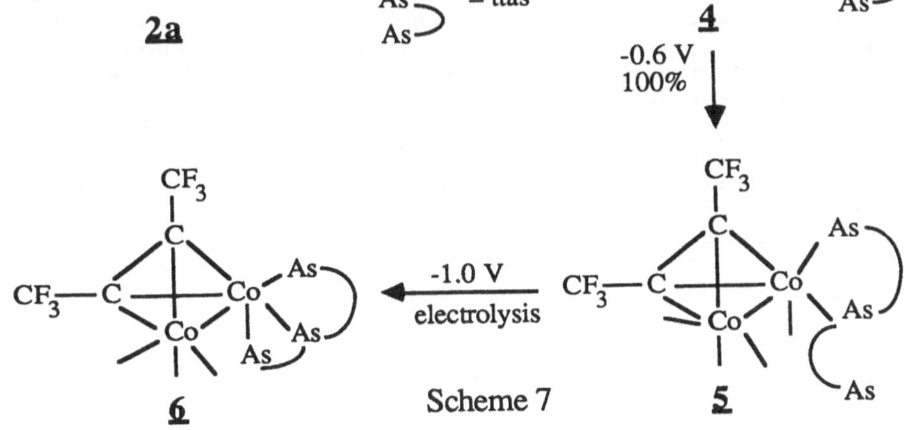

Scheme 7

classified as electron induced. It should be noted that a major structural rearrangement accompanies this final substitution step with destruction of the traditional sawhorse geometry of the C_2Co_2 system and the adoption of a semi-bridging carbonyl configuration across the Co-Co bond.

Similar investigations of the reaction of 1a and 2a with dppm and dppe further demonstrate that greater stereochemical control of the substitution process is achieved when the ETC reaction is initiated electrochemically.[29,32] Factors such as the "bite" of the ligand determine the ultimate product stereochemistry for a given mode of initiation but the preliminary catalytic coordination of a single donor atom, to give a monodentate or "dangling" product, is a common factor in the substitution reactions and the catalytic efficiencies are ligand independent. Sequential coordination of the phosphorus atoms of the dppe ligand to individual substrate molecules to give linked cluster products such as 7 were a feature of the ETC reactions of both 1a and 2a with the more flexible dppe ligand.

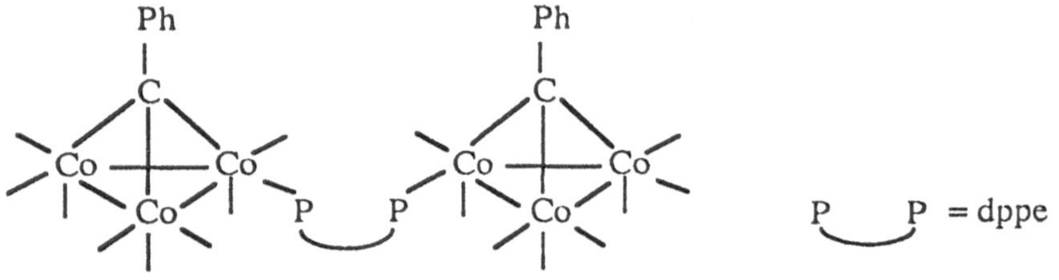

7

3. ETC ISOMERISATION.

Isomeric conversion resulting from electron transfer is one of the more spectacular manifestations of structural change resulting from an electron transfer reaction.[46] The "flyover" complex, 3, forms an extremely stable radical anion[47] and ETC substitution by phosphite ligands results in the formation of mono and bis-substituted complexes in high yield with a blue form of the bis-phosphite complexes predominating. Electrolysis of the the blue form at the potential of the first observed reduction process produces an instantaneous colour change from blue to purple leading to the isolation of the alternative isomers 8a, 8b in high yield. The structures of 8a and 8b have been determined crystallographically, Figure 2.[26] Alternatively, the purple isomer can be prepared stereospecifically from 3 by electrolysis at the potential of the blue isomer in the presence of the appropriate phosphine. The use of BPK as the reductant provided an additional route to the isomerisation.

Attempts to emulate the isomerisation in thermal or photolytic reactions proved fruitless. The blue isomers could be prepared in very low yield as the sole disubstituted products of thermal reactions. It is also not possible to reverse the transformation by thermal, chemical

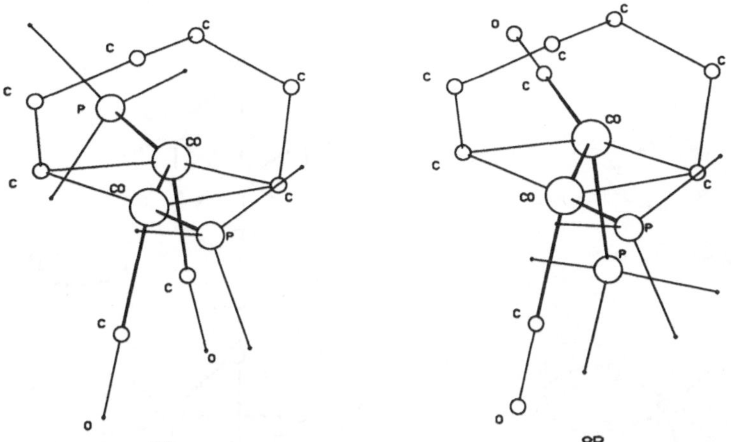

8A 8B

Figure 2. Structures of the blue 8a and purple 8b isomeric forms of $(CF_3)_6C_6Co_2(CO)_2[P(OMe)_3]_2$. The CF_3 groups of the "flyover" ligand and the Me groups of the phosphites have been omitted for clarity.

or electrochemical means. The isomerisation reaction occurs at a rate which defies attempts at coulometric measurements and the isomerisation may be characterised as a genuine bulk ETC process for which the catalytic chain is extremely long. Establishing the criteria for the ETC process is complicated in this instance by the fact that both isomeric forms exhibit identical redox behaviour, a phenomenon

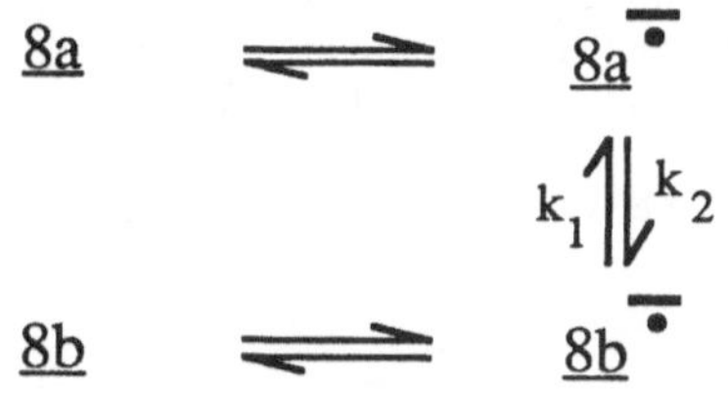

Scheme 8

characteristic of a system in which isomerisation occurs with zero current flow. The gross features of the isomerisation are represented by a limiting case of the square Scheme 8, for which $k_1 = k_2 = \infty$; that is the isomerisation is essentially instantaneous.[46]

Catalytic isomerisation is also observed in the electrochemistry of dppm derivatives of 1a and 2a. Thus isomers 9a, 9b differing by the mode of attachment of the bidentate ligand with respect to the cobalt triangle were found for the complexes $RCCo_3(CO)_7(dppm)$. Both isomeric forms gave identical i-E responses and the conversion 9a → 9b could be achieved electrochemically.[18] The isomerisation from metal chelation to metal-metal bond bridging was similarly catalysed for 10a and 10b.[32]

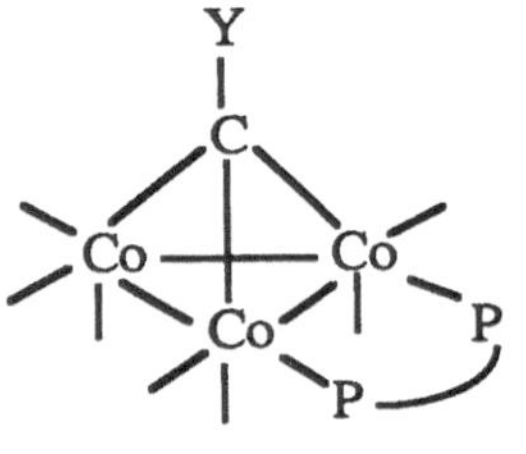

9a

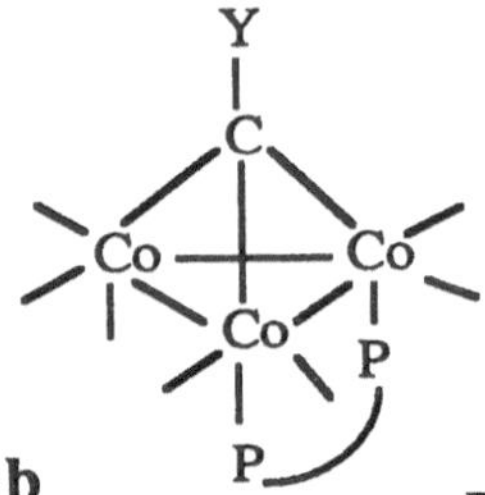

9b

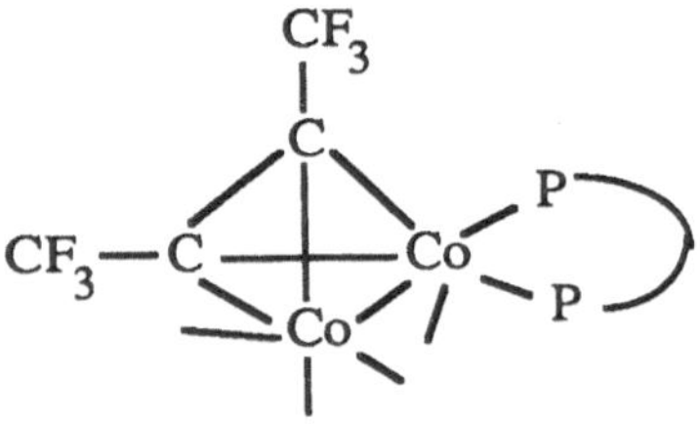

10a

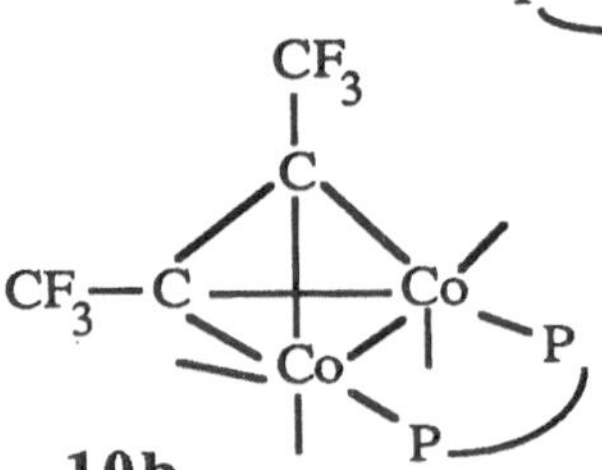

10b

4. ISOLOBAL REPLACEMENT REACTIONS.

Recognition of the isolobal relationships that exist between organometallic and organic fragments has not only produced a significant ordering of the structural diversity that characterises organometallic chemistry, but has also provided new synthetic strategies based on the extension of facile organic reactions to their organometallic isolobal analogues.[48] The concept of ETC isolobal replacement reactions begins with the recognition that metal carbonyl anions are Lewis bases, albeit of widely varying nucleophilicity.[49] This fact, coupled with the isolobal relationships

$$RC \longleftrightarrow Co(CO)_3 \longleftrightarrow CpM(CO)_2 \longleftrightarrow CpM'(CO)$$

M = Cr, Mo, W; M' = Fe, Ru

that exist between a carbyne, the $Co(CO)_3$ fragment and a variety of metal carbonyl moieties prompted an investigation of whether metal centred anions could replace more traditional nucleophiles in ETC substitution reactions and, if so, whether additional metal-metal bond formation would follow the substitution step. Compound 1a reacted in a matter of minutes with the carbonyl anions $CpM(CO)_3^-$ (M = Mo, W) and $CpFe(CO)_2^-$ with either BPK or electrochemical initiation to produce the heteronuclear tri-metallic compounds 11a-c in good yields.[27] Thus the ETC nucleophilic substitution of the metal carbonyl anion is followed by expulsion of the good leaving group $Co(CO)_4^-$ and incorporation of the heterometal atom into the metal triangle Scheme 9. The mercury salts $Hg(anion)_2$ provided an excellent source of the reactant carbonyl anions or, alternatively, these may be produced in situ by electrochemical reduction of the metal carbonyl dimers.[50] The high yields and short reaction times of the ETC reactions are in sharp contrast to an analogous thermal synthesis of 11a from 1a and the dimer $[CpMo(CO)_3]_2$ which gave a yield of 20% after 3 days in refluxing benzene.[51] The mild reaction conditions of the ETC process also allowed the use of clusters such as 1c, which have limited thermal stability, as substrates for these replacement reactions.

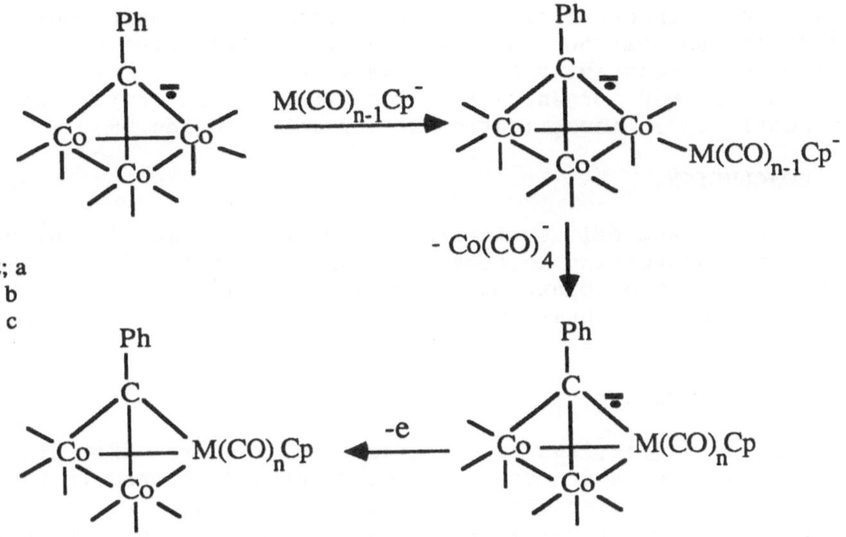

M = Mo, n = 2; a
M = W, n = 2, b
M = Fe, n = 1, c

11

Scheme 9

5. ELECTROGENERATION OF REACTIVE FRAGMENTS.

Synthesis of heteronuclear and/or large clusters has been achieved using metal-fragment replacement or redox-condensation reactions.[52] Since exhaustive reduction or oxidation of clusters leads in many instances to fragmentation, electrosynthesis using the generated species is an attractive possibility. An example[53] of electrochemically initiated cluster synthesis is that of $Pt_2Co_2(\mu\text{-}CO)_3(CO)_5(PPh_3)_2$ from $PtCo(\mu\text{-}CO)(CO)_7PPh_3$ and radicals have been implicated in mixed-metal cluster synthesis[54]. As Shriver et al comment, many reported syntheses of cluster systems may include radical intermediates, especially those involving BPK reduction.[55] Some examples from our work are given below.

$$Rh_4(CO)_{12} \xrightarrow{e} Rh_6(CO)_{16} \quad \text{(pure product in high yield)} \quad (50)$$

$$Fe_6(CO)_{16}C^{2-} \xrightarrow{-e} Fe_5(CO)_{15}C \quad \text{(good yield)} \quad (56)$$

$$Ru_3(CO)_{12} \xrightarrow{e} Ru_6(CO)_{18}^{2-} \text{ or } Ru_4(CO)_{13}^{2-} \qquad (16,50)$$

$$\mathbf{2a} + CpMo(CO)_3^- \xrightarrow{e} (CF_3)_2C_2CoMo(CO)_5Cp \qquad (57)$$

Electrochemically generated reactive fragments could, in principle, be a means of overcoming an endogenic electron transfer (with respect to the cluster). Cluster-mediated intramolecular electron transfer to oxygen with incorporation of oxygen in the organic moiety has been reported by Darchen et al.[58] However, this area has yet to be fully explored as reactivity may involve the metal or the ligand;[59] the design of new organic electrosyntheses and catalysis is an area currently under investigation in our laboratories.[60]

6. CONCLUSION.

Electrochemical synthesis incorporating metal clusters is in its infancy and exciting developments can be expected in the next few years. The method opens up a new dimension in the analogy between metal surfaces and clusters and a study of the electron transfer properties of clusters gives an insight into their electronic properties.

ACKNOWLEDGEMENTS.

The work described in this article results from the skill and dedication of a generation of co-workers mentioned in the references. We also thank Professors A.M. Bond, R.S. Dickson, P.H. Rieger and H. Vahrenkamp and Drs R.G. Cunninghame, L.R. Hanton, B.M. Peake and W.T. Robinson all of whom have made significant contributions to these developments. Acknowledgement is also made to the donors of the Petroleum Research Fund, administered by the American Chemical Society, for partial support of this work.

REFERENCES.

1. Penfold,B.R; Robinson,B.H. Accounts Chem. Research, 1973, 6, 73
2. Seyferth,D. Adv. Organometal. Chem. 1976, 16, 97
3. Manning,M.C; Trogler,W.C. Coord. Chem. Rev. 1981, 38, 89
4. Peake,B.M; Robinson,B.H; Simpson,J; Watson,D.J. J. Chem. Soc. Chem. Commun. 1974, 945
5. Dessy,R.E; Bares,L.A; Accounts Chem. Res 1972, 5, 415
6. Lemoine,P. Coord. Chem. Rev. 1982, 47, 55; Geiger,W.E; Connelly,N.G. Adv. Organometal. Chem. 1985 ,24, 87.
7. Peake,B.M; Rieger,P.H; Robinson,B.H; Simpson,J. J. Amer. Chem. Soc. 1980, 102, 156; Dickson,R.S; Peake,B.M; Rieger,P.H; Robinson,B.H; Simpson,J. J. Organometal. Chem. 1979, 172, C63; Peake,B.M; Rieger,P.H; Robinson,B.H; Simpson,J. Inorg. Chem. 1981, 20, 2540; Watson,D.J.; Peake,B.M; Robinson,B.H; Simpson,J. Inorg. Chem. 1977, 16, 405

8. Beurich,H; Madach,T; Richter,F; Vahrenkamp,H. Angew. Chem. Int. Ed. 1979, **18**, 690; Delley,B; Manning,M.C; Ellis,D.E; Berkowitz,J; Trogler,W.C. Inorg. Chem. 1982, **21**, 2247; Beringhelli,T; Gervasini,A; Morazzoni,S; Strumolo,D; Martinengo, S Zandererighi,L. J. Chem. Soc. Faraday Trans. 1. 1984, **86**, 1479; Rimmelin,J; Lemoine,P; Gross,M; de Montauzon,S. Nouv. J. Chim. 1983, **7**, 453
9. Ohst,H.H; Kochi,J.K. Organometallics, 1986, **5**, 1359, ibid, J. Chem. Soc. Chem. Commun. 1986, 121
10 Holland,G.F; Ellis,D.E; Trogler,W.C. J. Amer. Chem. Soc. 1986, **108**, 1884
11. Bond,A.M; Honrath,U; Lindsay,P.N.T; Peake,B.M; Robinson,B.H; Simpson,J; Vahrenkamp,H. Organometallics, 1984, **3**, 413
12. Bond,A.M; Dawson,P.A; Peake,B.M; Robinson,B.H; Simpson,J. Inorg. Chem., 1977, **16**, 2199; Bond,A.M; Dawson,P.A; Peake,B.M; Rieger,P.H; Robinson,B.H; Simpson,J. Inorg. Chem. 1979, **18**, 1413; Worth,G.H. unpublished work, University of Otago,1984
13. Miholova,D; Klima,J; Vlcek,A.A. J. Electroanal. Chem. Interfac. Electrochem. 1983, **143**, 195
14. Dawson, P.A; Peake,B.M; Robinson,B.H; Simpson,J. Inorg.Chem. 1980, **19**, 465; Krusic,P.J; San Filippo,J; Hutchinson,B; Hance,R.L; Daniels,L.M. J. Amer. Chem. Soc. 1981, **103**, 2129
15. Liddell,M. B.Sc.(Hons),report. University of Otago, 1982.
16. Downard,A.J; Robinson,B.H; Simpson,J. J. Organometal. Chem. 1987, **320**, 363; Cyr,J.E; De Gray,J.A; Gosser,D.K; Lee,E.S; Rieger,P.H. Organometallics, 1985, **4**, 950
17. Bonny,A; Crane,T.J; Kane-Maquire,N.A.P. Inorg. Chim. Acta, 1982, **65**, L83
18. Downard,A.M; Robinson,B.H; Simpson,J. Organometallics, 1986, **5**, 1132; Downard,A.M. Ph.D. thesis, University of Otago, 1984
19. Hinkelmann,K; Mahlendorf,M; Heinze,J; Schact,H-T; Field,J.S; Vahrenkamp,H. Angew. Chem. Int. Ed. 1987, **26**, 352
20. Arewgoda,C.M; Rieger,P.H; Robinson,B.H; Simpson,J; Visco,S. J. Amer. Chem. Soc. 1982, **104**, 5633
21. McGillen,B. unpublished work, University of Otago
22. Kirk,C.M; Peake,B.M; Robinson,B.H; Simpson,J; Huffadine,A.S. Aust. J. Chem. 1983, **36**, 441; Lindsay,P.N.T. Ph.D. Thesis, University of Otago, 1982
23. Elder,S.M; Worth,G.H. unpublished work, University of Otago
24. Alder,R.W; J. Chem. Soc. Chem. Commun. 1980, 1184; Saveant,J.M. Accounts Chem. Res. 1980, **13**, 323; Chanon,M. Bull. Soc. Chim. Fr. 1982, 197; Chanon,M; Tobe,M.L. Angew. Chem. Int. Ed. 1982, **21**, 1
25. Arewgoda,C.M; Robinson,B.H; Simpson,J. J. Amer. Chem. Soc. 1983, **105**, 1893
26. Arewgoda,C.M; Robinson,B.H; Simpson,J. J. Chem. Soc. Chem. Commun. 1982, 284
27. Jensen,S.D.; Robinson,B.H; Simpson,J. J. Chem. Soc. Chem. Commun. 1983, 1081
28. Cunninghame,R.G; Downard,A.J; Hanton,L.R; Jensen,S.D; Robinson,B.H; Simpson,J. Organometallics, 1984, **3**, 180
29. Downard,A.J;Robinson,B.H;Simpson,J. Organometallics, 1986, **5**, 1122
30. Downard,A.J;Robinson,B.H;Simpson,J. Organometallics, 1986, **5**, 1140

31. Robinson,B.H; Simpson,J; Trounson,M.E. Aust.J.Chem. 1986, **39**, 1435
32. Cunninghame,R.G; Hanton,L.R; Jensen,S.D; Robinson,B.H; Simpson,J. Organometallics, 1987, **6**, 1470
33. Jensen,S.D; Robinson,B.H; Simpson,J. Organometallics, 1987, **6**, 1479
34. Borgdorff,J; Shand,T. unpublished work, University of Otago
35. Bezems,G.J; Rieger,P.H; Visco,S. J. Chem. Soc. Chem. Commun. 1981, 265; Bruce,M.I. Coord. Chem. Revs. 1987, **76**, 1
36. Darchen,A; Mahe,C; Patin,H. Nouv. J. Chim. 1983, **7**, 453; Lhadi,E.K; Mahe,C; Patin,H; Darchen,A. J. Organometal. Chem. 1983, **146**, C61 and references therein.
37. Richmond,M.G; Kochi,J.K. Inorg. Chem. 1986, **25**, 656
38. Ohst,H.H; Kochi,J.K. Organometallics, 1986, **5**, 1359
39. Kochi,J.K. J. Organometal. Chem. 1986, **300**, 139
40. Worth,G.H. B.Sc(Hons) report, University of Otago, 1983.
41. Lavigne,G; Kaesz, H.D. J.Amer.Chem.Soc. 1984, **106**, 4647
42. Casagrande,L.G; Chen,T; Rieger,P.H; Robinson,B.H; Simpson,J; Visco, S. Inorg. Chem. 1984, **23**, 2019
43. Cartner,A; Cunninghame,R.G; Robinson,B.H. J. Organometal. Chem. 1975, **92**, 49
44. Vahrenkamp,H. Adv. Organometal. Chem. 1983, **22**, 169
45. Wei,C.H; Dahl,L.F. Inorg. Chem. 1967, **6**, 1229
46. Arewgoda,C.M;Robinson,B.H;Simpson,J. J.Organometal.Chem. in press Geiger, W.E. Prog. Inorg. Chem 1985 **33**, 275
47. Arewgoda,C.M; Bond,A.M; Dickson R.S; Mann,T.F; Moir,J.E; Rieger,P.H; Robinson,B.H; Simpson,J. Organometallics, 1985, **4**, 1077
48. Stone,F.G.A. Angew. Chem. Int. Ed. 1984, **23**, 89
49. King,R.B. Accounts Chem. Res. 1970, **3**, 417
50. Jensen,S. M.Sc thesis, University of Otago, 1984.
51. Beurich,H; Vahrenkamp,H. Chem. Ber. 1982, **115**, 2409; Richter,F; Beurich,H; Vahrenkamp,H. J. Organometal. Chem. 1976, **166**, C5
52. Johnson,B.F.G.,Ed. Transtion Metal Clusters, Wiley, NY.1980
53. Lemoine,P; Giraudeau,A; Gross,M; Bender,R; Braunstein,P. J. Chem. Soc. Dalton Trans. 1981, 2059
54. Horwitz,C.P; Holt,E.M; Shriver,D.F. Organometallics, 1985, **4**, 1117
55. Bhattacharyya,A.A; Nagel,C.C; Shore,S.G. Organometallics, 1983, **2**, 1187
56. Dawson,P.A. Ph.D. Thesis, University of Otago, 1978.
57. Jensen,S.D; Robinson,B.H; Simpson,J. Organometallics, 1986, **5**, 1690
58. Darchen,A; Mahe,C; Patin,H. J. Organometal. Chem. 1983,**251**, C9
59. Richmond,M.G; Kochi,J.K. Organometallics, 1987 **6**, 777
60. Headford,C.F.M. work in progress, University of Otago, 1987
61. Shand,T; Paris,S. B.Sc(Hons) reports, University of Otago, 1983
62. Robinson,B.H; Simpson,J. unpublished work

ELECTRON SPIN RESONANCE STUDIES OF ORGANOTRANSITION METAL REACTIVE INTERMEDIATES

Philip H. Rieger
Department of Chemistry
Brown University
Providence, Rhode Island 02912
U. S. A.

ABSTRACT. Electron spin resonance spectroscopy is a powerful tool for the study of organotransition metal radicals and radical ions. The technique can be used for the identification of radicals, measurement of radical concentrations, characterization of the molecular orbital containing the unpaired electron, and measurement of the rate of fluxionality in stereochemically nonrigid radicals. In this review, examples of ESR spectra of some simple radicals derived from manganese, iron, ruthenium, and cobalt carbonyls are discussed in the context of related electrochemical and chemical studies.

1987 marks the tenth anniversary of the first detailed study of the tricobalt carbon radical anions by Robinson, Simpson, and co-workers [1,2] and of my entry into the organometallic radical field as a sabbatical visitor at the University of Otago. I was not alone in my shift in research interests as several other ESR spectroscopists turned their attention in the same direction at about the same time, most notably Paul Krusic, Takashi Kawamura, John Morton, Keith Preston, and Martyn Symons. It seems appropriate to take the opportunity of a meeting devoted to the reactivities of organometallic radicals to review the role of ESR spectroscopy in the study of such species.

1. *WHAT CAN BE LEARNED FROM ESR SPECTRA?*

Isotropic ESR spectra can be used to identify radicals, to measure their concentrations, and to study the rates of fluxional processes. ESR spectra of radicals in solid matrices can be used to characterize the semi-occupied molecular orbital (SOMO) which is important in determining their reactivities. Sections 2 - 6 will review some examples of these applications. While the point of this paper is that ESR spectra can give valuable insights, ESR results by themselves rarely tell the whole chemical story. Thus ESR data will be correlated with those from cyclic voltammetry and other electrochemical experiments, from IR spectroscopy, from ion cyclotron resonance (ICR) studies of organometallic radicals in the gas phase, and from reaction chemistry, particularly studies of electron-transfer chain (ETC) catalysis of nucleophilic substitution reactions.

Identification of Radicals. Isotropic ESR spectra of radicals with one unpaired electron ($S = \frac{1}{2}$) are characterized by a g-value and hyperfine coupling constants for the

M. Chanon et al. (eds.), Paramagnetic Organometallic Species in Activation / Selectivity, Catalysis, 375–389.

magnetic nuclei. All transition metals have at least one isotope with a nuclear spin and a large enough magnetic moment to give rise to hyperfine coupling, although some, like ^{57}Fe, are low in natural abundance. The g-values of organotransition metal radicals differ from the free electron value ($g_e = 2.0023$) because of spin-orbit coupling and cover a wider range than those of organic radicals (analogous to the NMR chemical shift range for heavier atoms compared with hydrogen). When both the g-value and the hyperfine coupling pattern are known, it is often possible to identify a radical quite uniquely.

Information from ESR Spectra of Radicals in Solids. Although isotropic parameters are useful for radical identification, there is a lot more information available from ESR spectra of radicals in dilute single crystals or in frozen solutions. In particular, analysis of such spectra provides the magnitudes of the g- and nuclear hyperfine tensor components. Single crystal spectra also provide the orientations of the tensor principal axes; this information is largely lost in frozen solution spectra, although the orientations of the hyperfine tensor axes relative to those of the g-tensor sometimes can be extracted [3,4]. There are many pitfalls in powder pattern analysis, but, with care, spectra usually can be interpreted with some confidence [5].

An anisotropic hyperfine tensor usually is mostly due to electron-nuclear magnetic dipole-dipole interaction and reflects the odd electron distribution in the vicinity of the nucleus. ESR spectroscopists have had a lot of experience in the interpretation of dipolar tensors for simple radicals [6] and classical transition metal complexes [7]. Although complications from spin-orbit coupling and nd/(n+1)p mixing often have to be taken into account [8,9], this experience usually can be extended to organotransition metal radicals to extract the metal d-orbital contributions to the SOMO and sometimes the ligand contributions as well. Spin-orbit coupling contributions to the g-tensor contain information about the nature of the nearby filled and empty orbitals. The orientations of the principal axes of the g- and hyperfine tensors often can provide valuable clues to molecular structure.

Information from Linewidths of Solution Spectra. The time scale of ESR spectroscopy lies about midway between those of NMR and IR; processes with rates roughly in the range 10^7 to 10^{11} s^{-1} can influence linewidths. By far the most common contribution to line broadening arises from incomplete averaging of anisotropies in the g- and hyperfine tensors; the rate of tumbling in solution can be extracted from a linewidth analysis [10,11]. Intermolecular chemical processes are generally too slow to influence ESR linewidths, but intramolecular processes, e.g., ligand exchange in fluxional molecules, can be studied.

Measurement of Radical Concentrations. Since the integrated ESR intensity is proportional to radical concentration, ESR spectroscopy can be used as an analytical tool, e.g., to follow the concentration of a radical involved in a reaction. The principal limitations are those of sensitivity and response time. If the radical concentration is low, data acquisition will be slow, and only a relatively slow reaction can be followed. There has been little use made of the quantitative capabilities of ESR spectroscopy. This is unfortunate since the results of ESR studies are often held in some suspicion; the sensitivity of the technique allows detection of tiny amounts of side product radicals which may have little to do with a reaction involving spin-paired species. Chang's recent review of calibration methods [12] should be of value in quantitative applications of ESR spectra.

Generation of Radicals for ESR Study. Organometallic radicals are most commonly generated by chemical or electrochemical oxidation or reduction of a diamagnetic substrate in liquid solution. When an electrochemical oxidation or reduction involves one electron and is chemically reversible, it is usually assumed that a radical cation or anion is formed, and ESR spectroscopy is often used to verify that assumption. Ideally, the experiment should be done using a controlled-potential electrochemical cell mounted in the ESR cavity so that the ESR signal can be correlated to the passage of current at the potential thought to produce the radical species. Several cells have been described which are particularly suitable for this purpose [13,14]. Even when the expected radical is not detected, an ESR experiment can provide important clues to mechanism, particularly when an unexpected radical species is detected. Radicals can also be produced by photolytic bond cleavage, but they are often too reactive to study in solution. Important insights sometimes can be obtained from experiments where the reduction or oxidation is performed in the solid state, for example, by photolysis or γ-radiolysis of a dilute solution of the diamagnetic precursor in a frozen matrix (or single crystal) known to produce electrons or holes on irradiation. The behavior of the ESR spectrum on annealling the irradiated solid can provide information on the chemical properties of the radicals.

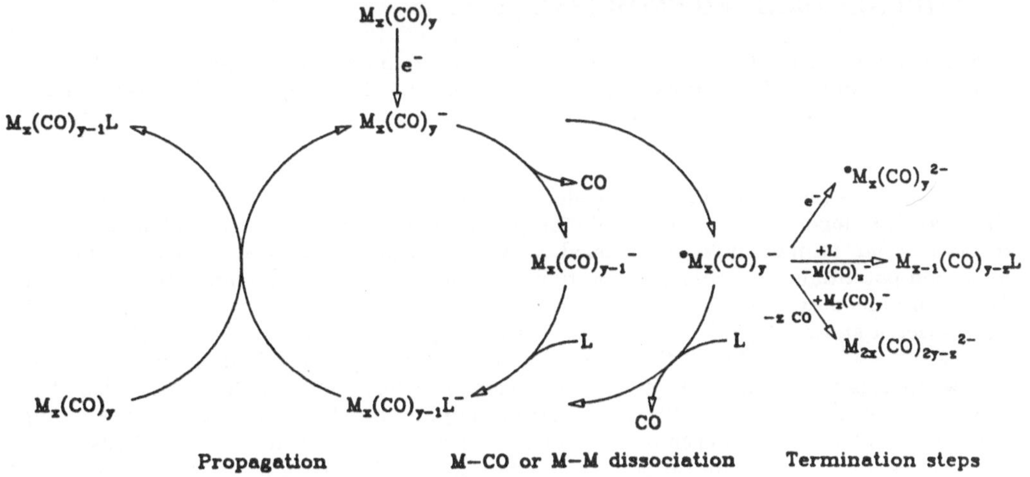

Scheme 1. The $S_{RN}1$ Mechanism

Electron-Transfer Chain Catalysed Nucleophilic Substitution. ESR spectroscopy is particularly relevant to the electron-transfer chain (ETC) catalysis mechanism for nucleophilic substitution since the chain carriers are radicals or radical ions which may be ESR-detectable. The mechanism has been studied extensively since the first organotransition metal examples were discovered in 1981 [15-17]. The general mechanism, which has ample precedent in organic and classical coordination chemistry [18-22], can be subdivided into reactions initiated by reduction with a unimolecular rate-limiting step ($S_{RN}1$), and by oxidation with a bimolecular rate-limiting step ($S_{ON}2$). In either case, initiation can be either chemical or electrochemical, and nucleophilic substitution is at a labile 17-electron metal center [23]. To be truly catalytic, it is necessary, of course, that the electron-transfer step which completes the chain be exoergic. In general, this condition is satisfied for $S_{RN}1$ displacement of CO by a phosphine or phosphite and this has been the most commonly encountered case. The $S_{ON}2$ mechanism has been demonstrated for a number of mononuclear organotransition metal systems [24], but most of the $S_{RN}1$

examples have been polynuclear clusters, e.g., $Fe_3(CO)_{12}$ [25] and other iron clusters [26,27], $Ru_3(CO)_{12}$ [25,28–30], $YCCo_3(CO)_9$ [15,25,31], and $(R_2C_2)Co_2(CO)_6$ [15,32,33]. For an $S_{RN}1$ reaction of a mononuclear complex, the mechanism must involve dissociation of a ligand [16], but, as shown in Scheme 1, two modes of reaction are possible for a cluster: (i) metal–ligand bond dissociation and (ii) metal–metal bond dissociation to produce an activated intermediate (designated by * in the scheme). Both modes have been postulated [15,27], and one of the goals of mechanistic studies of ETC catalysed reactions is the elucidation of this detail. A variety of terminating side reactions are possible, many of which have analogs in organic systems [21]. Some of the side reactions known to occur are: (1) Further reduction of the 17–electron intermediate; (2) Cluster fragmentation to produce, e.g., a carbonylmetallate; and (3) Production of clusters of higher nuclearity through radical dimerization or reaction of the radical anion (or a dianion) as a nucleophile with the activated radical anion.

2. *RADICALS DERIVED FROM $YCCo_3(CO)_9$*

The tricobaltcarbon clusters were among the first organotransition metal systems to receive a systematic ESR study [1]. In this section, the ESR results will be reviewed, looking particularly for light shed on the electrochemistry and $S_{RN}1$ reactions of these clusters.

ESR Results. Alkali metal or electrochemical reduction [1] of $YCCo_3(CO)_9$ (Y = H, alkyl, aryl, halogen, etc.) in THF solution produces anion radicals, most of which are quite stable at room temperature. In general, they are not particularly air-sensitive since the reduction potentials are more positive than that of oxygen [2,34]; indeed the anion radicals can be produced by reduction of the neutral cluster with potassium superoxide [35]. The ESR spectra show 22 lines as expected for three equivalent spin $\frac{7}{2}$ ^{59}Co nuclei, $\langle g \rangle$ = 2.01–2.03, $\langle a^{Co} \rangle$ = 34–37 G [1]. Several P- and Ge–capped cobalt clusters gave similar radical species with nearly identical ESR spectra [36–38]. With the sole exception of the Y = F anion, $\langle a^F \rangle$ = 12 G [39], no hyperfine coupling has been detected to the capping atom or to substituents thereon. Somewhat surprisingly, controlled-potential electrolysis of the phosphine- or phosphite-substituted derivatives, $YCCo_3(CO)_8L$, does not give ESR-detectable radical anions.

ESR spectra of $YCCo_3(CO)_9^-$ in frozen THF solution [39] show well-resolved parallel features, $g_\parallel$ = 1.996, $a_\parallel$ = 79.0 G for Y = Ph. The perpendicular features are not resolved, but the parameters can be estimated from the isotropic values to give $g_\perp$ = 2.022, $a_\perp$ = 14.2 G for Y = Ph. The ^{59}Co hyperfine tensors are very similar to those obtained by Strouse and Dahl [40] from ESR spectra of the neutral radicals, $SCo_3(CO)_9$ and $SeCo_3(CO)_9$, diluted in single crystals of the diamagnetic clusters, $SFeCo_2(CO)_9$ and $SeFeCo_2(CO)_9$. A variety of arguments [39,40] lead to the conclusion that the SOMO in these radicals is a trimetal antibonding orbital with approximately 25% d_{xy} contributions from each cobalt (taking the three-fold axis as z). Apparently the SOMO is very similar for the phosphorus- and germanium–capped tricobalt nonacarbonyl radical anions.

Correlation with Electrochemical and Chemical Data. As expected from the ESR results, cyclic voltammograms of $YCCo_3(CO)_9$ show chemically reversible one–electron reductions in acetone [2] or CH_2Cl_2 [34] solutions.

The tricobaltcarbon clusters are generally good substrates for ETC–catalysed nucleophilic substitution reactions [15,25,31]. Because the odd electron occupies a metal-metal antibonding orbital in the anion radical, it has been assumed that the dissociative step consists of Co–Co bond cleavage to produce a bicyclobutane structure with a formal 17-electron center. This interpretation is supported by the 60 cm^{-1} shift to lower frequency observed in the C-O stretching bands of the IR spectrum of the Y = Ph anion radical [1], which suggests a weakening of the C–O bond possibly accompanied by a small increase in the Co–C bond strength. The form of the rate law expected for an ETC reaction with Co–Co bond dissociation does not differ significantly from that expected for Co–CO dissociation, and kinetic studies [31,35] lead to empirical rate laws which are consistent with either mechanism. However, the rate constants seem to make most sense when applied to the Co–Co bond dissociation model.

Production of di- or trisubstitution products via the ETC reaction is much less efficient, especially when the reaction is electrochemically initiated. This fact, together with the failure to observe ESR spectra of phosphine-substituted radical anions and the overall two-electron reduction found for $PhCCo_3(CO)_8PPh_3$ [31], can be understood if we make the reasonable assumptions that phosphine substitution (i) increases the rate of ring-opening, (ii) shifts the equilibrium toward the open structure, and (iii) slows the substitution step; thus substitution no longer competes effectively with further electrochemical reduction.

3. *RADICALS DERIVED FROM $(R_2C_2)Co_2(CO)_6$*

The acetylene-bridged dicobalt clusters, $(R_2C_2)Co_2(CO)_{6-n}L_n$, can be reduced electrochemically to anion radicals in THF solution [41,42]. The stability of these radicals depends critically on the acetylene substituents and on Lewis base substitution with no evidence for long-lived radicals for n > 0. For R = CF_3, the anions are stable for hours, but for R = Ph, the anion lifetime is about one minute at -60° C.

Identification of Radicals. Isotropic ESR spectra show coupling to two equivalent ^{59}Co nuclei, $\langle g \rangle$ = 2.009–2.015, $\langle a^{Co} \rangle$ = 24–29 G, consistent with the expected anion radicals [3]. However, when the R = Ph cluster is reduced in the range -50° to -30° C, an eight-line spectrum is obtained, $\langle g \rangle$ = 2.06, $\langle a^{Co} \rangle$ = 50 G, suggesting a mononuclear radical species. Reduction of a Lewis base-substituted cluster or reduction of the unsubstituted cluster in the presence of an excess of Lewis base results in spectra showing coupling to one ^{59}Co, $\langle a^{Co} \rangle$ = 39–48 G, and, depending on the conditions, to one, two or three ^{31}P nuclei, $\langle a^{P} \rangle$ = 111–170, 67–105, and 77–79 G, respectively. The ESR parameters depend on the R group, and, since up three Lewis bases can be substituted, these spectra have been attributed to the mononuclear acetylene complexes, $(R_2C_2)Co(CO)_{3-n}L_n$ [42,43].

Correlation with Electrochemical Results. The identification of the mononuclear radicals allowed interpretation of the complex electrochemical behavior of the binuclear parents [42,43]. Cyclic voltammograms show electrochemically reversible one-electron reductions which are chemically reversible at room temperature for R = CF_3, but only at -30° C for R = Ph [42]. For R = Ph, the reduction involves approximately two electrons under 1 atm CO and 0.3-0.5 electrons in the presence of Lewis bases. In either case, the characteristic oxidation of $Co(CO)_4^-$ is observed in cyclic voltammograms, as well as features attributable to the reversible reduction of the mononuclear radicals and (with Lewis bases) to the reduction of substituted binuclear clusters. These results can be understood in terms of a competition between ETC substitution and displacement of $Co(CO)_4^-$ by the

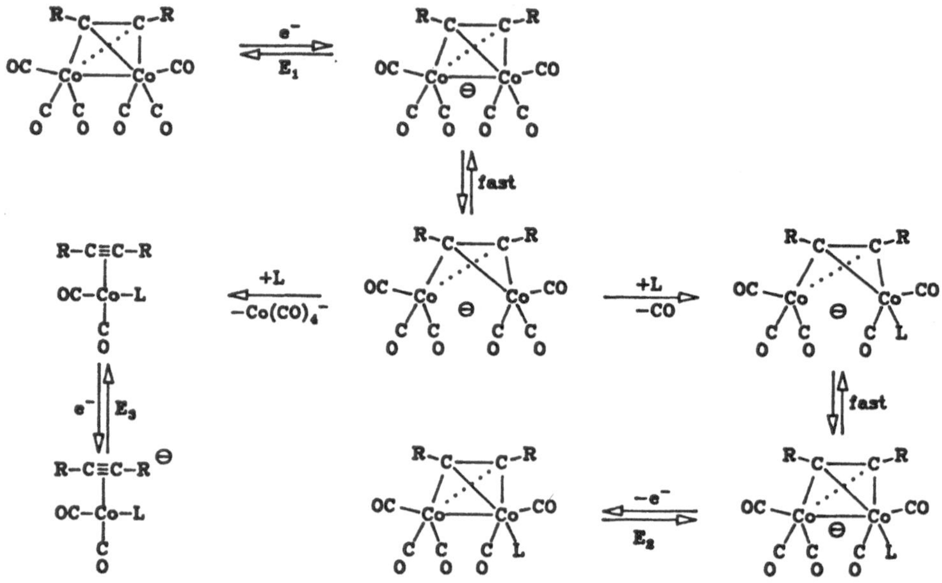

Scheme 2. Substitution/Fragmentation Competition

nucleophile (CO or Lewis base) as shown in Scheme 2. Since $E_3 > E_1$ for L = CO, the major reduction product under CO is the mononuclear anion (two electrons overall). When L is a phosphine or phosphite, $E_3 \cong E_2 < E_1$, so that the net current is reduced to the extent that substitution of the binuclear cluster occurs. The fragmentation/substitution ratio increases with further substitution so that, for example, a mixture of $(R_2C_2)Co(CO)_2L^-$ and $(R_2C_2)Co(CO)L_2$ is formed when the binuclear cluster is reduced in the presence of excess Lewis base. Tri-substituted radicals are obtained when $(R_2C_2)Co_2(CO)_3L_3$ is reduced in the presence of excess L.

When one or both of the acetylene substituents is strongly electron-withdrawing, e.g., $-CF_3$, the fragmentation pathway is largely suppressed and the $S_{RN}1$ reaction gives excellent yields [15,32,33,42]. Because the SOMO is strongly metal-metal antibonding, it has been assumed that Co-Co bond cleavage constitutes the dissociative step. This interpretation receives circumstantial support from the fragmentation side reaction, since Co-Co bond cleavage is obviously a necessary step in this path.

Solid State ESR Spectra. Frozen THF solutions of $(R_2C_2)Co_2(CO)_6^-$ give well-resolved ESR spectra. Detailed analysis of a typical spectrum shows that the g- and ^{59}Co hyperfine tensors do not share principal axes [3]. Accordingly, the SOMO can be described as a bent Co-Co σ^* orbital with about 64% of the spin density in Co d_{z^2} orbitals. The local z-axes are oriented approximately toward axial carbonyl ligands and thus are tilted away from the Co-Co internuclear vector by 15-20°.

Frozen THF solution spectra of the mononuclear acetylene radicals are mostly poorly resolved, but two examples have been found with sufficient resolution to permit at least partial analyses. Spectra of $(Ph_2C_2)Co(CO)_2PBu_3$ and $(Ph_2C_2)Co(CO)[P(OEt)_3]_2$ [43,44] are consistent with a d_z2 SOMO and a pseudo-C_{3v} structure, analogous to that of the isoelectronic $Co(CO)_4$ radical [45-47]. The *bis*-phosphite radical exists in two isomeric forms, one with axial and equatorial phosphite ligands, the other with both phosphites equatorial [44].

Rates of Fluxionality from ESR Linewidths. The isotropic spectrum of the *bis*-phosphite radicals show two equivalent phosphorus nuclei. However, rapid interconversion of the ax,eq and eq,eq isomers (and interchange of the phosphite ligands in the ax,eq isomer) leads to broadening of the central lines of the 1:2:1 ^{31}P triplets. Given the average phosphorus couplings for the static conformations from the frozen solution ESR spectrum, measured linewidths lead to an estimate of the conformation lifetime, 5×10^{-11} s at 290 K [43,44]. Similar analyses at other temperatures give an activation energy of 17 kJ mol^{-1}, consistent with the rather small distortion required to convert a pseudo-C_{3v} structure to the tetrahedral transition state.

4. *RADICALS DERIVED FROM IRON CARBONYLS*

The rich carbonyl chemistry of iron has produced a family of radical species identified by ESR spectroscopy. This section will examine the evidence for radical identification, the correlation between ESR and electrochemical data, and ESR results for more reactive species which have been observed only in the solid state.

1	$Fe(CO)_5$	$\mathbf{2}^{2-}$	$Fe(CO)_4{}^{2-}$	$\mathbf{3}^{2-}$	$Fe_2(CO)_8{}^{2-}$
4	$Fe_3(CO)_{12}$	$\mathbf{5}^{2-}$	$Fe_3(CO)_{11}{}^{2-}$	$\mathbf{6}^{2-}$	$Fe_4(CO)_{13}{}^{2-}$

Iron carbonyl radical precursors

Identification of Radicals. Krusic, et al. [48], found that alkali metal reduction of $Fe(CO)_5$ (**1**) in THF solution gave four resonances at $\langle g \rangle = 2.0016$, 2.0134, 2.0385, and 2.0497, attributed to anion radicals $\mathbf{4}^-$, $\mathbf{6}^-$, $\mathbf{3}^-$, and $\mathbf{5}^-$, respectively. ^{57}Fe and ^{13}C labelling studies provided a unique identification of $\mathbf{4}^-$ (three equivalent iron nuclei, 12 equivalent carbons) [48]. The same resonance was observed when **4** was reduced in CH_2Cl_2 solution at low temperature [49,50]. (When the reduction was at room temperature or in THF solution [51], the $\langle g \rangle = 2.038$ and 2.050 resonances were observed as well.) $\mathbf{4}^-$ is also formed on acidification of THF solutions of $NaHFe_2(CO)_8$ [52] and has been produced by Symons and co-workers by γ-radiolysis of **4** in frozen methyltetrahydrofuran [53]. Labelling studies similarly identified $\mathbf{6}^-$ (three equivalent and one unique iron, four sets of carbons in groups of six, three, three, and one) [48]. The spectrum of $\mathbf{3}^-$, which showed coupling to two equivalent ^{57}Fe nuclei but unresolved ^{13}C couplings, could also be obtained by chemical oxidation of $\mathbf{3}^{2-}$ [48]. This resonance was observed when **4** was reduced at room temperature [49,50] and when $NaHFe_2(CO)_8$ was dissolved in THF [52]. The $\mathbf{5}^-$ resonance showed coupling to a single iron nucleus and ^{13}C hyperfine structure was unresolved. However, the same signal was obtained on chemical oxidation of $\mathbf{5}^{2-}$ [48]. Additional evidence supporting this identification was provided by Chini [54] who reported that $\mathbf{5}^-$ is obtained from the redox redistribution reaction,

$$Fe_3(CO)_{11}^{2-} + Fe_3(CO)_{12} \rightarrow 2\,Fe_3(CO)_{11}^{-} + CO$$

Correlation with Electrochemical Experiments. Dawson, et al. [50], who observed the <g> = 2.050 resonance in reduced THF or CH_2Cl_2 solutions of **4**, claimed that $\mathbf{5}^{2-}$ is electro-inactive and suggested that the resonance is due to an Fe(I) species, possibly $\mathbf{1}^{+}$ or a derivative thereof. Reduction of phosphine- and phosphite-substituted clusters, $Fe_3(CO)_{12-n}L_n$ (n = 1-3) produced spectra which resembled those of $Fe(CO)_3L_2^{+}$ (see below), lending some support to this hypothesis. However, recent work by Amatore and Krusic [58] has shown that $\mathbf{5}^{2-}$ is indeed reducible, and the assignment of the <g> = 2.050 resonance to $\mathbf{5}^{-}$ now seems secure. Electrochemical studies are otherwise consistent with the ESR results. Reduction of **4** in acetone, THF or CH_2Cl_2 solutions involves one electron and is both chemically and electrochemically reversible [49,55-57], consistent with $\mathbf{4}^{-}$ as the primary reduction product. A second reduction, about 0.5 V more negative than the first, is thought to generate the dianion which either reproportionates to $\mathbf{4}^{-}$ or reacts to give unidentified products (anodic peaks in cyclic voltammograms). Controlled potential electrolysis at the potential of the first wave [57] resulted in uptake of one Faraday and production of two moles of **1** per mole of **4**, but the other products were not identified. **4** slowly disproportionates in THF, yielding small amounts of $\mathbf{3}^{-}$ and $\mathbf{5}^{-}$, which are ESR-detectable.

The main shortcoming of electrochemical studies of **4** has been that the radical anion is stable on the cyclic voltammetry time scale. Thus the various products have either been generated by sweeping into the second wave or by controlled potential electrolysis. In the first instance, the properties of the radical anion are not probed and the second approach misses the short-lived intermediates. Robinson [50] has argued that the principal decay mode of $\mathbf{4}^{-}$ is via disproportionation and decomposition of the dianion,

$$2\,Fe_3(CO)_{12}^{-} \rightleftarrows Fe_3(CO)_{12} + Fe_3(CO)_{12}^{2-}$$

$$Fe_3(CO)_{12}^{2-} \rightarrow \text{Products}$$

However, since K_{eq} for the disproportionation is on the order of 10^{-9} at room temperature, this mechanism seems unlikely. Miholová, et al. [59], have shown that the reproportionation reaction is quite important and have estimated a rate on the order of 10^{4}-10^{5} $M^{-1}s^{-1}$, suggesting a disproportionation rate constant of 10^{-5}-10^{-4} $M^{-1}s^{-1}$, probably too slow to materially influence the chemistry. Relying on the ESR results, it would seem that $\mathbf{4}^{-}$ decays via two pathways: (1) CO loss to form $\mathbf{5}^{-}$ (Fe-C bond cleavage); and (2) fragmentation to mono- and binuclear species (Fe-Fe bond cleavage). However, whether these are parallel or sequential pathways is unknown; confirmation from electrochemistry awaits further experiments.

4 is an indifferent substrate for ETC-catalysed nucleophilic substitution [25], requiring relatively large amounts of BPK initiator and giving only fair yields. This is consistent with a slow rate of bond cleavage in $\mathbf{4}^{-}$. It is tempting to suppose that the propagation steps include unimolecular CO loss followed by nucleophilic attachment and electron transfer, i.e., that $\mathbf{5}^{-}$ is the key intermediate. However, the evidence now available is insufficient to warrent this conclusion.

Until recently, the electrochemistry of **1** was replete with unanswered questions and seemed to show little connection to the ESR results. Many of these connections have now been completed [58]. A chemically irreversible reduction is seen in THF or CH_3CN

solutions [55,57]; controlled-potential electrolysis showed that one electron is involved and that the ultimate product is 4^{2-}. Addition of water apparently does not affect the primary reduction process, but the ultimate product then is $HFe(CO)_4^-$, and the total electron consumption increases from one to two [57], presumably via the process,

$$Fe(CO)_5 + 2\,e^- + H^+ \rightarrow HFe(CO)_4^- + CO$$

Bond, et al. [56], found a two-electron reduction in acetone, presumably the ECE process,

$$Fe(CO)_5 \xrightarrow{e^-} [Fe(CO)_5^-] \xrightarrow{-CO} Fe(CO)_4^- \xrightarrow{e^-} Fe(CO)_4^{2-}$$

Amatore and Krusic [58] found that the reduction does indeed involve two electrons at fast scan rates, but at slower scans, the reproportionation reaction,

$$Fe(CO)_4^{2-} + Fe(CO)_5 \rightarrow Fe_2(CO)_8^{2-} + CO$$

reduces the net electron uptake to one. There is evidence for the intermediacy of the $\mathbf{2}^-$ in the reduction of **1**: Reduction in acetone in the presence of PPh_3 shows the reduction peak characteristic of $Fe(CO)_4PPh_3$ [15], presumably formed through ETC catalysed nucleophilic substitution of **1**. It is noteworthy that neither $\mathbf{1}^-$ nor $\mathbf{2}^-$ is observed when **1** is reduced in liquid solution [48]. There is no evidence for $\mathbf{1}^-$ from ICR studies of **1** [60,61]; $\mathbf{2}^-$ is the major product of gas-phase electron attachment. Thus CO loss from $\mathbf{1}^-$ must be very rapid. $\mathbf{2}^-$ is apparently very reactive, either attacking **1** to produce $\mathbf{3}^-$ or dimerizing to give $\mathbf{3}^{2-}$.

Solid State ESR Experiments. Morton and Preston [62] produced $\mathbf{1}^-$ by γ-radiolysis at 77 K of a single crystal of $Cr(CO)_6$ doped with **1**. The ESR spectrum, which was well resolved below 20 K, gave a detailed picture of the structure. The anion is best described as an acyl radical, $(OC)_4FeCO^-$, with the odd electron shared between the iron and a bent carbonyl group. A spectrum thought to be that of $\mathbf{1}^-$ was observed by Symons and co-workers [53] on γ-radiolysis of **1** in a MeTHF glass at 77 K, but this spectrum was very different from that of Morton and Preston and is most likely due to $\mathbf{2}^-$. The spectrum has one g-component near the free electron value, suggesting a d_{z^2} SOMO. By analogy with the isoelectronic $Co(CO)_4$ radical [45-47], a C_{3v} structure is expected for $\mathbf{2}^-$. $Co(CO)_4$ has a d_{z^2} SOMO and a unique carbonyl group giving a large ^{13}C hyperfine coupling. The presumed spectrum of $\mathbf{2}^-$ [53] showed an increase in line width when enriched in ^{13}C, but it is unclear how many carbon nuclei are coupled. On annealling the sample at 120 K, a new resonance appeared. This signal is almost certainly due to $\mathbf{3}^-$ since the g-tensor is in excellent agreement with that obtained for $\mathbf{3}^-$ in MeTHF [58]. Thus the annealling process must correspond to

$$Fe(CO)_4^- + Fe(CO)_5 \rightarrow Fe_2(CO)_8^- + CO$$

Cation Radicals. Morton and Preston also found the cation radical, $\mathbf{1}^+$, in a γ-irradiated crystal of $Cr(CO)_6$ doped with **1** [63] and in γ-irradiated Kr matrix containing **1** [64]. The spectra differed significantly, possibly through medium effects. $Cr(CO)_6$ may impose a square pyramidal structure and the ^{83}Kr hyperfine coupling observed in Kr matrices suggests pseudo-octahedral geometry. These spectra were interpretable in terms of a SOMO about 55% d_{z^2} in character, consistent with a C_{4v} d^7 species. Symons and co-workers [53] obtained a spectrum assigned to the same radical cation from γ-irradiated frozen solutions of **1** in $CFCl_3$ with ESR parameters rather similar to those obtained in the

Kr matrix (though without the ^{83}Kr coupling, of course).

1 is oxidized electrochemically in trifluoroacetic acid by a reversible one-electron process [65], but the lifetime of $\mathbf{1}^+$ from cyclic voltammetry is less than a second at room temperature. On the other hand, Morton and Preston [63] obtained the ESR spectrum of $\mathbf{1}^+$ by dissolving **1** in concentrated sulfuric acid.

A large number of phosphine-substituted derivatives of $\mathbf{1}^+$ have been observed, e.g., $Fe(CO)_3(PPh_3)_2^+$ [66-69]. These show coupling to two equivalent ^{31}P nuclei, $<a^P> =$ 17-25 G, and a g-tensor at least somewhat similar to those reported for $\mathbf{1}^+$. The neutral precursors are known to be trigonal bipyramidal with *trans* phosphine ligands, and Therien and Trogler [69] have used IR spectra of ^{13}C labelled compounds to show that the stereochemistry is retained on oxidation to the radical cation. These authors also presented SCF-Xα-DV calculations on $Fe(CO)_3(PH_3)_2^+$ which show the SOMO (for D_{3h}) to have e' symmetry with 54% metal d_{xy}, $d_{x^2-y^2}$ character. However, the g-tensor reported [67] for $Fe(CO)_3(PPh_3)_2^+$ suggests a d_{z^2} SOMO similar to that of $\mathbf{1}^+$. Furthermore, the dppm- and dppe-substituted cations, which are necessarily *cis*, show ^{31}P couplings similar to those of the *bis*-phosphine derivatives [67]; the magnitudes of these couplings are consistent with equatorial phosphine ligands and a d_{z^2} SOMO or with axial phosphines and a $d_{xy}/d_{x^2-y^2}$ SOMO. These inconsistencies have yet to be resolved.

5. *RADICALS DERIVED FROM $Ru_3(CO)_{12}$*

A weak ESR signal is obtained when $Ru_3(CO)_{12}$ in THF solution is reduced by sodium at room temperature and the solution quickly cooled [51,70]. The single line, $<g> = 1.986$, has ^{99}Ru,^{101}Ru satellites with amplitudes consistent with three equivalent Ru atoms, $<a^{Ru}> = 24$ G. This signal was originally assigned to the anion radical [51], but low temperature electrochemical or chemical reduction of $Ru_3(CO)_{12}$ or electrochemical oxidation of $Ru_3(CO)_{11}^{2-}$ gave no ESR signal so that assignment of the observed resonance to $Ru_3(CO)_{12}^-$ has been somewhat doubtful.

Electrochemical reduction of $Ru_3(CO)_{12}$ in acetone, THF, or CH_2Cl_2 is a chemically and electrochemically irreversible two-electron process [70-72],

$$Ru_3(CO)_{12} \xrightarrow{e^-} [Ru_3(CO)_{12}^-] \xrightarrow{fast} {}^*Ru_3(CO)_{12}^- \xrightarrow{e^-} {}^*Ru_3(CO)_{12}^{2-}$$

Ru-Ru bond cleavage is either concerted with the first electron transfer step or very rapidly follows (the cyclic voltammogram remains irreversible down to -100° C [72] and up to scan rates of 10^4 V s^{-1} at a Pt microelectrode [71]). Reduction of the open chain radical anion, $^*Ru_3(CO)_{12}^-$, is highly exoergic at the reduction potential of $Ru_3(CO)_{12}$. This behavior is not unexpected since the $Ru_3(CO)_{12}$ LUMO is strongly metal-metal antibonding [73]. The open-chain dianion is quite unstable and rapidly decays to other products. Among the decomposition products is $Ru_3(CO)_{11}^{2-}$, the formation of which Robinson and co-workers [71] have shown to be catalysed by traces of moisture. At higher concentrations of the substrate, the octahedral cluster, $Ru_6(CO)_{18}^{2-}$, is also formed, probably through the redox condensation reaction,

$$Ru_3(CO)_{12}^{2-} + Ru_3(CO)_{12} \rightarrow Ru_6(CO)_{18}^{2-} + 6\,CO$$

In general, the electrochemistry of $Ru_3(CO)_{12}$ is very sensitive to solvent, supporting electrolyte, temperature, and substrate concentration.

In marked contrast to the complex electrochemistry and the apparent instability of $Ru_3(CO)_{12}^-$, $Ru_3(CO)_{12}$ is one of the best substrates yet found for the ETC catalysed nucleophilic substitution reaction, provided that the reaction is initiated chemically in THF solution [25,28–30]. When the ETC reaction is initiated by an electrode, the chain lengths are short and yields low, an understandable result since reduction of the radical anion competes effectively with nucleophilic substitution. When the reaction is initiated with a small concentration of a one–electron reducing agent such as BPK, on the other hand, a radical anion is more likely to encounter a nucleophile than another reducing agent and, provided that electron transfer to an unreacted substrate molecule is rapid (as it apparently is in THF, but less so in other solvents [71]), the ETC chain is completed and the reaction is fast and efficient. This suggests, of course, that the open chain radical anion is stable in the absence of nucleophiles and electron donors and that the observed ESR signal might correspond to this species after all. If this is the case, the spectrum, which showed hyperfine coupling to three equivalent Ru nuclei, would suggest rapid scrambling of the Ru atoms, presumably via a *triangulo* intermediate or transition state.

6. RADICALS DERIVED FROM $Mn_2(CO)_{10}$

The quest for ESR spectra of radicals formed on oxidation or reduction of $Mn_2(CO)_{10}$ has been particularly frustrating with many misleading and/or misinterpreted results. This species undergoes a chemically irreversible two–electron reduction in CH_2Cl_2, CH_3CN, and THF to give $Mn(CO)_5^-$, which in turn can be oxidized back to $Mn_2(CO)_{10}$. Several careful studies [55,74,75] have led to the conclusion that the $Mn(CO)_5$ radical is the key intermediate in an ECE electrode process. The first electron–transfer step is electrochemically irreversible, suggesting that Mn–Mn bond cleavage is concerted with, or very quickly follows, electron transfer,

$$Mn_2(CO)_{10} \xrightarrow{e^-} [Mn_2(CO)_{10}^-] \xrightarrow{fast} Mn(CO)_5 + Mn(CO)_5^-$$

$$Mn(CO)_5 \xrightarrow{e^-} Mn(CO)_5^-$$

$Mn(CO)_5$ is also thought to be the key intermediate in the two–electron oxidation; however, since the first electron–transfer step appears to be electrochemically reversible, the binuclear cation radical may have a finite lifetime. The ultimate product, $Mn(CO)_5^+$, is solvated by coordinating solvents such as CH_3CN, and the solvent is probably intimately involved in chemical or electrochemical steps involving this species,

$$Mn_2(CO)_{10} \xrightarrow{-e^-} [Mn_2(CO)_{10}^+] \xrightarrow[+S]{fast} Mn(CO)_5 + Mn(CO)_5S^+$$

$$Mn(CO)_5 \xrightarrow[+S]{-e^-} Mn(CO)_5S^+$$

It seems likely that the $Mn(CO)_5$ radical is also weakly solvated.

Solution photochemistry of $Mn_2(CO)_{10}$ is also dominated by $Mn(CO)_5$ production. ESR spin–trapping experiments have produced a variety of stable radical derivatives of $Mn(CO)_5$ and have provided informative probes of the mechanism [76–82]. The $Mn(CO)_5$ radical proved an elusive species for ESR study, consistent with its considerable reactivity. Early on, it was thought that the resonance obtained on photolysis of THF solutions of

$Mn_2(CO)_{10}$ might be due to $Mn(CO)_5$ or some closely related Mn(0) species [83,84], but the signal was shown [85,86] to be due to Mn(II) produced by photo-disproportionation

$$\tfrac{3}{2}\ Mn_2(CO)_{10} + 6\ THF \xrightarrow{h\nu} 2\ Mn(THF)_6^{2+} + Mn(CO)_5^- + 5\ CO$$

Early attempts to produce $Mn(CO)_5$ in solid matrices were unsuccessful [47], but the radical has since been produced and the ESR spectrum obtained by reaction of $Mn(CO)_5Br$ with silver atoms in a rotating cryostat [87], by co-condensation of Mn, CO and Ar on a cold finger [88], and by photolysis of $HMn(CO)_5$ in an argon matrix [89]. The odd electron in $Mn(CO)_5$ occupies an orbital which is about 60% $3d_{z^2}$, 8% 4s in character, consistent with the C_{4v} structure established by infrared spectroscopy [90].

In contrast to liquid solution electrochemistry, ICR studies show that roughly equal amounts of $Mn_2(CO)_9^-$ and $Mn(CO)_5^-$ are formed on gas-phase electron attachment of $Mn_2(CO)_{10}$. Interestingly, $Re_2(CO)_{10}$, which behaves like $Mn_2(CO)_{10}$ electrochemically, undergoes only CO loss on gas phase electron attachment [91]. Symons and co-workers [92,93] reduced $Mn_2(CO)_{10}$ via γ-radiolysis and obtained a powder spectrum thought to be that of the radical anion, $Mn_2(CO)_{10}^-$. However, a single crystal ESR study by Morton and Preston [94] showed that the ^{55}Mn hyperfine tensor principal axes are not colinear, suggesting that the observed radical is actually $Mn_2(CO)_9^-$ with a bridging CO ligand. The very similar spectrum obtained by γ-radiolysis of $Mn_2(CO)_8(P\text{-}n\text{-}Bu_3)_2$ and assigned [95] to the parent radical anion may thus correspond to $Mn_2(CO)_7(P\text{-}n\text{-}Bu_3)_2^-$. These results are consistent with the electrochemistry in that the binuclear radical anion is apparently never seen. Either Mn–Mn or Mn–CO bond cleavage occurs rapidly. It remains to be established, however, whether $Mn_2(CO)_9^{2-}$ is ever formed in solution, and, if not, why metal–metal bond cleavage dominates.

Although the electrochemical results described above suggest that the radical cation, $Mn_2(CO)_{10}^+$, might be more stable than the corresponding anion, there seem to have been no attempts to generate the species for ESR study. A related radical cation derived from $Mn_2(CO)_6(dmpm)_2$ [96] is apparently stable in CH_2Cl_2 solution ($\langle g \rangle = 2.009$, $\langle a^{Mn} \rangle = 24$ G), but the anisotropic parameters have not been determined.

Abbreviations used: ESR = electron spin resonance; NMR = nuclear magnetic resonance; IR = infrared; UV = ultraviolet; ICR = ion-cyclotron resonance; SOMO = semi-occupied molecular orbital; ETC = electron-transfer chain; THF = tetrahydrofuran; MeTHF = 2-methyltetrahydrofuran; BPK = benzophenone ketyl; dppm = *bis*(diphenylphosphino)methane; dppe = 1,2-(diphenylphosphino)-ethane; dmpm = *bis*(dimethylphosphino)methane.

References:

[1] B. M. Peake, B. H. Robinson, J. Simpson, and D. J. Watson, *Inorg. Chem.* **1977,** *16,* 405.
[2] A. M. Bond, B. M. Peake, B. H. Robinson, J. Simpson, and D. J. Watson, *Inorg. Chem.* **1977,** *16,* 410.
[3] B. M. Peake, P. H. Rieger, B. H. Robinson, and J. Simpson, *J. Am. Chem. Soc.* **1980,** *102,* 156.
[4] W. E. Geiger, P. H. Rieger, B. Tulyathan, and M. D. Rausch, *J. Am. Chem. Soc.* **1984,** *106,* 7000.
[5] P. H. Rieger, *J. Mag. Reson.* **1982,** *50,* 485; J. A. DeGray and P. H. Rieger, *Bull.*

Mag. Reson. **1987,** *8,* 95.
[6] P. W. Atkins and M. C. R. Symons, *The Structure of Inorganic Radicals,* Amsterdam: Elsevier, 1967.
[7] B. R. McGarvey, in *Transition Metal Chemistry,* Vol. 3, R. L. Carlin, ed, New York: Dekker, 1967; B. R. McGarvey, in *Electron Spin Resonance of Metal Complexes,* T. F. Yen, ed, New York: Plenum, 1969.
[8] A. J. Stone, *Proc. Roy. Soc. (London)* **1963,** *A271,* 424.
[9] T. Kawamura, S. Hayashida, and T. Yonezawa, *Chem. Phys. Lett.* **1981,** *77,* 348; T. Sowa, T. Kawamura, T. Yamabe, and T. Yonezawa, *J. Am. Chem. Soc.* **1985,** *107,* 6471.
[10] R. Wilson and D. Kivelson, *J. Chem. Phys.* **1964,** *44,* 154.
[11] B. M. Peake, P. H. Rieger, B. H. Robinson, and J. Simpson, *Inorg. Chem.* **1979,** *18,* 1000.
[12] T.-T. Chang, *Mag. Reson. Rev.* **1984,** *9,* 65.
[13] R. D. Allendoerfer, G. A. Martinchek, and S. Bruckenstein, *Anal. Chem.* **1975,** *47,* 890.
[14] K. R. Fernando, A. J. McQuillan, B. M. Peake, and J. Wells, *J. Magn. Reson.* **1986,** *68,* 551.
[15] G. J. Bezems, P. H. Rieger, and S. J. Visco, *J. Chem. Soc., Chem. Commun.* **1981,** 265.
[16] J. Grobe and H. Zimmermann, *Z. Naturforsch.* **1981,** *36b,* 301, 482.
[17] D. P. Summers, J. C. Luong, and M. S. Wrighton, *J. Am. Chem. Soc.* **1981,** *103,* 5238.
[18] J. F. Bunnett, *Accts. Chem. Res.* **1978,** *11,* 413.
[19] I. P. Beletskaya and V. N. Drozd, *Russ. Chem. Rev.* **1979,** *48,* 431.
[20] R. W. Alder, *J. Chem. Soc., Chem. Commun.* **1980,** 1184.
[21] J. M. Saveant, *Accts. Chem. Res.* **1980,** *13,* 323.
[22] M. Chanon and M. L. Tobe, *Angew. Chem., Intl. Ed.* **1982,** *21,* 1.
[23] B. H. Byers and T. L. Brown, *J. Am. Chem. Soc.* **1977,** *99,* 2527; T. L. Brown, *Ann. N. Y. Acad. Sci.* **1980,** *333,* 80.
[24] J. K. Kochi, *J. Organomet. Chem.* **1986,** *300,* 139.
[25] M. I. Bruce, D. C. Kehoe, J. G. Matisons, B. K. Nicholson, P. H. Rieger, and M. L. Williams, *J. Chem. Soc., Chem. Commun.* **1982,** 442.
[26] H. H. Ohst and J. K. Kochi, *J. Am. Chem. Soc.* **1986,** *108,* 2897; *Inorg. Chem.* **1986,** *25,* 2066.
[27] A. Darchen, C. Mahe, and H. Patin, *J. Chem. Soc., Chem. Commun.* **1982,** 243.
[28] M. I. Bruce, T. W. Hambley, B. K. Nicholson, and M. R. Snow, *J. Organomet. Chem.* **1982,** *235,* 83.
[29] M. I. Bruce, J. G. Matisons, B. K. Nicholson, and M. L. Williams, *J. Organomet. Chem.* **1982,** *236,* C57.
[30] M. I. Bruce, J. G. Matisons, and B. K. Nicholson, *J. Organomet. Chem.* **1983,** *247,* 321.
[31] A. J. Downard, B. H. Robinson, and J. Simpson, *Organometallics* **1986,** *5,* 1122, 1132.
[32] M. Arewgoda, B. H. Robinson, and J. Simpson, *J. Am. Chem. Soc.* **1983,** *105,* 1893.
[33] R. B. Cunninghame, L. R. Hanton, S. D. Jensen, B. H. Robinson, and J. Simpson, *Organometallics* **1987,** *6,* 1470; S. D. Jensen, B. H. Robinson, and J. Simpson, *ibid.* **1987,** *6,* 1479.
[34] J. C. Kotz, J. V. Petersen, and R. C. Reed, *J. Organomet. Chem.* **1976,** *120,* 433.
[35] E. S. Lee and P. H. Rieger, unpublished work.
[36] H. Beurich, T. Madach, R. Richter, and H. Vahrenkamp, *Angew. Chem.* **1979,** *91,*

760; U. Honrath. L. Shu–Tang, and H. Vahrenkamp, *Chem. Ber.* **1985,** *118,* 132.

[37] J. A. Christie, D. N. Duffy, K. M. MacKay, and B. K. Nicholson, *J. Organomet. Chem.* **1982,** *226,* 165.

[38] P. N. Lindsay, B. M. Peake, B. H. Robinson, J. Simpson, U. Honrath, H. Vahrenkamp, and A. M. Bond, *Organometallics* **1984,** *3,* 413.

[39] B. M. Peake, P. H. Rieger, B. H. Robinson, and J. Simpson, *Inorg. Chem.* **1981,** *20,* 2540.

[40] C. E. Strouse and L. F. Dahl, *Disc. Faraday Soc.* **1969,** *47,* 93. C. E. Strouse and L. F. Dahl, *J. Am. Chem. Soc.* **1971,** *93,* 6032.

[41] R. S. Dickson, B. M. Peake, P. H. Rieger, B. H. Robinson, and J. Simpson, *J. Organomet. Chem.* **1979,** *172,* C63.

[42] M. Arewgoda, P. H. Rieger, B. H. Robinson, J. Simpson, and S. J. Visco, *J. Am. Chem. Soc.* **1982,** *104,* 5633.

[43] L. V. Casagrande, T. Chen, P. H. Rieger, R. H. Robinson, J. Simpson, and S. J. Visco, *Inorg. Chem.* **1984,** *23,* 2019.

[44] J. A. DeGray, Q.–J. Meng, and P. H. Rieger, *J. Chem. Soc., Faraday Trans. I,* in press.

[45] H. J. Keller and H. Wawersik, *Z. Naturforsch.* **1965,** *20b,* 938.

[46] L. A. Hanlan, E. P. Kundig, B. R. McGarvey, and G. A. Ozin, *J. Am. Chem. Soc.* **1975,** *97,* 7054.

[47] S. A. Fieldhouse, B. W. Fulham, G. W. Neilson, and M. C. R. Symons, *J. Chem. Soc., Dalton Trans.* **1974,** 567.

[48] P. J. Krusic, J. San Filippo, Jr., B. Hutchinson, R. L. Hance, and L. M. Daniels, *J. Am. Chem. Soc.* **1981,** *103,* 2129.

[49] D. Miholová, J. Klima, and A. A. Vlček, *Inorg. Chim. Acta* **1978,** *27,* L67.

[50] P. A. Dawson, B. M. Peake, B. H. Robinson, and J. Simpson, *Inorg. Chem.* **1980,** *19,* 465.

[51] B. M. Peake, B. H. Robinson, J. Simpson, and D. J. Watson, *J. Chem. Soc., Chem. Commun.* **1974,** 945.

[52] J. P. Collman, R. G. Finke, P. L. Matlock, R. Wahren, R. G. Komoto, and J. L. Brauman, *J. Am. Chem. Soc.* **1978,** *100,* 1119.

[53] B. M. Peake, M. C. R. Symons, and J. L. Wyatt, *J. Chem. Soc., Dalton Trans.* **1983,** 1171.

[54] P. Chini, *J. Organomet. Chem.* **1980,** *200,* 37.

[55] C. J. Pickett and D. Pletcher, *J. Chem. Soc., Dalton Trans.* **1975,** 879.

[56] A. M. Bond, P. A. Dawson, B. M. Peake, B. H. Robinson, and J. Simpson, *Inorg. Chem.* **1977,** *16,* 2199.

[57] N. El Murr and A. Chaloyard, *Inorg. Chem.* **1982,** *21,* 2206.

[58] C. Amatore and P. J. Krusic, private communication.

[59] D. Miholová, J. Fiedler, and A. A. Vlček, *J. Electroanal. Chem.* **1983,** *143,* 195.

[60] R. C. Dunbar, J. P. Ennever, and J. P. Fackler, Jr., *Inorg. Chem.* **1973,** *12,* 2734.

[61] J. H. Richardson, L. M. Stephenson, and J. L. Brauman, *J. Am. Chem. Soc.* **1974,** *96,* 3671.

[62] S. A. Fairhurst, J. R. Morton, and K. F. Preston, *J. Chem. Phys.* **1982,** *77,* 5872.

[63] T. Lionel, J. R. Morton, and K. F. Preston, *J. Chem. Phys.* **1982,** *76,* 234.

[64] S. A. Fairhurst, J. R. Morton, R. N. Perutz, and K. F. Preston, *Organometallics* **1984,** *3,* 1389.

[65] C. J. Pickett and D. Pletcher, *J. Chem. Soc., Dalton Trans.* **1976,** 636.

[66] N. G. Connelly and K. R. Somers, *J. Organomet. Chem.* **1976,** *113,* C39.

[67] P. K. Baker, N. G. Connelly, B. M. R. Jones, J. P. Maher, and K. R. Somers, *J. Chem. Soc., Dalton Trans.,* **1980,** 579.

[68] R. N. Bagchi, A. M. Bond, C. L. Heggie, T. L. Henderson, E. Mocellin, and R. A. Seikel, *Inorg. Chem.* **1983,** *22,* 3007.
[69] M. J. Therien and W. C. Trogler, *J. Am. Chem. Soc.* **1986,** *108,* 3697.
[70] J. E. Cyr, J. A. DeGray, D. K. Gosser, E. S. Lee, and P. H. Rieger, *Organometallics* **1985,** *4,* 950.
[71] A. J. Downard, B. H. Robinson, J. Simpson, and A. M. Bond, *J. Organomet. Chem.* **1987,** *320,* 363.
[72] J. E. Cyr and P. H. Rieger, to be published.
[73] M. C. Manning and W. C. Trogler, *Coord. Chem. Rev.* **1981,** *38,* 89.
[74] P. Lemoine, A. Giraudeau, and M. Gross, *Electrochim. Acta* **1976,** *21,* 1.
[75] D. A. Lacombe, J. E. Anderson, and K. M. Kadish, *Inorg. Chem.* **1986,** *25,* 2074.
[76] A. Vlček, Jr. *J. Organomet. Chem.* **1986,** *306,* 63.
[77] A. Alberti, M. C. Depew, A. Hudson, W. G. McGimpsey, and J. K. S. Wan, *J. Organomet. Chem.* **1985,** *280,* C21.
[78] G. A. Abakumov, V. K. Cherkasov, K. G. Shalnova, I. A. Teplova, and G. A. Razuvaev, *J. Organomet. Chem.* **1982,** *236,* 333.
[79] A. Alberti and C. M. Camaggi, *J. Organomet. Chem.* **1980,** *194,* 343.
[80] L. Pasimeni, P. L. Zanonato, and C. Corvaja, *Inorg. Chim. Acta* **1979,** *37,* 241.
[81] A. S. Huffadine, B M. Peake, B. M. Robinson, J. Simpson, and P. A. Dawson, *J. Organomet. Chem.* **1976,** *121,* 391.
[82] A. Hudson, M. G. Lappert, P. W. Lednor, and B. K. Nicholson, *J. Chem. Soc., Chem. Commun.* **1974,** 966.
[83] S. A. Hallock and A. Wojcicki, *J. Organomet. Chem.* **1973,** *54,* C27.
[84] C. L. Kwan and J. K. Kochi, *J. Organomet. Chem.* **1975,** *101,* C9.
[85] A. Hudson, M. F. Lappert, and B. K. Nicholson, *J. Organomet. Chem* **1975,** *92,* C11.
[86] A. Hudson, M. F. Lappert, J. J. MacQuitty, B. K. Nicholson, H. Zainal, G. R. Luckhurst, C. Zannoni, S. W. Bratt, and M. C. R. Symons, *J. Organomet. Chem.* **1976,** *110,* C5.
[87] J. A. Howard, J. R. Morton, and K. F. Preston, *Chem. Phys. Lett.* **1981,** *83,* 226.
[88] H. Huber, E. P. Kündig, G. A. Ozin, and A. J. Poë, *J. Am. Chem. Soc.* **1985,** *97,* 308. ESR result quoted by Huffadine, *et al.* [81].
[89] M. C. R. Symons and R. L. Sweany, *Organometallics* **1982,** *1,* 834.
[90] S. P. Church, M. Poliakoff, J. A. Timney, and J. J. Turner, *J. Am. Chem. Soc.* **1981,** *103,* 7515.
[91] W. K. Meckstroth and D. P. Ridge, *J. Am. Chem. Soc.* **1985,** *107,* 2281.
[92] O. P. Anderson and M. C. R. Symons, *J. Chem. Soc., Chem. Commun.* **1972,** 1020.
[93] S. W. Bratt and M. C. R. Symons, *J. Chem. Soc., Dalton Trans.* **1977,** 1314.
[94] T. Lionel, J. R. Morton, and K. F. Preston, *Inorg. Chem.* **1983,** *22,* 143.
[95] M. C. R. Symons, J. Wyatt, B. M. Peake, J. Simpson, and B. H. Robinson, *J. Chem. Soc., Dalton Trans.* **1982,** 2037.
[96] G. R. Lemke and C. P. Kubiak, *Inorg. Chim. Acta* **1986,** *113,* 125.

IMPORTANCE OF PARAMAGNETIC ORGANOMETALLIC SPECIES IN CLUSTER ELECTROCHEMISTRY

P. LEMOINE
Laboratoire d'Electrochimie et de Chimie-Physique du Corps Solide, U.A. au C.N.R.S. n°405, UNIVERSITE LOUIS PASTEUR
4, rue Blaise Pascal 67000 STRASBOURG - FRANCE -

ABSTRACT : This report aims to summarize the important new trends in cluster electrochemistry that have been reported in the literature. In particular, the reactivity of paramagnetic cluster species, a relatively new area of electrochemical research, will be addressed. Important contributions and examples of recent cluster synthesis will be given. In addition, mechanistic studies of electron-transfer catalysed reactions involving paramagnetic cluster species intermediates will be presented.

1. INTRODUCTION

Many research teams work in this area ; those directed by KOCHI (1), RIEGER (2), GEIGER (3), ROBINSON-SIMPSON (4), GRAY (5), DAHL (6), DARCHEN (7), are among the most known. Clusters as models for nitrogenase and non-heme iron proteins are under investigation. The catalytic reduction of dinitrogen to NH_3 and N_2H_4 at ambient temperature and pressure has been achieved by sodium amalgam or at a mercury cathode (8). Some data suggest that the active form of the catalyst has a composition <u>derived from a cluster</u> containing a Mo_4 - O MgO - Mo_4 framework (9) . Furthermore, using reduced clusters of $[Fe_4S_4L_4]^{2-}$ and $[Fe_6Mo_2S_8L_9]^{3-}$ (L = SPh , SCH_2CH_2OH) as homogeneous catalysts, TANAKA et al have reduced N_2H_4 into NH_3 . When L = SPh, these species were also able to reduce CH_3NC into CH_4 and CH_3NH_2 , whereas CH_3CN could be reduced into C_2H_6 and NH_3 (10,11). Also, the anionic cluster $[Fe_4S_4(SR)_4]^{2-}$ was shown to be active in the electrocatalytic reduction of CO_2 to CO and formate at a facilitated potential of about 0.7 V lower than that of CO_2 (12). These few examples show that clusters are important in nitrogenase as well as in the design of models for non-heme iron proteins reactions. <u>Synthetic molecular compounds intermediate between molecules and crystals containing zero-valent or low-valent metals</u> offer a promising research field. Two examples have been reported in the literature. Firstly, the solid state structure of $[Nb_3Cl_6(C_6Me_6)_3]^{2+}$

M. Chanon et al. (eds.), Paramagnetic Organometallic Species in Activation / Selectivity, Catalysis, 391–405.

alternating with $[TCNQ]_2^{2-}$ dimers in a bent chain configuration, which is reminiscent of organic charge-transfer complexes [13] and secondly, the organic donor-hexanuclear cluster halide $[(TMTTF)_2Mo_6Cl_{14}]$ (TMTTF = tetramethyl-tetrathiafulvalene) has been recently prepared [14] by electrocrystallization of (TMTTF) in the presence of the tetraethylammonium salt of the discrete metal anion cluster $(Mo_6Cl_{14})^{2-}$ in CH_3CN Clusters with new electric or magnetic properties have been synthesized. The pioneering work in the superconductor materials field was carried out by CHEVREL et al [15], and has since been the subject of extensive study. These phases display superconductivity properties at high temperatures and high fields, which is of interest due to their potential applications e.g. electricity transport [16]. A series of these compounds of formula M Mo_6X_8 (M = Pb, Sn, Ba, Cu, Au, Li ; X = S, Se, Te) [17] displays the structural unit Mo_6X_8 . Mixed-valence cluster compounds, in which the cluster is linked to another redox center, have recently been synthesized and studied electrochemically. As an example, the complex [18] $[FcCCo_3(CO)_9]$ (Fc = ferrocene) displays a reduction centered on the cluster framework (CCo_3) which forms a radical anion as detected by esr , whereas the oxidation is centered on the Fc moiety [18]. Thus, in these clusters, there are localized valence sites for which the Hush model holds.

To date there is only limited information available in the literature. LEMOINE [19] has reviewed the works published between 1966 and 1982 [20]. A more general view was compiled by GEIGER [21] while a recent review that involves work carried out between 1982-1986 is in press in Coordination Chemistry reviews [22]. In contrast, there are numerous general reviews on clusters (see for instance references 1-30 in ref. 23). Experimental conditions for electrochemical investigation of clusters have been described in several reports [19,24]. In the following text, we will summarize our contribution to cluster electrochemistry and give a synthetic view on the principal results obtained by several research teams. Lastly, we will comment on further developments which might be important in future investigations.

2. SUMMARY OF OUR CONTRIBUTION TO CLUSTER ELECTROCHEMISTRY [19,22]

About ten years ago, we started with simple carbonyl metals of the type $M(CO)_n$ (M = Cr, Mo, W ; n = 6) and derivatives in order to obtain mechanistic informations on their redox behaviour [25]. Later, the electrochemistry of compounds containing a metal-metal bond such as $M_2(CO)_{10}$ (M_2 = Mn_2, MnRe, Re_2) were investigated [26]. In both studies, the products resulting from the electrochemical reactions (oxidation and reduction) were markedly different when compared to the starting complexes. Thus, electron transfers were followed by chemical reactions such as metal-metal or metal-ligand bond rupture, ligand rearrangement or solvent interactions. Together with PICKET and PLETCHER [27], we were among the first to suspect the prominent role of radicals in the subsequent chemical reactions. At the same time, a study of photolysis of identical compounds was completed which showed that paramagnetic intermediates are involved indiscriminately in photochemical as well as in electrochemical reactions.

The existence of radical species, resulting from metal-metal bond cleavage in carbonyl compounds is now currently recognized in spit of the difficulty in their unambiguous characterization by in-situ esr spectroscopy. As an example of electrochemical rupture of metal-metal bonds, the electrochemical and esr results concerning complexes $Mn_2(CO)_{10}$ (28) and $[M(CO)_3(\eta^5 - C_6H_6)]_2$ (29) (M = Mo, W) are consistent with an ECE mechanism (28,29). Thus, in spite of their rapid rate of redimerization, the radicals M· are key intermediate species in electrochemical as well as in photochemical mechanisms (30).

These preliminary results prompted us to study more sophisticated metal-metal bonded molecules i.e. clusters. A cluster is a molecule in which each metallic atom is directly bonded to at least two other metallic atoms, the simplest compound being the triangular M_3L_n (M = metal, L = ligands) (31). Initial investigations commenced in the early 80's with linear new clusters containing three metals in their core bonded by two metal-metal bonds M – Pt – M (M = Cr, Mo, W, Mn, Fe, Co . The redox behaviour of M – Pt – M complexes was compared (32,33,34) to that of the corresponding complexes with the following core : M – Hg – M and $[M - Au - M]^-$. The M· radical has been postulated as an intermediate during electroreduction, based on peak potentials for oxidation and reduction of the M· fragments as well as the presence of esr spectra (32). The latter could also be due to a trapped species such as $M(O)_2$, since $Mn(CO)_5(O)_2$ has been generated by reaction of $Mn(CO)_5$ with O_2 (28,29). In the absence of hyperfine structure, conclusive remarks cannot be made.

We also succeeded in using the reactivity of radical intermediates to accomplish electrochemical syntheses in situ, as shown on the following example (35) : The cluster $[Pt_2Co_2(\mu - CO)_3(CO)_5(PPh_3)_2]$ was isolated in good yields as one of the electrochemical reduction products of the trinuclear starting cluster $[PtCo_2(\mu - CO)(CO)_7(PPh_3)]$. The following reaction scheme was proposed :

$$[PtCo_2(\mu - CO)(CO)_7(PPh_3)] + e^- \xrightarrow{\text{Electro-chemical step}} [PtCo_2(\mu - CO)(CO)_7(PPh_3)]^-$$

$$\downarrow \text{Chemical step}$$

$$[Pt_2Co_2(\mu - CO)_3(CO)_5(PPh_3)_2] \xleftarrow{\text{Dimerization}} [Co(CO)_4]^- + [PtCo(CO)_4(PPh_3)]^{\cdot}$$

At the same time, the chemical reduction of $[Ru_3(CO)_{12}]$ was demonstrated to yield $[Ru_4(CO)_{13}]^{2-}$ through partial or total fragmentation of the triangular starting species (36).

Ligand-bridged polymetallic complexes were also studied. Thus, four trimetallic Ir complexes were studied in CH_3CN containing 0.1M TEAP at a platinum electrode (37).

Later work was centered on the redox behaviour of trimetallic and tetrametallic clusters of Pt and Co together with new planar mixed compounds containing the Pd_2M_2 or Pt_2M_2 (M = Cr, Mo, W) core [38]. All these compounds show irreversible reduction steps due to subsequent chemical reactions taking place after the electron transfer, leading to very reactive fragments which may rearrange in solution. No electrochemical investigation has been published on the typical tetrahedral cluster $[Co_4(CO)_{12}]$ in spit of the fact that this very reactive compound shows interesting catalytic properties. In contrast, the tetrahedral King's cluster [39] $[Fe(CO)(\eta^5 - C_6H_6)]_4$ investigated in 1971 by FERGUSON and MEYER [40] was shown to be stable in several oxidation states. Later, DAHL and coworkers [41] obtained similar results with clusters derivatives of the prototypical $[Fe(\eta^5 - C_6H_6)(\mu^3 - S)]_4$. A series of synthetic analogues of the active sites of the iron-sulfur proteins also behave as electron sponges i.e. they undergo several one-electron reversible redox steps without breakdown [42]. On the basis of these results, we investigated the cluster series $[Co_4(CO)_{12-2n}(dppm)_n]$ (n = 0, 1, 2 ; dppm = bis(diphenyl phosphino) methane. It was shown [43] that $[Co_4(CO)_{12}]$ undergoes a facile one-electron reversible reduction, followed by cluster fragmentation. In order to avoid cluster fragmentation, chelating ligands dppm ($Ph_2PCH_2PPh_2$) were substituted for CO ligands in $[Co_4(CO)_{12}]$ and a comparative study of the electrochemistry of the cluster series $[Co_4(CO)_{12-2n}(dppm)_n]$ was undertaken. Whereas the clusters with n = 0,1 exhibit the reversible 0/-1 redox couple, the tetrasubstituted cluster $[Co_4(CO)_8(dppm)_2]$ shows the reversible redox couples +1/0 , 0/-1 , -1/-2 (the latter only under drastic experimental conditions). The redox potentials of the 0/-1 redox couples correlate with the number of phosphorus ligands bonded to the cluster and with the electronic frequencies of the metal-metal transition. This was also the case for the series $Fe_3(CO)_{12-n}L_n$ (n = 0,2 ; L = PPh_3) in which each CO replaced by a PPh_3 ligand caused a cathodic shift of ca. 200 mV for the first reduction potential [44]. The effect of the tripod ligand $HC(PPh_2)_3$ capping one face of the tetrahedron $[Co_4(CO)_{12}]$ on the electrochemical properties of the cluster has been investigated [45]. The substitution of three CO by the chelating ligand tripod would be expected to shift the first reduction potential by 600 mV cathodically. The experimental value of 590 mV is in agreement with the predicted value. Thus, the ligand effects appear to be additive. This conclusion is important as (i) <u>chelating ligands may reinforce the cohesion of the cluster and (ii) the excess negative charge in clusters may be delocalized from the metal core on the ligands</u>.

Not only chelating phosphine ligands are able to stabilize oxidized or reduced cluster species. For instance, the ligand C_2Ph_2 may increase the stability of the reduced cluster as shown by comparison of the redox behaviour of the series of anionic clusters $[M\ Co_3(CO)_{12}]^-$ and the corresponding substituted species $[M\ Co_3(CO)_{10}(C_2Ph_2)]^-$ (M = Fe, Ru) [46]. These effects of different ligand environments on the electrochemical behaviour of $[Co_4(CO)_{12}]$ and related clusters have recently be confirmed (47).

Having established the influence of the ligands bonded to the Co_4 core on their redox behaviour, we next investigated the metal influence on the electrochemistry of $M_4(CO)_{12}$ (M_4 = Co_4, Co_2Rh_2, Rh_4) (45). Whereas the $E_{1/2}$ of the first reductions of $[Co_4(CO)_{12}]$ and $[Co_2Rh_2(CO)_{12}]$ are close, the complex $[Rh_4(CO)_{12}]$ displays a one-electron reduction wave at a more cathodic potential. The LUMO energy level is displaced compared with that of $[Co_4(CO)_{12}]$ and $[Co_2Rh_2(CO)_{12}]$. Thus, as the basicity of the metal increases from Co to Rh (i) the $E_{1/2}$ are cathodically shifted in agreement with an increase of metal basicity and (ii) the stability of the radical anions formed decreases. Similar conclusions (45) were reached from the investigation of the parent series $[M_4(CO)_9tripod]$ (M_4 = Co_4, Co_2Rh_2, Rh_4) . Not only metals from the same column of the peridic table, but also atoms like Fe , Ru can be substituted to the Co atom in the Co_4 tetrahedron. The resulting anionic clusters $[M\ Co_3(CO)_{12}]^-$ (M = Fe, Ru) exhibit two close one-electron reduction steps at more cathodic potentials than $[Co_4(CO)_{12}]$ (46). As the basicity of the metal increases from Fe to Ru , the cluster becomes more difficult to reduce. The donor ability of the solvent (48) is also of importance. In donor solvents such as CH_3CN , DMF or DMSO , the two waves are separated by about 200 mV , whereas in poor donor solvent, such as 1,2 $C_2H_4Cl_2$, the two waves overlap. The esr spectrum (46) of the radicals $[M\ Co_3(CO)_{12}]^{-}$ exhibit a 22 lines hyperfine structure. The corresponding g values g = 2.036 for M = Fe and g = 2.085 for M = Ru were measured.

2.1. Application of the enhanced reactivity of cluster radicals in the ETC reaction

Photochemistry, sonochemistry and redox reagents, together with electrochemistry are able to increase cluster reactivity. The reactivity of radical species electrochemically generated is of great interest. The electron-transfer catalysed reaction (ETC reaction) was first discovered for the aqueous $[Cr(CN)_6]^{3-}$ anion reduction (49). Analogous to this reaction is the aromatic substitution S_NR_1 investigated by BUNNETT (50), SAVEANT (51) and others (52). BEZEMS, RIEGER and VISCO published in 1981 (2) the first example of an ETC reaction in cluster. In the presence of nucleophilic reagents, phosphines or phosphites, the first reduction peaks of the clusters $[Y\ CCo_3(CO)_9]$ (Y = Ph, Cl) drop to zero, whereas a new peak expected for the substitution products $[Y\ CCo_3(CO)_8$ phosphine$]$ appears at more cathodic potentials. Thus, carbonyl groups around the clusters may be replaced stepwise by nucleophilic reagents, depending on the basicity and the steric hindrance of the latter. Thermal reactions offer conditions of lower selectivity and yields than ETC reactions. Chemical reagents able to generate radicals may replace (53) the electrode for example sodium diphenylketyl ca. 0.025 M in THF .

Nucleophilic reagents used in the ETC reaction include phosphines and phosphites, arsine and stilbines, isocyanides etc... . Other very interesting reagents include metal carbonyl anions which may increase the number of metal-metal bonds, for instance in the synthesis of the cluster $[RCCo_2M(\eta^5 - C_6H_6)(CO)_8]$ (R = Me, Ph, H ; M = Cr, Mo, W) (54). Bulk conditions i.e. solvent, supporting electrolyte, residual water, oxygen, ion-pairing effects and temperature are required for the chain process (55). The necessary potential must also be achieved (50,51) in reduction, the product must reduce at a more negative potential than the starting material. Good leaving groups such as CO , $Co(CO)_4^-$ and Cl^- contribute to facilitate the reaction. KOCHI et al (56) have recently investigated a certain number of ETC reactions, and proposed mechanisms in agreement with our interpretations. Our contribution in this area concerns the ETC reaction on the prototypical $[Co_4(CO)_9tripod]$ cluster (45). This cluster was convenient to investigate for ETC reaction because of (i) its stability and (ii) the stability of its radical anion the esr spectrum of which is known. The thermal substitution of $[Co_4(CO)_9(tripod)]$ by PR_3 (R = Me_3, Me_2Ph, $MePh_2$, Ph_3, $(C_6H_{11})_3$) is known. We also verified independently that the neutral cluster parent $[Co_4(CO)_6(tripod)(\pi\text{-toluene})]$, for which all apically bonded CO ligands are substituted by π-toluene, and its radical anion were equally inert towards substitution by all investigated phosphines. This observation is in agreement with a substitution occuring on the apical cobalt atom as shown independently by IR ν_{CO} data. We proposed an ETC mechanism on the basis of our experimental results. The slowest reaction in the chain propagation process is the loss of the apical CO ligand as the substitution rate is slowed down considerably by the presence of ca. 2 atm of CO gas. We also verified on the cluster radical anion $[Rh_4(CO)_9tripod]^{-}$ that the decomposition reaction of that anion involves the reversible loss of CO , a step which can be stopped by the presence of gaseous. Another interesting observation made in this laboratory is the reactivity of the cluster radical cation $[Co_4(CO)_9tripod]^+$ towards nucleophilic reagent (57).

2.2. **Redox behaviour of mixed-metal clusters**

Whereas one-electron processes are generally common for homometallic close-packed structure clusters, mixed metal clusters often undergo multi-electron transfers leading to the rupture of their metallic core into fragments of lower nuclearity. We have investigated (38) the cluster series : $[M^1M^2(\mu^3 - CO)(\mu^2 - CO)_2(C_6H_6)(PPh_3)]_2$. M^1 = Pd, Pt ; M^2 = Cr, Mo, W) with the planar $M^1_2M^2_2$ metallic core. These neutral clusters are built around the Pd-Pd or the Pt-Pt bonds. The Pd or Pt atoms are in the formal +1 oxidation state (46). A unique two-electron reduction step, irreversible in cyclic voltammetry, is observed. Cluster fragments of lower nuclearity have been identified among the reduction products. A reaction mechanism has been proposed (58) for both reduction and oxidation. Esr experiments, as well as the geometry of the curves obtained in differential and normal pulse polarography (current log transforms) and in cyclic voltammetry ($E_p - E_{p/2}$), also indicate that the reduction is a two-electron step within the time scale of the electrochemical experiments. Thus, these clusters behave in reduction like M^1-M^1 bimetallic carbonyl complexes rather than sinks of electrons. In oxidation in contrast, they exhibit two one-electron oxidation steps and behave as sources of electrons.

When a single two-electron reduction step occurs, bielectronic processes can be interpreted in terms of two successive one-electron processes, a conclusion which is supported by our studies (46) on the following example. The cluster anion $[RuCo_3(CO)_{12}]^-$ undergoes a single two-electron reduction in 1,2 $C_2H_4Cl_2$, a poor donor solvent. In contrast in donor solvents such as DMF or DMSO, two narrow one-electron redox steps were obtained.

Clusters built around the Pd-Pd, Pd-Pt or Pt-Pt bond, where the metals are in the formal + 1 oxidation state, have been the subject of numerous investigations by this research group. In these heterometallic compounds, the role of bridged PPh_2 ligands may be demonstrated by the following examples. The triangular cluster : $[Co\,Pt_2\,(\mu - PPh_2)(\mu - CO)\,(CO)\,(PPh_3)_3]$ exhibits an electrochemically reversible one-electron reduction leading to an unstable reduced species which is converted into the $[Co(CO)_3PPh_3]^-$ anion, identified as one of the decomposition products (59). An irreversible one-electron oxidation also occurs. The cluster is easier to oxidize and more difficult to reduce than the clusters with Pt Co_2 metal core (35) : $[Co_2Pt\,(\mu - CO)(CO)_7(PPh_3)]$ and $[Co_2Pt\,(\mu - CO)(CO)_6(Ph_2PCH_2CH_2PPh_2)]$ in agreement with a strengthening effect of the bridging group.

The nature of the metal atoms Pd or Pt in the metal-metal bond is of importance on the electrochemical reduction (60,61).

2.3. Cluster cohesion reinforced by an heteroatom inside their core

The electrochemical irreversible two-electron reduction of compounds $[M_5(CO)_{16}C]$ (M = Fe, Ru, Os) led to $[M_5(CO)_{14}C]^{2-}$ dianion (62). We have reported (63,) new results on carbido-carbonyl clusters $[Co_8(CO)_{18}C]\,(R_4N)_2$, $[Co_6(CO)_{15}C]\,(R_4N)_2$, $[Rh_6(CO)_{15}C]\,(R_4N)_2$ (R = Et or Bu) and $[Fe_6(CO)_{16}C]\,(Et_4N)_2$. In these compounds, the carbide inside the metal core may contribute to the strength of the cluster, as it donates four electrons to the cluster without occupying the peripheric coordination sites. Other heteroatoms may maintain the metal core, for instance N, S, P. (Example $[Co_6S_8(PEt_3)_6]$ or $[Co_8S_6(SPh)_8]^{5-}$) (64). Interpretation of our results involved primarily consideration of the geometry and shape of the clusters. For instance the "close packed" cluster $[Co_8(CO)_{18}C]^{2-}$ undergoes three reversible one-electron steps, whereas the cluster with "open geometry" $[Co_6(CO)_{15}C]^{2-}$ exhibits irreversible reduction and oxidation steps (63).

3° SYNTHETIC VIEW OF THE PRINCIPAL RESULTS ON CLUSTER ELECTROCHEMISTRY

For general cluster species, according to the notation of DESSY (20) and LEHMKUHL (65), a general reaction scheme may occur, as follows :

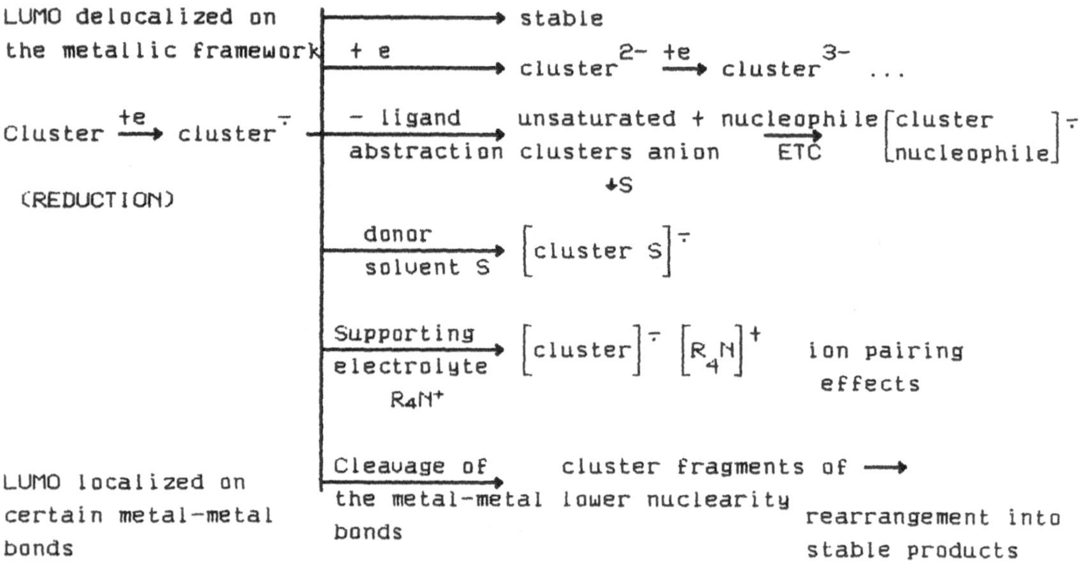

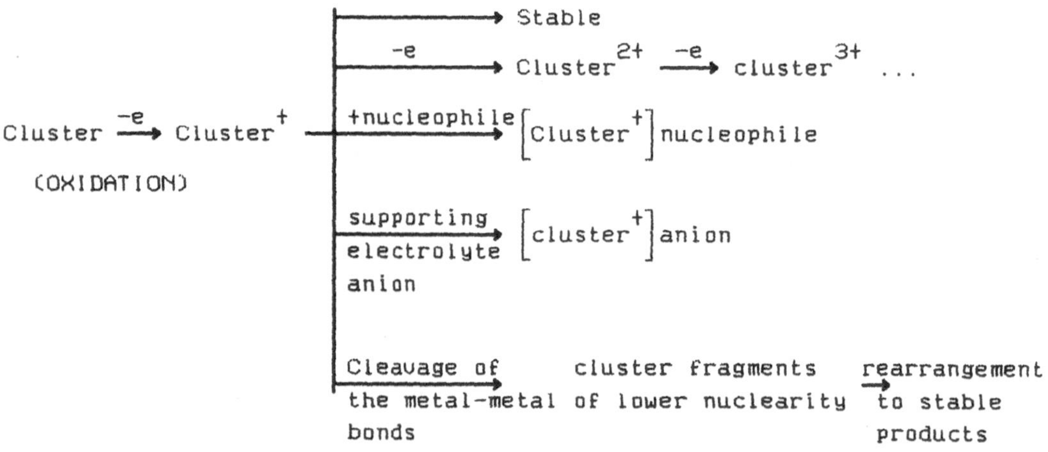

In contrast, for heteronuclear clusters with open structures the ionic character of certain metal-metal bonds - i.e. Pt(+1) - Co(-1) for instance in the Pt_2Co core- may be responsible for their fragmentation after reduction. In these complexes, the reduction center is localized on the Pt(+1) which is reduced to Pt(O) . These clusters behave as metal-metal bonded complexes rather than delocalized sources and sinks of electrons.

Clusters with the following metallic cores have been investigated electrochemically :

Trinuclear compounds	:	see references 13,20,32,33,35,59 to 61, 66 to 83,97,98.
Tetranuclear compounds	:	see references 1,10-12,38,40-43,45-47,56,58,60, 61,75,80,82,84-94.
Pentanuclear compounds	:	see references 5,62,95,96.
Hexanuclear compounds	:	see references 3,5,64,99-101,104.
Octanuclear compounds	:	see references 63,102
Nonanuclear clusters	:	see reference 103

Although this list is not exhaustive, it is interesting to note that the more studied clusters are trinuclear compounds. For other studies, references [19] and [22] cover the field.

4. CONCLUSION - Future developments in cluster electrochemistry

Our contribution in cluster electrochemistry has established two types of behaviour. (i) A wide range of homonuclear compounds with "close-packed structures" may undergo reversible one-electron reactions without notable alteration. Thus, they are considered as electron "reservoirs" i.e. sources and sinks of electrons. However, as the LUMO fronter orbital involved in reduction is antibonding and the HOMO involved in oxidation is often bonding in nature, both oxidation and reduction will tend to weaken the metal-metal interactions i.e. to destabilize the cluster. Hence, electrochemically generated cluster radical species are often not stable. We have found conditions to enhance their stability : chelating ligands, presence of an inside heteroatom, ion-pairing effects. Thus, cluster radical anions under appropriate conditions can exchange ligands with nucleophilic reagents. Whereas the reactivity of radical anion cluster has now ample precedent in the ETC reaction, no information is available on the reactivity of radical cation clusters toward nucleophilic or electrophilic reagents. In general, while the electroreduction of clusters has attracted numerous studies, there is comparatively limited information available on electrooxidation. Thus, systematic investigations, in particular of electrooxidation, will provide further information on the reactivity of radical cation clusters. A basic application of these reactions lies in the electrosynthesis of new or modified cluster system which offers a new field of research in electrochemistry. As it is now possible to bridge two clusters together, the synthesis of polymeric cluster systems leading to new materials can also be envisaged. Relatively few systems in which clusters alternate with organic or organometallic molecules are known. A further application of cluster electrochemistry is electrocatalysis by active cluster species. While neutral clusters are often unreactive, in comparison paramagnetic cluster species exhibit enhanced reactivity.

The efficiency of these clusters in the reduction of N_2 and CO_2 has yet to be explored. Similarly, the role of clusters in the bioelectrochemistry for nitrogenase and their suitability as models for non heme iron proteins requires further investigations. ii) A second class of "open structured" clusters often with heteronuclear compounds may undergo irreversible multielectron processes associated with cluster breakdown into species of lower nuclearity. For the case of a single two-electron reduction, a stepwise process with a detectable intermediate may occur. Often intermediate species are not detectable. Thus the electrochemical study of heteronuclear compounds is more complex. Identification of fragments obtained after cluster breakdown is not straightforward. It has been shown that certain clusters behave like binuclear metal-metal bonded carbonyl complexes rather than delocalized bonded compounds, in particular those with Pt(+1) or Pd(+1) atoms bonded to Mo(-1) or Co(-1) atoms. In spite of the fact that it is possible to enhance cluster cohesion, many oxidation and reduction reactions remain irreversible in clusters. Fragments have often not been identified and their reactivity has not been investigated. Another potential area of research is the modification induced by electron-transfer in clusters, initial studies of which have been carried out by DAHL and GEIGER's teams.

ACKNOWLEDGMENTS

I am grateful for the collaboration with Pr. M. GROSS and the following co-workers Jean RIMMELIN, Rodolphe JUND, Ghinwa NEMRA. The C.N.R.S. is gratefully acknowledged for financial support of this research.

REFERENCES

1. H.H. OHST, J.K. KOCHI, J. Am. Chem. Soc., **108** (1986) 2897 ; Inorg. Chem, **25** (1986) 2066.
2. G.J. BEZEMS, P.H. RIEGER, S. VISCO, J. Chem. Soc. Chem. Comm., (1981) 265.
3. B. TULYATHAN, W.E. GEIGER, J. Am. Chem. Soc., **107** (1985) 5960.
4. S.B. COLBRAN, B.H. ROBINSON, J. SIMPSON, Organometallics, **5** (1986) 1122.
5. T.C. ZIETLOW, H.B. GRAY, Inorg. Chem., **25** (1986) 631.
6. W.L. OLSON, L.F. DAHL, J. Am. Chem. Soc., **108** (1986) 7657.
7. A. BENOIT, A. DARCHEN, J.Y. LE MAROUILLE, C. MAHE, H. PATIN, Organometallics, **2** (1983) 555.
8. L.P. DIDENKO, A.B. GAVRILOV, A.K. SHILOVA, V.V. STRELETS, V.N. TSAREV, A.E. SHILOV, V.D. MAKHAEV, A.K. BANERJEE, L. POSPISIL, Nouv. J. Chim., **10** (1986) 583.
9. L. POSPISIL, L.P. DIDENKO, A.E. SHILOV, J. Electroanal. Chem., **197** (1986) 305.
10. Y. HOZUMI, Y. IMASAKA, K. TANAKA, T. TANAKA, Chem. Lett., (1983) 897.

11. K. TANAKA, Y. IMASAKA, M. TANAKA, M. HONJO, T. TANAKA, J. Am. Chem. Soc., **104** (1982) 4258.
12. M. TEZUKA, T. YAJIMA, A. TSUCHIYA, Y. MATSUMOTO, Y. UCHIDA, M. HIDIA, J. Am. Chem. Soc., **104** (1982) 6834.
13. S.Z. GOLDBERG, B. SPIVACK, G. STANLEY, R. EISENBERG, D.M. BRAITSCH, J.S. MILLER, M. ABKOWITZ, J. Am. Chem. Soc., **99** (1977) 110.
14. L. OUAHAB, P. BATAIL, C. PERRIN, C. GARRIGAU-LAGRANGE, Mat. Res. Bull., **21** (1986) 1223.
15. R. CHEVREL, M. SERGENT, J. PRIGENT, J. Solid State Chem., **3** (1971) 515.
16. R. CHEVREL in *Superconductor Materials Science : Metallurgy, Fabrication and Applications*. Eds. FONERS, Schwartz B.B. Plenum Press : New York 1981, Chapter 10 ; O. FISHER Appl. Phys., **16** (1978) 1.
17. A. PERRIN, M. SERGENT, O. FISHER, Mater. Res. Bull., **13** (1978) 259
A. PERRIN, R. CHEVREL, M. SERGENT, O. FISHER, J. Solid. state Chem., **33** (1980) 43.
M. POTEL, R. CHEVREL, M. SERGENT, J.C. ARMICI, M. DEROUX, O. FISHER, J. Solid State Chem., **35** (1980) 286 and references therein
18. S. COLBRAN, B.H. ROBINSON, J. SIMPSON, J.C.S. Chem. Comm., **1361** (1982).
19. P. LEMOINE, Coord. Chem. Rev., **47** (1982) 56.
20. R.E. DESSY, P.M. WEISSMAN, R.J. POHL, J. Am. Chem. Soc., **88** (1966) 5112 ; **88** (1966) 471 ; **88** (1966) 5117 ; **90** (1968) 1995.
21. W.E. GEIGER in *Progress in Inorganic Chemistry*, S.J. LIPPARD Ed. J. WILEY, NEW YORK (1985) Vol. **33**, p. 275.
22. P. LEMOINE, Coord. Chem. Rev., in the press.
23. P. BRAUNSTEIN, Nouv. J. Chim., **10** (1986) 365 and references therein.
24. D. De MONTAUZON, R. POILBLANC, P. LEMOINE, M. GROSS, Electrochimica Acta, **23** (1978) 1247.
L.I. DENISOVICH, S.P. GUBIN, Uspekhi Khimii, **46** (1977) 50.
A.J. BARD, Electroanalytical Chemistry, Vol. **7**, Marcel Dekker, New York 1974 and vol. **4**, New York, 1970.
A.J. FRY, Synthetic Organic Electrochemistry, Haper and Row, New York 1972.
J.B. HEADRIDGE, Electrochemical Techniques for Inorganic Chemists, Acad. Press, New York 1967, Chap. 5.
L. MEITES, Electrochemical Data, Organic Organometallic and Biochemical Substances, Wiley Interscience, New York 1974.
25. P. LEMOINE, M. GROSS, C.R. Acad Sci., Paris, Série C, **280** (1975) 797 ; Electrochim. Acta **23** (1978) 1219.
26. P. LEMOINE, M. GROSS, Electrochim. Acta, **21** (1976) 1.
M. DIOT, J. BOUSQUET, P. LEMOINE, M. GROSS, J. Organomet. Chem., **112** (1976) 79.
P. LEMOINE, M. GROSS, M. DIOT, J. BOUSQUET, J. Organomet. Chem., **104** (1976) 221.
P. LEMOINE, M. GROSS, J. Organomet. Chem., **133** (1977) 193.

27. C.J. PICKETT, D. PLETCHER, J.C.S. Dalton, **749** (1976) ; **636** (1976); J.C.S. Chem. Comm., **660** (1974) ; J. Org. Chem., **102** (1975) 327.

28. K.M. KADISH, D.A. LACOMBE, J.E. ANDERSON, Inorg. Chem., **25** (1986) 2074.

29. K.M. KADISH, D.A. LACOMBE, J.E. ANDERSON, Inorg. Chem., **25** (1986) 2246.

30. M. CHANON, M.L. TOBE, Angew. Chem., Int. Ed., **21** (1982) 1. M. CHANON, Bull. Soc. Chim. Fr., **2** (1985) 209 and references therein.

31. D. LABROUE, Soc. Chim. Fr., Actual. Chim., (1980) 7.

32. P. LEMOINE, A. GIRAUDEAU, M. GROSS, P. BRAUNSTEIN, J.C.S. Chem. Comm. (1980) 77.

33. P. LEMOINE, A. GIRAUDEAU, M. GROSS, P. BRAUNSTEIN, J. Organomet. Chem., **202** (1980) 447.

34. A. GIRAUDEAU, P. LEMOINE, M. GROSS, P. BRAUNSTEIN, J. Organomet. Chem., **202** (1980) 455.

35. P. LEMOINE, A. GIRAUDEAU, M. GROSS, R. BENDER, P. BRAUNSTEIN, J.C.S. Dalton Trans. (1981) 2059.

36. C.C. NAGEL, K.E. INKROTT, S.G. SHORE, Abstract of papers, 11th Central Regional Meeting of American Chemical Society, Columbus, Ohio May 1979.

37. P. LEMOINE, M. GROSS, D. De MONTAUZON, R. POILBLANC, Inorg. Chim. Acta, **71** (1983) 15.

38. R. JUND, P. LEMOINE, M. GROSS, R. BENDER, P. BRAUNSTEIN, J.C.S. Chem. Comm., (1983) 86.

39. R.B. KING, Inorg. Chem., **5** (1966) 2227.

40. J.A. FERGUSON, T.J. MEYER, J.C.S. Chem. Comm., (1971) 623. J. Am. Chem. Soc., **94** (1972) 3409.

41. TRINH-TOAN, B.K. TEO, J.A. FERGUSON, T.J. MEYER, L.F. DAHL, J. Am. Chem. Soc., **99** (1977) 408.

42. J.A. IBERS, R.H. HOLM, Science, **209** (1980) 223. C.T.W. CHU, Y.K. LO, L.F. DAHL, J. Am. Chem. Soc., **104** (1982) 3409

43. J. RIMMELIN, P. LEMOINE, M. GROSS, D. De MONTAUZON, Nouv. J. Chim., **7** (1983) 453.

44. A.M. BOND, P.A. DAWSON, B.M. PEAKE, B.H. ROBINSON, J. SIMPSON, Inorg. Chem., **16** (1977) 2199.

45. J. RIMMELIN, P. LEMOINE, M. GROSS, A.A. BAHSOUN, J.A. OSBORN, Nouv. J. Chim., **9** (1985) 184.

46. R. JUND, Ph.D. Thesis, Strasbourg University (1987).

47. G.F. HOLLAND, D.E. ELLIS, W.C. TROGLER, J. Am. Chem. Soc., **108** (1986) 1884.

48. V. GUTMANN, Electrochim. Acta, **21** (1976) 661.

49. S.W. FELDBERG, L. FEFTIC, J. Phys. Chem., **76** (1972) 2439.

50. J.F. BUNNETT, Acc. Chem. Res., **11** (1978) 413.

51. J.M. SAVEANT, Acc. Chem. Res., **13** (1980) 323.

52. I.P. BELETSKAYA, V.N. DROZD, Russ. Chem. Rev., **48** (1979) 431.

53. M.I. BRUCE, D.C. KEHOE, J.G. MAISONS, B.K. NICHOLSON, P.H. RIEGER, M.L. WILLIAMS, J.C.S. Chem. Comm., (1982) 442.

54. S.D. JENSEN, B.H. ROBINSON, J. SIMPSON, J. Chem. Soc. Chem. Comm., (1983) 1081.

55. C.M. KIRK, B.M. PEAKE, B.H. ROBINSON, J. SIMPSON, Aust. J. Chem., **441** (1983) 36.
56. H.H. OHST, J.K. KOCHI, Organometallics, **5** (1986) 1359.
57. J. RIMMELIN, PhD Thesis Strasbourg University (1987).
58. R. JUND, P. LEMOINE, M. GROSS, R. BENDER, P. BRAUNSTEIN, J.C.S. Dalton, (1985) 711.
59. R. BENDER, P. BRAUNSTEIN, B. METZ, P. LEMOINE, Organometallics, **3** (1984) 381.
60. G. NEMRA, P. LEMOINE, P. BRAUNSTEIN, C. DE MERIC DE BELLEFON, M. RIES, J. Organomet. Chem., **304** (1986) 245.
61. G. NEMRA, P. LEMOINE, M. GROSS, C. DE MERIC DE BELLEFON, M. RIES, Electrochim. Acta, **31** (1986) 1205.
62. A. GOURDON, T. JEANNIN, J. Organomet. Chem., **290** (1985) 199.
B.F.G. JOHNSON, J. LEWIS, W.J. NELSON, J.N. NICHOLLS, J. PUGA, P.R. RAITHBY, M.J. ROSALES, M. SCHRODER, M.D. VARGAS, J.C.S. Dalton, (1983) 2447.
63. J. RIMMELIN, P. LEMOINE, M. GROSS, R. MATHIEU, D De MONTAUZON, J. Organomet. Chem., **309** (1986) 355.
64. Z. ZANELLO, F. CECCONI, C.A. GHILARDI, S. MIDOLLINI, A. ORLANDINI, Journées d'Electrochimie, 28-31 Mai, Firenze 1986. See also Inorg. Chim. Acta, **64** (1981) L47 ; **76** (1983) L183.
65. H. LEHMKUHL, in Organic Electrochemistry, M. BAIZER, Ed. M. DEKKER, New York 1973, p. 621.
66. A.A. VLCEK, Special Pub ; The Chemical Society, London 1959 N° **13** p. 121.
67. R.E. DESSY, R.B. KING, W. WALDROP, J. Am. Chem. Soc., **88** (1966) 5112.
68. A.M. BOND, P.A. DAWSON, B.M. PEAKE, B.H. ROBINSON, J. SIMPSON, Inorg. chem., **16** (1977) 2199.
69. D. MIHOLOVA, J. KLIMA, A.A. VLCEK, Inorg. Chim. Acta, **27** (1978) L67.
70. A.A. VLCEK, Coll. Czech. Chem. Comm., **24** (1959) 1748.
71. D. De MONTAUZON, Thèse de Doctorat 3ème Cycle, Paris 1974.
72. D. De MONTAUZON, R. POILBLANC, J. Organomet. Chem., **104** (1976) 99.
73. L.I. DENISOVICH, S.P. GUBIN, Y.A. CHAPOVSKII, N.A. USTYNYUK, Izv. Akad. Nauk SSSR, Ser. Khim., **4** (1968) 924 ; **2** (1969) 258.
74. J.J. HABEEB, D.G. TUCK, S. ZHANDIRE, Can. J. Chem., **57** (1979) 2196.
75. L.R. BYERS, V.A. UCHTMAN, L.F. DAHL, J. Am. Chem. Soc., **103** (1981) 1942.
T. MADACH, H. VAHRENKAMP, Chem. Ber, **114** (1981) 505 ; **114** (1981) 513.
76. H. BEURICH, T. MADACH, F. RICHTER, H. VAHRENKAMP, Angew. Chem. Int. Ed., **18** (1979) 9.
77. A.D. YOUNG, Inorg. Chem., **20** (1981) 2049.
78. **a)** B.M. PEAKE, B.H. ROBINSON, J. SIMPSON, D.J. WATSON, Inorg. Chem., **16** (1977) 405.
b) A.M. BOND, B.M. PEAKE, B.H. ROBINSON, J. SIMPSON, D.J. WATSON, Inorg. Chem., **16** (1977) 410.

c) A.S. HUFFADINE, B.M. PEAKE, B.H. ROBINSON, J. SIMPSON, P.A. DAWSON, J. Organomet. Chem., **121** (1976) 391.
d) A.M. BOND, P.A. DAWSON, B.M. PEAKE, P.H. RIEGER, B.H. ROBINSON, J. SIMPSON, Inorg. Chem., **18** (1979) 1413.
e) B.M. PEAKE, P.H. RIEGER, B.H. ROBINSON, J. SIMPSON, Inorg. Chem., **18** (1979) 1000.
f) B.M. PEAKE, P.H. RIEGER, B.H. ROBINSON, J. SIMPSON, J. Organomet. Chem., **C 63** (1979) 172.
g) P.A. DAWSON, B.M. PEAKE, B.H. ROBINSON, J. SIMPSON, Inorg. Chem., **19** (1980) 465.
h) B.M. PEAKE, P.H. RIEGER, B.H. ROBINSON, J. SIMPSON, J. Am. Chem. Soc., **102** (1980) 156.
i) C.M. KIRK, B.M. PEAKE, B.H. ROBINSON, J. SIMPSON, J. Chem. Res. **S 106** (1982) ; **M 1201** (1982).
j) B.M. PEAKE, P.H. RIEGER, B.H. ROBINSON, J. SIMPSON, Inorg. Chem., **20** (1981) 2540.
k) C.M. AREWGODA, P.H. RIEGER, B.H. ROBINSON, J. SIMPSON, J. Am. Chem. Soc., **104** (1982) 5633.
l) C.M. AREWGODA, B.H. ROBINSON, J. SIMPSON, J. Am. Chem. Soc., **105** (1983) 1893.
m) L.V. CASAGRANDE, T. CHEN, P.H. RIEGER, B.H. ROBINSON, J. SIMPSON, S.J. VISCO, Inorg. Chem., **23** (1984) 2019.
n) P.N. LINDSAY, B.M. PEAKE, B.H. ROBINSON, J. SIMPSON, U. HONRATH, H.Z. VAHRENKAMP, A.M. BOND, Organometallics, **3** (1984) 413.
o) U. HONRATH, H.Z. VAHRENKAMP, Z. Naturforsch., **39 b** (1984) 545.
p) C.M. AREWGODA, A.M. BOND, R.S. DICKSON, T.F. MANN, J.E. MOIR, P.H. RIEGER, B.H. ROBINSON, J. SIMPSON, Organometallics, **4** (1985) 1077.
q) S.B. COLBRAN, B.H. ROBINSON, J. SIMPSON, Organometallics, **2** (1983) 943.
r) S.B. COLBRAN, B.H. ROBINSON, J. SIMPSON, Organometallics, **2** (1983) 952.
s) S.B. COLBRAN, B.H. ROBINSON, J. SIMPSON, J. Organomet. Chem., **265** (1984) 199.
t) R.G. CUNNINGHAME, A.J. DOWNARD, L.R. HANTON, S.D. JENSEN, B.H. ROBINSON, J. SIMPSON, Organometallics, **3** (1984) 180.
u) S.B. COLBRAN, B.H. ROBINSON, J. SIMPSON, Organometallics, **3** (1984) 1344.
v) A.J. DOWNARD, B.H. ROBINSON, J. SIMPSON, Organometallics, **5** (1986) 1132.
w) A.J. DOWNARD, B.H. ROBINSON, J. SIMPSON, Organometallics, **5** (1986) 1140.
x) S.D. JENSEN, B.H. ROBINSON, J. SIMPSON, Organometallics, **5** (1986) 1690.
y) C.M. AREWGODA, B.H. ROBINSON, J. SIMPSON, J. Chem. Soc. Chem. Comm., **284** (1982).

79. a) R.L. BEDARD, Ph.D., The University of Wisconsin, Madison 1985.
b) R.L. BEDARD, L.F. DAHL, J. Am. Chem. Soc., **108** (1986) 5924 ; **108** (1986) 5933.

c) W.L. OLSON, L.F. DAHL, J. Am. Chem. Soc., **108** (1986) 7657.
d) W.L. OLSON, A.M. STACY, L.F. DAHL, J. Am. Chem. Soc., **108** (1986) 7646.

80. A.M. BONNY, T.J. CRANE, N.A.P. KANE-MAGUIRE, J. Organomet. Chem., **289** (1985) 157.

81. M.I. BRUCE, M.G. HUMHREY, O.S. SHAWKATALY, M.R. SNOW, E.R.T. TIEKINK, J. Organomet. Chem., **C 51** (1986) 315.

82. M.G. RICHMOND, J.K. KOCHI, Inorg. Chem., **25** (1986) 656.

83. J.C. CYR, J.A. de GRAY, D.K. GOSSER, E.S. LEE, P.H. RIEGER, Organometallics, **4** (1985) 950.

84. K. TANAKA, M. HONJO, T. TANAKA, J. Inorg. Biochem., **22** (1984) 1873.

85. R. JOHNSON, R.H. HOLM, J. Am. Chem. Soc., **100** (1978) 5338.

86. D. COUCOUVANIS, M. KANATZIDIS, E. SIMHON, N.C. BAENZIGER, J. Am. Chem. Soc., **104** (1982) 1874.

87. M.G. KANATZIDIS, M. RYAN, D. COUCOUVANIS, A. SIMOPOULOS, A. KOSTIKAS, Inorg. Chem., **22** (1983) 181.

88. B.A. AVERILL, T. HERSKOVITZ, R.H. HOLM, J.A. IBERS, J. Am. Chem. Soc., **95** (1973) 3523.

89. C.J. PICKETT, J. Chem. Soc. Chem. Comm., (1985) 323.

90. P.K. MASCHARAK, K.S. HAGEN, J.T. SPENCE, R.H. HOLM, Inorg. Chim. Acta, **80** (1983) 157.

91. J.M. BERG, R.H. HOLM in *"Metal Ions In Biology"* ; T.G. SPIRO Ed., Interscience, New York, 1982, Vol. **4**, Chapter 1.

92. F.A. ARMSTRONG, H.A.O. HILL, N.J. WALTON, FEBS Letters, **145** (1982) 241 ; **150** (1982) 214.

93. T. SHIBAHARA, H. KUROYA, K. MATSUMOTO, S. OOI, J. Am. Chem. Soc., **106** (1984) 789.
P. KATHIRGAMANATHAN, M. MARTINES, A.G. SYKES, J. Chem. Soc. Chem. Comm., (1985) 953.

94. P.K. MASCHARAK, W.H. ARMSTRONG, Y. MIZOBE, R.H. HOLM, J. Am. Chem. Soc., **105** (1983) 475.

95. C.M. HAY, B.F.G. JOHNSON, J. LEWIS, R.C.S. McQUEEN, P.R. RALTHBY, R.M. SORRELL, M.J. TAYLOR, Organometallics, **4** (1985) 202.

96. H.P. KLEIN, U. THEWALT, G. HERRMANN, G. SUSS-FINK, C. MOINET, J. Organomet. Chem., **286** (1985) 225.

97. D. OSELLA, R. GOBETTO, P. MONTANGERO, P. ZANELLO, A. CINQUANTINI, Organometallics, **5** (1986) 1247.

98. D. De MONTAUZON, R. MATHIEU, J. Organomet. Chem., **252** (1983) C 83.

99. G. TAINTURIER, B. GAUTHERON, C. DEGRAND, Organometallics, **5** (1986) 942.

100. D.D. KENDWORTH, R.A. WALTON, Inorg. Chem., **20** (1981) 1151.

101. M.S. PAQUETTE, L.F. DAHL, J. Am. Chem. Soc., **102** (1980) 6622.

102. R.E. PALERMO, R.H. HOLM, J. Am. Chem. Soc., **105** (1983) 4310 and references therein.

103. J.G.M. VAN DER LINDEN, M.L.H. PAULISSEN, J.E.J. SCHMITZ, J. Am. chem. Soc., **105** (1983) 1903.

104. A.W. MAVERICK, S.J. NAJDZIONEK, D. MACKENZIE, D.G. NOCERA, H.B. GRAY, J. Am. Chem. Soc., **105** (1983) 1878 ; **106** (1984) 824.

ELECTRON TRANSFER-CATALYSED SUBSTITUTION REACTIONS OF METAL CLUSTER CARBONYLS

Michael I. Bruce
Department of Physical and Inorganic Chemistry, University of Adelaide, Adelaide, South Australia 5001

ABSTRACT
This article summarises the scope of the radical anion (alkali metal diphenylketyl) initiated substitution reactions of metal cluster carbonyls. The method has been applied mainly to $Ru_3(CO)_{12}$, but substitution of several other homo- and hetero-metallic clusters containing Fe, Ru, Os, Co, Rh, or Ir has been reported. Some suggestions are made concerning the mechanism of these reactions, which probably proceed by a radical chain process, but it is clear that present information does not provide an unequivocal answer to this problem.

INTRODUCTION

Cluster complexes containing transition metals have an extensive chemistry, development of this area being particularly intensive over the last two decades as a result of expectations that the behaviour of ligands in clusters might model that of molecules adsorbed on metal surfaces, and also because of their potential as catalysts.[1] Early studies of the chemistry of metal cluster carbonyls were hampered by the high reactivity of cluster-bound ligands coupled with the harsh conditions often required for initial substitution of CO. These two factors result in complex reaction mixtures, products containing significantly altered ligands, or fragmentation of the cluster. The reactions of the Group 8 $M_3(CO)_{12}$ compounds with PPh_3 illustrate some of this diversity of behaviour:

M. Chanon et al. (eds.), Paramagnetic Organometallic Species in Activation / Selectivity, Catalysis, 407–422.

$$Fe_3(CO)_{12} + PPh_3 \rightarrow Fe(CO)_{5-n}(PPh_3)_n \quad (n = 1, 2)$$
$$Ru_3(CO)_{12} + PPh_3 \rightarrow Ru_3(CO)_9(PPh_3)_3$$
$$Os_3(CO)_{12} + PPh_3 \rightarrow Os_3(CO)_{12-n}(PPh_3)_n \quad (n = 1, 2, 3)$$
$$+ \text{ six other complexes}$$

The solvents employed for these reactions were refluxing tetrahydrofuran or dioxan (Fe),[2] refluxing benzene or cyclohexane (Ru),[3] and refluxing xylene (Os).[4] In the latter case, the other complexes isolated are formed by C-H and C-P bond cleavage and rearrangement reactions, some of which are only now beginning to be unravelled.[5] The rate of reaction of the ruthenium cluster is governed by the introduction of the first ligand: substitution of the second and third CO groups is fast, and results in the exclusive formation of the trisubstituted product. Another example which may be cited to illustrate the difficulty of controlling the reaction products is the formation of a mixture of all five complexes $Ru_4(\mu\text{-}H)_4(CO)_{12-n}\{P(OMe)_3\}_n$ (n = 0-4), each isolated in only small amounts after extensive chromatography after the initial reaction between $Ru_3(CO)_{12}$ and $P(OMe)_3$.[6]

ELECTRON TRANSFER-CATALYSED REACTIONS

It is now apparent that one approach to simple substituted cluster carbonyls which does not suffer from these difficulties is via the appropriate carbonylmetal anion radical, in which the lability of CO towards substitution is considerably enhanced. This property has led to the development of a mild and stereospecific method of synthesis of metal cluster carbonyl derivatives, namely electron- or radical anion-induced substitution. The first report of such a reaction applied to metal cluster carbonyls was the substitution of the radical anion obtained electrolytically from $Co_3(\mu_3\text{-}CR)(CO)_9$ (R = Ph, Cl) by tertiary phosphines or phosphites; the reaction afforded a quantitative yield of product at room temperature in less than five minutes.[7] The method has been applied extensively to several bi-, tri- and poly-nuclear metal carbonyl complexes by Darchen,[8] Robinson,[9] Bruce[10] and their coworkers. The reactions are examples of substitution reactions that occur by electron-transfer (ET) catalysis.[11]

The development of this reaction applied to $Ru_3(CO)_{12}$ has been recounted elsewhere.[12] We have used, for preference, a solution of sodium diphenylketyl, obtained by stirring excess sodium with benzophenone in dry tetrahydrofuran for about an hour under nitrogen.[13] The solution is deep purple in colour, and it should be noted that the first-formed deep blue charge-transfer complex is not an efficient catalyst. The active solution prepared as above is stable for about two days; regeneration is a simple matter of adding more benzophenone and stirring as above. The optimum concentration is about 0.025 M. The method employed for the substitution reactions consists of adding a few drops of the radical-anion solution to a stoichiometric mixture of the cluster carbonyl complex and the ligand. The reactions generally occur immediately upon addition of the radical anion, as evidenced by TLC and IR ν(CO) examination of the reaction mixture: isolated yields of products are generally good to excellent.

This mild substitution reaction has enabled many cluster complexes which are not accessible by thermal reactions (substitution of $Ru_3(CO)_{12}$ does not usually commence until the mixture is heated to 60-80 °C) to be synthesised (Table 1).

Some examples chosen from our own studies follow:

(i) The reaction between $Ru_3(CO)_{12}$ and $PPh_2CH_2CH_2PPh_2$ (dppe)* afforded good yields of the complexes $Ru_3(CO)_{11}$(dppe-*P*) (with only one P atom coordinated), $\{Ru_3(CO)_{11}\}_2$(μ-dppe) (with the dppe ligand bridging two Ru_3 clusters), and $Ru_3(\mu\text{-dppe})_n(CO)_{12-2n}$ (n = 1, 2) (in both of which the dppe ligand(s) bridges Ru-Ru bonds).[14]

(ii) The olefinic tertiary phosphine $PPh_2(C_6H_4CH{=}CH_2\text{-}2)$ afforded the complex $Ru_3(\mu\text{-}\eta^2,P\text{-}CH_2{=}CHC_6H_4PPh_2)(CO)_{10}$, which

---

* Abbreviations used: dpae $AsPh_2CH_2CH_2AsPh_2$; dpam $CH_2(AsPh_2)_2$; dppa $C_2(PPh_2)_2$; dppe $PPh_2CH_2CH_2PPh_2$; dppm $CH_2(PPh_2)_2$; ebdp *cis* -CH(PPh_2)=CH(PPh_2); ppn $[N(PPh_3)_2]^+$; xy 2,6-$Me_2C_6H_3$.

TABLE 1 Radical anion-induced reactions of $Ru_3(CO)_{12}$

Ligand	Product $Ru_3(CO)_{12-n}(L)_n$ n^a	Yield	Reference
$CNBu^t$	1	78	10, 13
	2	61	10, 13
CNCy	1	89	13
(R)-(+)-CNCHMePh	1	32	13
$CNCH_2SO_2C_6H_4Me$-4	2	87	13
NCMe	1	65	9a
$PHPh_2$	1	n.g.	34, 35
$PHPh\{CH_2CH_2Si(OEt)_3\}$	1	n.g.	35
PMe_3	1	75	13
	2	60	13
	3	76	10, 13
$PMe_2(CH_2Ph)$	3	44	36
PMe_2Ph	1	76	13
PPh_3	1	81	10, 13
	2	96	10, 13
	3	76	10, 13
$P(C_6H_4Me\text{-}2)_3$	1	37	13
$P(C_6H_4Me\text{-}4)_3$	1	79	13
	3	87	13
$PPh_2(C_6H_4CH{=}CH_2\text{-}2)$	1	74	15a
$PPh_2\{CH_2CH_2Si(OEt)_3\}$	1	n.g.	37
PCy_3	1	89	13
	2	55	13
$P(CH_2CH_2CN)_3$	2	91	13
	3	72	13
$PPh(OMe)_2$	1	74	13
	2	63	13
	3	44	10, 13
	4	91	10, 13

$P(OMe)_3$	1	81	9a,10,13, 38
	2	81	13, 38
	3	70	13, 38
$P(OEt)_3$	3	15	36
	4	16	
$P(OCH_2CF_3)_3$	1	11	36
	2	21	
	3	7	
$P(OCH_2)_3CEt$	1	50	13
	2	57	13
	3	92	13
$P(OPh)_3$	1	80	13
$P(OC_6H_4Me\text{-}4)_3$	1	64	13
	2	26	13
	3	66	13
dppm	(2)	90	9a, 10, 39
	(4)	26	13
	(1)	98	13
dppe	(2)	52	14
	(4)	53	10, 14
	(1)-μ	43	10, 14
	(1)	n.g.	14
dppa	(1)-μ	71	16
$AsPh_3$	1	66	10, 13
	2	48	16
dpam	(2)	53	10, 13
	(4)	8	13
dpae	(2)	78	13
$SbPh_3$	1	44	10, 13
$CNBu^t/CNCy$	2	85	13
$CNBu^t/PMe_2Ph$	2	42	13, 19a
$CNBu^t/P(C_6H_4Me\text{-}4)_3$	2	88	13
$PMe_3/P(OMe)_3$	2	87	13
$P(C_6H_4Me\text{-}4)_3/AsPh_3$	2	39	13
$CNBu^t/\{P(OCH_2)_3CEt\}_2$	3	34	13

$CNBu^t/PPh_2(C_6H_4CH{=}CH_2\text{-}2)$	(3)	18	19a
$PMe_3/PPh_3/P(OMe)_3$	3	41	13, 19a
PMe_3/dppe	(3)	81	13
$P(OMe)_3$/dpam	(3)	51	13, 19a

[a] Number of CO groups displaced; ( ) indicates bidentate ligand, ( )-μ indicates bidentate ligand bridging two Ru_3 clusters. n.g. not given.

requires warming to only 40 °C to undergo dehydrogenation of the olefinic function wiuth formation of the hydrido-alkyne cluster $Ru_3(\mu\text{-}H)_2(\mu_3\text{-}\eta^2,P\text{-}HC{\equiv}CC_6H_4PPh_2)(CO)_8$; at higher temperatures, further condensation to an Ru_4C_2 cluster occurs.[15]

(iii) A second example of a complex containing two Ru_3 nuclei linked by a bidentate tertiary phosphine ligand is provided by $\{Ru_3(CO)_{11}\}_2(\mu\text{-dppa})$. On heating to 90 °C, rapid formation of the pentanuclear complex $Ru_5(\mu_5\text{-}\eta^2,P\text{-}C{\equiv}CPPh_2)(\mu\text{-}PPh_2)(CO)_{13}$ occurs, containing an open Ru_5 cluster.[16] The fascinating chemistry of this molecule, already partially explored in its reactions with CO and with H_2,[17] continues to provide unusual complexes, including a bis-phosphinidene-vinylidene derivative, in which cleavage of both P-C bonds present in dppa, and migration of one phenyl group from each phosphorus to the same carbon atom, have occurred.[18]

A further application of this new synthetic method has been to the synthesis of cluster complexes containing two or more different ligands, other than CO. Successive application of the reaction after adding the appropriate ligand to the reaction mixture, offers a stepwise, high yield route which may be carried out in one pot if suitable monitoring of the reaction mixture can take place.[19] By this route, not only were complexes containing three different phosphorus ligands obtained, such as $Ru_3(CO)_9(PMe_3)(PPh_3)\{P(OMe)_3\}$, but also some containing mixed ligand types, such as the isocyanide-olefin-tertiary phosphine combination found in $Ru_3(CO)_9(CNBu^t)(\eta^2,P\text{-}CH_2{=}CHC_6H_4PPh_2)$.

Although most of our studies have been directed at the synthesis of substituted derivatives of $Ru_3(CO)_{12}$, both we and others have found that the method is applicable to many other polynuclear complexes (Table 2). Of course, if the thermal reactions proceed readily and without complication, as found for $M_4(CO)_{12}$ (M = Co, Rh, Ir), $M_6(CO)_{16}$ (M = Rh, Ir), $Ru_5C(CO)_{15}$ or $Ru_6C(CO)_{17}$, for example, then the ET-catalysed reaction offers little advantage. However, radical-anion induced stoichiometric substitution has been found to occur with many systems which are thermally sensitive, and low-temperature conditions may be advantageous in these cases: substitution of $Fe_3(CO)_{12}$ can be achieved at -30 °C with relatively little cluster fragmentation.[9a] Other cluster carbonyls reported to undergo the reaction include $Os_3(CO)_{12}$, $Rh_6(CO)_{16}$ and $Ir_4(CO)_{12}$.[9a,10] In our experience, stoichiometric substitution does not usually occur with the osmium complex, a mixture of CO substitution products being obtained; nevertheless, replacement of CO occurs readily at room temperature and the products are readily separated by chromatography. In the case of $Ru_4(\mu\text{-H})_4(CO)_{12}$, we have noted increasing difficulty in replacing more than three CO groups with phosphorus ligands, when deprotonation of the cluster hydride competes effectively with electron transfer. Another class of cluster complex which appears to be well-suited to this type of reaction are those in which one or more faces are capped by Main Group atoms, such as $Co_3(\mu_3\text{-CR})(CO)_9$, $Ru_3(\mu\text{-H})_2(\mu_3\text{-E})(CO)_9$ (E = S, PR), $Ru_3(\mu_3\text{-NR})_2(CO)_9$ and $Co_4(\mu_4\text{-PR})_2(CO)_{10}$.

An exciting extension of the method is its application to the exchange of ML_n moieties in metal cluster complexes. This route has been used in the preparation of $Co_2Mo(\mu_3\text{-CPh})(CO)_8(\eta\text{-}C_5H_5)$ from $Co_3(\mu_3\text{-CPh})(CO)_9$ and $\{Mo(CO)_3(\eta\text{-}C_5H_5)\}_2$, and of $CoFeM(\mu_3\text{-E})(CO)_8(\eta\text{-}C_5H_5)$ (M = Mo or W, E = S or Se; M = Mo, E = $PNEt_2$) from $Co_2Fe(\mu_3\text{-E})(CO)_9$ and $Na[M(CO)_3(\eta\text{-}C_5H_5)]$, the reactions being carried out in the presence of sodium diphenylketyl.[9a,20] Again, however, the scope of these reactions remains to be established.

TABLE 2 Radical anion-induced reactions of other cluster carbonyl complexes

Cluster	Ligand	n[a]	Yield	Reference
$Fe_3(CO)_{12}$	$CNBu^t$	1	53	10, 40
	PPh_3	1	60	9a, 10
	$P(OMe)_3$	1	75	9a
	$P(OC_6H_4Me\text{-}4)_3$	1	25	10
$Fe_2Ru(CO)_{12}$	PPh_3	1	50	41
	$P(OMe)_3$	1	90	41
$FeRu_2(CO)_{12}$	PPh_3	1	80	41
		2	90	41
	$P(OMe)_3$	1	80	41
		2	90	41
$Ru_3(\mu\text{-H})(\mu_3\text{-}C_2Bu^t)(CO)_9$	$CNBu^t$	1	n.g.	42
	PPh_3	1	n.g.	42
$Ru_3(\mu_3\text{-NPh})_2(CO)_9$	CNxy	1	40	33
	PPh_3	1	52	33
	$P(OMe)_3$	1	37	33
	dppm	(2)	80	33
	dppe	(2)	7	33
	dppa	(1)-μ	31	33
$Ru_3(\mu\text{-H})_2(\mu_3\text{-S})(CO)_9$	dppm	(2)	76	43
	ebdp	(2)	70	43
	dpam	(2)	56	43
$Ru_4(\mu\text{-H})_4(CO)_{12}$	$CNBu^t$	1	63	10,13
		2	34	10,13
	PPh_3	1	55	10,13
		2	53	10
	$PPh_2\{CH_2CH_2\text{-}Si(OEt)_3\}$	1	n.g.	37
	$P(OMe)_3$	1	90	10,19a
		3	47	10,19a
	$P(OPh)_3$	1	76	10,13

	$P(OC_6H_4Me\text{-}4)_3$	1	65	10
	$PMe_2Ph/$ $P(OCH_2)_3CEt$	2	66	13
	$P(OMe)_3/$ $PPh(OMe)_2$	2	70	19
	$PMe_2Ph/P(OMe)_3/$ $PPh(OMe)_2$	3	75	19
	$PMe_2Ph/P(OCH_2)_3CEt/$ $P(OC_6H_4Me\text{-}4)_3$	3	64	19
$RuCo_3(\mu\text{-}H)(CO)_{12}$	PPh_3	2	25	44
	dppe	(2)	47	45
$RuRh_3(\mu\text{-}H)(CO)_{12}$	PPh_3	2	n.g.	44
$Ru_3Co(\mu\text{-}H)_3(CO)_{12}$	dppe	(2)	70	45
$Os_3(CO)_{12}$	$CNBu^t$	1	17	10
	NCMe	1	70	9a
	PPh_3	1	24	10
		2	21	
		3	17	
	$P(OMe)_3$	1	90	9a
$Co_3(\mu_3\text{-}CPh)(CO)_9$	PPh_3	1	90	9a
	dppm	(2)	90	9a
$Co_3(\mu_3\text{-}CCl)(CO)_9$	$CNBu^t$	3	62	42
	$P(OC_6H_4Me\text{-}4)_3$	1	78	19a
	dppe	(2)	51	10
$Co_4\{\mu_3\text{-}(PPh_2)_3CH\}(CO)_9$	PPh_3	1	n.g.	46
	$PMePh_2$	1	n.g.	46
	PMe_2Ph	1	n.g.	46
$Co_4(\mu_4\text{-}PPh)_2(CO)_{10}$	PPh_3	1	60	32a
	$P(OMe)_3$	1	40	32a
	dppe	(2)	50	32a
$Rh_6(CO)_{16}$	$CNBu^t$	5	30	10
$Ir_4(CO)_{12}$	PPh_3	1	90	9a

[a] See footnote to Table 1.

There are many reactions of cluster carbonyls which take place under mild conditions and which either do not require, or cannot be carried out in the presence of, a radical anion initiator. In the case of $Os_3(CO)_{12}$, ready substitution of CO by acetonitrile occurs on addition of Me_3NO to a solution of the carbonyl in acetonitrile, and the complexes $Os_3(CO)_{12-n}(NCMe)_n$ (n = 1, 2) can be either isolated or used directly; the disubstituted complex can also be obtained from $Os_3(CO)_{10}$(*cis*-cyclooctene)$_2$.[21] The reaction can be modified to afford the analogous ruthenium complexes by working at low temperatures, or by carrying out the reaction in dichloromethane containing about 10% MeCN and the ligand.[22] Facile reactions of $Ru_3(CO)_{12}$ with amines, such as $NHMe_2$, to give μ-carboxamido complexes (at -30 °C),[23] or with Group 14 ligands (isocyanides or carbene precursors)[24] have also been reported.

Of more practical import are the findings of Kaesz and coworkers,[25] who noted the specific site activation on ruthenium by bridging acyl or halide groups, as found in $Ru_3(\mu\text{-H})(\mu\text{-O=CMe})(CO)_{10}$, $Ru_3(\mu\text{-H})(\mu\text{-X})(CO)_{10}$ or $Ru_3(\mu\text{-X})(\mu\text{-COMe})(CO)_{10}$ (X = Cl, Br, I), which results in progressively faster substitution reactions being found along this series of complexes. These results are probably relevant also to the nucleophile-induced substitution of $Ru_3(CO)_{12}$ found by Kaesz and Lavigne,[26] and by Ford.[27] Thus, addition of catalytic amounts of [ppn][X] (X = Cl, CN, OAc, BH_4) to solutions of $Ru_3(CO)_{12}$ containing the ligand, promotes substitution of CO by tertiary phosphines, phosphites or arsines. Similarly, addition of methoxide ion accelerates substitution of CO. In all cases, formation of a labile anionic $[Ru_3(Nu)(CO)_{11}]^-$ (Nu = Cl, OAc, CN, BH_4, CO_2Me, etc) species is likely. More recently, it has been found that addition of a few drops of a tetrahydrofuran solution of $K[BHBu^s_3]$ (*K* -Selectride) achieves the same result,[28] probably via the hydrido anion $[Ru_3(\mu\text{-H})(\mu\text{-CO})(CO)_{10}]^-$.

MECHANISM

There has been much discussion of the mechanism of the radical anion-induced reactions. Electron transfer from sodium diphenylketyl to $Ru_3(CO)_{12}$ can occur via addition of the electron to the LUMO, which is metal-metal antibonding.[29] This results in cleavage or weakening of an Ru-Ru bond to give a site which is both sterically and electronically favoured for nucleophilic attack by the entering ligand. We suppose that the radical anion so formed is less stable than that of the parent cluster carbonyl, so that electron transfer back to the unsubstituted carbonyl occurs:

$$Ru_3(CO)_{12} + e^- \rightarrow [Ru_3(CO)_{12}]^{-\cdot}$$
$$[Ru_3(CO)_{12}]^{-\cdot} + L \rightarrow [Ru_3(CO)_{11}(L)]^{-\cdot} + CO$$
$$[Ru_3(CO)_{11}(L)]^{-\cdot} + Ru_3(CO)_{12} \rightarrow Ru_3(CO)_{11}(L) + [Ru_3(CO)_{12}]^{-\cdot}$$

The reaction cycle so established continues until either carbonyl or ligand is used up, or another pathway removes the anion from the reaction mixture. Electrochemical studies have confirmed the relative stabilities of the parent and substituted cluster anions to be in accord with this proposal;[30] however, lifetime measurements and findings that the formation of the reduced species is irreversible appears to suggest that they cannot be intermediates in these reactions.

In the absence of added ligand, $Ru_3(CO)_{12}$ reacts with sodium diphenylketyl to give anionic Ru_4 and Ru_6 clusters; it has been shown independently that condensation of $Ru_3(CO)_{12}$ with $[Ru_3(CO)_{11}]^-$ affords $[Ru_6(CO)_{18}]^{2-}$.[31] Detailed examination[30] of the reduction of $Ru_3(CO)_{12}$ suggests the formation of the short-lived $[Ru_3(CO)_{12}]^{-\cdot}$ radical anion, which may either isomerise to an open cluster, or lose CO to give $[Ru_3(CO)_{11}]^{-\cdot}$. The open cluster can accept a second electron to form $[Ru_3(CO)_{12}]^{2-}$. In the ligand substitution reactions, however, this further reduction is generally slower than formation of the substituted product.

The formation of $[M_3(CO)_{12-n}(PPh_3)_n]^{2-}$ (M = Ru, Os) has been confirmed electrochemically in several solvents, although the half-life of $[M_3(CO)_{12}]^{-\cdot}$ was determined to be $< 10^{-6}$ sec. in CH_2Cl_2. The 'open' $[M_3(CO)_{12}]^{2-}$ can be detected in low concentrations, and the reactions are sensitive to the presence of water. Accordingly, it has been suggested that the radical anion-catalysed substitution is actually another example of a nucleophile-assisted substitution, with Nu = OH.[30] The electrode-initiated ET-catalysed substitution occurs within the electrochemical timescale for $Ru_3(CO)_{12}$, but is found to be inefficient and non-specific, in contrast with the sodium diphenylketyl-initiated reactions; the latter are highly solvent dependent, proceeding best in ethers such as tetrahydrofuran. The differences observed between the two systems suggest that two different processes are occurring, and that our present understanding does not enable a meaningful connection to be made between them.

That the reactions initiated by sodium diphenylketyl are genuine ET-catalysed reactions proceeding by a radical chain process seems to be confirmed, at least in the cases of the phosphinidene-capped complexes $Fe_3(\mu_3\text{-}PPh)_2(CO)_9$ and $Co_4(\mu_4\text{-}PPh)_2(CO)_{10}$, by the elegant work of Kochi and coworkers.[32] They used electrochemical and ESR techniques to identify both the radical ions produced by successive one-electron reductions, and their Group 15 ligand substitution products. An important result found for the iron complex is the opening of the Fe_2P triangle to generate a 17-e centre; by analogy, this mechanism probably operates also with the analogous $Ru_3(\mu_3\text{-}NPh)_2(CO)_9$ complex.[33] In the case of the tetranuclear cobalt complex, a key step is the dissociative loss of CO from the anion radical to form the unsaturated intermediate $[Co_4(PPh_2)_2(CO)_9]^{-\cdot}$, which then adds the ligand in a rapid reaction. The efficiency of the substitution reaction depends on the kinetic stability of the anion radical, which in turn is a function of other routes by which it can react.

CONCLUSION

Our studies have shown that sodium diphenylketyl-initiated reactions of several metal cluster carbonyls with a variety of ligands are efficient synthetic routes to a wide variety of carbonyl-substituted complexes. For $Ru_3(CO)_{12}$, it is presently the method of choice, particularly if the resulting complexes are thermally labile. The scope of the reactions involving other clusters still requires to be established. Further studies are needed to establish the precise nature of these reactions, particularly to determine whether they are ET-initiated or induced by nucleophiles. At present, however, we favour a radical anion chain process, which can afford the specific products observed.

ACKNOWLEDGEMENTS

It is a pleasure to acknowledge the essential contributions of Drs John Matisons and Brian Nicholson to this work, and the further contributions of the coworkers listed in the references. Much of this work was supported by grants received under the Australian Research Grants Scheme.

REFERENCES

1. (a) B.F.G. Johnson, ed., **Transition Metal Clusters**, Wiley, New York, 1980; (b) M. Moskovits, ed., **Metal Clusters**, Wiley-Interscience, New York, 1986; (c) B.C. Gates, L. Guczi and H. Knozinger, eds, **Metal Clusters in Catalysis**, Elsevier, Amsterdam, 1986
2. A.F. Clifford and A.K. Mukherjee, *Inorg.Chem.*, 1963, **2**,151.
3. (a) B.F.G. Johnson, R.D. Johnston, P.L. Josty, J. Lewis and I.G. Williams, *Nature*, 1967, **213**, 901; (b) F. Piacenti, M. Bianchi, E. Benedetti and G. Sbrana, *J.Inorg.Nuclear Chem.*, 1967, **29**, 1389; (c) J.P. Candlin and A.P. Shortland, *J.Organomet.Chem.*, 1969, **16**, 289; (d) M.I. Bruce, G. Shaw and F.G.A. Stone, *J.Chem.Soc., Dalton Trans.*, **1972**, 2094.
4. (a) C.W. Bradford, R.S. Nyholm, G.J. Gainsford, J.M. Guss, P.R. Ireland and R. Mason, *J.Chem.Soc., Chem.Commun.*, **1972**, 87; (b) G.J. Gainsford, J.M. Guss, P.R. Ireland, R. Mason, C.W. Bradford and R.S. Nyholm, *J.Organomet.Chem.*, 1972, **40**, C70; (c) C.W. Bradford and R.S. Nyholm, *J.Chem.Soc., Dalton Trans.*, **1973**, 529.

5. A.J. Deeming, S.E. Kabir, N.I. Powell, P.A. Bates and M.B. Hursthouse, *J.Chem.Soc., Dalton Trans.*, **1987**, 1529.
6. S.A.R. Knox and H.D. Kaesz, *J.Am.Chem.Soc.*, 1971, **93**, 4594.
7. G.J. Bezems, P.H. Rieger and S. Visco, *J.Chem.Soc., Chem.Commun.*, **1981**, 265.
8. (a) A. Darchen, C. Mahé and H. Patin, *J.Chem.Soc., Chem.Commun.*, **1982**, 243; (b) E.K. Lhardi, C. Mahé, H. Patin and A. Darchen, *J.Organomet.Chem.*, 1983, **246**, C61; (c) A. Darchen, C. Mahé and H. Patin, *J.Organomet.Chem.*, 1983, **251**, C9; (d) A. Darchen, E.K. Lhardi and H. Patin, *J.Organomet.Chem.*, 1983, **259**, 189.
9. (a) M. Arewgoda, B.H. Robinson and J. Simpson, *J.Chem.Soc., Chem.Commun.*, **1982**, 284; *J.Am.Chem.Soc.*, 1983, **105**, 1893; (b) R.G. Cunninghame, A.J. Downard, L.R. Hanton, S.D. Jensen, B.H. Robinson and J. Simpson, *Organometallics*, 1984, **3**, 180.
10. M.I. Bruce, D.C. Kehoe, J.G. Matisons, B.K. Nicholson, P.H. Rieger and M.L. Williams, *J.Chem.Soc., Chem.Commun.*, **1982**, 442.
11. (a) M. Chanon and M.L. Tobe, *Angew.Chem.*, 1982, **94**, 27; *Angew.Chem., Int.Ed.Engl.*, 1982, **21**, 211; (b) M. Juillard and M. Chanon, *Chem.Rev.*, 1983, **83**, 425.
12. M.I. Bruce, *Coord.Chem.Rev.*, 1987, **76**, 1.
13. M.I. Bruce, J.G. Matisons and B.K. Nicholson, *J.Organomet.Chem.*, 1983, **247**, 321.
14. M.I. Bruce, T.W. Hambley, B.K. Nicholson and M.R. Snow, *J.Organomet.Chem.*, 1982, **235**, 83.
15. (a) M.I. Bruce, B.K. Nicholson and M.L. Williams, *J.Organomet.Chem.*, 1983, **243**, 69; (b) M.I. Bruce and M.L. Williams, *J.Organomet.Chem.*, 1986, **314**, 323; (c) M.I. Bruce, E. Horn, M.R. Snow and M.L. Williams, *J.Organomet.Chem.*, 1983, **255**, 255.
16. M.I. Bruce, M.L. Williams, J.M. Patrick and A.H. White, *J.Chem.Soc., Dalton Trans.*, **1985**, 1229.
17. (a) M.I. Bruce and M.L. Williams, *J.Organomet.Chem.*, 1985, **282**, C11; (b) M.I. Bruce, M.L. Williams, B.W. Skelton and A.H. White, *J.Organomet.Chem.*, 1985, **282**, C53; (c) M.I. Bruce, B.W. Skelton, A.H. White and M.L. Williams, *J.Chem.Soc., Chem.Commun.*, **1985**, 744.
18. M.I. Bruce and M.J. Liddell, unpublished results.

19. (a) M.I. Bruce, J.G. Matisons, B.K. Nicholson and M.L. Williams, *J.Organomet.Chem.*, 1982, **236**, C57; (b) M.I. Bruce, B.K. Nicholson, J.M. Patrick and A.H. White, *J.Organomet.Chem.*, 1983, **254**, 361.
20. (a) S. Jensen, B.H. Robinson and J. Simpson, *J.Chem.Soc., Chem.Commun.*, **1983**, 1081; (b) U. Honrath and H. Vahrenkamp, *Z.Naturforsch., Ser.B*, 1984, **39b**, 559.
21. B.F.G. Johnson, J. Lewis and D. Pippard, *J.Organomet.Chem.*, 1978, **160**, 263;1981, **213**, 249. See also K. Burgess, *Polyhedron*, 1984, **3**, 1175.
22. (a) G.A. Foulds, B.F.G. Johnson and J. Lewis, *J.Organomet.Chem.*, 1985, **294**, 123; 1985, **296**, 147; M.I. Bruce and P.A. Humphrey, unpublished results.
23. R. Szostak, C.E. Strouse and H.D. Kaesz, *J.Organomet.Chem.*, 1980, **191**, 243.
24. (a) M.I. Bruce, J.G. Matisons, R.C. Wallis, J.M. Patrick, B.W. Skelton and A.H. White, *J.Chem.Soc., Dalton Trans.*, **1983**, 2365; (b) M.F. Lappert and P.L. Pye, *J.Chem.Soc., Dalton Trans.*, **1977**, 2172; (c) M.M. Singh and R.J. Angelici, *Inorg.Chim.Acta*, 1985, **100**, 57.
25. C.E. Kampe and H.D. Kaesz, *Inorg.Chem.*, 1984, **23**, 4646.
26. G. Lavigne and H.D. Kaesz, *J.Am.Chem.Soc.*, 1984, **106**, 4647.
27. M. Anstock, D. Taube, D.C. Gross and P.C. Ford, *J.Am.Chem.Soc.*, 1984, **106**, 3696.
28. G. Lavigne, N. Lugan and J-J. Bonnet, *Inorg.Chem.*, 1987, **26**, 2345.
29. (a) D.R. Tyler, R.A. Levenson and H.B. Gray, *J.Am.Chem.Soc.*, 1978, **100**, 7888; (b) B.E.R. Schilling and R. Hoffman, *J.Am.Chem.Soc.*, 1979, **101**, 3456.
30. A.J. Downard, B.H. Robinson, J. Simpson and A.M. Bond, *J.Organomet.Chem.*, 1987, **320**, 363.
31. A.A. Bhattacharyya, C.C. Nagel and S.G. Shore, *Organometallics*, 1983, **2**, 1187.
32. (a) M.G. Richmond and J.K. Kochi, *Inorg.Chem.*, 1986, **25**, 656, 1334; H.H. Ohst and J.K. Kochi, *Inorg.Chem.*, 1986, **25**, 2066.
33. M.I. Bruce, M.G. Humphrey, O. bin Shawkataly, M.R. Snow and E.R.T. Tiekink, *J.Organomet.Chem.*, 1986, **315**, C51; 1987, in press.
34. S.A. MacLaughlin, N.J. Taylor and A.J. Carty, *Organometallics*, 1984, **3**, 392.
35. S.L. Cook and J. Evans, *J.Chem.Soc., Chem.Commun.*, **1983**, 713.

36. M.I. Bruce, M.J. Liddell and O. bin Shawkataly, unpublished results.
37. J. Evans and B.P. Gracey, *J.Chem.Soc., Chem.Commun.*, **1983**, 247.
38. D.C. Cross and P.C. Ford, *J.Am.Chem.Soc.*, 1985, **107**, 585.
39. J.A. Ladd, H. Hope and A.L. Balch, *Organometallics,* 1984, **3**, 1838.
40. M.I. Bruce, T.W. Hambley and B.K. Nicholson, *J.Chem.Soc., Dalton Trans.*, **1983**, 2385.
41. T. Venäläinen and T. Pakkanen, *J.Organomet.Chem.*, 1984, **266**, 269.
42. M.I. Bruce, J.G. Matisons and B.K. Nicholson, unpublished results.
43. M.I. Bruce, M.G. Humphrey and O. bin Shawkataly, unpublished results.
44. J. Pursiainen, T.A. Pakkanen and J. Jääskeläinen, *J.Organomet.Chem.*, 1985, **290**, 85.
45. J. Pursiainen and T.A. Pakkanen, *J.Organomet.Chem.*, 1986, **309**, 187.
46. J. Rimmelin, P. Lemoine, M. Gross, A.A. Bahsoun and J.A. Osborn, *Nouv.J.Chim.*, 1985, **9**, 181.

THE ROLE OF FREE RADICALS IN ORGANOMETALLIC CATALYSIS

Jack Halpern
Department of Chemistry
The University of Chicago
Chicago, Illinois 60637
U. S. A.

ABSTRACT. A variety of reactions of metal complexes, reflecting the characteristic weakness of many metal-carbon, metal-hydrogen and metal-metal bonds, result in the formation of carbon-centered and/or metal-centered free radicals. Such reactions play important roles in many catalytic processes including coenzyme B_{12}-dependent enzymic reactions. The mechanistic aspects of such processes are discussed.

I. INTRODUCTION

The connection between free radicals and catalysis is an old one as manifested, for example, by the familiar catalytic roles of metal ions (cobalt, manganese, etc.) in the autoxidation of organic compounds. However, the scope of such catalysis has been greatly expanded in recent years, notably through the recognition of many examples of organometallic free radical catalytic processes (1,2). These now are recognized as being far more widespread than once suspected not only in new contexts but also in the context of familiar catalytic reactions, such as hydrogenation and hydroformylation, that previously had been interpreted in terms of alternative non-radical mechanisms. Important manifestations of free radical catalytic mechanisms, having their origin in organometallic chemistry, also have been identified in biological systems, notably the enzymic rearrangement reactions for which coenzyme B_{12} (5'-deoxyadenosylcobalamin) serves as a cofactor (3).

Metal-centered and/or carbon-centered free radicals ($L_nM^{\bullet}$ and $R^{\bullet}$, respectively) arise through a variety of reactions of metal complexes, among them the following:

<u>Metal-centered radicals</u>:

$$L_{2n}M_2 \longrightarrow 2L_nM^{\bullet} \quad (1)$$

$$L_{2n}M_2 + X^{\bullet} \longrightarrow L_nM\text{-}X + L_nM^{\bullet} \quad (2)$$

$$L_nM\text{-}H + R^{\bullet} \longrightarrow L_nM^{\bullet} + RH \quad (3)$$

M. Chanon et al. (eds.), Paramagnetic Organometallic Species in Activation / Selectivity, Catalysis, 423–433.

<u>Carbon-centered radicals:</u>

$$L_nM^{\bullet} + R\text{-}X \longrightarrow L_nM\text{-}X + R^{\bullet} \qquad (4)$$

$$L_nM\text{-}R \xrightarrow{e^-} [L_nM\text{-}R]^- \longrightarrow [L_nM]^- + R^{\bullet} \qquad (5)$$

$$L_nM\text{-}R \xrightarrow{-e^-} [L_nM\text{-}R]^+ \longrightarrow [L_nM]^+ + R^{\bullet} \qquad (6)$$

<u>Both types of radicals:</u>

$$L_nM\text{-}H + \ \rangle C{=}C\langle \ \longrightarrow L_nM^{\bullet} + M\text{—}\rangle C\text{-}C\langle^{\bullet} \qquad (7)$$

$$L_nM\text{-}R \longrightarrow L_nM^{\bullet} + R^{\bullet} \qquad (8)$$

The facile occurrence of reactions 1 and 2 reflect the characteristic weakness of many transition metal-metal single bonds, while reactions 3 and 7 and reaction 8 reflect the characteristic weakness of many transition metal-hydrogen and transition metal-alkyl bonds, respectively. The metal-alkyl bond dissociation reactions that frequently follow electron transfer (eq. 5 and 6) reflect the stabilities typically associated with closed shell (usually 18 electron) configurations of organotransition metal complexes.

II. CONEZYME B_{12}-DEPENDENT REARRANGEMENTS

Coenzyme B_{12} (5'-deoxyadenosylcobalamin, abbreviated $AdCH_2$-B_{12}), whose structure is depicted in Fig. 1, serves as a cofactor for several enzymic reactions, a common feature of which is the 1,2-interchange of a hydrogen atom and a substituent on adjacent carbon atoms as depicted by eq. 9, where X = OH, NH_2, $CH(NH_2)COOH$, C(=O)SCoenzyme A or $C(=CH_2)COOH$ (3,4).

$$\rangle \overset{H}{\overset{|}{C_1}}\text{—}\overset{X}{\overset{|}{C_2}}\langle \ \underset{\text{Coenzyme } B_{12}}{\overset{\text{Enzyme}}{\rightleftharpoons}} \ \rangle \overset{X}{\overset{|}{C_1}}\text{—}\overset{H}{\overset{|}{C_2}}\langle \qquad (9)$$

Examples of three such reactions are depicted by eq. 10-12.

$$HOOC\text{-}CH_2\text{-}CH_2\text{-}\overset{O}{\overset{\|}{C}}SCoA \longrightarrow HOOC\text{-}\overset{CH_3}{\overset{|}{CH}}\text{—}CSCoA \qquad (10)$$

$$\overset{NH_2}{\overset{|}{CH_2}}\text{-}CH_2\text{-}CH_2\text{-}CH_2\text{-}\overset{NH_2}{\overset{|}{CH}}\text{-}COOH \longrightarrow CH_3\text{-}\overset{NH_2}{\overset{|}{CH}}\text{-}CH_2\text{-}CH_2\text{-}\overset{NH_2}{\overset{|}{CH}}\text{—}COOH \qquad (11)$$

$$\overset{OH}{\overset{|}{CH_2}}\text{-}CH_2OH \longrightarrow [CH_3\text{-}CH(OH)_2] \longrightarrow CH_3\overset{O}{\overset{\|}{C}}H + H_2O \qquad (12)$$

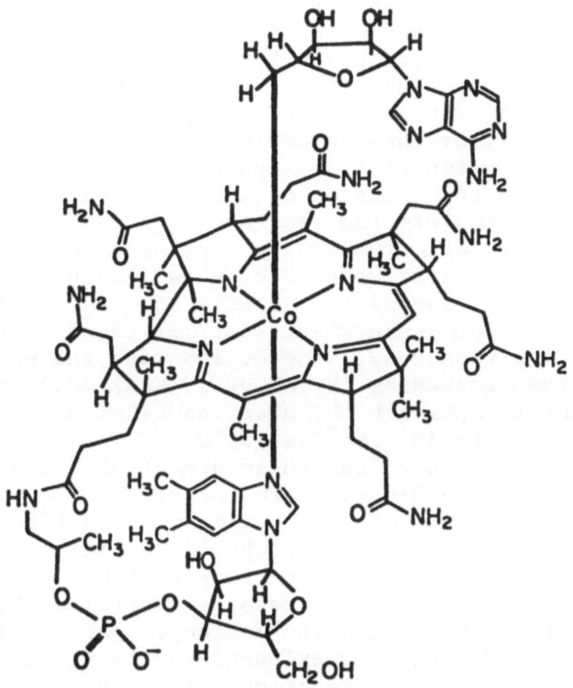

Figure 1. 5'-Deoxyadenosylcobalamin (Coenzyme B_{12})

Various enzymatic studies have provided convincing evidence that these reactions proceed through free radical chain mechanisms depicted schematically by eq. 13 and 14, where B_{12r} is cob(II)alamin (3-8).

$$AdCH_2\text{-}B_{12} \underset{}{\overset{\text{Enzyme}}{\rightleftharpoons}} AdCH_2^{\bullet} + B_{12r} \tag{13}$$

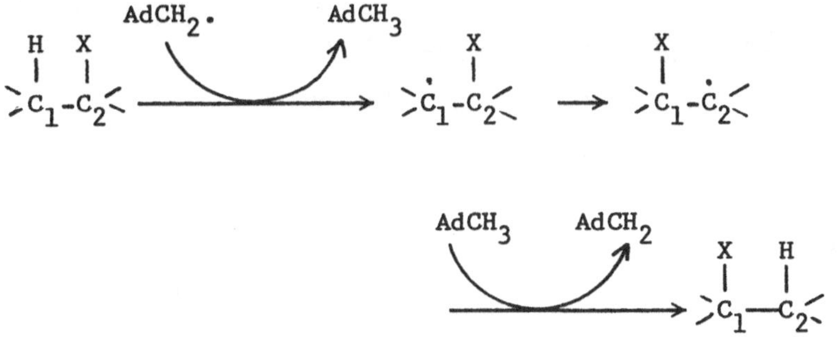

(14)

This mechanism encompasses the following sequence of steps, (1) enzyme-induced homolytic dissociation of the coenzyme B_{12} cobalt-carbon bond to generate B_{12r} and a 5'-deoxyadenosyl radical, (2) abstraction of a hydrogen atom from the substrate to generate a substrate radical and 5'-deoxyadenosine ($AdCH_3$), (3) substrate rearrangement (either directly or through additional intermediate steps) involving the 1,2-migration of X and generation of the corresponding product radical, and (4) transfer of a hydrogen atom from $AdCH_3$ to the product radical to complete the rearrangement reaction.

The following recent studies are pertinent to a knowledge and understanding of the chemical aspects of this mechanistic scheme.

1. The Co-C bond dissociation energy of coenzyme B_{12} has been determined to be in the range 26-30 kcal/mol (9,10). While this does indeed constitute a weak bond, interaction with the enzyme must induce significant further weakening (by about 10 kcal/mol) for the rate of bond dissociation to approach the observed rates of coenzymeB_{12}-dependent reactions (ca 10^2 sec^{-1} at 30°C).

2. Extensive studies on the influence of electronic and steric factors on the Co-C bond dissociation energies of B_{12} model systems suggest that the mechanism of this enzyme-induced bond weakening and dissociation is steric in origin, probably resulting from upward conformational distortion of the corrin ring (10,11).

3. Studies on model radicals have demonstrated the spontaneous 1,2-migrations of thioester and vinyl groups, paralleling the methylmalonyl-CoA (12) and α-methyleneglutarate rearrangements (13).

Thus, at least for these reactions it appears that the only role of coenzyme B_{12} is to serve as a precursor of a reactive primary 5'-deoxyadenosyl radical and that the only distinctive property of the coenzyme that is directly utilized in the enzymic reaction is the weakness of the cobalt-carbon bond. There is a striking parallel between this role of coenzyme B_{12} as a "reversible radical carrier" and that of myoglobin as a reversible oxygen carrier (3).

It has been suggested that the weakness of the Co-C bond in coenzyme B_{12}, which extends also to many other transition metal-alkyl bonds (14), has its origin in steric crowding around the metal (3). Facile dissociation of such bonds gives rise to other free radical processes, for example the carbon-hydrogen bond-forming reductive elimination process depicted by eq. 15-18 (where R = $p\text{-}CH_3OC_6H_4CH_2$ and P = $(p\text{-}CH_3OC_6H_4)_3P$). The R-Mn bond dissociation energy of $R\text{-}Mn(CO)_4P$ has been estimated to be 25-30 kcal/mol (15).

$$R\text{-}Mn(CO)_4P \rightleftharpoons R^{\bullet} + {}^{\bullet}Mn(CO)_4P \qquad (15)$$

$$R^{\bullet} + HMn(CO)_4P \longrightarrow RH + {}^{\bullet}Mn(CO)_4P \qquad (16)$$

$$2\,{}^{\bullet}Mn(CO)_4P \rightleftharpoons Mn_2(CO)_8P_2 \qquad (17)$$

$$\text{Overall:}\quad R\text{-}Mn(CO)_4P + H\text{-}Mn(CO)_4P \longrightarrow RH + Mn_2(CO)_8P_2 \qquad (18)$$

III. CATALYTIC HYDROGENATION AND RELATED REACTIONS

Many homogeneously-catalyzed hydrogenation and related olefin addition reactions (e.g., hydroformylation) proceed through non-radical mechanisms involving organometallic intrmediates formed, typically, by concerted insertion of olefins into transition metal-hydrogen bonds (16).

However, in recent years it has become increasingly clear that alternative mechanisms for such reactions, involving non-organometallic free radical intermediates, also are widespread. These pathways have their origins in the characteristic weakness of transition metal-hydrogen bonds (typical dissociation energies ca 60 kcal/mol) as a result of which H-atom transfer from a metal hydride to an unsaturated substrate such an olefin (eq. 19) is only moderately endothermic (ca 15-20 kcal/mol in the case of styrene) (1).

$$L_nM\text{-}H + {>}C{=}C{<} \longrightarrow L_nM^{\bullet} + H\text{-}C{-}\dot{C}{<} \qquad (19)$$

Convincing evidence for such a mechanism (eq. 20-23) has been provided by the results of a study of the stoichiometric hydrogenation of α-methylstyrene by $HMn(CO)_5$ (eq. 23) (17). Supporting this mechanistic scheme are (a) the kinetics of the reactions ($-d[PhC(CH_3){=}CH_2]/dt = k[PhC(CH_3){=}CH_2][HMn(CO)_5]$, $\Delta H^{\ddagger} = 21.4$ kcal mol^{-1}, $\Delta S^{\ddagger} = -12$ cal mol^{-1} K^{-1}, no inverse CO dependence), (b) an inverse kinetic isotope effect $[k(HMn(CO)_5)/k(DMn(CO)_4) \sim 0.4]$, interpreted in terms of the increase in the stretching frequency when H(D) is transferred from Mn to C [$\nu_{Mn-H} \sim 1800$ cm^{-1}, $\nu_{C-H} \sim 3000$ cm^{-1}], (c) formation of $HMn(CO)_5$ accompanying the reaction of $DMn(CO)_5$ with α-methylstyrene, consistent with the reversibility of the first step of eq. 20 and, most convincingly, (d) the observation of CIDNP signals that could be quantitatively interpreted in terms of this mechanistic scheme.

$$PhC(CH_3){=}CH_2 + HMn(CO)_5 \rightleftharpoons \overline{Ph\dot{C}(CH_3)_2,\ \dot{M}n(CO)_5}$$

$$\longrightarrow Ph\dot{C}(CH_3)_2 + \dot{M}n(CO)_5 \quad \text{(rate-determining)} \qquad (20)$$

$$Ph\dot{C}(CH_3)_2 + HMn(CO)_5 \longrightarrow PhCH(CH_3)_2 + \dot{M}n(CO)_5 \qquad (21)$$

$$2\dot{M}n(CO)_5 \longrightarrow Mn_2(CO)_{10} \qquad (22)$$

$$\text{Overall:} \quad PhC(CH_3){=}CH_2 + 2HMn(CO)_5 \longrightarrow PhCH(CH_3)_2 + Mn_2(CO)_{10} \qquad (23)$$

Such free radical mechanisms extend also to the hydrogenation of aromatic hydrocarbons. In 1975, Feder and Halpern (18) proposed that the $HCo(CO)_4$-catalyzed hydrogenation of arenes, which previously had been interpreted in terms of a concerted insertion of the arene into the Co-H bond of $HCo(CO)_4$ (19), proceeds instead by a free radical mechanism, depicted for the case of anthracene by eq. 24-27, in which the rate-determining step (eq. 25) involves the transfer of an H atom from $HCo(CO)_4$ to the substrate. Neither coordination of the substrate nor formation of organometallic intermediates plays a role in this mechanism.

$$Co_2(CO)_8 + H_2 \rightleftharpoons 2HCo(CO)_4 \qquad (24)$$

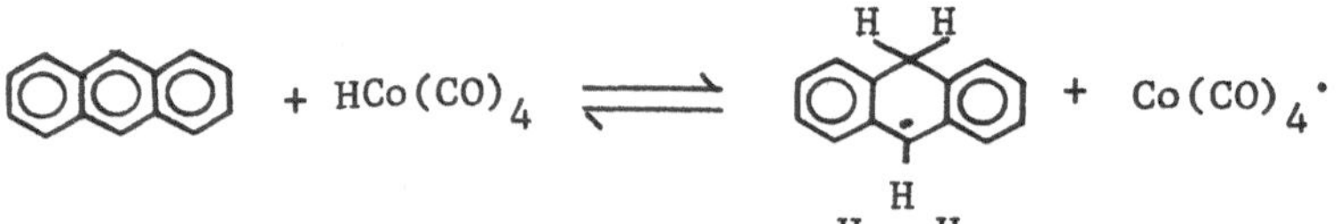

(25)

(26)

$$2Co(CO)_4^{\bullet} \rightleftharpoons Co_2(CO)_8 \qquad (27)$$

In view of evidence that the reverse of reaction 25 and related H atom transfers from organic radicals to metal-centered radicals are characterized by very low activation barriers (exhibiting near diffusion-controlled kinetics) it may be concluded that $\Delta H^{\ddagger}$ for a reaction such as eq. 25 can be approximated by the endothermicity of the reaction. On this basis it can be predicted that the mechanistic scheme of eq. 24-27 should be widespread and should extend particularly to the catalysis by metal hydrides of the hydrogenation of conjugated olefins such as styrene which yield stabilized radicals after H atom transfer (1). Indeed, analogous mechanisms had been advanced earlier for the $Co(CN)_5^{3-}$-catalyzed hydrogenation of cinnamate and related substrates (20-22).

Based on these considerations we also proposed (1) that the $HCo(CO)_4$-catalyzed hydroformylation of styrene and related conjugated substrates (but not of simple alkenes such as propylene) proceeds by an analogous free radical mechanism depicted by Scheme 1 rather than by the conventional Heck-Breslow mechanism involving coordination and migratory insertion of the olefin (i.e.,

$HCo(CO)_4 \xrightarrow{RCH=CH_2} HCo(CO)_3(RCH=CH_2) \longrightarrow RCH_2CH_2Co(CO)_3$, etc.).

Convincing evidence for this proposal now has been advanced through a number of studies on the stoichiometric and catalytic hydrogenation and hydroformylation reactions of styrene and related substrates with $HCo(CO)_4$ (23-29).

A mechanism analogous to that of Scheme 1, but involving $HFe(CO)_4^-$ instead of $HCo(CO)_4$, also has been invoked for the hydroformylation and hydrogenation of styrene by CO and water, catalyzed by iron carbonyls (30).

We also have proposed that the product-forming step in the cobalt carbonyl-catalyzed hydroformylation reaction (eq. 31) may proceed through a free radical mechanism, namely that depicted by eq. 28-31) (1).

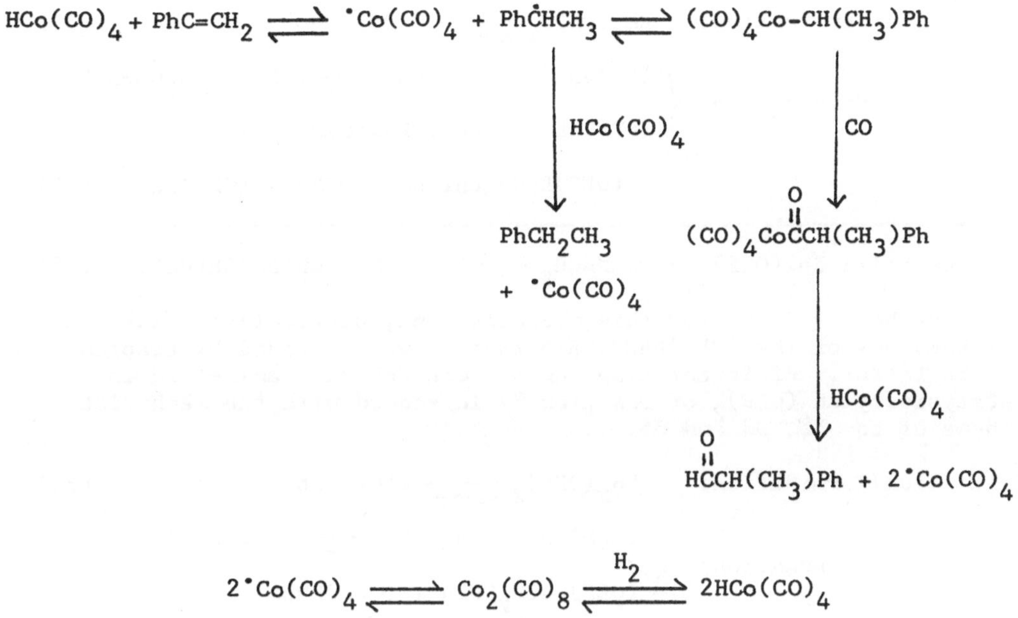

Scheme 1. Proposed mechanism of the hydroformylation of styrene

$$\mathrm{RC(=O)Co(CO)_4} \longrightarrow \mathrm{R\dot{C}O} + {}^{\bullet}\mathrm{Co(CO)_4} \quad (28)$$

$$\mathrm{R\dot{C}O} + \mathrm{HCo(CO)_4} \longrightarrow \mathrm{RCHO} + {}^{\bullet}\mathrm{Co(CO)_4} \quad (29)$$

$$2\,{}^{\bullet}\mathrm{Co(CO)_4} \longrightarrow \mathrm{Co_2(CO)_8} \quad (30)$$

Overall: $$\mathrm{RC(=O)Co(CO)_4} + \mathrm{HCo(CO)_4} \longrightarrow \mathrm{RC(=O)HO} + \mathrm{Co_2(CO)_8} \quad (31)$$

This scheme parallels that of eq. 15-18 and is supported by other recent studies (31).

IV. FREE RADICAL CHAIN MECHANISMS OF INSERTION AND OXIDATIVE ADDITION

Free radical pathways also may be initiated by homolytic dissociation of characteristically weak metal-metal single bonds.

Illustrative of this is the insertion of $PhCH{=}CH_2$ into the Rh-Rh bond of $Rh_2(OEP)_2$ (OEP = octaethylporphyrin) depicted by eq. 35. On the basis of kinetic evidence it was concluded that reaction 35 proceeds through the free radical chain mechanism depicted by eq. 32-34 (32).

$$\text{Initiation/Termination:} \quad Rh_2(OEP)_2 \rightleftharpoons 2(OEP)Rh^{\bullet} \qquad (32)$$

$$\text{Propogation:} \begin{cases} (OEP)Rh^{\bullet} + PhCH{=}CH_2 \rightleftharpoons (OEP)RhCH_2\dot{C}HPh & (33) \\ (OEP)RhCH_2\dot{C}HPh + Rh_2(OEP)_2 \longrightarrow & \\ \qquad (OEP)RhCH_2CH(Ph)Rh(OEP) + (OEP)Rh^{\bullet} & (34) \end{cases}$$

$$\text{Overall:} \quad Rh_2(OEP)_2 + PhCH{=}CH_2 \longrightarrow (OEP)RhCH_2CH(Ph)Rh(OEP) \qquad (35)$$

Further evidence for this mechanism and, particularly, for the intermediacy of the $(OEP)RhCH_2\dot{C}HPh$ radical was provided by trapping of the latter. Efficient trapping by (OEP)RhH was manifested in catalysis by $Rh_2(OEP)_2$ of reaction 37 in accord with the mechanistic scheme of eq. 32, 33 and 36.

$$\text{Initiation/Termination:} \quad Rh_2(OEP)_2 \rightleftharpoons 2(OEP)Rh^{\bullet} \qquad (32)$$

$$\text{Propogation:} \begin{cases} (OEP)Rh^{\bullet} + PhCH{=}CH_2 \rightleftharpoons (OEP)RhCH_2\dot{C}HPh & (33) \\ (OEP)RhCH_2\dot{C}HPh + (OEP)RhH & \\ \qquad \longrightarrow (OEP)RhCH_2CH_2Ph + (OEP)Rh^{\bullet} & (36) \end{cases}$$

$$\text{Overall:} \quad (OEP)RhH + PhCH{=}CH_2 \longrightarrow (OEP)RhCH_2CH_2Ph \qquad (37)$$

A free radical mechanism, analogous to that of eq. 32, 33 and 36, also accommodates the previously reported (33) insertion of CO into the Rh-H bond of (OEP)RhH (eq. 33, 38-40).

$$\text{Initiation/Termination:} \quad Rh_2(OEP)_2 \rightleftharpoons 2(OEP)Rh^{\bullet} \qquad (32)$$

$$\text{Propogation:} \begin{cases} (OEP)Rh^{\bullet} + CO \rightleftharpoons 2(OEP)Rh\dot{C}O & (38) \\ (OEP)Rh\dot{C}O + (OEP)RhH & \\ \qquad \longrightarrow (OEP)RhCHO + (OEP)Rh^{\bullet} & (39) \end{cases}$$

$$\text{Overall:} \quad (OEP)RhH + CO \longrightarrow (OEP)RhCHO \qquad (40)$$

This mechanism parallels the microscopic reverse of pathways previously postulated for the decarbonylation of metal formyl complexes (34). Reaction 39 also finds a parallel in the mechanism recently proposed for the generation of metal formyl complexes by H-atom transfer to electrochemically generated metal carbonyl radicals (35).

The oxidative addition of benzyl bromide to $Rh_2(OEP)_2$ (eq. 42) also can be accommodated by a free radical chain mechanism, depicted by eq. 33, 41 and 42 (32).

Initiation/Termination: $Rh_2(OEP)_2 \rightleftharpoons 2(OEP)Rh^{\bullet}$ (32)

Propogation:

$(OEP)Rh^{\bullet} + C_6H_5CH_2Br \longrightarrow (OEP)RhBr + C_6H_5CH_2^{\bullet}$ (41)

$C_6H_5CH_2^{\bullet} + Rh_2(OEP)_2 \longrightarrow (OEP)RhCH_2C_6H_5 + (OEP)Rh^{\bullet}$ (42)

Overall: $Rh_2(OEP)_2 + C_6H_5CH_2Br \longrightarrow (OEP)RhCH_2C_6H_5 + (OEP)RhBr$ (43)

Several factors may be identified as contributing to the distinctive free radical chain processes that we have identified in these systems, namely (a) the weak Rh-Rh bond in $Rh_2(OEP)_2$ which is reponsible for the accessibility of the $(OEP)Rh^{\bullet}$ radical (32), (b) unusually strong Rh-C bonds which contribute to the driving force for the chain propogating steps 32 and 38 and (c) the absence of axial ligands in $Rh_2(OEP)_2$ which renders feasible the homolytic displacement steps 34 and 42. Several features of these free radical chain mechanisms, notably the chain propogation sequences involving addition of metal free radicals to olefins and to CO and the S_H2 displacement of metal radicals at metal-metal bonds, are without direct precedent in transition metal chemistry.

ACKNOWLEDGMENT: This research was supported by grants from the National Science Foundation (CHE82-17950) and the National Institutes of Health (DK-13339).

REFERENCES

(1) J. Halpern, Pure Appl. Chem. 51, 2171 (1979).

(2) J. Halpern, Pure Appl. Chem. 58, 575 (1986).

(3) J. Halpern, Science 227, 869 (1985) and references cited therein.

(4) D. Dolphin, Ed., "B_{12}", Wiley, New York, 1982 Vol. I and II, and references cited therein.

(5) B. Zagalak and W. Friedrich, Eds., "Vitamin B_{12}", de Gruyter, Belin, 1979.

(6) B. M. Babior, Acc. Chem. Res. 8, 376 (1975).

(7) R. H. Abeles, Ref. 5, p. 373.

(8) B. T. Golding, Ref. 4, p. 543.

(9) J. Halpern, S. H. Kim and T. W. Leung, J. Am. Chem. Soc. 106, 8317 (1984).

(10) B. P. Hay and R. G. Finke, J. Am. Chem. Soc. 108, 4820 (1986).

(11) N. Bresciani-Pahor, M. Forcolin, L. G. Marzilli, L. Randaccio, M. F. Summers and P. J. Toscano, Coord. Chem. Revs. 63, 1 (1985).

(12) S. Wollowitz and J. Halpern, J. Am. Chem. Soc. 106, 8319 (1984).

(13) (a) B. Maillard, D. Forest, K. U. Ingold, J. Am. Chem. Soc. 98, 7024 (1976). (b) A. Effio, D. Griller, K. U. Ingold, A. L. J. Beckwith and A. K. Serelis, J. Am. Chem. Soc. 102, 1734 (1980).

(14) J. Halpern, Accts. Chem. Res. 15, 238 (1982).

(15) M. J. Nappa, R. Santi and J. Halpern, Organometallics, 4, 34 (1985).

(16) J. Halpern, Inorg. Chem. Acta 50, 11 (1981) and references cited therein.

(17) R. L. Sweany and J. Halpern, J. Am. Chem. Soc. 99, 8335 (1977).

(18) H. M. Feder and J. Halpern, J. Am. Chem. Soc. 97, 7186 (1975).

(19) P. D. Taylor and M. Orchin, J. Org. Chem. 37, 3913 (1972).

(20) (a) J. Kwiatek, Catalysis Revs. 1, 37 (1968). (b) J. Kwiatek and J. K. Seyler, Adv. Chem. Ser. 70, 207 (1968).

(21) J. Halpern and L. Y. Wong, J. Am. Chem. Soc. 90, 6665 (1968).

(22) L. Simandi and F. Nagy, Acta Chim. Hung. 46, 137 (1968).

(23) F. Ungvary and L. Marko, Organometallics 1, 1120 (1982).

(24) J. A. Roth and M. Orchin, J. Organomet. Chem. 182, 299 (1979).

(25) (a) T. E. Nalesnik and M. Orchin, J. Organomet. Chem. 199, 265 (1980). (b) T. E. Nalesnik, J. H. Frendenberger and M. Orchin, J. Molec. Catalysis 16, 43 (1982).

(26) J. A. Roth, P. Wiseman and L. Ruzzala, J. Organomet. Chem. 240, 271 (1983).

(27) F. Ungvary and L. Marko, J. Organomet. Chem. 249, 411 (1983).

(28) M. Orchin, Ann. N.Y. Acad. Sci., 415, 129 (1983).

(29) T. M. Bockman, J. F. Garst, R. B. King, L. Marko and F. Ungvary, J. Organomet. Chem. **279**, 165 (1985).

(30) J. Palagyi and L. Marko, J. Organomet. Chem. **236**, 343 (1982).

(31) J. Azran and M. Orchin, Organometallics **3**, 197 (1984).

(32) R. S. Paonessa, N. C. Thomas and J. Halpern, J. Am. Chem. Soc. **106**, 8319 (1984)

(33) (a) B. B. Wayland and B. A. Woods, J. Chem. Soc. Chem. Commun. 700 (1981). (b) B. B. Wayland, B. A. Woods and R. Pierce, J. Am. Chem. Soc. **104**, 302 (1982).

(34) (a) B. A. Naraganan, C. Amatore, C. P. Casey and J. K. Kochi, J. Am. Chem. Soc. **105**, 6351 (1983). (b) C. E. Summer and G. O. Nelson, J. Am. Chem. Soc. **106**, 432 (1984).

(35) B. A. Narayanan and J. K. Kochi, J. Organomet. Chem. **272**, C49 (1984).

ORGANOMETALLIC AND RADICAL CHEMISTRY OF B_{12} COENZYMES

Bernhard Kräutler
Laboratory of Organic Chemistry
ETH-Zürich
Universitätstr. 16
CH-8092 Zürich
Switzerland

ABSTRACT. Vitamin B_{12}-derivatives act as organometallic catalysts in nature and as one-electron redox centers. Besides their participation in the complex coenzyme B_{12}-catalyzed enzymatic isomerization reactions, an essential biological role of the corrinoids also is methyl group transfer catalysis. Methyl group transfer reactions between corrinoid cobalt complexes were now found to take place with ease in aqueous solution and at room temperature, and the corresponding equilibria were studied. In this way, the thermodynamic effect of the intramolecular coordination of the unique nucleotide function of the cobalamins on the Co-C bond was studied. Secondly, the structure of a paramagnetic Co(II)corrin was analyzed by X-ray analysis. When compared with the corresponding diamagnetic Co(III)corrin precursor, the cobalt corrin portions were found to be highly similar, geometrically. This X-ray analysis points to the existence of a long axial bond in Co(II)corrins, but fails to reveal a deformation of the cobalt-corrin part that might represent a mode of activation of the protein bound coenzyme B_{12} towards Co-C bond homolysis.

- . -

In nearly all spheres of life, vitamin B_{12} derivatives are essential catalysts in biosynthesis[1,2]. Instead of the originally isolated cyano-complex vitamin B_{12} (1, cyanocobalamin), however, the organometallic derivatives methylcobalamin (2) and coenzyme B_{12} (3, adenosylcobalamin, see Fig. 1) have been identified as metabolically active B_{12} coenzymes, as well as the Co(II)corrin cob(II)alamin (4)[1-3]. The biological functions of the corrinoids are associated

M. Chanon et al. (eds.), Paramagnetic Organometallic Species in Activation / Selectivity, Catalysis, 435–446.

largely with the unusual Co-C bond, which is stable towards cleavage by physiological acids, and which was discovered in the coenzyme 3 by X-ray analysis (see Fig. 2 [4]).

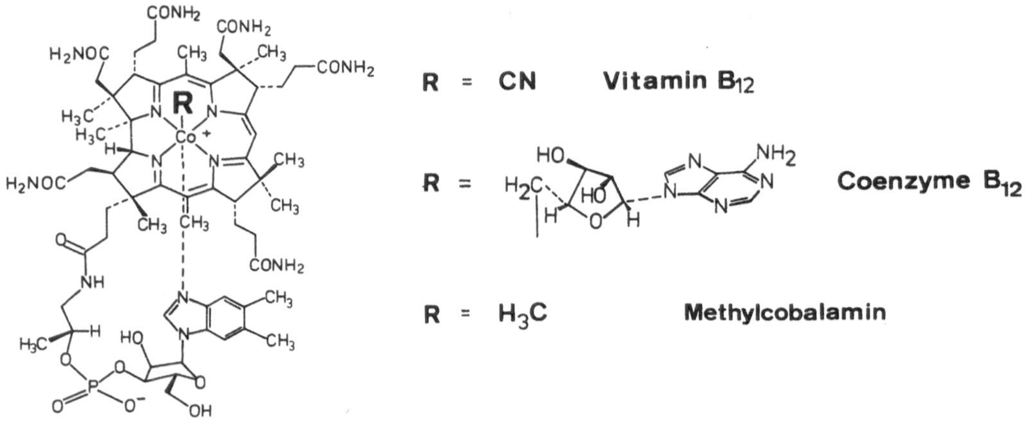

Figure 1: Structural formulae of vitamin B_{12} (1), of methylcobalamin (2) and of coenzyme B_{12} (3).

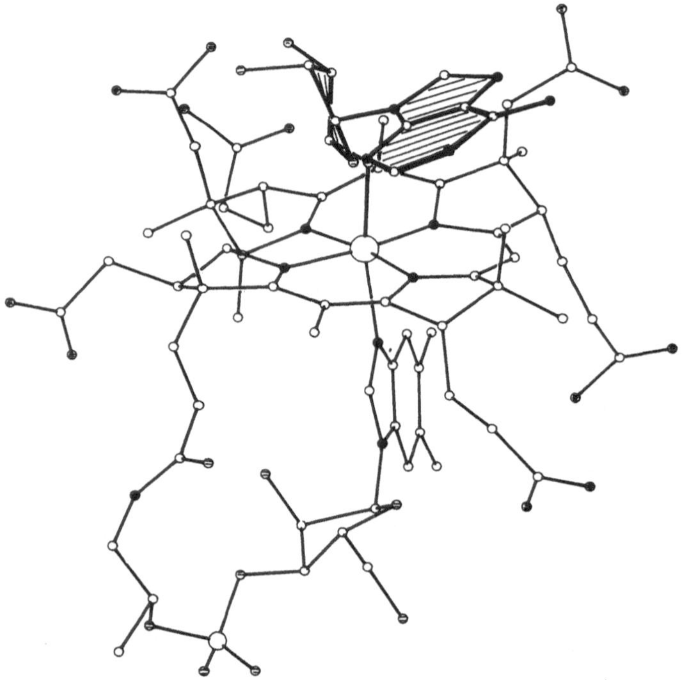

Figure 2: Three dimensional structure of coenzyme B_{12} (3), according to X-ray analysis [4].

Coenzyme B_{12} (3), and related adenosyl-cobamides, are cofactors in some ten complex enzymatic isomerization reactions, one of which is the (R)-methylmalonyl-coenzyme A to succinyl-coenzyme A rearrangement (see Fig. 3 [5]), which also occurs and is essential in human metabolism. Secondly, they participate in one form of reduction of ribonucelotides to 2′-deoxy-ribonucleotides [6]. The unique coenzyme B_{12} catalyzed skeletal 1,2 isomerizations have attracted considerable scientific interest [1,7]: the concomitant 1,2 hydrogen migration is not truly intramolecular, but occurs via intermediate binding of the "migrating hydrogen" to the 5′-position of the adenosyl group of the coenzyme 3 [8a].

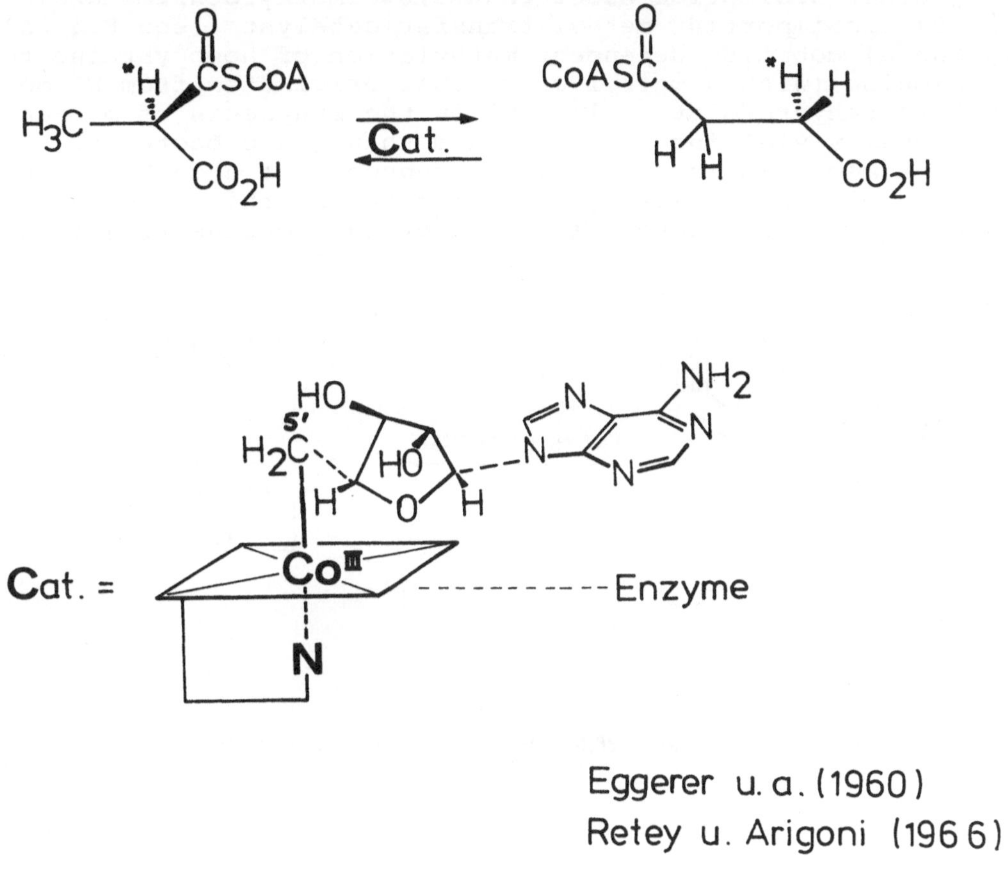

Figure 3: The coenzyme B_{12}-catalyzed enzymatic rearangement of (R)-methylmalonyl-coenzyme A to succinyl-coenzyme A.

Since related skeletal isomerizations, as in the enzymatic (R)-methylmalonyl-CoA to succinyl-CoA isomerization, can be observed in radical reactions [7,8b], and radical species and Co(II)corrins have indeed been detected in the enzymatic reactions [3], the homolysis of the Co-C bond of protein-bound 3 is taken as the initial step of the coenzyme B_{12} catalyzed enzymatic rearragement reactions [7]. Correspondingly, the homolysis of the Co-C bond is considered the biologically relevant reactivity of the coenzyme 3 (see Fig. 4 [7]). Indeed, the bond dissociation energy is estimated to be close to 30 kcal/mol only, based on the rate of homolytic decomposition of 3 in homogeneous solution [9].

Methylcobalamin (2) and other methylcobamides (i.e. having other nucleotide bases than 5,6-dimethylbenzimidazole) [2,10] are important methyl transfer catalysts (see Fig. 5) in the a) cobamide dependent methylation of homocysteine to methionine (with a methyl group that originates from N^5-methyl-tetrahydrofolate) [11]; b) in the synthesis of methane (the energy yielding process) in methanogenic bacteria, where a mechanistically similar transport of a methyl group from an N^5-methyl-tetrahydropterine to the thiol coenzyme M appears to be a relevant function of the corrins [12] (and

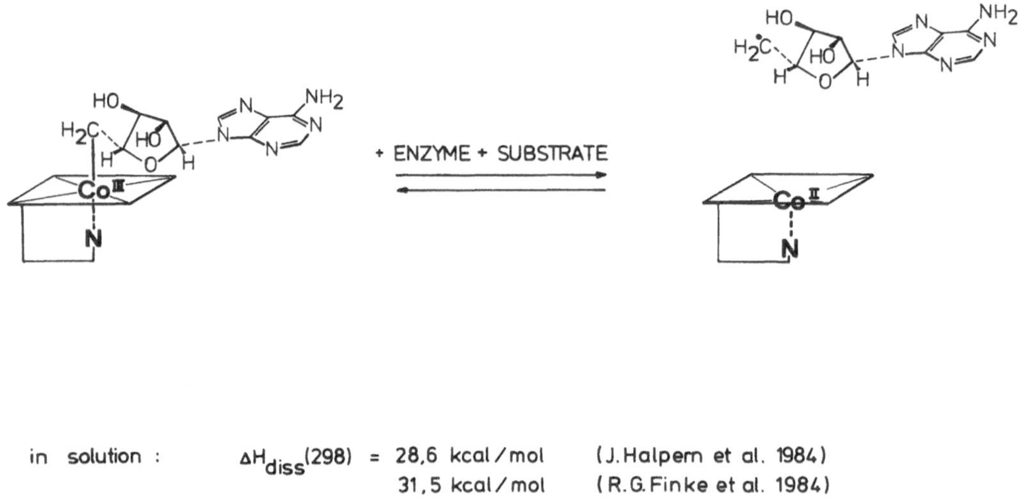

Figure 4: Coenzyme B_{12} (3), as biological source of 5′-deoxyadenosyl radicals (top) and Co-C bond dissociation enthalpies, based on kinetic estimates [9].

not the reductive liberation of methane itself, now recognized to be accomplished with the help of the porphinoid nickel-complex coenzyme F430 [13]; c) in the bacterial "acetate" - synthesis and carbon dioxide fixation via acetyl-coenzyme A, in a chemically still little understood sequence of events [14]; (and d) in the methylation of mercuric ions and of other metal ions [15]).

The relevant reactivity of corrinoids in the metabolic methyl group transfer processes is the high nucleophilicity (and the nucleofugicity) of Co(I)corrins, as well as the resistance of methylcorrinoids towards physiological Brønsted acids. To what extent the radical reactivity of organocorrinoids is involved (e.g. in the bacterial synthesis of the coenzyme A bound acetyl group [16]) remains to be clarified.

Indeed, the rapid transfer of the cobalt-bound methyl group from CH_3-corrinoids (such as 2) to Co(II)corrinoids [17], indicates considerable reactivity for CH_3-transfer in radicaloid reactions. Secondly, in equilibration experiments, it also allows the determination of relative Co-CH_3

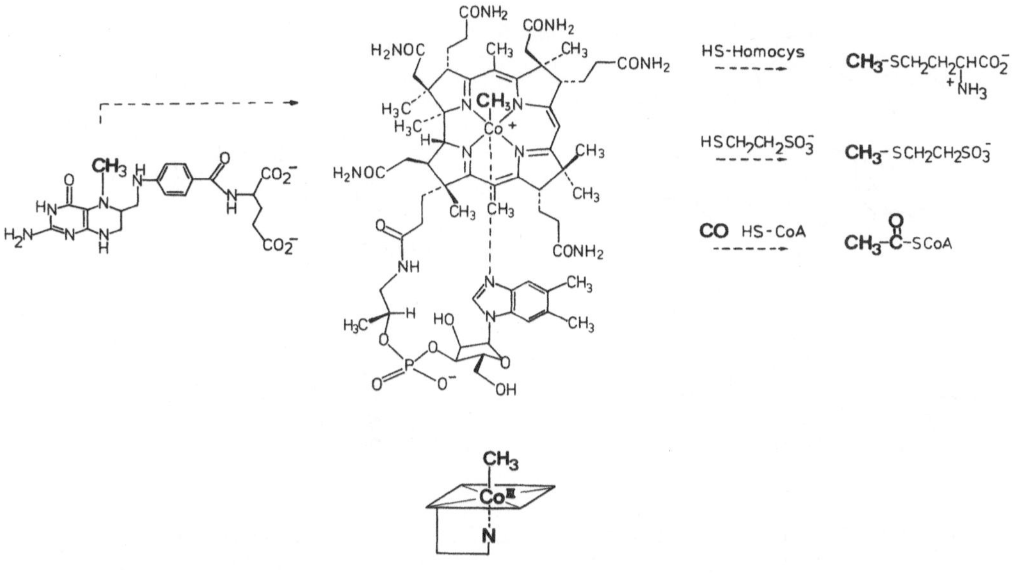

Figure 5: Methyl corrinoids as biological methyl group transfer catalysts: methyl group transfer from N^5-methyl-tetrahydropterins to thiols (synthesis of methionine, of methyl coenzyme M) and formation of the acetyl group of acetyl-coenzyme A, with "carbonylation" of a methylated intermediate [14].

bond dissociation energies: this way, by comparison between the nucleotide containing cobalamins (such as 2) and the nucleotide free cobinamides (such as 5, see Fig. 6), the thermodynamic trans-effect of the nucleotide base on the Co-CH_3 bond strength in 2 can be determined [17].

Equilibration experiments starting with 2 and the Cob(II)inamide 6 or with 4 and 5 (see Fig. 7) indicated an equilibrium constant $K_{II/III}$ = 0.60: the nucleotide function of 2 hardly affects the Co-C bond strength in 2, and, in fact, slightly stabilizes it. In analogy to experiments carried out earlier by Johnson and co-workers [18], the CH_3 group transfer presumably proceeds via backside attack of the radicaloid Co(II) center on the Co-CH_3 bond, i.e. without free CH_3 radicals [17].

Similar equilibration experiments, but on the Co(I) oxidation level of the CH_3 group acceptor, can also be analyzed to reveal the effect of the nucleotide coordination in 2 on the heterolytic, nucleofugic Co-C bond dissociation: the nucleotide coordination is found to stabilize the Co-CH_3 bond by ca. 4 kcal/mol against displacement of the CH_3 group

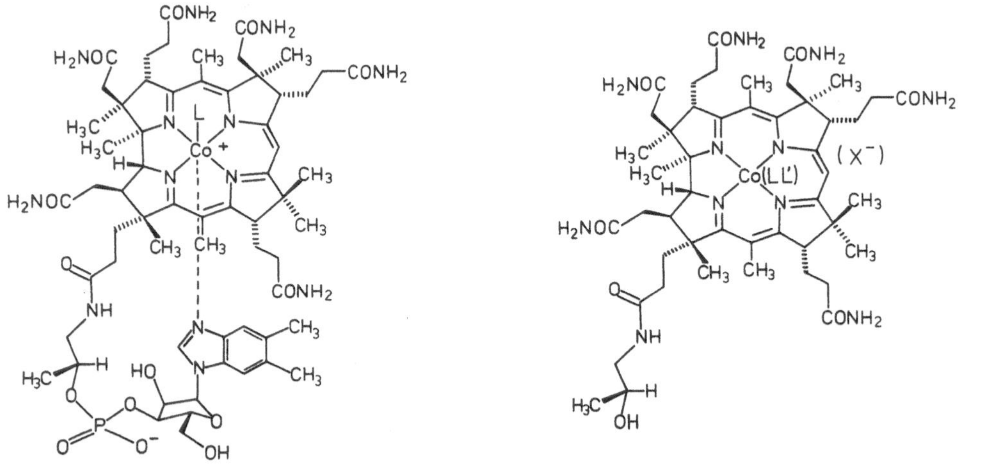

Figure 6: Structural formulae left: of methylcobalamin (2, Co-L=Co(III)-CH_3), of cob(II)alamin (4, Co-L = Co(II)), of aquocobalamin (7, Co-L = Co(III)-H_2O) and of cob(I)alamin (8, Co-L = Co(I), nucleotide not coordinating) and right: of methyl-cobinamide (5, Co(L,L')= Co(III)(CH_3,H_2O)), of cob(II)inamide (6, Co(L,L')= Co(II)(H_2O)), of aquo-cobinamide (9, Co(L,L')= Co(III)(OH,H_2O)) and of cob(I)inamide (10, Co(L,L')= Co(I)).

by nucleophiles. In contrast, as earlier experiments have shown [15,19], the nucleotide coordination in 2 enhances the readiness of 2 to give up its cobalt bound methyl group to electrophiles, such as Hg^{2+} ions [15] and other cobalt(III) corrinates [19].

A quantitative correlation can be established between the homolytic and heterolytic $Co-CH_3$ bond dissociation energies and the stabilization of the complete corrinoids by the nucleotide coordination in aqueous solution, as expressed by the (nucleotide) pKa-values. This way, the unique nucleotide function of the cobalamin is seen to influence thermodynamically their organometallic chemistry, of e.g. 2 and 3 (see Fig. 8 [17]).

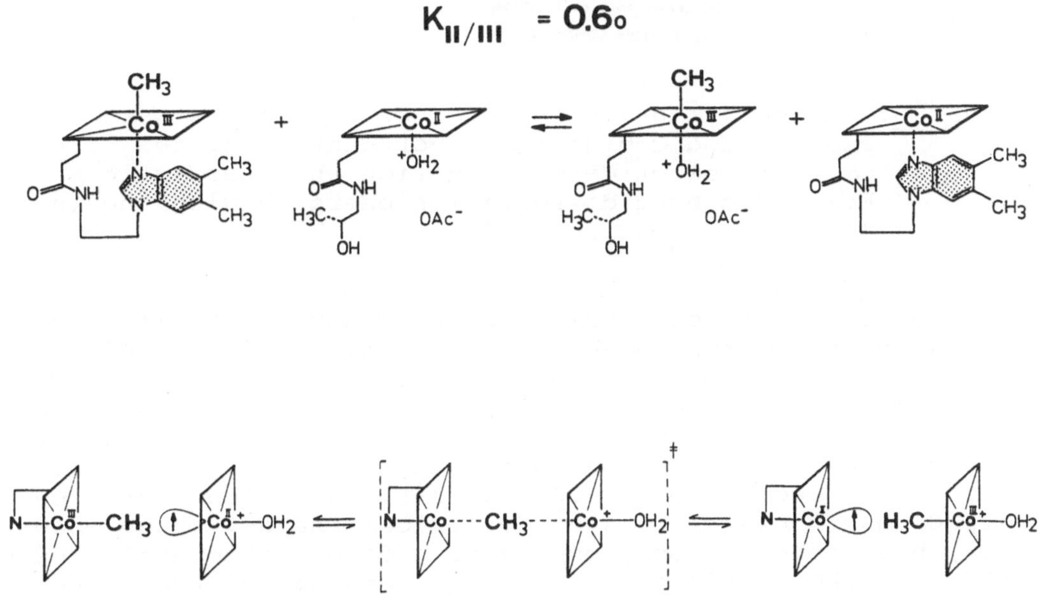

Figure 7: The methyl group transfer equilibrium between the CH_3-Co(III)- and Co(II)-forms of cobalamins and cobinamides: determination of the effect of the nucleotide coordination on the homolytic $Co-CH_3$ bond dissociation energy in methylcobalamin [17].

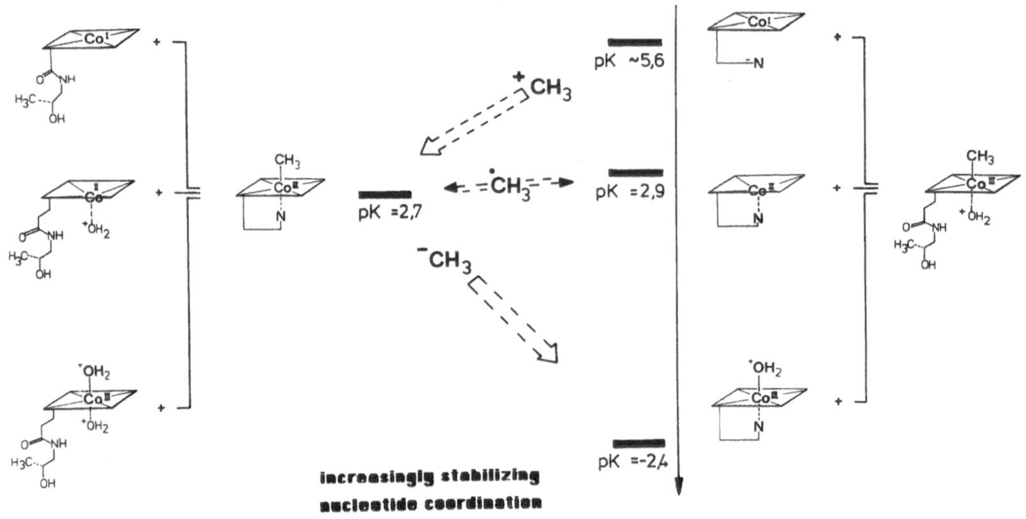

Figure 8: The methyl group transfer equilibria between cobalamins and cobinamides: correlation of the effect of the nucleotide coordination on heterolytic and homolytic bond dissociation energies with the (nucleotide) pK_a-values [17].

For the tasks of corrinoids as coenzymes, their reactivity in the protein-bound state is critical. To explain the high rates of coenzyme B_{12}-catyzed isomerization reactions, presumed to be induced by the homolysis of the Co-C bond in the protein-bound coenzyme, an as yet structurally little understood Co-C bond weakening effect by steric distortion has been invoked [20]. So far, these considerations relied on structural data of Co(III)corrinates, which, however, have revealed only one major mode of structural flexibility of the corrines, the "upward-bending" of the "northern" and "southern" half of the molecule in cob(III)alamins [21]. We have recently succeeded in crystallizing a monomeric Co(II)cobyrinate, derived from vitamin B_{12}, and in determining the three-dimensional structure of a paramagnetic Co(II)corrinate by x-ray analysis [22]: formic acid reduction of the lipophilic "cobester" (11, dicyano-heptamethyl-cob(III)yrinate [23]), followed by aqueous work-up with addition of sodium perchlorate produced "Co(II)-cobester" (12, perchlorato-heptamethyl- cob(II)yrinate, see Fig.9). Its three-dimensional structure, determined by X-ray analysis, is shown in Fig.9b.

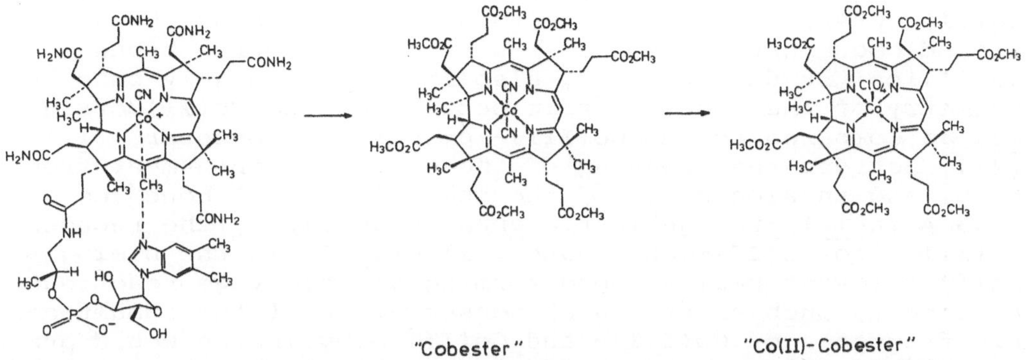

"Cobester"
R.Keese, L.Werthemann
& A.Eschenmoser
(1965)

"Co(II)-Cobester"
C. Caderas & M. Hughes

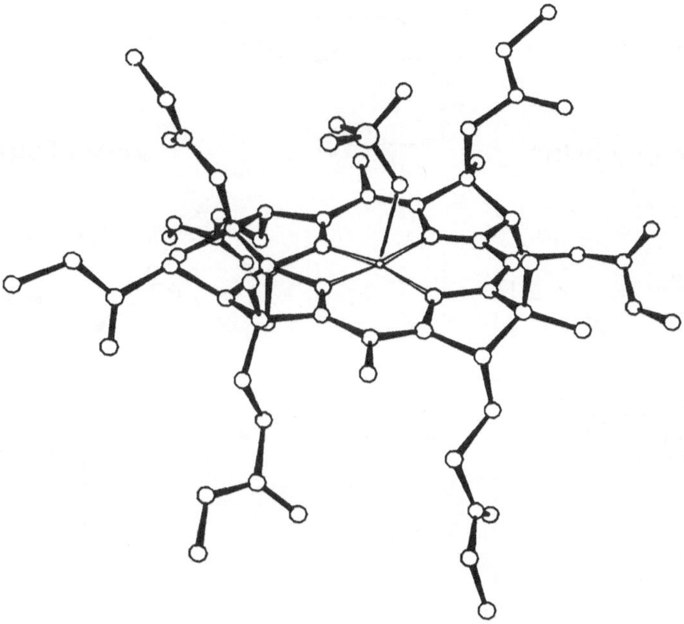

Figure 9: Top: Structural formulae of vitamin B_{12} (1), of the heptamethyl cobyrinate "cobester" (11) [23] and of "Co(II)cobester" (12) [22], obtained from 11 by reduction with formic acid. Bottom: Three-dimensional structure of a Co(II)corrinate derived from vitamin B_{12}: X-ray crystal structure of "Co(II)cobester" [22].

The X-ray analysis of 12 revealed a pentacoordinate Co(II)center, as expected, with an axially-bound perchlorate ligand coordinating to the β-side of the corrin-bound Co(II)ion. It did, however, also indicate a highly conserved geometry of the cobalt corrin portion in the Co(II)corrin 12, as compared to the Co(III)- precursor "cobester" (11) [24]. Besides the lowering of the coordination number from 6 to 5, and an alongation of the remaining axial bond (Co-O= 2,33 A in 12), few geometric changes accompany the reduction Co(III)- to Co(II)-cobyrinate (11⟶12). In the biologically relevant radical bond forming and cleavage reactions at cobalt, such a structural conservation of the cobalt-corrin fragments of Co(III)- and Co(II)-cobyrinates would presumably be a factor in support of particularly low activation barriers [22].

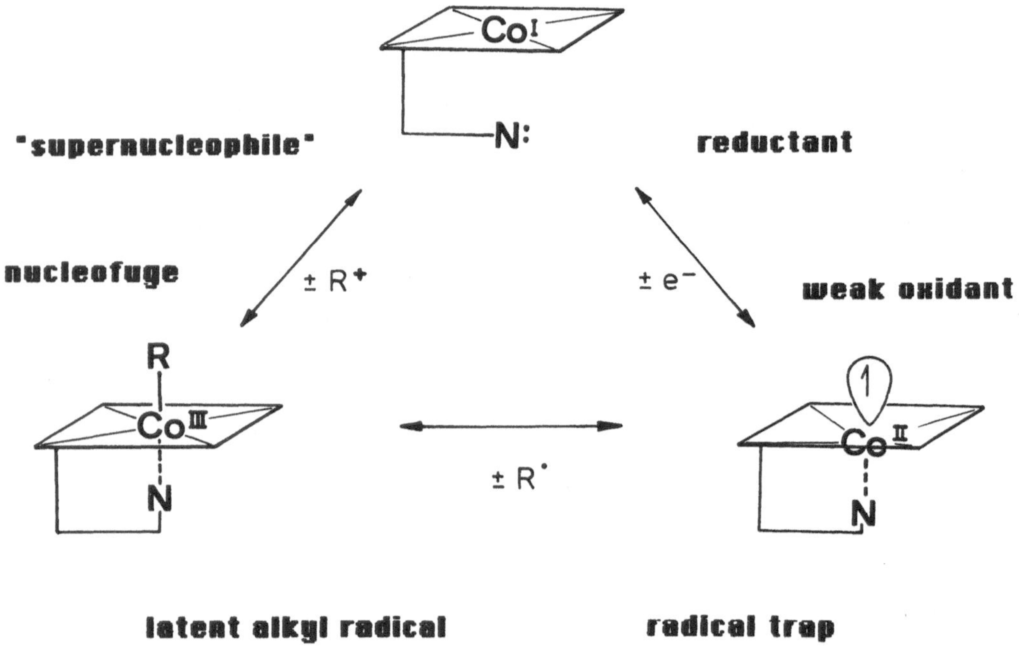

Figure 10: Patterns of reactivity in the bioorganometallic chemistry of nucleotide containing corrinoids.

In view of the biological functions of the corrinoids, but also with respect to their use as catalysts in organic synthesis [25], the radical chemistry of organo- and Co(II) cobyrinates, the two-electron nucleophilic/nucleofugic reactivity of Co(I) cobyrinates and the Co(II)/Co(I) corrin redox properties, appear to be the most relevant organometallic properties of corrinoids (see Fig. 10): Alkyl-Co(III)-corrins represent latent alkylradicals, but contain a hexacoordinate, diamagnetic Co(III)center, that gives the weak Co-C bond considerable inertness against Brønsted acids.The Co(II)corrins are persistent radicaloid species, that are highly efficient traps for organic radicals. The strongly reducing Co(I)corrins, finally contain a highly nucleophilic (although formally coordinatively unsaturated) tetracoordinate Co(I) d^8-ion.

The extent to which these unusual reactivities of the vitamin B_{12}-derivates are a specific consequence of binding the cobalt-ion to a corrin ligand still is unknown. The unique nucleotide function of the "complete" corrinoids however is seen to modulate thermodynamically and kinetically [26] the basic reactivities of the cobaltcorrin. The three-dimensional structure of the Co(II)corrin <u>12</u> so far does not point to a particular flexibility of the corrin ligand that is effective in the transition Co(III)- to Co(II)-corrin, in particular. This is of interest, since steric perturbations have been suggested to be a factor allowing the promotion of Co-C bond homolysis in the protein-bound coenzyme B_{12} (see e.g. [20]).

<u>Acknowledgements:</u> I would like to thank Dr.C.Kratky & W.Keller (University of Graz, Austria) for their engaged X-ray analytical work, and the Swiss National Science Foundation for financial support of this work.

REFERENCES.

[1] D.Dolphin (ed.) B_{12}, J.Wiley & Sons, New York 1982.
[2] W.Friedrich, Vitamin B_{12} und verwandte Corrinoide, R. Ammon & W.Dirscherl (eds.), G.Thieme Verlag, Stuttgart 1975.
[3] J.R.Pilbrow in [1], Vol. 1, p.431.
[4] P.G.Lenhert & D.C.Hodgkin, Nature (London), **192**, 937 (1961).
[5] J.Retey, in [1], Vol.2, p.357.
[6] R.L.Blakley, in [1], Vol.2, p.381.
[7] J.Halpern, Science, **227**, 869 (1985).
[8] a. J.Retey & D.Arigoni, Experientia, **22**, 783 (1966); b. H.Eggerer, P.Overath, F.Lynen & E.R.Stadtman, J.Am.Chem. Soc., **82**, 2643 (1960).
[9] a. J.Halpern, S.H.Kim & T.W.Leung, J.Am.Chem.Soc. **106**, 8317 (1984); b. B.P.Hay & R.G.Finke, ibid, **108**, 4820 (1986).
[10] B.Kräutler, Chimia, **41**, 277 (1987).
[11] R.T.Taylor, in [1], Vol.2, P.307.
[12] P.van der Meijden, B.W.te Brömmelstroet, C.M.Poirot, C. van der Drift & G.D.Vogels, J.Bacteriol. **160**, 629 (1984).
[13] W.L.Ellefson, W.B.Whitman & R.S.Wolfw, Proc.Natl.Acad. Sci.USA, **79**, 3707 (1982).
[14] H.G.Wood, S.W.Ragsdale & E.Pezacka, Trends Biochem.Sci. **11** , 14 (1986).
[15] J.S.Thayer & F.E.Brinckman, Adv.Organometall.Chem. **20**, 313 (1982).
[16] B.Kräutler, Helv.Chim.Acta, **67**, 1053 (1984).
[17] B.Kräutler, Helv.Chim.Acta, **70**, 1268 (1987).
[18] D.Dodd, M.D.Johnson & B.L.Lochmann, J.Am.Chem.Soc. **99**, 3664 (1977).
[19] Y.T.Fanchiang, G.T.Bratt & H.P.C.Hogenkamp, Proc.Natl. Acad.Sci.USA, **81**, 2698 (1984).
[20] a.J.M.Pratt, in [1], Vol.1, p.325; b. M.K.Geno & J.Halpern, J.Am.Chem.Soc., **109**, 1238 (1987).
[21] V.B.Pett, M.N.Liebman, P.Murray-Rust, K.Prasad & J.P.Glusker, J.Am.Chem.Soc. **109**, 3207 (1987).
[22] B.Kräutler, W.Keller, M.Hughes, C.Caderas & C.Kratky, J.Chem.Soc.,Chem.Commun., **1987**, in press.
[23] R.Keese, L.Werthemann & A.Eschenmoser, published in L.Werthemann, Dissertation ETH Zürich, no. 4097, Juris Druck+Verlag, Zürich 1968.
[24] A.Fischli & J.J.Daly, Helv.Chim.Acta, **63**, 1628 (1980).
[25] R.Scheffold, Chimia, **39**, 203 (1985).
[26] B.Kräutler, in The Bioalkylation of Heavy Elements, F. Glockling & J.P.Craig (eds.), Roy.Soc.Chem.(London), in press.

STRUCTURE AND REACTIONS OF SOME ORGANOMETALLIC RADICALS

M. C. R. Symons
Department of Chemistry
The University
Leicester LE1 7RH
U.K.

ABSTRACT. The effect of ionizing radiation on some selected organometallic systems is discussed, both for the pure compounds and for dilute solid solutions. Results obtained from e.s.r. studies show that, frequently, electron-gain and electron-loss centres can be prepared in this way, the e.s.r. spectra providing identification and much structural detail. The systems described in detail are tetra-alkyl-tin and -lead derivatives, dialkyl mercury derivatives, and alkyl-cobalt complexes including methyl cobalamine and vitamin B12.

1. INTRODUCTION

1.1. General Outline

This paper is designed to illustrate how the combined use of ionizing radiation and electron spin resonance (ESR) spectroscopy can give useful information about electron-loss (e-l) and electron-gain (e-g) by a range of compounds. In other words, the redox chemistry of various compounds can be conveniently studied by this method. Because space is limited, attention is given to solid-state work, in which, often, primary species are detected, and, at times, their unimolecular break-down products.

After a brief outline of the main techniques (2), three systems are discussed in some depth (3) tetraalkyl-tin and -lead derivaties, (4) dialkyl-mercury derivatives, and (5) alkyl-cobalt derivatives.

2. METHODS

2.1. Irradiation of Pure Compounds

Exposure of solids to ionizing radiation can, from a chemical viewpoint, be treated as a series of ionization events. In over 30 years of study I have never needed to use other concepts, and this approach will be used herein. This results in the formation of <u>electron-loss</u> (e-1) centres (radical cations if the substrate is neutral) and electrons. Once the latter are thermalised, they <u>may</u> react to give

M. Chanon et al. (eds.), Paramagnetic Organometallic Species in Activation / Selectivity, Catalysis, 447–461.

electron-gain (e-g) centres (radical-anions for neutral substrates). These primary centres may be trapped in the solid, in which case they can generally be studied by e.s.r. spectroscopy, which will normally provide clear identification, and give useful structural information, as I show below. Normally the electrons migrate far enough that spin-spin interactions between the centres cannot be detected, but occasionally triplet-states are formed by proximal pairs.

Sometimes either the e-l or e-g centres are mobile, in which case electron-return may dominate. This is usually the case for hydrocarbons, for example. For alkanes, the resulting excited molecules undergo homolysis and neutral radicals are detected. For aromatic compounds however, light is usually emitted, and radical yields are consequently low.

2.2. Dilute Solutions: Electron-gain Centres

Methods have been devised such that specific e-g centres can be prepared, free of the corresponding e-l centres. These methods remove some of the complexities and ambiguities associated with the spectra from pure materials. If the substrate is a cation, it is best to use an ionic matrix in which the matrix cation has one less charge. For example, M^{2+} ions in alkali-halides will often give $M^{\cdot +}$ centres. Conversely, anionic substrates incorporated in salts whose anions have one extra charge will usually capture electrons efficiently, an example being PO_4^{3-} in silicates, which gives $\cdot PO_4^{4-}$ on irradiation.[1]

If the substrate is neutral, we usually use solid-solutions in methanol (CD_3OD generally) or methyltetrahydrofuran (MTHF), since these form good glasses so that phase-separation is generally avoided. Alternatively, tetramethylsilane or adamantane often give e-g centres, and these solvents have the adantage of being rotator solids, so that narrow-line isotropic e.s.r. spectra may be obtained, which are easier to interpret than solid-state 'powder' spectra.

An example of the mechanism of radiolysis of these solutions is given for substrate S in CD_3OD, in equations [1]-[6]. Of these

$$CD_3OD \rightarrow CD_3OD^{\cdot +} + e^- \quad .... [1]$$

$$CD_3OD^+ + CD_3OD \rightarrow \underline{D_2\dot{C}OD} + CD_3OD_2^+ \quad .... [2]$$

$$e^- + nCD_3OD \rightarrow \underline{e^-_{solv}} \quad .... [3]$$

$$e^- + CD_3OD \rightarrow \underline{\dot{C}D_3} + OH^- \quad .... [4]$$

$$e^- + S \rightarrow \underline{S^{\cdot -}} \quad .... [5]$$

$$S^{\cdot -} + CD_3OD \rightleftharpoons \cdot SH + CD_3O^- \quad .... [6]$$

reactions, [2] is fast, so the e-l centre is rapidly trapped. It gives a narrow central e.s.r. quintet (species underlined are detected by ESR spectroscopy). Reaction [3] is fast, but these trapped electrons, responsible for the violet colour of the glasses, can still readily migrate to S, as in [5], to give $S^{\cdot -}$ may be partially or wholly converted into its conjugate acid, $\cdot SH$. [Thus, for example,

CN^- ions give $\cdot CN^{2-}$, but these are converted into $H\dot{C}N^-$ and even some $H_2\dot{C}N$.[2]] This can be avoided by using other solvents, but they are, otherwise, generally less satisfactory than CD_3OD for technical reasons.

2.3. Dilute Solutions: Electron-loss Centres

A range of solvents, including CCl_4, $CFCl_3$, $CFCl_2$-CF_2Cl, C_2F_6 and SF_6 have been used for the specific radiolytic conversion of S into $S^{\cdot+}$ at low temperatures.[3] The reactions leading to $S^{\cdot+}$ formation are thought to be [7]-[10], in which the most popular solvent, $CFCl_3$, is used for

$$CFCl_3 \rightarrow CFCl_3^{\cdot+} + e^- \quad .... \quad [7]$$

$$CFCl_3 + e^- \rightarrow CFCl_3^{\cdot-} \rightarrow \dot{C}FCl_2 + Cl^- \quad .... \quad [8]$$

$$CFCl_3^{\cdot+} + CFCl_3 \rightleftharpoons (CFCl_3^{+})_2^{\cdot} \quad .... \quad [9]$$

$$CFCl_3^{\cdot+} + S \rightarrow CFCl_3 + S^{\cdot+} \quad .... \quad [10]$$

illustration. The 'hole' is transferred from one solvent to another, possibly *via* a dimeric intermediate [9],[4] until it reaches S. The primary radical cation, $S^{\cdot+}$ may exhibit weak interaction with a solvent molecule,[5] or it may undergo some form of unimolecular break-down, such as a rearrangement, or loss of a diamagnetic fragment.

The use of these two types of solvent system is illustrated in the three topics selected for detailed discussion herein.

3. RESULTS AND DISCUSSION

3.1. Tetraalkyl-Tin and -Lead Centres

3.1.1. Tin Centres. Exposure of Me_4Sn to ionizing radiation at 77 K might have been expected to give $Me_4Sn^{\cdot-}$ and $Me_4Sn^{\cdot+}$. Evidence for the former was indeed forthcoming, but the latter was not detected.[6] Subsequently, we studied Me_4Sn in $CFCl_3$[7] and indeed detected $Me_4Sn^{\cdot+}$ the results confirming that no such centre was formed in the pure material. However, on annealing, this centre decomposed to give methyl radicals, in accord with reaction [11], previously predicted by

$$Me_4Sn^{\cdot+} \rightarrow Me^{\cdot} + Me_3Sn^{+} \quad .... \quad [11]$$

Kochi.[8] This reaction was even more facile for $Me_3SnR^{\cdot+}$ (R = Et-, Me_2CH-) although for the Me_3Sn-$CMe_3^{\cdot+}$ cation the reaction was thermally reversible. We envisage a cage reaction [12] to explain this unusual result.

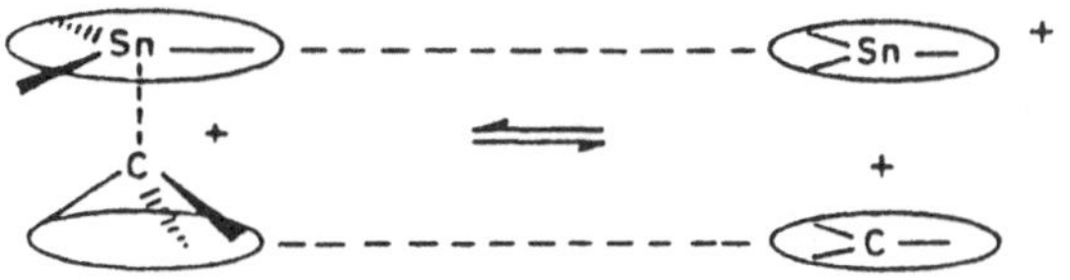

.... [12]

In pure Me_4Sn, breakdown of the primary cations to give $\cdot CH_3$ radicals could be responsible for the formation of $H_2\dot{C}SnMe_3$ radicals, formed in high abundance, and for the addition product $Me_5Sn^{\cdot}$ whose formation was tentatively deduced from the e.s.r. spectra. The only species whose formation in irradiated Me_4Sn is not accommodated by these reactions is $Me_3Sn^{\cdot}$, which was unambiguously detected. We suggest that electron capture by Me_4Sn is a slow process because of the major distortion required, and that some electron return may well occur with consequent homolysis, giving $\cdot CH_3$ and $Me_3Sn^{\cdot}$ radicals.

The structural detail which can be deduced from the e.s.r. spectra, especially the hyperfine coupling from ^{119}Sn and ^{117}Sn (both have $I=\frac{1}{2}$) is considerable. The large tin isotropic coupling shows that $Me_3Sn^{\cdot}$ radicals are pyramidal, the deduced 5*s*-character being *ca.* 15%.

Results for $Me_4Sn^{\cdot -}$ show that, as expected, the isotropic coupling, and hence the 5*s* orbital contribution is considerably enhanced, to *ca.* 18%. By analogy with isostructural phosphorayl radicals, $Me_4Sn^{\cdot -}$ is expected to have a "trigonal-bipyramidal" structure, with two axial and two equatorial ligands. These arguments have been strongly supported by studies of $\cdot SnH_4^-$ in neopentane[9] and tetramethylsilane.[10] Two strongly coupled (axial) protons were detected (A ~140 G), together with ^{119}Sn coupling of *ca.* 2000 G, which corresponds to a 5*s* population of *ca.* 19.8%. It is of particular structural importance that, rather than utilising the 'vacant' 5*d*-manifold, these molecules accept electrons into antibonding σ-levels, with major structural distortion to minimise the antibonding effect. Results for similar closed-shell ML_4 species suggest that this is a general property provided M is not a transition metal.

Interestingly, the corresponding radical-cation, $\cdot SnH_4^+$ also has a SOMO with large 5*s* character (*ca.* 17%) but there is a dramatic fall on going to $Me_4Sn^{\cdot +}$ (*ca.* 1%). It is interesting that $\cdot CH_4^+$ cations have a SOMO comprising only 2*p* orbitals on carbon, just as does $\cdot CH_3$, whereas both $\cdot SnH_3$ and $\cdot SnH_4^+$ have SOMO's which include considerable 5*s*-character.[11]

Two isomers of $\cdot SnH_4^+$ were detected at 77 K in $CFCl_3$, one having two strongly coupled protons (A = 85 G) and one having only one strongly coupled proton (A = 175 G). These species are thought to have distorted as indicated in Inserts I and II. It is noteworthy that the total spin-density on hydrogen remains about the same for these two SOMO's.

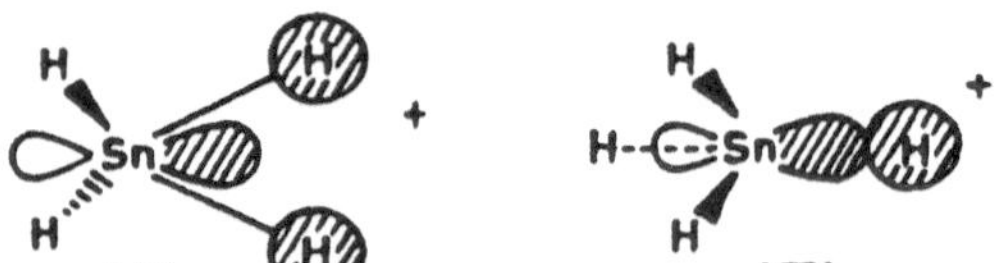

(I) (II)

For $Me_4Sn^{\cdot +}$ the Me_3Sn unit has become almost planar, and, from the ^{13}C splitting, which was very small,[12] the methyl group is also thought to be considerably flattened. Thus the cation has already moved some way along the reaction coordinate which leads to planar

$SnMe_3^+$ cations and planar $\cdot CH_3$ radicals.

For $R_3Sn-SnR_3$ species, our expectation was that both the radical-anions and radical-cations would have SOMO's largely confined to tin, the former being mainly the $Sn\dot{-}Sn$ σ^* orbital and the latter the $Sn\cdot Sn$ σ orbital. However, pure $Me_3Sn-SnMe_3$ failed to give clear evidence for either, the major tin-centred radical being $Me_2\dot{S}n-SnMe_3$,[6] which is probably a homolysis product. It is probable that $Ph_3Sn-SnPh_3$ gave the required anion,[6] and certain that dilute solutions of $Me_3Sn-SnMe_3$ in $CFCl_3$ gave the corresponding cation.[13] On an historical note, my suggested structure for the first alkane cation ever studied by e.s.r. spectroscopy, $Me_3C\cdot CMe_3^+$,[14] was one in which the two Me_3C- units had become almost planar. Whilst this was not a very accurate description of the alkane cation, the ^{119}Sn hyperfine coupling for $Me_3Sn\cdot SnMe_3^+$ shows that it is, in fact, a good description for this cation, the two Me_3Sn- units being apparently almost planar. In fact, of course, it would not be easy for the two Me_3C- units to become planar in $Me_3C\cdot CMe_3^+$ and still retain a significant C-C bond, for steric reasons.

3.1.2. <u>Lead Centres</u>. Exposure of pure Me_4Pb and Et_4Pb to ^{60}Co γ-rays gave no clear indication for the dominant formation of the parent cations and anions.[15] The former have been prepared in $CFCl_3$, being very similar to $Me_4Sn^{\cdot +}$,[16] whilst the latter should have a very large coupling to ^{207}Pb. Such a centre was detected in irradiated Me_4Pb and tentatively identified as the parent anion, but it was in low abundance relative, mainly, to $H_2\dot{C}PbMe_3$ and $Me_3Pb^{\cdot}$ radicals, which were unambiguously identified. It is worth noting that both $H_2\dot{C}SnMe_3$ and $H_2\dot{C}PbMe_3$ have only a small, almost isotropic hyperfine coupling to the metal nuclei. This accords with a spin-polarisation of the C-M σ electrons and rules out any major contribution from $p(\pi)-d(\pi)$ delocalisation, as does the magnitude of the ^1H coupling constants (*ca*. 21.5 G). It seems that electron-return is probably the major mechanism, giving $Me^{\cdot}$ + $\cdot PbMe_3$, the former going on to give $H_2\dot{C}PbMe_3$ radicals.

The e.s.r. spectra for $\cdot PbEt_3$ radicals were better defined than those for the methyl derivative. One aspect of these spectra has never been satisfactorily explained, namely, that the spin-density on Pb, estimated from calculated atomic values, is *ca*. 218%.[15] Proper corrections were applied in deriving this value. Even using the newer A° and $2B^\circ$ values given by Morton and Preston,[17] I find *ca*. 164% 6*p*-character, but only 6.6% 6*s*-character. Possibly these atomic values are still at fault. However, the $2B^\circ$ values that I recommend[18] tie in well with experimental values for atoms in several cases, and I therefore think that the values for $2B^\circ$ given by Morton and Preston are over, rather than underestimates. I mention these observations to show that not all results are as easy to interpret as I may have implied herein.

Results for irradiated R_3Pb-hal derivatives are also of interest.[15] I start by considering results for the corresponding alkyl halides, Rhal. These react with electrons to give $R^{\cdot}$---hal^- "adducts" which seem to be held together only by weak charge-transfer forces, and are detected only when the anion is unsolvated.[19] The cations, which have a SOMO largely confined to the halogen,[20] dimerise in the pure solids to give the σ* species $(Rhal\dot{-}halR)^+$.[21]

For R_3Pbhal, we expect to find both electron addition and possibly halide loss. Addition can give a trigonal bipyramidal complex or a σ* complex $R_3Pb\dot{-}hal^-$, the latter being the expected precursor for dissociation. In fact, the results imply an intermediate structure, which is certainly not the C_{3v} σ* structure because of differences between the halogen hyperfine structure for the non-magnetic lead isotopic species and the features for the ^{207}Pb species. These centres were well defined for $Ph_3PbCl^{\cdot -}$ and Ph_3PbBr^-.

The corresponding electron-loss centres were less well defined, but gave coupling to only one halogen nucleus (Cl, Br) and a small coupling to lead. There was, however, no large positive *g*-shift apparent, such as would be expected if the SOMO were largely confined to the halogen ligand. A careful search for $(R_3Pbhal\dot{-}halPbR_3)^+$ cations has so far been unsuccessful. Thus it seems that the electron-loss centres display large hyperfine coupling to halogen only, and yet have *g*-values close to the free-spin value, which is not expected for SOMO's largely centred on the halogen.

3.2. Dialkyl-Mercury Centres

3.2.1. Pure Compounds. These again gave high yields of carbon centred radicals. For $H_2\dot{C}HgMe$ and $Me\dot{C}HHgEt$ the coupling to mercury was low and almost isotropic, implying σ-electron spin-polarisation rather than direct delocalisation, despite the possibility of $p(\pi)$-$p(\pi)$ overlap.[22] Results for $H_2\dot{C}CH_2HgEt$ suggest that, as with $H_2\dot{C}CH_2SnEt_3$ and the lead derivative, the heavy atom prefers the out-of-plane site with some hyperconjugative delocalisation.[23] In fact, the β-Hg coupling of 750-800 G is much greater than the α-Hg coupling (*ca.* 220 G) in accord with these models.

Our results for the radical-cations (see on) show that no radical cation species were trapped as such in the pure compounds. Thus electron-return may have been important, giving $R^{\cdot}$ + $RHg^{\cdot}$ centres, the former reacting further to give the carbon-centred radicals.

The radicals $RHg^{\cdot}$ are expected to have large ^{199}Hg coupling constants, but two such species were detected, one having $A(^{199}Hg)$ ~3700 G and the other ~2700 G. We tentatively suggested that the former was $RHg^{\cdot}$ and the latter, $R_2Hg^{\cdot -}$. If the latter is correctly identified, we expect that it is bent. Electron-capture, for the linear R_2Hg molecules should occur in the Hg ($6p_\pi$) levels, and bending is expected to occur thereby putting 6*s* character into the SOMO. Unfortunately, the spectra were too poorly defined for a good estimate of the anisotropic coupling to be made, but the results certainly accord with a model having *ca.* 22% 6*s* and up to 78% 6*p* character for the SOMO.[22]

3.2.2. Solutions in CCl_4 and $CFCl_3$. Dilute solutions of $(CD_3)_2Hg$ in CCl_4 gave well defined e.s.r. spectra, with no hyperfine splitting other than that from the mercury isotopes (Fig. 1a).[25] Using a computer programme with no approximations, we have derived the following data: $g_\parallel$ = 1.995, $g_\perp$ = 1.850, $A_\parallel(^{199}Hg)$ = (-)807 G and $A_\perp(^{199}Hg)$ = (-)1568 G. Using these values the computed spectrum in Fig. 1b was obtained. [Note that the low-field line displays an extra turning point, arising from the large *g*- and A-anisotropies.] These results accord well with expectation for a SOMO having major 6*p*(σ) character with low spin-density on the methyl groups. The unusual observation that $|A_\perp|>|A_\parallel|$ is understandable if A_{iso} is negative. This, in turn, requires that spin-polarisation of the inner *s*-electrons on mercury dominates over that of the carbon-mercury σ_1-electrons, which is somewhat surprising.

Results for $(CD_3)_2Hg$ in $CFCl_3$ were complicated by extra hyperfine coupling, which we ascribe to ^{19}F from one solvent molecule. This is an interaction sometimes observed for substrates with relatively low ionization potentials, and does not imply any specific bonding as is the case when chlorine coupling is observed.[5] For $(CH_3)_2Hg$ in CCl_4 lines were broadened but not resolved. However, in $CFCl_3$, there was a splitting of *ca*. 5 G in addition to the ^{19}F splitting, which has been analysed in terms of six equivalent protons. When ^{13}C was introduced extra broadening was seen, but the coupling was not resolved and, therefore, must be small.

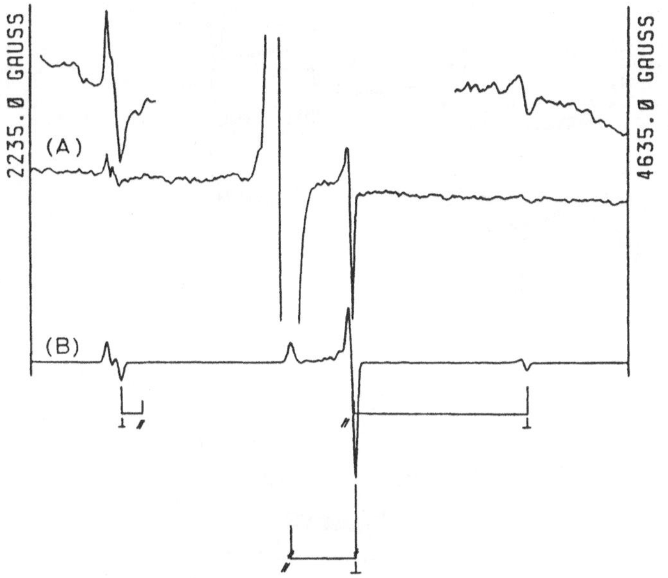

Figure 1. First derivative X-band e.s.r. spectrum for $(CD_3)_2Hg$ in CCl_4 after exposure to ^{60}Co γ-rays at 77 K, showing features assigned to $(CD_3)_2Hg^{\cdot+}$ cations (A) together with a computer simulation based on the data in the text. [Note that the main parallel feature is taken to be concealed under the intense central line, whilst the M_I = -½ (^{199}Hg) parallel feature fortuitously coincides with the main perpendicular feature.

In marked contrast, Et_2Hg gave a 42 G triplet splitting in both solvents, with no extra ^{19}F splitting. This large coupling must come from two unique methyl protons. The structure can be compared with that of the ethane cation, which also shows major coupling to only two of the six protons.[26]

4. METHYL COBALAMIN AND COENZYME B_{12}

Animals and man contain three major types of cobalamin: the methyl derivative, the adenosyl derivative (coenzyme B_{12}) and the hydroxo derivative (Fig. 2). The first two have labile Co-C bonds which are

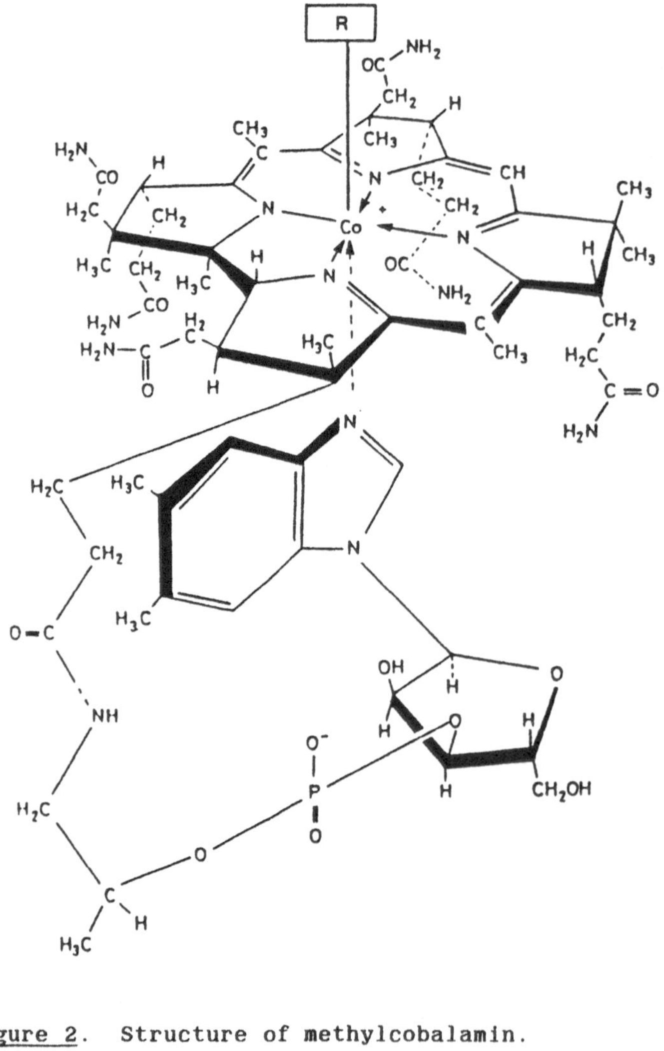

Figure 2. Structure of methylcobalamin.

highly sensitive. All three contain Co(III) in their normal states. The photochemistry of the first two derivatives has been widely studied[27-34] but little attention has been given to their radiolyses. This is a complex area, and my attempt to draw general conclusions may well be an over-simplification. It is helpful to cover both topics to see what useful information can be forthcoming.

4.1. Electron Addition

In our radiolysis studies attention was focused on electron capture. Hence, mainly, solutions in CD_3OD or $CD_3OD + D_2O$ were used. These give good glasses and were therefore also useful in photochemical work.

For methyl cobalamine, electron addition at 77 K did not result in initial formation of Co(II): instead, a radical species (g=2) was obtained, which we identified as a corrin anion (π^*) having undetectably low spin-density on Co.[35] That the added electron was bound to the cobalamin was established by the appearance of Co(II) signals on annealing. However, analysis of the g- and hyperfine tensor components showed that the centre was not a normal $\cdots 3\underline{d}_{z^2}^1$ Co(II) species but probably a $\cdots 3\underline{d}_{x^2-y^2}^1$ species (Table 1). Further annealing gave the well known $\cdots 3\underline{d}_{z^2}^1$ centre. Thus the nett reaction [13] appears to

$$-Co^{III} - CH_3 + e^- \rightarrow -Co^{II} + CH_3^- \qquad \ldots. [13]$$

occur, but by an unusually complex route. We assume that electron addition is kinetically controlled, and that the path with least activation is π-addition. Apparently a switch to the σ^* (x^2-y^2) orbital is then favoured perhaps because of the strongly covalent nature of the Co-C bond. Possibly the methyl group moves away at this stage, prior to another electron-switch to give B_{12r}, or these may coincide. It is highly unlikely that this reaction involves CH_3^- formation. We suggest that simultaneous protonation occurs, but this step is difficult to formulate. Loss of the *trans*-ligand [a dimethyl imidazole derivative] is surely more reasonable, but this is covalently linked to the corrin ring and hence will be able to return as the methyl group leaves. We stress that loss of $\cdot CH_3$, as occurs on photolysis, leaves a Co(I) derivative which would not give any e.s.r. signal].

It is interesting to contrast this result with that for electron-addition to the structurally similar compound, methyl(pyridine)cobaloxime.[36] In this case neither the π^* nor the $\underline{d}_{x^2-y^2}$ structures were detected, the electron apparently going directly into the $\underline{d}_{z^2}$ orbital. However, the absence of any ^{14}N hyperfine coupling and the well defined doublets obtained for the $^{13}CH_3$ derivative showed that in contrast with methyl cobalamin, the axial pyridine ligand was lost and the methyl group was retained. This is the more obvious pathway, and the contrast with methyl cobalamin is remarkable.

Similar studies of electron addition to coenzyme B_{12} gave the same broad central 'radical' feature, which we assign to the π^* anion derivative. On annealing we were unable to detect any Co($\underline{d}_{x^2-y^2}^1$) type species, but normal B_{12r} was eventually detected. Again it seems that the Co-C bond breaks in preference to the *trans* Co-N bond. Presumably

TABLE 1. - Some Experimental E.S.R. Parameters for Derivatives of Vitamin B_{12} and Related Compounds

Compound	Nucleus	Hyperfine Coupling Constants/G[a]			g-values		
		A_z	A_x	A_y	g_z	g_x	g_y
$(B_{12})^{-b}$	^{59}Co	87	*ca.* 40	*ca.* 40	2.000	*ca.* 2.22	*ca.* 2.22
	^{14}N	*ca.* 13	-	-	-	-	-
	^{13}C	70	*ca.* 60	*ca.* 60	-	-	-
$(B_{12}CN)^{-c}$	^{59}Co	85	*ca.* 12	*ca.* 12	2.004	*ca.* 2.17	*ca.* 2.17
	^{13}C	65	*ca.* 55	*ca.* 55	-	-	-
$Co(CN)_5{}^{2-d}$	^{59}Co	81	(-) 26	(-) 26	2.004	2.17	2.17
	$^{13}C_{ax}$	37	31	31	-	-	-
$Co(^{13}CH_3)^e$	^{59}Co	62.5	*ca.* 42	*ca.* 42	2.002	*ca.* 2.099	*ca.* 2.099
	^{13}C	19	*ca.* 17	*ca.* 17			
$B_{12}r$	^{59}Co	108	NR	NR	2.003	2.32	2.32
	^{14}N	18	NR	NR	-	-	-
$B_{12}(d_{x^2-y^2})$	^{59}Co	87	30	30	2.077	2.016	2.016
	^{14}N	10	NR	NR	-	-	-
$(CH_3CO\ B_{12})^-$	^{59}Co	106	NR	NR	2.006	2.28	2.28
	^{14}N	17	NR	NR	-	-	-

[a] G = 10^{-4} T; corrected when necessary; NR, not resolved.
[b] The anion of B_{12}.
[c] The monocyanide derived from the dicyanide by electron addition.
[d] Ref. 35.
[e] $^{13}CH_3$ cobaloxime, see Ref. 36.

the same factors that appear to facilitate CH_3^- generation also favour displacement of the deoxy adenosyl 'anion'.

Yet another variation was found with vitamin B_{12} (the cyano derivative) which, using ^{13}CN, was shown to form an unstable anion with a $\cdots d_{z^2}^1$ configuration, whereas the dicyano derivative lost one CN^- ligand even at 77 K.[35]

4.2. Photolyses

A very interesting low temperature e.s.r. study of the photolysis of coenzyme B_{12} is that of Lowe *et al.*[27] Using dilute aqueous propane-1,2-diol as a suitable glass, they found that photolysis at 77 K gave no detectable e.s.r. features, so the thermal back-reaction must be efficient in this rigid glass matrix. However, on photolysis at higher temperatures, but below the softening point of the glass, they observed e.s.r. spectra for the 'normal' Co(II), 5-coordinated species (B_{12r}) together with an unusually broad (~30 G) free radical signal. On warming to room temperature and re-freezing, the radical signal was lost but the B_{12r} signal was enhanced considerably. We have confirmed these results using both (D_2O + CD_3OD) and (H_2O + glycol) glasses.[28] It seems that after Co-C bond fission occurs, the adenosyl radical unit swings away, given sufficient thermal energy, and becomes trapped over an ill-defined range. This results in variable spin-spin interactions which greatly broaden the features. Ultimately the radicals move away completely and react, leaving B_{12r} as a relatively stable product.

Greater insight can be obtained by considering results for methylcobalamin.[28] Using extensive precautions to prevent photolysis prior to freezing, we found that dilute glassy solutions gave 1:3:3:1 quartet features characteristic of $\dot{C}H_3$ radicals (the $-CD_3$ derivative gave $\cdot CD_3$ and the $^{13}CH_3$ derivative gave $^{13}CH_3$ features). When CD_3OD glasses were used, $D_2\dot{C}OD$ radicals were also detected (Fig. 3a). These, we presume, were formed by attack of 'hot' $\cdot CH_3$ radicals on nearby solvent molecules. Using higher microwave powers these signals were saturated and clear features characteristic of a Co(II) species could be seen (Fig. 3b). This species is not the normal Co(II) B_{12r} derivative, but was converted into B_{12r} on annealing above *ca*. 180 K (Fig. 3c). Note that for B_{12r}, the parallel ^{59}Co octet features are all split into triplets, characteristic of ^{14}N coupling from the axial ligand. Triplet splitting is not resolved for the spectrum in Fig. 3b, but the features are broad and a triplet (1:1:1) with half the normal splitting fits well. The $^{59}Co(A_\parallel)$ splitting (56 G) is almost half that for B_{12r} (108 G) (Fig. 3c) and there is an extra doublet splitting ($2D_\parallel$) of *ca*. 100 G. Although the perpendicular features are poorly defined, the approximate $g_\perp$-shift [g(exp) -2.0023] (*ca*. 0.16) is half that for B_{12r} (*ca*. 0.32). All this evidence suggests that the species is in a triplet-state, with fast electron exchange, so that any one electron is only 50% on cobalt and 50% on some radical, R$^\cdot$. This explains the factor of two for $A(^{14}N)$, $A_\parallel(^{59}Co)$ and $\Delta g_\perp$. It also explains the 100 G doublet splitting, which is largely due to dipolar spin-spin coupling.

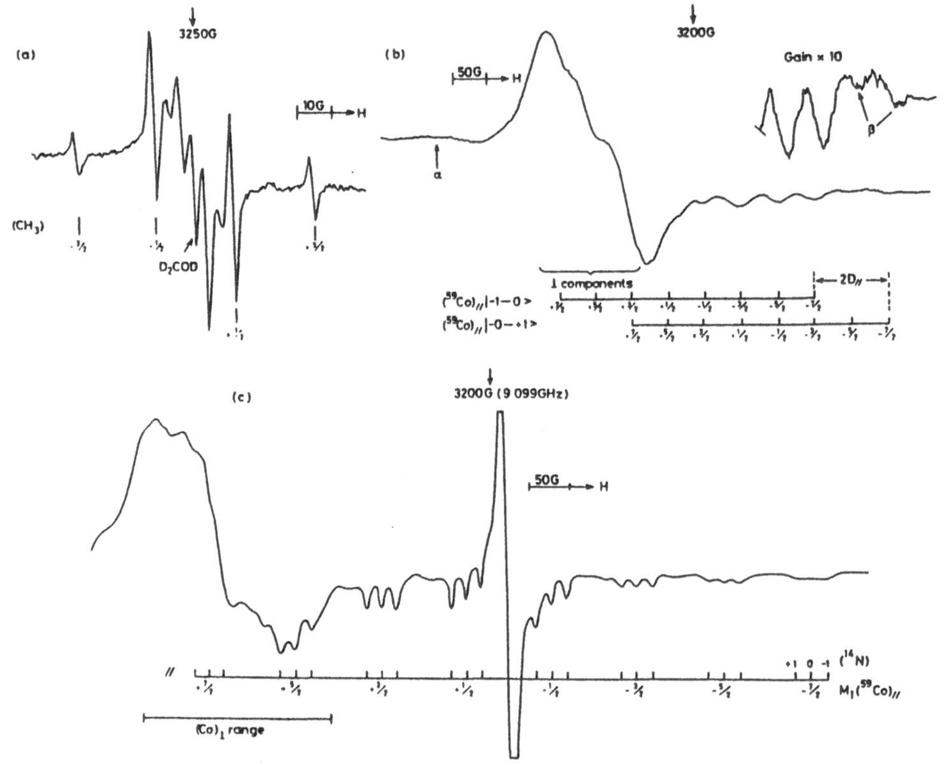

Figure 3. First derivative X-band e.s.r. spectrum for methylcobalamin in D_2O-CD_3OD solution after exposure to u.v. light, (a) at low microwave power, showing features assigned to CH_3 and CD_2OD radicals; (b) at high power, showing features assigned to $B_{12}(\alpha)$ (⊥ features only) and species A [note the extra pair of parallel features at high field (β)]; (c) a typical e.s.r. spectrum for $Co^{II}B_{12r}$, as obtained, for example, on annealing the above system to *ca.* 180 K.

Had the species been the expected Co^{II}---$\cdot CH_3$ pair, there should have been an additional 1:3:3:1 splitting of each parallel component of half the normal methyl coupling (i.e. ~11 G). This was definitely absent and the spectra obtained from the Co^{II}-CD_3 complex were unchanged in this region. It therefore seems that the R˙ component is $D_2\dot{C}OD$ rather than $\cdot CH_3$, since the small ^2H coupling (predicted to be *ca.* 1.8 G) would not contribute significantly to the line-width. We suggest that as the 'hot' $\cdot CH_3$ radical moves away from cobalt, it encounters a solvent molecule, extracting D, and leaving $D_2\dot{C}OD$ trapped 8.3 Å away (Scheme I). This scheme nicely explains why trapping occurs at *ca.* 8 Å. In our view, if and when reaction with CD_3OD does not occur, the $\cdot CH_3$ radical is most unlikely to be permanently trapped

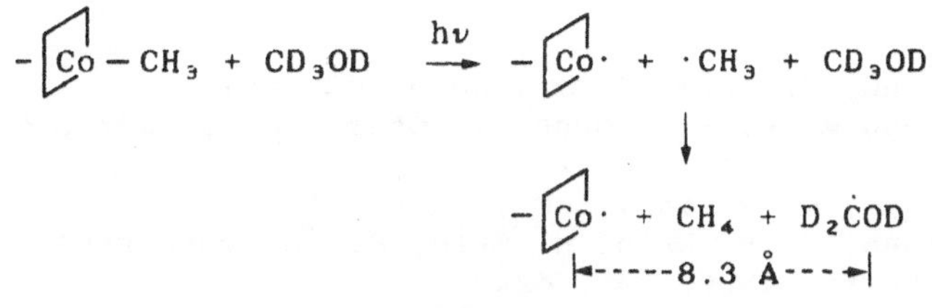

Scheme I

at, say, ~5 Å from cobalt, since there is nothing to prevent thermal return. For this, and other reasons, the suggestion that 5 Å trapping occurs in photolysed coenzyme B_{12}[37] is, in our view, unlikely to be correct.

Note that the concentration of pairs is far greater than that for 'free' $\dot{C}H_3$ and $D_2\dot{C}OD$ radicals. Nevertheless, there should be some normal B_{12r} present and a weak perpendicular feature was indeed detectable (α in Fig. 3b).

5. CONCLUSIONS

The results of these studies support our contention[38] that the combined use of rigid systems plus ionizing radiation and e.s.r. spectroscopy is an unusual but powerful method for studying electron-capture and electron-loss centres formed from organometallic compounds. For transition-metal derivatives this usually amounts to gain or loss of electrons from the *d*-manifold,[39] and hence the results are perhaps less exceptional than those for the non-transition metal derivatives described herein. The resulting radical intermediates have low stability, but well defined structures, as judged by e.s.r. spectroscopy. In favourable cases it is possible to study unimolecular decompositions of these radical-ions.

Acknowledgements
I thank Professor R. L. Petersen for checking the manuscript and Miss V. Orson-Wright for preparing the typescript.

REFERENCES

[1] M. C. R. Symons, *J. Chem. Phys.*, 1970, **53**, 857.
[2] I. S. Ginns and M. C. R. Symons, *J. Chem. Soc., Dalton Trans.*, 1972, 185.
[3] M. C. R. Symons, *Chem. Soc. Rev.*, 1984, **13**, 393.
[4] M. C. R. Symons, B. W. Wren, H. Muto, K. Toriyama and M. Iwasaki, *Chem. Phys. Lett.*, 1986, **127**, 424.
[5] T. Clark, A. Hasegawa and M. C. R. Symons, *Chem. Phys. Lett.*, 1985, **116**, 79.
[6] A. Hasegawa, S. Kaminaka, T. Wakabayashi, M. Hayashi and M. C. R. Symons, *J. Chem. Soc., Chem. Commun.*, 1983, 1199.
[7] M. C. R. Symons, *J. Chem. Soc., Chem. Commun.*, 1982, 869.
[8] J. K. Kochi, *"Organometallic Mechanisms and Catalysis"* Academic Press, New York, 1978.
[9] J. R. Morton and K. F. Preston, *Mol. Phys.*, 1975, **30**, 1213.
[10] A. Hasegawa, T. Yamaguchi and M. Hayashi, *Chem. Lett.*, 1980, 611.
[11] S. A. Fieldhouse, A. R. Lyons, H. C. Starkie and M. C. R. Symons, *J. Chem. Soc., Dalton Trans.*, 1974, 1966.
[12] A. Hasegawa, S. Kaminaki, T. Wakabayashi, M. Hayashi, M. C. R. Symons and J. Rideout, *J. Chem. Soc., Dalton Trans.*, 1984, 1667.
[13] M. C. R. Symons, *J. Chem. Soc., Chem. Commun.*, 1981, 1251.
[14] I. G. Smith and M. C. R. Symons, *J. Chem. Res. (S)*, 1979, 382; M. C. R. Symons, *Chem. Phys. Lett.*, 1980, **69**, 198.
[15] R. J. Booth, S. A. Fieldhouse, H. C. Starkie and M. C. R. Symons, *J. Chem. Soc., Dalton Trans.*, 1976, 1506.
[16] B. W. Walther, F. Williams, W. Lau and J. K. Kochi, *Organometallics*, 1983, **2**, 688.
[17] J. R. Morton and K. F. Preston, *J. Magn. Reson.*, 1978, **30**, 577.
[18] M. C. R. Symons, *"Chemical and Biochemical Aspects of Electron Spin Resonance Spectroscopy"* Van Nostrand Reinhold Co. Ltd., Wokingham, Berkshire, 1978.
[19] I. G. Smith and M. C. R. Symons, *J. Chem. Soc., Perkin Trans. 2*, 1979, 1362.
[20] G. W. Eastland, S. P. Maj, A. Hasegawa, C. Glidewell, M. Hayashi and T. Wakabayashi, *J. Chem. Soc., Perkin Trans. 2*, 1984, 1439.
[21] S. P. Mishra and M. C. R. Symons, *J. Chem. Soc., Perkin Trans. 2*, 1975, 1492.
[22] B. W. Fullam and M. C. R. Symons, *J. Chem. Soc., Dalton Trans.*, 1974, 1086.
[23] M. C. R. Symons and M. M. Aly, *J. Organomet. Chem.*, 1979, **166**, 101.
[24] A. Hasegawa, J. Rideout and M. C. R. Symons, unpublished results.
[25] J. Rideout and M. C. R. Symons, *J. Chem. Soc., Chem. Commun.*, 1985, 129.
[26] K. Toriyama, K. Nunome and M. Iwasaki, *J. Chem. Phys.*, 1982, **77**, 5891; 1983, **79**, 2499.
[27] D. J. Lowe, K. N. Joblin and D. J. Cardin, *Biochim. Biophys. Acta*, 1978, **539**, 398.
[28] D. N. R. Rao and M. C. R. Symons, *J. Chem. Soc., Perkin Trans. 2*, 1983, 187.

[29] J. F. Endicott and T. L. Netzell, *J. Am. Chem. Soc.*, 1979, **101**, 4000.
[30] K. N. Joblin, A. W. Johnson, M. F. Lappert and B. K. Nicholson, *J. Chem. Soc., Chem. Commun.*, 1975, 441.
[31] Ph. Maillard and C. Giannotti, *J. Organomet. Chem.*, 1979, **182**, 225.
[32] Ph. Maillard, J. C. Massot and C. Giannotti, *J. Organomet. Chem.*, 1978, **159**, 219.
[33] C. Giannotti and J. R. Bolton, *J. Organomet. Chem.*, 1974, **80**, 379.
[34] H. P. C. Hogenkamp, *Biochemistry*, 1966, **5**, 417.
[35] D. N. R. Rao and M. C. R. Symons, *J. Chem. Soc., Faraday Trans. I*, 1983, **79**, 269.
[36] D. N. R. Rao and M. C. R. Symons, *J. Chem. Soc., Faraday Trans. I*, 1984, **80**, 423.
[37] A. Pezeshk and R. E. Coffman, *J. Chem. Soc., Dalton Trans.*, 1985, 891.
[38] M. C. R. Symons, *Pure and Appl. Chem.*, 1981, **53**, 223.
[39] M. C. R. Symons, J. R. Morton and K. F. Preston, *Amer. Chem. Soc., Symp. Ser.*, 1987, **333**, 169.

COMPETITION BETWEEN ODD- AND EVEN-ELECTRON PROCESSES

Joseph San Filippo
Department of Chemistry
Rutgers University
New Brunswick, NJ 08903, USA

Introduction

The competition between odd- and even-electron processes has been generally recognized in chemistry for many years. One of our earlier interests in this area arose in the study of the reactions of the superoxide radical anion. The obvious question as to whether this

$$RX \xrightarrow{O_2^-} \longrightarrow R\text{-}OH$$

carbon-oxygen bond formation involved an odd- or even-electron process was answered by examining the reaction's stereochemistry.[1]

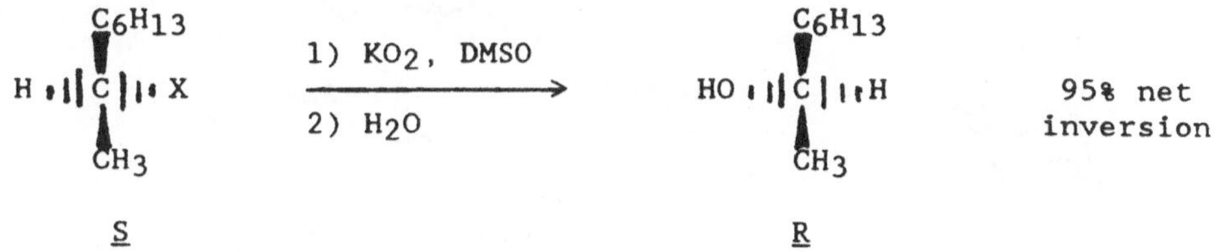

Traditional stereochemical studies are useful diagnostic probes of reaction mechanism provided the product is sufficiently stable to isolate and its maximum rotation and configuration are known. These conditions are frequently not met in many organometallic reactions and a variety of alternative stereochemical probes have been developed to supplement entantiomeric studies. The diastereotopic probes introduced by Whitesides[2] and Osborn[3] are two such examples.

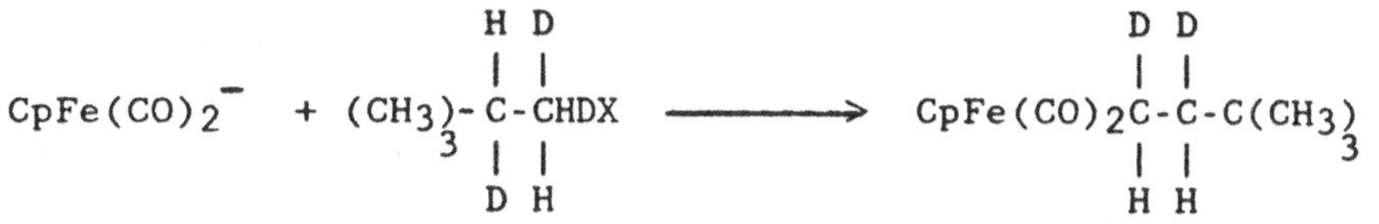

M. Chanon et al. (eds.), Paramagnetic Organometallic Species in Activation / Selectivity, Catalysis, 463–475.

$$IrL_2(CO)Cl \;+\; Ph\text{-}\overset{F}{\underset{H}{C}}\text{-}\overset{H}{\underset{D}{C}}\text{-}Br \longrightarrow IrL_2(CO)(Cl)CHDCFHPh$$

In certain reactions, however, product stability -- both thermal and stereochemical -- is adequate to permit the determination of the stereochemical consequences of a reaction by classical entantiomeric procedures, provided the maximum rotation of the product can be determined. One such system is that posed by the reaction of trimethyl- and triphenylstannate, $(CH_3)_3Sn^-$ and Ph_3Sn^-, with alkyl halides.

ALKYLATION with 4-*tert*-BUTYLCYCLOHEXYL and 2-OCTYL TOSLYATE and HALIDE

Because it is generally not possible to optically resolve the alkylation products of these metalate anions, we had to develop a reaction that would permit us to synthesize these products stereospecifically. Our studies[4] of the reaction of $(CH_3)_3Sn^-$ with cis- and trans-4-*tert*-butylcyclohexyl tosylate demonstrated that these reactions proceed with essentially complete (>99%) *inversion* of configuration. It seemed reasonable, therefore, to conclude that

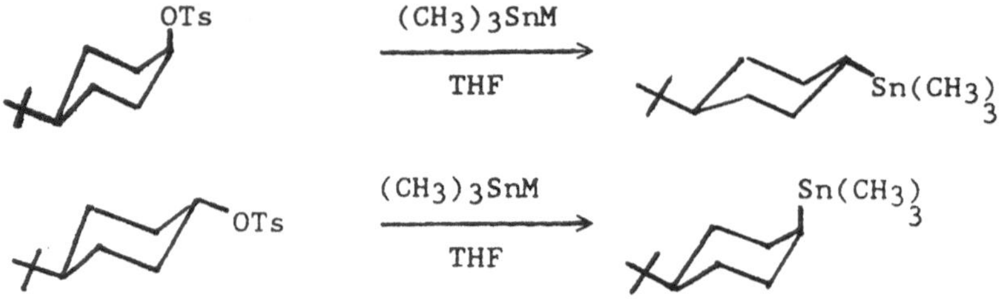

corresponding reaction with the tosylate of optically active 2-octanol would take place with equivalent stereochemical integrity. In an effort to establish the stereospecificity of this reaction, we carried out an independent but stereochemically equivalent synthesis. Collectively, these results sustain our contention that the reaction of $(CH_3)_3SnM$ with optically active 2-octyltosylate is stereospecific and that the rotation of optically pure trimethyl(2-octyl)tin is 26.1°.

Scheme I

$$(R)\text{-}2\text{-}C_8H_{17}OH \xrightarrow[\text{2. }(CH_3)_3SnLi\text{, THF, }-70°C]{\text{1. TsCl, py}} (S)\text{-}2\text{-}C_8H_{17}Sn(CH_3)_3$$

$[\alpha]_D^{17}$ -9.9° $[\alpha]_D^{25}$ −26.1° (neat)

↓ PBr_3

$$(S)\text{-}2\text{-}C_8H_{17}Br \xrightarrow[\text{THF, }-50°C]{\text{"}[(CH_3)_3Sn]_2CuLi\text{"}} (R)\text{-}2\text{-}C_8H_{17}Sn(CH_3)_3$$

$[\alpha]_D^{22}$ +43.6° $[\alpha]_D^{25}$ +26.1° (neat)

In a parallel study the reaction of lithium triphenyltin with the tosylate of optically active 2-octanol was carried out. The results, summarized below, suggest that optically pure product has a rotation of 23.3°.

$$(R)\text{-}2\text{-}C_8H_{17}OH \xrightarrow[\text{2.}Ph_3SnLi\text{, THF, 25°C}]{\text{1.TsCl,py}} (S)\text{-}2\text{-}C_8H_{17}SnPh_3$$

$[\alpha]_D^{17}$ -9.9° $[\alpha]_D^{22}$ +23.3°,(c 4.15, C_6H_6)

As seen below, the equivalent reaction involving optically active 2-octyl bromide is also stereospecific.

$$(S)\text{-}2\text{-}C_8H_{17}Br \xrightarrow[\text{THF, 25°C}]{Ph_3SnLi} (R)\text{-}2\text{-}C_8H_{17}SnPh_3$$

$[\alpha]_D$ +38.6° $[\alpha]_D$ −23.3°,(c 4.15, C_6H_6)

We extended these studies to include the alkylation of other group IV anions, examining specifically the alkylation of triphenylmethyl-, triphenylsilyl-, and triphenylgermyllithium. The stereo-chemical consequences of these reactions with optically active 2-octyltosylate are summarized below.

$$(R)\text{-}2\text{-}C_8H_{17}OH \xrightarrow[\text{py}]{\text{TsCl}} \xrightarrow[\text{THF, 25°C}]{Ph_3CLi} (S)\text{-}2\text{-}C_8H_{17}CPh_3$$

$[\alpha]_D^{17}$ -9.9° $[\alpha]_D^{25}$ +42.50°,(c 3.05, C_6H_6)

$$(R)\text{-}2\text{-}C_8H_{17}OH \xrightarrow[\text{py}]{\text{TsCl}} \xrightarrow[\text{THF, 25°C}]{Ph_3SiLi} (S)\text{-}C_8H_{17}SiPh_3$$

$[\alpha]_D^{25}$ +42.50°,(c 3.71, C_6H_6)

Table I. Normal-Addition Reaction of $(CH_3)_3SnM$ with Optically Active 2-Octyl Bromide and Chloride.

$$(CH_3)_3SnM \xrightarrow{R^*\text{-}X} (CH_3)_3SnR^* + MX$$

R^*-X	$(CH_3)_3SnM$ (concn, M)	Temp, °C	Solvent	Yield, % R-Sn$(CH_3)_3$	Entantiomeric Excess, %
X=Br	Li(0.2)	0	THF	74	49
	(0.4)	0		69	53
	(0.8)	0		74	40
	(0.4)	-70		60	82
	Na(0.2)	0		66	33
	(0.4)	0		55	34
	(0.8)	0		32	36
	(0.4)	-70		45	62
	K(0.2)	0		80	33
	(0.4)	0		54	33
	(0.8)	0		66	42
X=Cl	Li(0.4)	25		51	90
	(0.4)	0		62	100
X=Br	(0.4)	0	DME	79	34
	Na(0.4)	0	DME	56	31
	K(0.4)	0	DME	52	32

Table II. Inverse-Addition Reaction of $(CH_3)_3SnM$ with (*S*)-Octyl Bromide and Chloride.

$$(CH_3)_3SnM \xrightarrow[\text{THF, 0°C}]{R^*\text{-}X} (CH_3)_3SnR^* + MX$$

R^*-X	Me_3SnM (concn, *M*)	Yield, % $RSnMe_3$	Entantiomeric Excess, %
Cl	Na(0.4)	52	93
Br	Li(0.4)	63	97
Br	Na(0.4)	58	98
Br	Na(0.1)	68	100

Table III. Reaction of cis- and trans-4-*tert*-Butylcyclohexyl Bromide (0.5M) with $(CH_3)_3SnM$ (M = Li,Na,K).

Substrate (concn, M)	$(CH_3)_3SnM$ (concn, M)	Solvent	Temp °C	Isomer Distribution trans:cis
cis-Br	M=Li(0.04)	THF	0	68:32
	(0.8)	THF	0	73:27
	(0.4)	THF	-70	82:18
	(0.4)	DME	0	78:22
	(0.4)	DME	-70	85:15
trans-Br	M=Li(0.04)	THF	0	69:31
	(0.8)	THF	0	71:29
	(0.4)	THF	-70	70:30
	(0.4)	DME	0	80:20
	(0.4)	DME	-70	77:23
cis-Br	M=Na(0.04)	THF	0	72:28
	(0.8)	THF	0	53:47
cis-Br	M=Na(0.4)	THF	-70	64:36
	(0.4)	DME	0	48:52
	(0.4)	DME	-70	45:55
trans-Br	M=Na(0.04)	THF	0	74:26
	(0.8)	THF	0	55:45
	(0.4)	THF	-70	64:36
	(0.4)	DME	0	46:54
	(0.4)	DME	-70	49:51
cis-Br	M=K(0.04)	THF	0	76:24
	(0.8)	THF	0	62:38
	(0.4)	THF	-70	68:32
	(0.4)	DME	0	54:46
	(0.4)	DME	-70	50:50
trans-Br	M=K(0.04)	THF	0	76:24
	(0.8)	THF	0	63:37
	(0.4)	THF	-70	65:35
	(0.4)	DME	0	52:48
	(0.4)	DME	-70	52:48

$$(\underline{R})\text{-}2\text{-}C_8H_{17}OH \xrightarrow[py]{TsCl} \xrightarrow[THF,\ 25°C]{Ph_3GeLi} (\underline{S})\text{-}C_8H_{17}GePh_3$$

$[\alpha]_D^{25}$ +12.60°,(c 4.65, C_6H_6)

By comparison, the corresponding reaction of 2-octyl chloride and bromide with Ph_3CLi, Ph_3GeLi and Ph_3SnLi also proceeds with high stereoselectivity, whereas the corresponding alkylation of Ph_3SiLi is notably less stereospecific: 70-75% net inversion observed with 2-octyl chloride and 25-55% net inversion occurred with 2-octyl bromide.

$$Ph_3CLi + R^*\text{-}X \longrightarrow Ph_3C\text{-}R^*$$

X=Cl, > 99% ee
Br, > 97% ee

$$Ph_3GeLi + R^*\text{-}X \longrightarrow Ph_3Ge\text{-}R^*$$

X=Cl, > 94% ee
Br, > 99% ee

$$Ph_3SnLi + R^*\text{-}X \longrightarrow Ph_3Sn\text{-}R^*$$

X=Cl, > 97% ee
Br, > 99% ee

$$Ph_3SiLi + R^*\text{-}X \longrightarrow Ph_3Si\text{-}R^*$$

X=Cl, ~ 75% ee
Br, ~ 25-55% ee

By contrast, the alkylation of $(CH_3)_3SnM$ with 2-octyl chloride and 2-octyl bromide shows considerably greater variation in reaction stereoselectivity. Stereoselectivity in these instances varies with the order of reagent addition, reaction temperature, solvent, gegenion, concentration and additives. Thus, for example, at 0°C the normal addition (i.e. halide to organometallic reagent) reaction of 2-octyl chloride with $(CH_3)_3SnM$ in THF proceeds with 34% (M = Na), 33% (M = K) and 53% (M = Li) net *inversion*. The latter reactions are considerably more stereoselective when carried out at lower temperatures: at -70°C the corresponding values are 62% (M = Na) and 83% (M = Li) net *inversion*.

Stereoselectivity in this series is also dramatically influenced by the order of reagent addition. For example, under *inverse-addition* conditions, i.e. addition of metalate anion to halide, the reaction of $(CH_3)_3SnM$ with 2-octyl bromide in THF at 0°C occurs with high stereoselectivity (97% and 98% net *inversion*, respectively, for M = Li and Na) and with minor adjustment of other parameters can be made to occur stereospecifically. Tables I-II summarize some of these findings.

The loss of configuration associated with an alkyl radical center occurs primarily within the parent solvent cage and is complete for kinetically free radicals.[5a] Reasonably, such intermediates can be envisioned as leading to product though a radical-radical coupling reaction, i.e.

$$R\cdot + R'_3Sn\cdot \xrightarrow{k_c} R\text{-}SnR'_3 \qquad (4)$$

However, eq. 4 cannot account for the observed influences of solvent on product isomer distribution observed in the alkylation of $(CH_3)_3SnM$ with both cis- and trans-4-*tert*-butylcyclohexyl bromide (Table III): only quite minor variations in product stereochemistry ratios are expected for the nongeminate, nonionic coupling process outlined in eq. 4 when R = 4-*tert*-butylcyclohexyl. The substantial influence which solvent, temperature and concentration have on product isomer distributions in the reaction of $(CH_3)_3SnM$ with cis- and trans-4-*tert*-butylcyclohexyl bromide suggests a product-forming step involving ionic character.

DISCUSSION

A reasonable mechanistic interpretation of these data is that at least one competing, nonstereospecific alkylation reaction is operating in addition to any stereospecific reaction(s) that may be occurring. We propose the first step in this competing reaction to be a single electron transfer that leads to an alkyl radical, halide ion, and a cluster radical cation (Scheme II). Subsequent dissociation of a trimethyltin radical from these partially oxidized clusters followed by its combination with R'· (eq. 4) and/or the direct reaction of $[R_3SnM]^{+\cdot}$ with R'· (eq. 5) would provide a method of forming a carbon-tin bond.

Scheme II. Proposed Mechanism for the Free-Radical Component of the Reaction of $(CH_3)_3SnLi$, -Na, -K with Alkyl Halides

$$R'X + [R_3SnM]_n \longrightarrow \overline{R + X^- + [R_3SnM]^{+\bullet}_n} \qquad (1)$$

$$\overline{R'\bullet + X^- + [R_3SnM]^{+\bullet}_n} \xrightarrow{k_d} R'\cdot + [R_3Sn]^{+\bullet}_n + X^- \qquad (2)$$

$$[R_3SnM]^{+\bullet}_n + X^- \xrightarrow{k_3} [R_3SnM]_{n-1} + MX + R_3Sn\bullet \qquad (3)$$

$$R'\bullet + R_3Sn\bullet \xrightarrow{k_c} R'SnR_3 \qquad (4)$$

$$R'\bullet + [R_3SnM]^{+\bullet}_n \longrightarrow R'SnR_3 + [(R_3Sn)_{n-1}M]^{+}_n \qquad (5)$$

Eq. 5 or a combination of the processes represented by eq. 4 and 5 describe the product-forming step in the reaction of $(CH_3)_3SnLi$ with cis- and trans-4-*tert*-butylcyclohexyl bromide. Moreover, eq. 5

involves the bimolecular reaction of R'· with $[R_3SnM]_n^{+\cdot}$, whose structure, like that of organolithium reagents in general, but unlike that of a simple free radical, is likely to be solvent dependent.[5b] Solvent-induced structural changes of this sort could reasonably lead to changes in the steric factors that control the stereochemistry of the carbon-tin coupling product. Of course, factors other than solvent can influence the structure of organolithium reagents. Temperature and concentration, for example, can also play a role in this regard. Collectively, this interpretation of the difference in the product stereochemistries observed in the reaction of $(CH_3)_3SnLi$ with cis- and trans-4-*tert*-butylcyclohexyl bromide is appealing in its simplicity and internal consistency. It also points out a useful additional fact, viz, the sensitivity of the 4-*tert*-butylcyclohexyl radical as a diagnostic probe with which to explore radical-radical coupling reactions *in those instances* where thermodynamic and steric product control reinforce each other, resulting in a clear preference for the formation of the trans product isomer which, conseq.uently, permits an interpretation of isomer distribution that is not beclouded by the conflicting factors which characterize those reactions of the 4-*tert*-butylcyclohexyl radical that produce a predominance of the sterically and thermodynamically disfavored cis product isomer.[6]

In addition to solvent, temperature, and concentration, gegenions can also influence the extent of anion association and, hence, structure, solubility, basicity, nucleophilicity, etc., of these reagents in solution. Thus, for example, it has long been recognized that, in contrast to the general solubility of organolithium reagents in many organic solvents (a fact attributed to the substantially covalent nature of the C-Li bond), corresponding organosodium and -potassium compounds, are generally quite insoluble, presumably as a conseq.uence of the high degree of ionic character possessed by the carbon-metal bond. Indeed, there are numerous parallels between the physical and chemical properties of organosodium and -potassium compounds which clearly do not extend to organolithium reagents. It is, therefore, perhaps not surprising to find that the alkylations of $(CH_3)_3SnNa$ and $(CH_3)_3SnK$ exhibit reaction profiles which are similar to each other but distinctively different from that observed for $(CH_3)_3SnLi$. For example, as previously noted, the reaction of $(CH_3)_3SnNa$, -K with cis- and trans-4-*tert*-butylcyclohexyl bromide in THF yields essentially eq.uivalent product isomer distributions containing appreciably less of the trans product isomer than is produced in the eq.uivalent alkylation employing $(CH_3)_3SnLi$. This difference is even more pronounced in the reaction of $(CH_3)_3SnNa$, -K with cis- and trans-4-*tert*-butylcyclohexyl bromide in DME in which case an essentially statistical distribution of product diastereomers occurs. In light of earlier discussion, it follows that these differences reflect differences in the steric factors that control the stereochemistry of C-Sn bond formation, and which presumably occur as a conseq.uence of the differing influences that solvent, concentration, and temperature have on the structure of the reaction intermediates generated from $(CH_3)_3SnNa$, -K vs. those generated from $(CH_3)_3SnLi$.

Finally, we note that recent years have witnessed various attempts to develop non-stereochemical probes as aids in distinguishing between competing odd- and even-electron processes. The cyclization of the 5-hexyl radical is a well-documented example of one such process. The rate constant for this cyclization is ca. 1.07 x 10^5 s^{-1} at 25°C[7], while the cyclization of the corresponding anion is significantly slower.

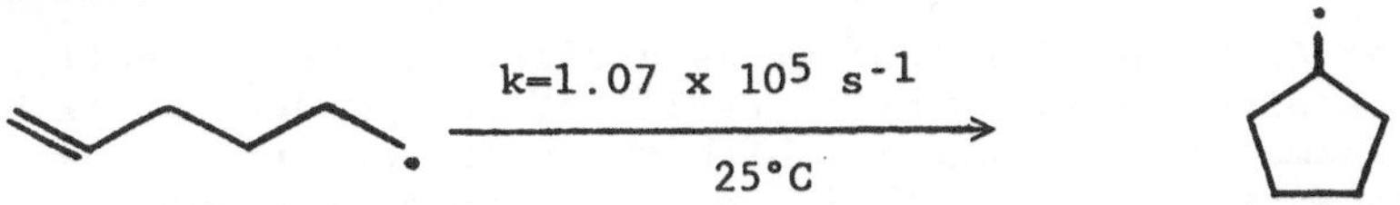

1 2 (1)

As a conseq.uence of this behavior, the 5-hexenyl system has become a standard probe, diagnostic of the intermediacy of (kinetically) free alkyl radicals.[8] The related 1-methyl-5-hexenyl system, 3, has been less studied; however, the rate of cyclization of 3 is known to be essentially the same as that observed for 1, i.e. k ≈ 1 x 10^5 s^{-1}.[7] It follows from a consideration of the magnitude of these rate

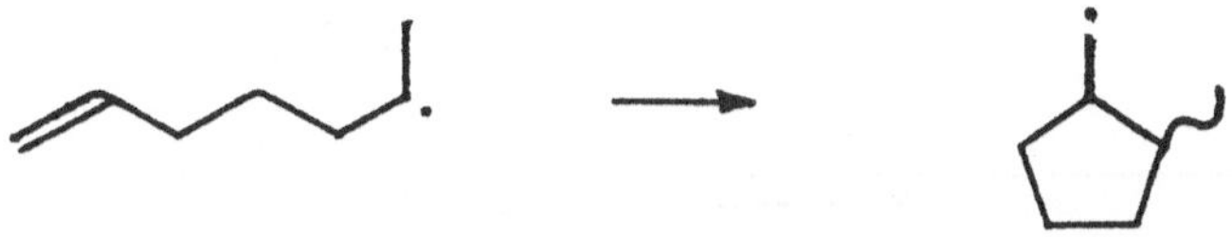

3 4 (6)

constants that the rearrangement of 1 and 3 is ≥10^4 times too slow to compete with geminate coupling processes which are generally recognized to proceed with effective rate constants $k_{effective} \geq_{\psi} 10^{10}$ s^{-1}.[9] Indeed, the virtue of both the 5-hexenyl and the 1-methyl-5-hexenyl systems as diagnostic probes in the elucidation of reaction mechanism rests squarely on the understanding that their characteristic rearrangement occurs during the time that the parent radical is kinetically free. It follows, therefore, that under eq.uivalent circumstances, the extent to which the rearrangement of 3 --> 4 takes place, can (at most) eq.ual but never exceed the extent of racemization associated with the eq.uivalent reaction involving a radical generated from a chiral precursor. We were surprised, therefore, to note that Ashby and DePriest[10], after examining the reaction of $(CH_3)_3SnNa$ with optically active 2-bromooctane and observing that this alkylation proceeds with overall net (~50% in their hands) *inversion* of configuration, went on to propose that this reaction proceeds by a carbon-tin bond-forming step involving a stereospecific reaction between $(CH_3)_3Sn\cdot$ and intermediate 2-octyl radicals, all with the same solvent cage. These authors based their conclusion on the fact that the corresponding reaction of $(CH_3)_3SnNa$ with 2-bromo-6-heptene, resulted (in their hands) in a 71% (absolute) yield of cyclized product (5), which they concluded could only arise through the general intermediacy of 1-methyl-5-hexenyl radicals.

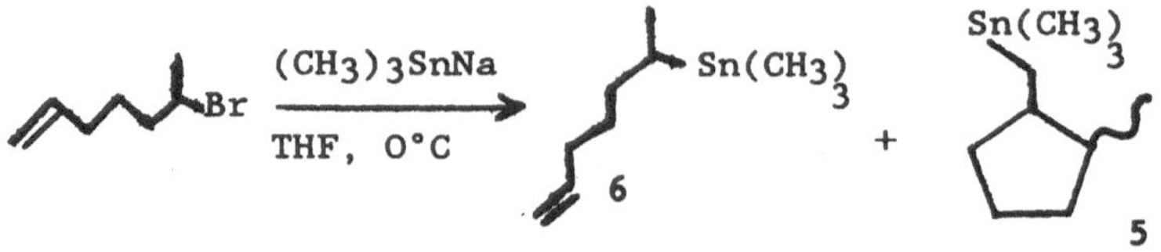

(7)

The proposal of Ashby and DePriest poses a dilemma. Specifically, they require that a significant fraction (≥33%) of the production of **5** must occur in the same solvent cage in which the electron-transfer step takes place.[11] However, since 1-methyl-5-hexenyl radicals do not have time to cyclize under such conditions, cyclization under these circumstances cannot be invoked as *prima facie* evidence for the intermediacy of 1-methyl-5-hexenyl radicals.

Table IV. Reactions of 2-Halohept-6-ene with $(CH_3)_3SnNa$ and Ph_3MLi M = C,Si,Ge,Sn) in THF

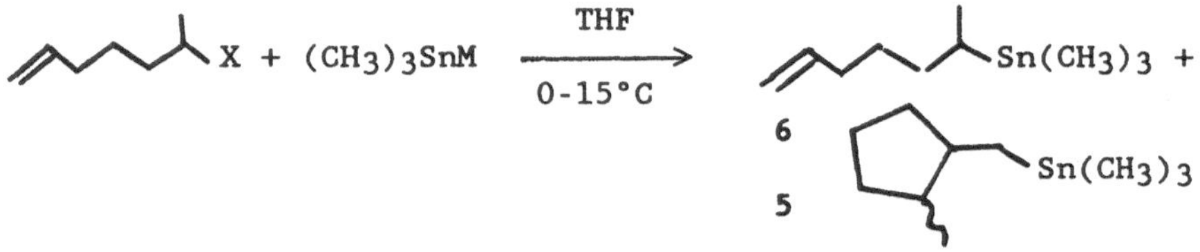

M	X	Additive	Order of Addition	yield,[a] % 6	5	cis/trans	Ref
Na	Cl		inverse	55(62)	33(38)	1.5	Ashby
Na	Br		normal	4(12)	71(88)	3.7	Ashby
Na	Br		inverse	11(13)	72(87)	3.2	Ashby
Na	Br	DCPH[b]	inverse	1(7)	14(93)	3.7	Ashby
Na	Br		normal	(6)	(94)		This work
Na	Br		inverse	(32)	(68)		This work
Li	Br		normal	12(21)	48(79)	2.8	Kitching

[a]Absolute yields; relative yields are in parentheses.
[b]Dicyclohexylphosphine.

Further evidence that 5 does not constitute *prima facie* evidence for the intermediacy of 1-methyl-5-hexenyl radicals in eq. 7 is recognized in the caveats posed by two additional observations: (i) the finding that the closely related rearrangement of **1** ⟶ **2** does not occur during the corresponding reaction of 1-bromohex-5-ene[12,13] (more sensitive probes[4,14] suggest that ca. 20% of the carbon-tin coupling product afforded by the reaction of $(CH_3)_3SnM$ with *primary* bromides arises via a pathway consistent with a free radical process) and (ii) the finding that the extent of rearrangement observed during the reaction of both Ph_3SiM and Ph_3GeM with 2-bromo-6-heptene exceeds the extent of racemization associated with the corresponding reaction between these same reagents and optically active 2-bromooctane (Table V).

Table V. A Comparison of the Reactions of 2-Bromohept-6-ene and (S)-2-Octyl Bromide with Ph_3MLi (M = C,Si,Ge,Sn).

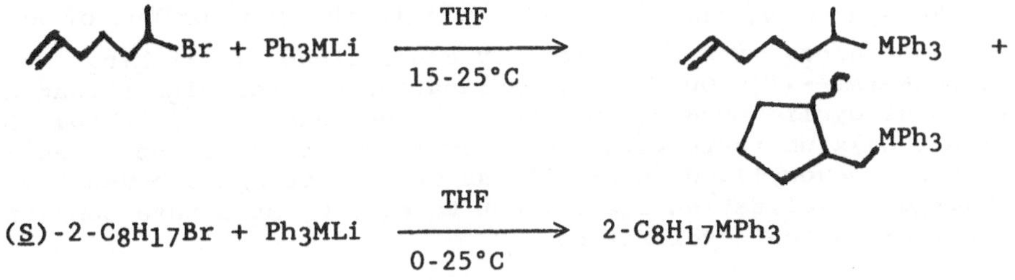

M	Order of Addition	Rel. yield, % 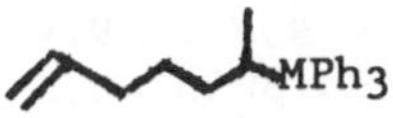	Rel. yield, % 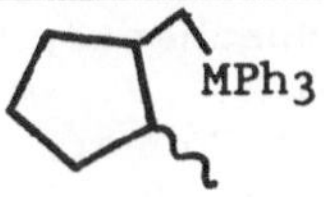	ee %
C	normal	>99	<1	94
	inverse	>99	<1	97
Si	normal	11	89	24
	inverse	37	63	50
Ge	normal	66	34	85
	inverse	84	16	99
Sn	normal	>99	<1	98

[a]Relative yields.

The reaction of $(CH_3)_3SnNa$ with 2-halo-6-heptene exhibits other qualities that are inconsistent with the intermediacy of 1-methyl-5-hexenyl radicals. Thus, prior studies[15] have established that the *cis*--trans ratio of the cyclic product produced from 3 is invariant: *cis*/*trans* = 2.3. Moreover, this value is, as it should be, independent of solvent, concentration, additives, and leaving group. By contrast, the *cis*/*trans* ratios seen in Table IV range between 1.5 and 3.7 and, in addition, are substantially influenced by the presence of the additive dicyclohexylphosphine and the nature of the leaving group.

In a follow-up report, Ashby and co-workers[16] acknowledged the inconsistencies caused by their interpretation, but dismissed these by imposing a series of unwarranted assumptions (e.g. "Since the tetralkyltin compounds containing the cyclized moeity devised from the alkyl halide probes are most reasonably attributed to the cyclization of intermediate radicals, the percentage of cyclized tetralkyltin compound can be assumed to indicate the minimal extent of reaction proceeding with radical...") and unconvincing arguments involving selective behavior (e.g. "If the 2-heptenyl radical escapes the solvent cage more readily than the 2-octyl radical, one could rationalize a high degree of cyclization in the 2-heptenyl system whereas predominant inversion of configuration of the 2-octyl system is explained by [backside] substitution of $(CH_3)_3Sn\cdot$ on RX within the [initial] solvent cage.") Unexplained, too, were the major differences between reaction product ratios cited in their preliminary report[10] and those presented in this later publication.

Collectively, these findings vitiate the conclusions of Ashby and DePriest along with the corollary conclusions of Kitching, Kuivila, and co-workers.[12,17] One likely explanation for the higher than expected yield of cyclic products produced in the reaction of 2-bromo-6-heptene with $(CH_3)_3SnM$ rests with the observation of Garst and Hines[18] that (1-methyl-5-hexenyl)sodium as well as other 1-methyl-5-hexenyl metallics undergo a cyclization eq.uivalent to eq.. 6, at a rate competitive with the free-radical cyclization.

REFERENCES

1. Chern, C.-I.; DiCosimo, R.; DeJesus, R.; and San Filippo, J. Jr., *J. Am. Chem. Soc.*, **1978**, *100*, 7317.
2. Bock, P.L.; Boschetto, D.J.; Rasmussen, J.R.; Demers, J.P.; and Whitesides, G.M., *J. Am. Chem. Soc.*, **1974**, *96*, 2814.
3. Bradley, J.S.; Connor, D.E.; Dolphin, D.E.; Labinger, J.A.; and Osborn, J.A., *J. Am. Chem. Soc.*, **1972**, *94*, 4043.
4. San Filippo, J. Jr.; Silbermann, J.; and Fagan, P.JU., *J. Am. Chem. Soc.*, **1978**, *100*, 4834.
5. (a) Greene, F.D.; Berwick, M.A.; Stowell, J.C., *J. Am. Chem. Soc.*, **1970**, *92*, 867. Kopecky, K.R.; Gillan, T., *Can. J. Chem. Soc.*, **1969**, *47*, 2371. (b) It is, of course, well known that the degree of association of organolithium reagents is strongly solvent dependent: the presence of basic solvents in an organolithium reagent solution generally predisposes the aggregate toward dissociation into smaller and presumably better solvated fragments (cf. Panek, E.J. and Whitesides, G.M., *J. Am. Chem. Soc.*, **1972**, *94*, 8768).
6. Jensen, F.R. and Rodgers, J.E., *J. Am. Chem. Soc.*, **1968**, *90*, 5793.
7. Cf. Griller, D. and Ingold, K.V., *Acc. Chem. Res.*, **1980**, *13*, 317.
8. (a) Garst, J.F. and Smith, C.D., *J. Am. Chem. Soc.*, **1976**, *98*, 1520 and references therein. (b) Wilt, J.W. in "Free Radicals"; Kochi, J.K., ed.; Wiley-Interscience: New York, 1973; Vol. 1, Chapter 8.
9. Noyes, R.M., *Prog. React. Kinet.*, **1969**, *1*, 129.
10. Ashby, E.C. and DePriest, R., *J. Am. Chem. Soc.*, **1982**, *104*, 6144.
11. The normalized yields of **5** and **6** (Table I) are, respectively, 88% and 12% (see ref 10). Since ~50% of the <u>total</u> yield of coupling product (i.e. **5** + **6**) must occur (according to ref 10) within the same solvent cage as the rate-limiting electron-transfer step, it follows (assuming all or less than all of trimethyl (2-octyl) tin is also derived in this initial solvent cage) that a *minimum* of (50-12)/88 or 33% of trimethyl (2-octyl) tin must also be formed in this initial solvent cage.
12. Kitching, W.; Olszowy, H.A.; and Harvey, K., *J. Org. Chem.*, **1982**, *47*, 1893.
13. Newcomb, M. and Courtney, A.R., *J. Org. Chem.*, **1980**, *45*, 1707.

14. Bock, P.L. and Whitesides, G.M., *J. Am. Chem. Soc.*, **1974**, *96*, 2826.
15. Beckwith, A.L.J.; Blair, I. and Philipou, G., *J. Am. Chem. Soc.*, **1974**, *96*, 1613 and references therein.
16. Ashby, E.C.; Su, W.-Y.; and Pham, T.N., *Organomet.*, **1985**, *4*, 1493.
17. Kuivila, H.G. and Alnajjar, M.S., *J. Am. Chem. Soc.*, **1982**, *104*, 6146.
18. Garst, J.F. and Hines, J.B., Jr., *J. Am. Chem. Soc.*, **1984**, *106*, 6443.

TRAPPING INTERMEDIATES IN ELECTRON TRANSFER REACTIONS OF SQUARE PLANAR ORGANOMETALLIC d-8 SPECIES

Gerard van Koten, David M. Grove, and Adolphus A.H. van der Zeijden
Laboratory of Organic Chemistry
Department of Metal-Mediated Synthesis
University of Utrecht
Padualaan 8
3584 CH Utrecht
The Netherlands

ABSTRACT. Chelating ligands are of a profound benefit in stabilizing paramagnetic or other transient organometallic species. This is illustrated by a series of examples in which *ortho*- and bis-*ortho* chelated, square-planar arylmetal- d^8 Ni^{II}, Pd^{II}, Pt^{II}, Rh^{I} and Ir^{I} compounds have been reacted with electrophiles H_2, MeI, halogens and Me_2SnBr_2.

1. INTRODUCTION

One of the most fundamental reactions in inorganic and organometallic chemistry is oxidative addition. It is a conceptually simple process in which a reagent, XY, reacts with a metal centre, M^{n+}, to form a M^{n+2} species with M-X and M-Y σ-bonds. This process and its reverse (reductive elimination), which have been much studied, particularly with metal d^8 complexes, are of fundamental importance in industrially employed catalytic systems.[1] However, there is still considerable discussion regarding the question of whether, in a given system, oxidative addition occurs as a concerted two-electron transfer process involving a three centre M(XY) transition state or whether it involves two separate (nonconcerted) one-electron transfers. In the latter case, the involvement of radical species is expected, and with metal d^8 substrates this should result in concomitant formation of paramagnetic metal species.[1,2]
Since the direct detection of short-lived metal radical species by ESR can be difficult, other ways of studying electron transfer reactions are more common. One approach, that can be quite successful, is to react electrophiles with a metal complex of known geometry and then to identify the mechanism by characterization of the product(s) together with interpretation of kinetic and spectroscopic evidence from the intervening step(s). Clearly, the identification or isolation of metal-containing intermediates would aid such studies significantly, but in most systems this is not feasible.

2. COMPLEXES OF ORTHO CHELATED ARYL LIGANDS

In our oxidative addition studies, we have found that specific chelating ligands afford metal complex substrates that are ideally suited for investigating intermediate species,

M. Chanon et al. (eds.), Paramagnetic Organometallic Species in Activation / Selectivity, Catalysis, 477–487.

which in a number of cases have been isolated and fully characterized. The present 'paper' is a short overview of some of our studies in this area using square-planar d^8 Ni^{II}, Pd^{II}, Pt^{II}, Rh^{I} and Ir^{I} substrates, and the unique organometallic species that we encountered. Research in our laboratory has been focussed on the chemistry of monoanionic aryl ligands having *ortho*- or bis-*ortho* dimethylaminomethyl substituents (*i.e.* $[C_6H_4(CH_2NMe_2)$-2] (A) and $[C_6H_3(CH_2NMe_2)_2$-2,6] (B), see Fig 1), that form organometallic compounds with a direct M-C_{aryl} σ-bond. Through coordination of the N donor site(s) of the *ortho* CH_2NMe_2 arm(s) five-membered M-C-C-C-N chelate ring

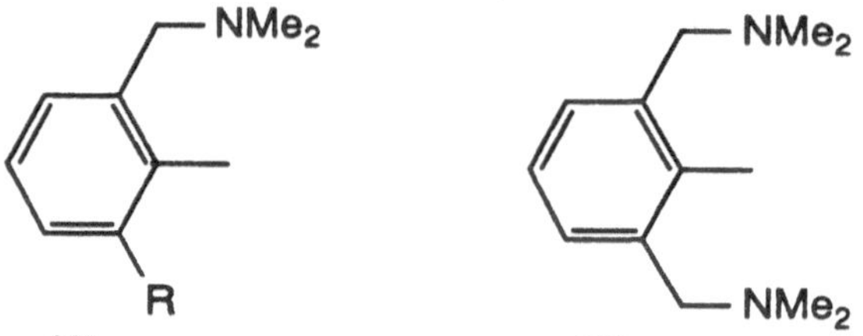

(A) (B)

Fig 1. The ortho chelating ligands A with R is H or Me and B abbreviated as NCN.

formation occurs. The use of a C-bonded aryl ring as a ligand prevents M-C bond breakage by α−elimination. If the aryl ligand is in addition bis-*ortho* substituted, decomposition via β-elimination pathways, although not very likely, is also prevented. Intramolecular coordination furthermore hampers other M-C bond breaking processes such as homolytic and heterolytic fission. In many reactions the rigid chelating coordination of the aryl ligand provides the metal centre with a set of unusual properties. Firstly, the chelating ligand restricts the available sites for incoming reagents and other ligands. This is a situation that, as we will see, can frustrate the normal course of an oxidative reaction and lead to "trapped" intermediates having considerable stability. Secondly, the hard C and N donor atom(s) increase the nucleophilicity of the metal centre. This destabilizes the d_{z^2} orbital that lies perpendicular to the coordination plane in square planar metal d^8 complexes and so makes them susceptible to attack by a variety of electrophiles - with startling results.

3. HALOGENS AS ELECTROPHILES

3.1. Nickel complexes

The square planar $[Ni\{C_6H_3(CH_2NMe_2)_2\text{-}2,6\}I]$ complex reacts with I_2 in benzene at room temperature, in a one-electron transfer, to afford a novel true organometallic Ni^{III} complex $[NiI_2\{C_6H_3(CH_2NMe_2)_2\text{-}2,6\}]$, see eqn. (1).[3] An X-ray structure of this iodo-complex, Fig 2a,

$$Ni^{II}(NCN)I \quad + \quad 1/2\, I_2 \quad \text{-------------->} \quad Ni^{III}(NCN)I_2 \qquad (1)$$

yellow-orange black

shows a square pyramidal metal coordination sphere (N,C,N-ligating atoms and one I atom defining the basal plane and the other I atom sited at the apex) that leaves no doubt as

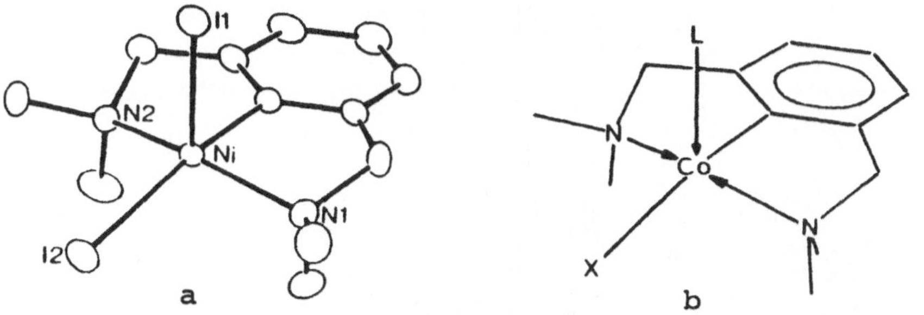

Fig 2. **a**, Molecular structure of $[NiI_2\{C_6H_3(CH_2NMe_2)_2\text{-}2,6\}]$; Ni-I1 2.613(1), Ni-I2 2.627(1), Ni-C 1.898(5) Å, I1-Ni-I2 103.0(1)°; **b**, Proposed structure of $[Co^{II}X\{C_6H_3(CH_2NMe_2)_2\text{-}2,6\}L]$.

to its nature. The ESR spectrum of this air-stable species and those of the bromo and chloro analogues (that are also easily prepared) is consistent with a d^7 electronic configuration having a primarily metal-centred unpaired electron that is localized in the d_{z^2} orbital , Fig 3a.[3]
The presence of a kinetically stable (aryl)C–Ni bond in these complexes is supporting evidence for the possibility of species with Ni^{III}–C σ-bonds being intermediates in the oxidative addition of aryl or alkyl halides to Ni^0-substrates. Interestingly the reaction of $BrC_6H_3(CH_2NMe_2)_2\text{-}2,6$ with $[Ni^0(COD)_2]$, that is the preferred route for the synthesis of $[Ni^{II}Br\{C_6H_3(CH_2NMe_2)_2\text{-}2,6\}]$, does yield small amounts of $[Ni^{III}Br_2\{C_6H_3(CH_2NMe_2)_2\text{-}2,6\}]$. This reaction is thus likely to have an electron transfer mechanism in which the following pathway dominates:[4]

$$Ni^0 + NCNBr \text{ -----> } Ni^I + NCNBr^{-\cdot} \text{ -----> } [Ni^{II}Br(NCN)] \quad (2)$$

Side reactions that then generate both Ni^{II} and Ni^{III} products can be envisaged:

$$NCNBr^{-\cdot} + Ni^0 \text{ -----------> } [Ni^I(NCN)] + Br^- \quad (3)$$

$$[Ni^I(NCN)] + NCNBr \text{ -----------> } [Ni^{II}Br(NCN)] + 1/2\,(NCN)_2 \quad (4)$$

$$Ni^I + NCNBr \longrightarrow [Ni^{III}Br(NCN)]^+ \xrightarrow{Br^-} [Ni^{III}Br_2(NCN)] \quad (5)$$

It is due to the special stabilizing properties of the NCN ligand that the otherwise transient Ni^{III} intermediates formed in reactions of $[Ni^0(COD)_2]$ with aryl halides are trapped and then even become isolable.
Other paramagnetic NCN-metal complexes, that recently emerged from our studies, are $[Fe^{II}Cl_2(NCN)]$ (SP, high spin)[5] and $[Co^{II}X(NCN)L]$ (see Fig 2b, L= *e.g.* pyridine, SP, low spin). The ESR spectrum of the latter species, Fig. 3b, shows nicely the hyperfine coupling with the Co^{59} nucleus and the superhyperfine coupling with the apicaly positioned pyridine-N^{14} nucleus.[6]

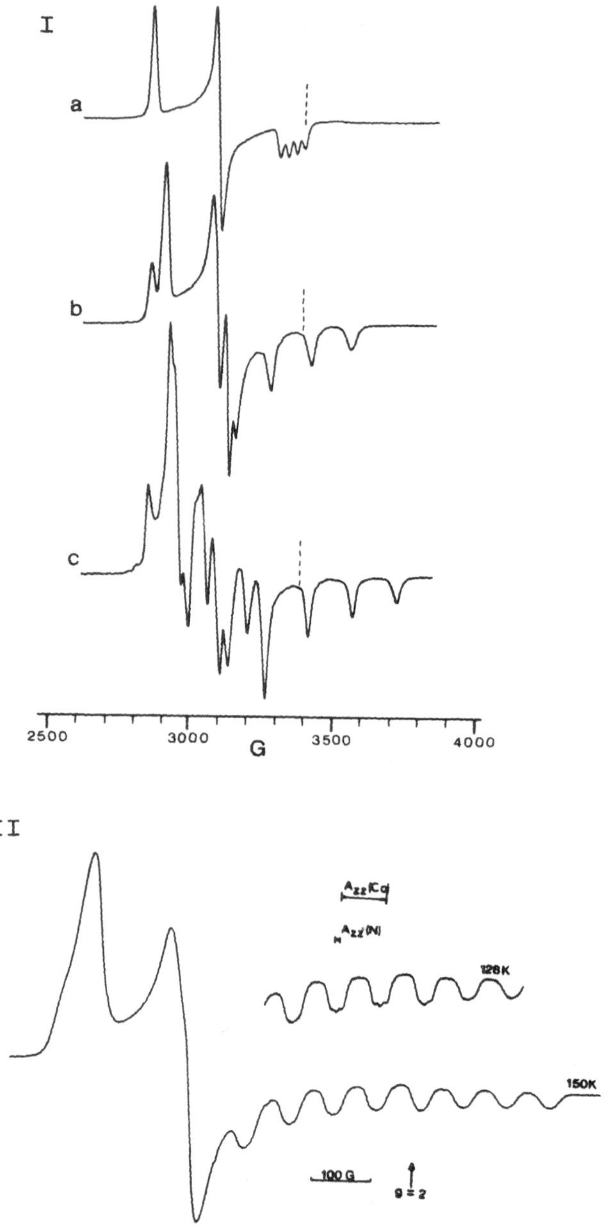

Fig 3. **I**, 9.50 GHz X-band ESR spectra of [$Ni^{III}X_2$(NCN)] species in toluene glasses at ≈150 K. g = 2.002 indicated by -------, (a) X = Cl (I = 3/2), (b) X = Br (I = 3/2), (c) X = I (I = 5/2); **II**, ESR spectrum of[$Co^{II}Cl\{C_6H_3(CH_2NMe_2)_2$-2,6}-Py] in toluene glass at ≈150 K.

3.2. Palladium Complexes

Analagous [PdI{$C_6H_3(CH_2NMe_2)_2$-2,6}] reacts with I_2 to form a dark material with a stoichiometry of [Pd{$C_6H_3(CH_2NMe_2)_2$-2,6}I_5}].[7] A crystallographic study showed discrete square planar [PdI{$C_6H_3(CH_2NMe_2)_2$-2,6}] units having Pd^{II} centres, with I_2 atoms forming weak interactions with each other and with the coordinated I atom of the complex, see Fig. 4. That interaction with the palladium centre from a direction perpendicular to the coordination plane is absent, is indicated by the opposite puckering of the two five-membered chelate rings in the [PdI{$C_6H_3(CH_2NMe_2)_2$-2,6}] unit. A clear example of zero electron transfer!

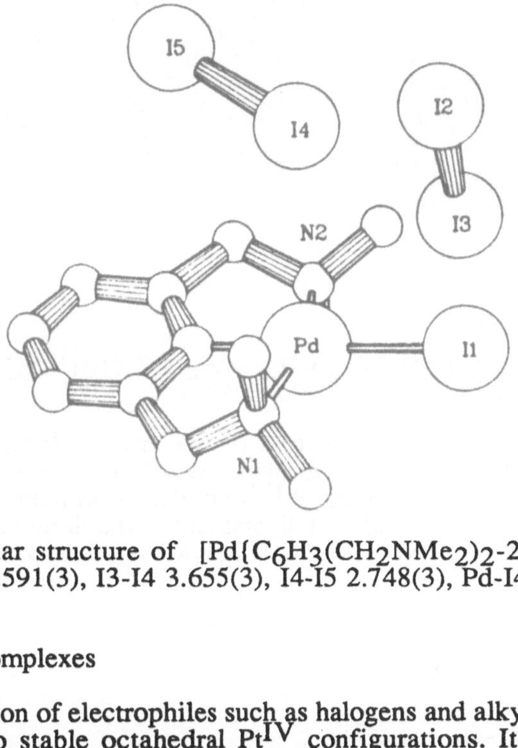

Fig 4. Molecular structure of [Pd{$C_6H_3(CH_2NMe_2)_2$-2,6}I_5}]; Pd-I1 2.7771(2), I1-I3 3.591(3), I3-I4 3.655(3), I4-I5 2.748(3), Pd-I4 3.992(2) Å.

3.3. Platinum complexes

Oxidative addition of electrophiles such as halogens and alkyl halides to Pt^{II} complexes usually leads to stable octahedral Pt^{IV} configurations. It was therefore not totally unexpected that [PtX{ $C_6H_3(CH_2NMe_2)_2$-2,6}] (X = Cl, Br) reacted with appropriate reagents X_2 or CuX_2 to afford [PtX_3{$C_6H_3(CH_2NMe_2)_2$-2,6}] which had properties consistent with an octahedral Pt^{IV} species.[8] Transfer of two electrons from Pt^{II} to the electrophilic halogen had been accomplished - though how was not known. In contrast to Cl_2 and Br_2, the reaction of I_2 with [PtI{$C_6H_3(CH_2NMe_2)_2$-2,6}] gave a dark purple diamagnetic complex of stoichiometry [PtI_3{$C_6H_3(CH_2NMe_2)_2$-2,6}] that had spectroscopic features which were not consistent with a Pt^{IV} configuration, and yet were totally unlike those of the Pt^{II} starting material.[9] The X-ray crystal structure, Fig 5, can only be described as unexpected. The [PtI{$C_6H_3(CH_2NMe_2)_2$-2,6}] substrate is caught in the intimate first step of its interaction with an intact I_2 molecule which has approached end-on along the z-axis and the platinum center attains a square pyramidal (5-coordinate) geometry. This platinum species can be truly called a "trapped" intermediate; there is a

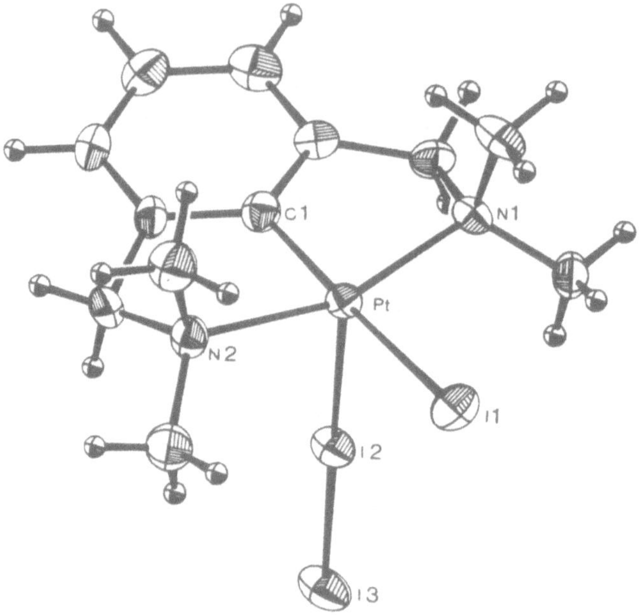

Fig 5. Molecular structure of $[PtI\{C_6H_3(CH_2NMe_2)_2\text{-}2,6\}(\eta^1\text{-} I_2)]$; Pt-I2 2.895(1), I2-I3 2.822(1) Å, Pt-I2-I3 179.43(4)°.

strong interaction between substrate and reagent (between the filled d_z2 orbital and an unoccupied σ^*-orbital on I_2) yet the path to oxidative addition by electron transfer is blocked. This failure to form an octahedral Pt^{IV} complex is probably due to the steric requirements of the $C_6H_3(CH_2NMe_2)_2$-2,6 ligand and of the three I atoms. However, reaction of I_2 with $[Pt\{C_6H_3(CH_2NMe_2)_2\text{-}2,6\}(C_6H_4Me\text{-}4)]$ does yield a Pt^{IV} species *cis*-$[PtI_2\{C_6H_3(CH_2NMe_2)_2\text{-}2,6\}(C_6H_4Me\text{-}4)]$, Fig 6.[9,10] In this product oxidative

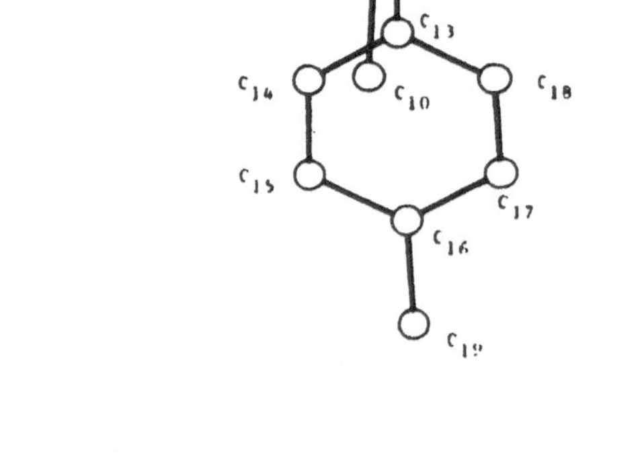

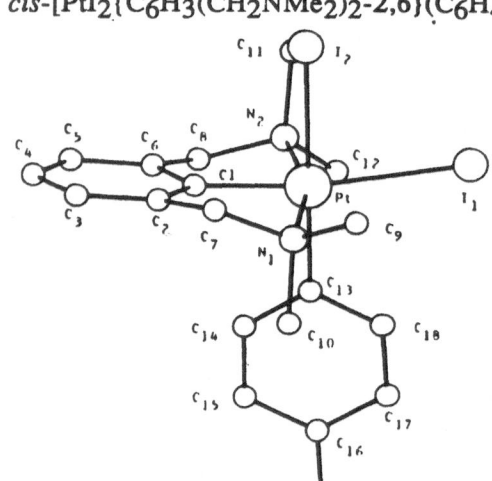

Fig 6. Molecular structure of $[PtI_2\{C_6H_3(CH_2NMe_2)_2\text{-}2,6\}(C_6H_4Me\text{-}4)]$.

addition has resulted in a mutually *cis* arrangement of the two Pt-I bonds with the fairly narrow flat *p*-tolyl group removed from the original $Pt(C_6H_3(CH_2NMe_2)_2$-2,6) coordination plane and now lying unable to rotate, between the arms of the $C_6H_3(CH_2NMe_2)_2$-2,6 ligand *trans* to an I atom. This result is most likely the consequence of a concerted two-electron transfer from Pt^{II} to I_2.

4. METHYL IODIDE AND MeOTf AS ELECTROPHILES.

Different aspects of the mechanism(s) are highlighted when MeY reagents are used in these oxidative reactions instead of halogens. For example, the reaction of MeOTf with $[Pt\{C_6H_3(CH_2NMe_2)_2$-2,6$\}X]$ (X = Cl, Br, I) provides the unusual cationic arenonium species [PtX(NCN-Me)]OTf and the corresponding reaction of MeI with $[Pt\{C_6H_3(CH_2NMe_2)_2$-2,6$\}(C_6H_4Me$-2)] affords related [Pt(NCN-Me)(C_6H_4Me-2)]$^+I^-$, Fig 7.[11] In both cases a methyl group, *i.e.* Me^+, has become bonded to the aryl carbon of $C_6H_3(CH_2NMe_2)_2$-2,6 that is σ–bonded to platinum and as a consequence the aryl ring has acquired a positive charge and can be considered as a metal-substituted Wheland intermediate.[12] Based on the results with I_2 one can envisage

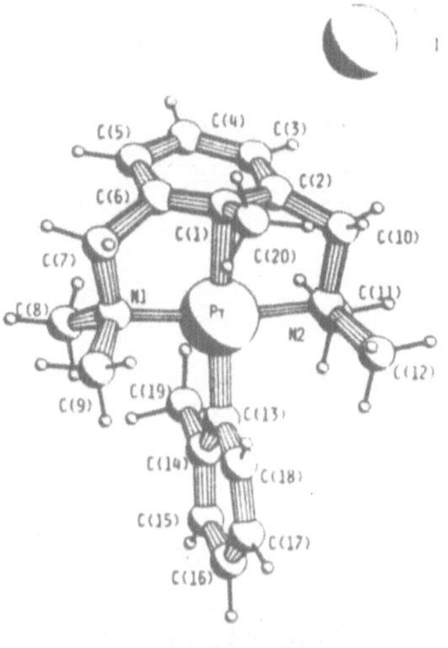

Fig 7. Molecular structure of [Pt(NCN-Me)(C_6H_4Me-2)]$^+I^-$.

that reactions of $Pt\{C_6H_3(CH_2NMe_2)_2$-2,6$\}$ substrates involve the d_{z^2} orbital functioning as a nucleophile to the carbon atom of the MeY molecules. An S_N2 reaction occurs providing a postulated five-coordinate platinum intermediate [PtMe($C_6H_3(CH_2NMe_2)_2$-2,6)(aryl/X)]$^+Y^-$. The latter is not stable and the methyl group transfers by a 1,2-shift to the metal-bonded C atom of $C_6H_3(CH_2NMe_2)_2$-2,6 in what appears to be an abortive attempt at reductive elimination (of 1,3-(Me_2NCH_2)-2-MeC_6H_3). However, this organic moiety is firmly held to the metal centre by the strongly bonded CH_2NMe_2 arms and another "intermediate" is trapped. Note that this species can also be seen as an intermediate in the oxidative addition of a C-C bond of a Me substituted aromatic (*e.g.* toluene) to a $[PtL_2(aryl)X]^+$ complex. The reaction of MeI with $[PtBr\{C_6H_3(CH_2NMe_2)_2$-2,6$\}]$ does not, however, afford such an arenonium product but results in an equilibrium between the starting materials, $[PtI(C_6H_3(CH_2NMe_2)_2\text{-}2,6)]$ and MeBr.[11,12] Although no intermediate(s) could be identified, it is likely that, as with the formation of the aforementioned arenonium complexes, a five coordinate Pt^{IV} cationic species is involved.

5. RHODIUM AND IRIDIUM COMPLEXES

Our studies of oxidative addition reactions using ortho chelated aryl ligands has also included square planar metal d^8 complexes $[M^I\{C_6H_3(CH_2NMe_2)\text{-}2\text{-}R\text{-}6\}(COD)]$ (M = Rh, Ir; R = H, Me or CH_2NMe_2). Thus, $[Rh^I\{C_6H_3(CH_2NMe_2)_2\text{-}2,6\text{-}C,N\}(COD)]$ reacts with MeI to afford square pyramidal $[Rh^{III}(Me)I\{C_6H_3(CH_2NMe_2)_2\text{-}2,6\}]$, Fig 8.[13] As postulated for related Pt^{II} systems above, the reaction probably proceeds by

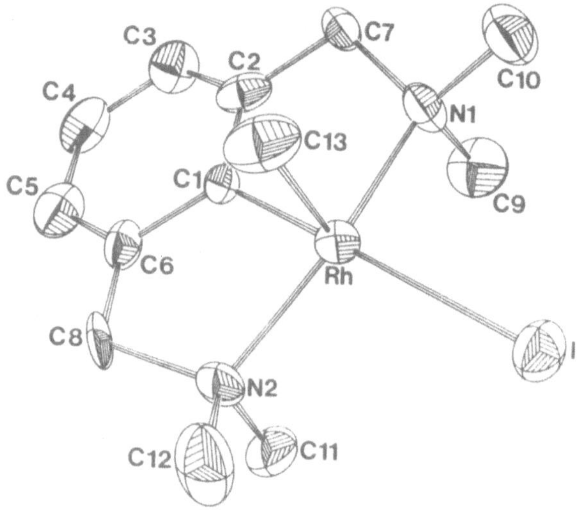

Fig 8. Molecular structure of $[Rh^{III}(Me)I\{C_6H_3(CH_2NMe_2)_2\text{-}2,6\}]$.

means of an S_N2- type mechanism, involving an initial two-electron transfer from the nucleophilic Rh^I centre to the electrophilic carbon atom of MeI. Notably, this oxidative addition is assisted by the initially uncoordinated CH_2NMe_2 group, that promotes electron transfer to afford a Rh^{III} centre. The Rh^{III} complex is inert towards subsequent reductive elimination, in contrast to the reaction products of the isoelectronic Pt^{II} complexes (*vide supra*). In fact, the square pyramidal Rh^{III} complex is the (stable) group 9 analogue of the isostructural (unstable) group 10 cationic intermediate $[Pt^{IV}(Me)X(C_6H_3(CH_2NMe_2)_2\text{-}2,6)]^+$.

6. METAL COMPLEXES AS ELECTROPHILES

The nucleophilic centres of the square metal d^8 complexes we have used containing ortho chelated ligands are also reactive towards various electrophilic metal compounds. Here two types of electron transfer, namely a transmetallation or a redox reaction, can occur. In these systems the ability to "trap" intermediates provides detailed information on the mechanistic pathways of these important reactions. For example, the reaction of *cis* -[$Pt^{II}\{C_6H_4(CH_2NMe_2)\text{-}2\}_2$] with $Hg(OAc)_2$ results in the formation of a stable dinuclear platinum-mercury adduct, Fig 9.[14] This adduct, with its square pyramidal

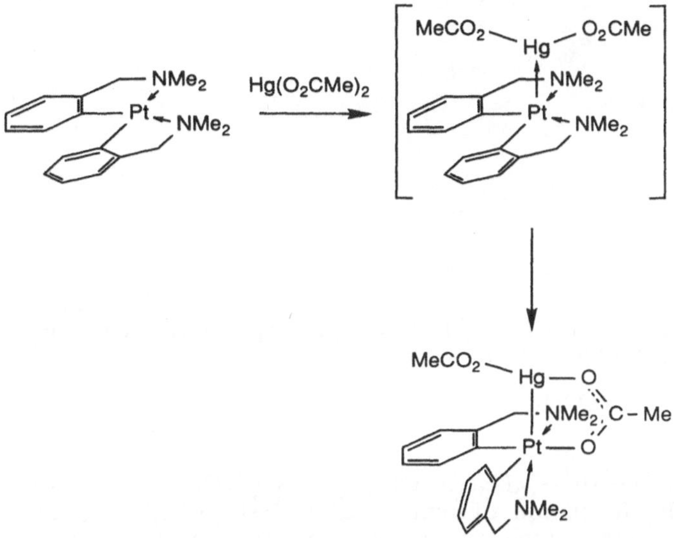

Fig 9. Formation and molecular structure of [$Pt\{C_6H_3(CH_2NMe_2)\text{-}2\}_2$ -(μ-OAc)HgOAc].

configuration around platinum, is probably formed by a donative action of the filled d_{z^2} orbital to the electrophilic mercury centre. The final product can be regarded as a "trapped" intermediate since it could potentially react further, either by a 1,2-shift of the Hg unit towards C(aryl), eventually leading to transmetallation, or by a two-electron transfer from platinum to mercury leading to a redox reaction. Although this adduct is stable towards both these reaction routes, *trans* -[$Pt^{II}\{C_6H_4(CH_2NMe_2)\text{-}2\}_2$], in contrast, does react further with $Hg(OAc)_2$, *via* the latter route, affording [$Pt^{IV}(OAc)_2\{C_6H_4(CH_2NMe_2)\text{-}2\}_2$] and Hg^0.[14]
The group 9 analogues [$M^I\{C_6H_3(CH_2NMe_2)\text{-}2\text{-}R\text{-}6\}(COD)$] (M = Rh, Ir; R = H, CH_2NMe_2) show both transmetallation and redox ractions with electrophilic metal compounds. Thus, [$Rh^I\{C_6H_3(CH_2NMe_2)_2\text{-}2,6\}(COD)$] reacts with AgOAc to yield [$Rh^{III}(OAc)_2\{C_6H_3(CH_2NMe_2)_2\text{-}2,6\}(H_2O)$], and with $SnMe_2Br_2$ to yield [$RhBr(COD)]_2$ and the very interesting five-coordinate ionic complex [$Sn(Me)_2\{C_6H_3(CH_2NMe_2)_2\text{-}2,6\}]^+Br^-$.[15] In another case, *i.e.* the reaction of [$Ir\{C_6H_4(CH_2NMe_2)\text{-}2\}(COD)$] with SnX_2RR', an unusual dinuclear 1:1 adduct was isolated, Fig 10.[16] This adduct, in which a tin-halide bond acts as a side-on coordinated ligand to Ir, can be regarded as a 'trapped' three-centre intermediate in the oxidative addition of a metal-halide bond to a metal d^8 centre. True oxidative addition in this complex (*i.e.* breakage of the Sn-X bond in the Ir-Sn-X triangle) is primarily prevented by steric crowding.

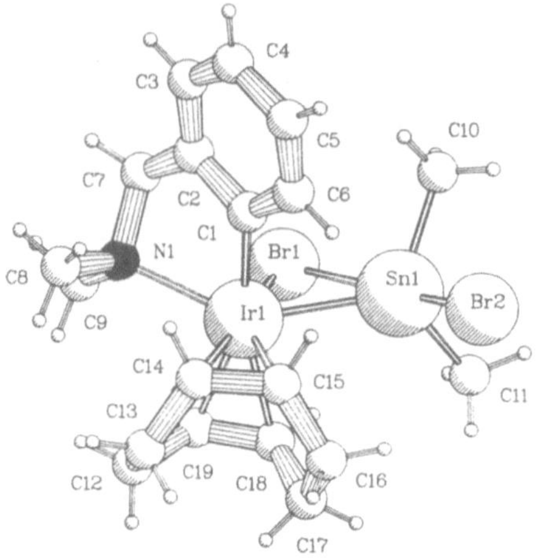

Fig 10. Molecular structure of [Ir{$C_6H_4(CH_2NMe_2)$-2}(COD)Br$SnMe_2$Br]; Ir-Sn 2.6297(8), Ir-Br1 2.582(1), Sn-Br1 2.934(1), Sn-Br2 2.608(1) Å.

7. HYDROGEN AS ELECTROPHILE

The reaction of square planar [Ir^I{$C_6H_3(CH_2NMe_2)$-2-R-6}(COD)] with dihydrogen results at -20 °C in the formation of octahedral [$Ir^{III}H_2${$C_6H_3(CH_2NMe_2)$-2-R-6}(COD)], Fig 11.[17] Despite the fact that two isomers are possible for these Ir^{III} dihydrides, there is exclusive formation of only one of them (governed by thermodynamic (steric) rather than by kinetic factors). They are rare examples of species containing both M-H, M-C and M-olefin bonds on one metal centre. Considering that the hydride atoms are in positions *cis* to either the Ir-C(aryl) or an Ir-olefin bond their stability (at -20°C) is remarkable. Both C-H reductive elimination or olefin insertion reactions, are probably hampered by the fact that both the C(aryl) atom and the olefin functions around the Ir centre are part of bidentate ligands. On raising the temperature of solutions of the Ir^{III} dihydride where R is H from -20 to 20 °C a reversible loss of dihydrogen and reformation of the Ir^I precursors occurs.

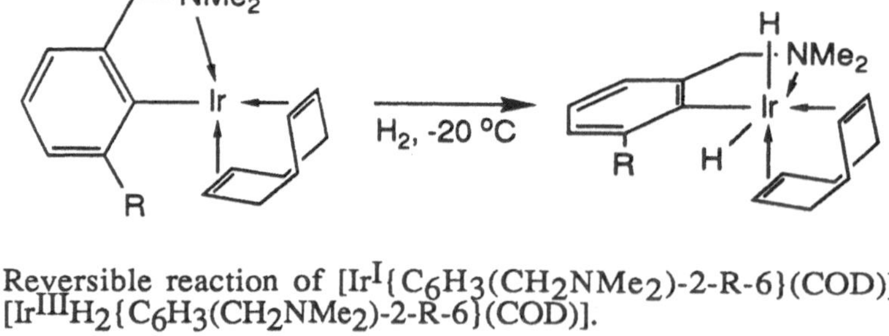

Fig 11. Reversible reaction of [Ir^I{$C_6H_3(CH_2NMe_2)$-2-R-6}(COD)] with H_2 to [$Ir^{III}H_2${$C_6H_3(CH_2NMe_2)$-2-R-6}(COD)].

ACKNOWLEDGEMENT

Thanks are due to my colleague Prof. dr. K. Vrieze and to the many coworkers who contributed to this work. In addition to my co-authors Drs. D.M. Grove and A.A.H. van der Zeijden, I would like to particularly thank Mr. J.A.M. van Beek, Dr. A.L. Spek, Dr. W.J.J. Smeets, Mr. J.T.B.H. Jastrzebski, Dr. J. Terheijden and Dr. C.H. Stam.

REFERENCES

1. R.S. Dickson, *Homogeneous Catalysis with compounds of Rhodium and Iridium*, Reidel, Dordrecht, Holland, 1985.
2. J.K. Kochi, *Organometallic Mechanisms and Catalysis*, Academic Press, New York, 1978.
3. a. D.M. Grove, G. van Koten, R. Zoet, N.W. Murrall, and A.J. Welch.*J. Am. Chem. Soc.* **1983**, *105*, 1379; b. D.M. Grove, G. van Koten, P. Mul, A.A.H. van der Zeijden, J. Terheijden, M.C. Zoutberg and C.H. Stam. *Organometallics*, **1986**, *5*, 322.
4. D.M. Grove, G. van Koten, P. Mul, R. Zoet, J.G.M. van der Linden, J. Legters, J.E.J. Schmitz, N.W. Murral, and A.J. Welch. *Inorganic Chemistry*, **1988**, 000.
5. A. De Koster, J.A. Kanters, A.L. Spek, A.A.H. van der Zeijden, G. van Koten, and K. Vrieze. *Acta Cryst*. **1985**, *C41*, 893.
6. A.A.H. van der Zeijden, and G. van Koten, *Inorg. Chem.*, **1986**, *25*, 4723.
7. J.A.M. van Beek, G. van Koten, W.J.J. Smeets, and A.L. Spek. *unpublished results*.
8. J. Terheijden, G. van Koten, J.L. de Booys, H.J.C. Ubbels, and C.H. Stam. *Organometallics*, **1983**, *2*, 1882.
9. J.A.M. van Beek, G. van Koten, W.J.J. Smeets, and A.L. Spek.*J. Am. Chem. Soc.* **1986**, *108*, 5010.
10. J.A.M. van Beek, G. van Koten, F. Muller, and C.H. Stam *to be published*.
11. J. Terheijden, G. van Koten, I.C. Vinke, and A.L. Spek. *J. Am. Chem. Soc.* **1985**, *107*, 2891.
12. D.M. Grove, G. van Koten, J.N. Louwen, J.G. Noltes, A.L. Spek, and H.J.C. Ubbels.*J. Am. Chem. Soc.* **1982**, *104*, 6609.
13. A.A.H. van der Zeijden, G. van Koten, J.-M. Ernsting, B. Krijnen, C. H. Stam. *submitted to J. Chem. Soc., Dalton Trans.*
14. A.F.M.J. van der Ploeg, G. van Koten, K. Vrieze, and A.L. Spek. *Inorg. Chem.* **1982**, *21*, 2014.
15. A.A.H. van der Zeijden, G. van Koten, R. A. Nordemann, B. Kojic-Prodic, and A.L. Spek *submitted to Organometallics*.
16. A.A.H. van der Zeijden, G. van Koten, J.M.A. Wouters, W.F.A. Wijsmuller, W.J.J. Smeets, and A.L. Spek *to be published*.
17. A.A.H. van der Zeijden, G. van Koten, R. Luijk, D.M. Grove. *submitted to Organometallics*.

POLAR VERSUS ELECTRON TRANSFER PATHWAY IN THE REACTION OF ALKYL LITHIUM AND ALKYL GRIGNARD REAGENTS WITH MONONITROARENES: FACTORS AFFECTING PRODUCT DISTRIBUTION

Giuseppe Bartoli [a*], Marcella Bosco [b], Renato Dalpozzo [b], Loris Grossi [b]
a.Dipartimento Scienze Chimiche,via S.Agostino 1,62032 Camerino (Mc),Italy.
b.Dipartimento Chimica Organica,viale Risorgimento 4,40136 Bologna,Italy.

ABSTRACT: The reaction of nitroarenes with alkyl lithium and alkyl Grignard reagents can mainly give 1,4 and 1,6-addition products, nitroarene radical anion and alkyl radical decay products. Employing magnesium instead of lithium derivatives and low temperatures, the yields in conjugate addition products increase to the detriment of the amount of nitroarene radical anion; while steric hindrances favour the redox process. These findings can be interpreted on the basis of a single electron transfer (SET) pathway in which conjugate addition derivatives and nitroarene radical anion can be considered as typical in-cage and out-of-cage products respectively. However, there are no conclusive evidences to exclude that the so-called cage products can arise from a polar pathway. Nevertheless, this question can be settled if the existence of a continuos mechanistic spectrum from 'full polar' to 'full SET' pathways is accepted on the basis of the Pross's electron shift theory.

1. INTRODUCTION

Until few years ago, many fundamental aspects of the action of strong basic carbanions (RM) such as lithium or magnesium compounds on nitroarenes were unknown or misinterpreted,[1,2] due to an incomplete or incorrect analysis of a very complex reaction product distribution. It is now established that in the initial phase of the reaction the following products may be formed (see Scheme 1):[3-6]

i) nitroarene radical anion **3** and decay products of radical **4**, originated via a single electron transfer (SET) from RM to nitroarene **1** (redox process);
ii) nitronate adducts **5** and **6** from 1,4 and 1,6-conjugate addition respectively of RM to the nitroarenic system (addition process);
iii) hydroxylamino derivative **9** from reductive addition of RM on nitro group. This reaction very likely proceeds via intermediate **7**. However, hitherto, this hypothesis is not supported by conclusive evidences.[7]

Reaction product distribution depends upon many factors, but, independently from the nature of the metallic counterion and from the solvent and temperature conditions employed, the reaction generally follows path (ii) and / or path (i) [1,4] when the negative charge lies on an sp^3 carbon as in alkyl carbanions. Viceversa when the negative charge occupies an sp^2 orbital as in aryl[3,5,7] and vinyl[8] carbanions the reaction generally follows path (iii).

M. Chanon et al. (eds.), Paramagnetic Organometallic Species in Activation / Selectivity, Catalysis, 489–502.

In this report, we shall discuss only on the reactivity of the alkyl lithium and magnesium compounds, which can give an irreversible attack to the nitroarenic system, to find out the better conditions favouring the conjugate addition product formation over the redox processes.

Scheme 1. General picture of the reaction between nitroarenes and strong basic carbanions .

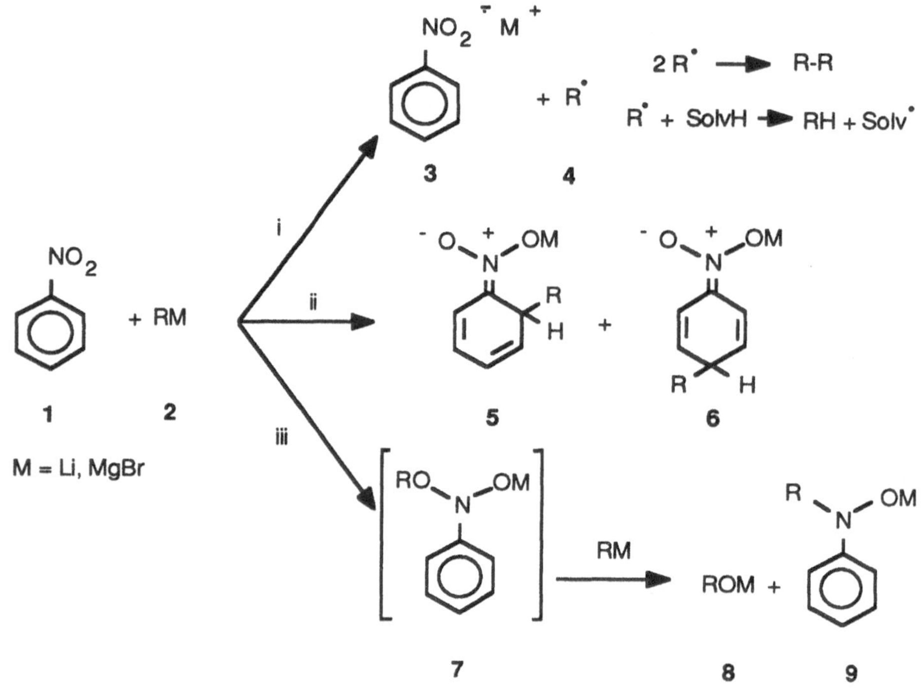

2. MATERIALS AND METHODS

THF was purified by refluxing it over sodium wires until the blue color of benzophenone ketyl persisted and then distilling it into a dry receiver under nitrogen atmosphere.

1,3-Dithianylmagnesium bromide was prepared by adding lithium dithiane (5 mmol) to a freshly prepared solution of anhydrous $MgBr_2$ (10 mmol; from 1,2-dibromoethane and magnesium turnings) at 0 °C and stirring the reaction mixture for about 30 minutes.[9]

The reaction between nitroarenes (4-chloronitrobenzene and 2-methoxy-1-nitronaphthalene) and organometallic reagents (BuLi, BuMgBr, lithium dithiane, lithium 1-methyldithiane dithianylmagnesium bromide, hex-5-enylmagnesium bromide) was carried out as follows: to a stirred solution of nitroarene (5 mmol) a solution of the organometallic reagent (5.5 mmol) in the same solvent was added dropwise under nitrogen atmosphere.

After about ten minutes a THF solution of 6 mmol of DDQ were added to the stirred reaction mixture and stirring was continued for one hour. The mixture was quenched with aqeous acetic acid (5%), extracted with ether, dried, evaporated under reduced pressure and flash-chromatographed to give the pure products which were identified by standard analytical methods[3,4].

ESR experiments were carried out by introducing a weighted amount of the two reactants under nitrogen atmosphere into a U-tube each in their separate chamber. The sealed sample tube was placed in a bath at the appropriate temperature, the reactants mixed, and then introduced in a cavity of the ESR spectrometer setted at the same temperature. Absolute radical anion concentrations were determined by double integration followed by calibration against a solution containing a known concentration of the diphenylpicrylhydrazyl free radical.

3.RESULTS AND DISCUSSION

A general picture, which can account for the formation of both conjugate addition and redox products, is represented in Scheme 2.[3]

This mechanism is similar to the one recently proposed to account for the closely related reactivity of strong electron acceptor ketones.[10,11]

Scheme 2. Mechanism of the reaction of nitrobenzene (**1**) with alkyl lithium and magnesium compounds (**2**).

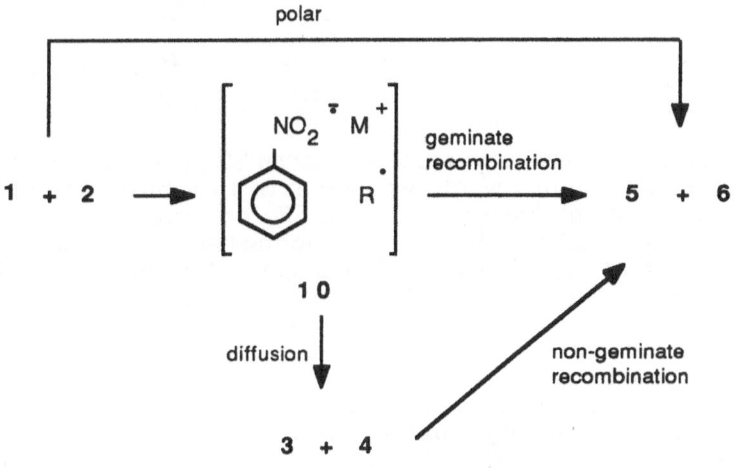

$$R^{\bullet} + R^{\bullet} \rightarrow R\text{-}R \qquad R^{\bullet} + SolvH \rightarrow RH + Solv^{\bullet}$$

According to this scheme, the conjugate addition products **5** and **6** can be supposed to be exclusively or in part formed by a polar pathway. Alternatively, we can suppose that all products arise from an initial electron transfer from RM (**2**) to nitroarenic system **1** to give the radical pair **10**. The alkyl radical can migrate within the solvent cage from the nitro group, near which it is formed, to the *ortho* or *para* position to give the addition products **5** and **6** respectively in a geminate recombination.[12] Alternatively, the two radicals of the pair **10** can diffuse out of the solvent cage. The free alkyl radicals **4** can collapse with the nitroarenic radical anion **3** to give the nitronate adducts **5** and **6** in a non-geminate recombination. Moreover, the radicals **4** can have an independent decay by dimerization and hydrogen abstraction. On the contrary, the radical anion **3** is generally stable and, therefore, its amount present at the end of the reaction corresponds to the amount of radicals **4** which have decayed by an independent route.

Radical anion formation can be detected by ESR spectroscopy. However, quantitative ESR measurements are usually affected by large experimental errors. Therefore, it is more convenient to oxidize the radical anion to starting material by an appropriate reagent. Dichloro-dicyanobenzoquinone (DDQ) is generally used, since, at the same time, it is able to quantitatively convert the unstable nitronate adducts into the corresponding nitro derivatives (see Scheme 3)[1].

Scheme 3. Conversion of unstable radical anion **3** and nitronate adducts **5** and **6** into stable compounds by oxidation with DDQ.

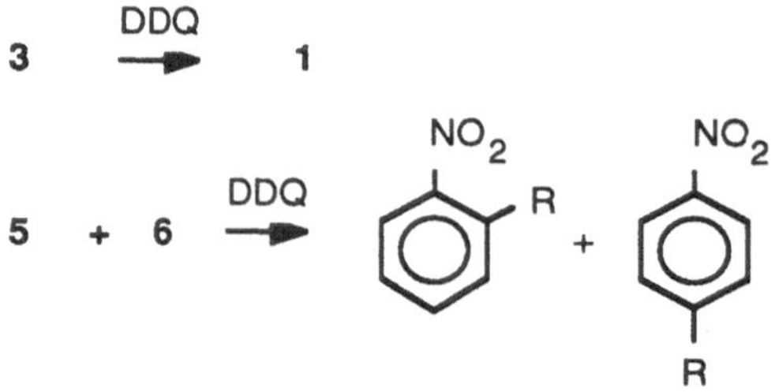

11 12

The main question to be answered is whether an SET pathway is responsible of the redox product **3** only. In other words, we must at first establish whether the addition product formation proceeds through a polar or an SET pathway.

The ability to 'trap' or 'observe' the intermediate radical species would be instrumental in establishing the integrity of the proposed SET mechanism. In the present system we chose the former approach by incorporating a 'radical probe' into the nucleophilic reagent, since it is experimentally difficult to observe by means of spectroscopic methods the formation and the decay of very short-living radical species such as the highly reactive alkyl radicals.

The hex-5-enylmagnesium bromide undoubtedly represents an unequivocal 'probe' to establish if the formation of alkyl radicals is involved in a reaction of a Grignard reagent with an electrophilic substrate.[10,13] In fact, if free hex-5-enyl radicals are formed during the reaction course, they may rapidly cyclize to cyclopentylmethyl radicals before decaying[14] and then products containing the cyclopentylmethyl framework may be obtained in a reaction proceeding with radical character.

Moreover, the use of a Grignard instead of a lithium reagent shows the advantage that the formation of redox products (nitroarene radical anion) is generally negligible (see later in the text).

'Probe' experiments were carried out on 2-methoxy-1-nitronaphthalene (**13**) in THF (see Scheme 4), since this system leads to only one addition product.[1] This feature facilitated the analysis of the reaction mixture.

As shown in Table I, at 0 °C we observed the formation of both uncyclized and cyclized addition product (**20** and **21**). These findings represent a clear evidence that a large part of the reaction proceeds through an SET pathway.

Scheme 4. Mechanism of the reaction between 2-methoxy-1-nitronaphthalene and hex-5-enylmagnesium bromide.

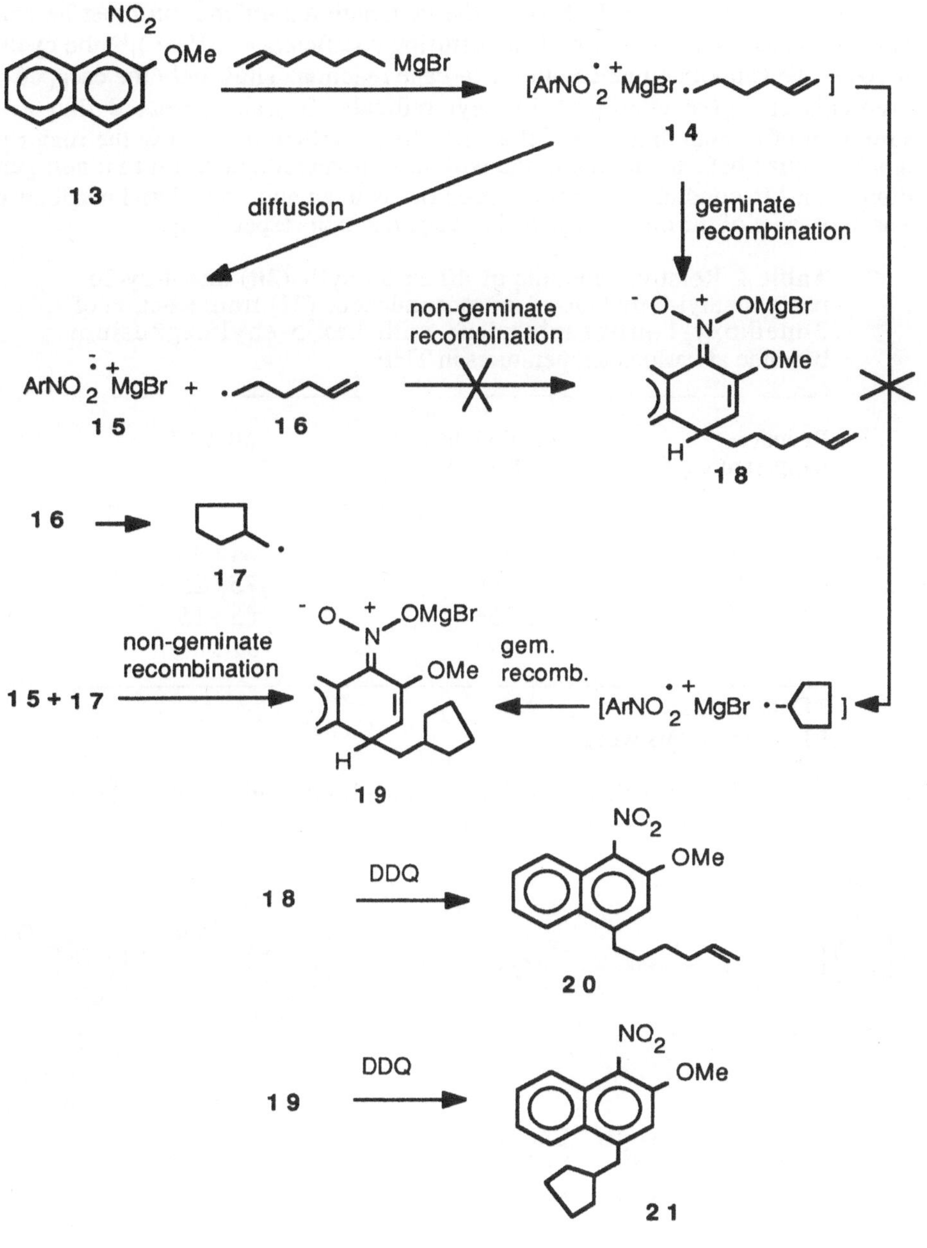

If addition products are formed from a geminate and a non-geminate recombination of the alkyl radicals with nitroarene radical anions, we can reasonably assume that uncyclized and cyclized products (**20** and **21**) arise from the in-cage and from the out-of-cage reaction respectively. In fact, since the geminate recombination must be complete before the radicals can escape their cage (diffusion coefficient ≈ 10^9 s^{-1}),[15] the cyclization (k_{cy} ≈ 10^5 s^{-1})[14] cannot compete with the in-cage reaction. Thus, only the escaped hex-5-enyl radicals can give cyclopentylmethyl radicals. It was estimated that a steady concentration of radical anions less than 10^{-3} M is sufficient to allow the major part of radicals to cyclize before recombinating with nitroarene radical anion in a non-geminate recombination.[3] In conclusion the proportion of cyclized and uncyclized products can be taken as a measure of the out-of-cage and in-cage reaction respectively.

Table I. Relative amounts of 4(hex-5-enyl)- (**20**) and 4-cyclo - pentylmethyl-2-methoxy-1-nitronaphthalene (**21**) from reaction of 2-methoxy-1-nitronaphthalene with hex-5-enylmagnesium bromide at various temperatures in THF

Reaction temperature (°C)	Overall yield (%) (**20** + **21**)	**20 : 21**
20	78[a]	64 : 36
0	76[a]	75 : 25
-30	75[a]	85 : 15
-70	77[b]	92 : 8

[a] Data from ref. 3
[b] Data from this work

Table II. Competitive reactions of 2-methoxy-1-nitronaphthalene (1 eq) with two Grignard reagents (5 eq each) in THF at 0°C.[a]

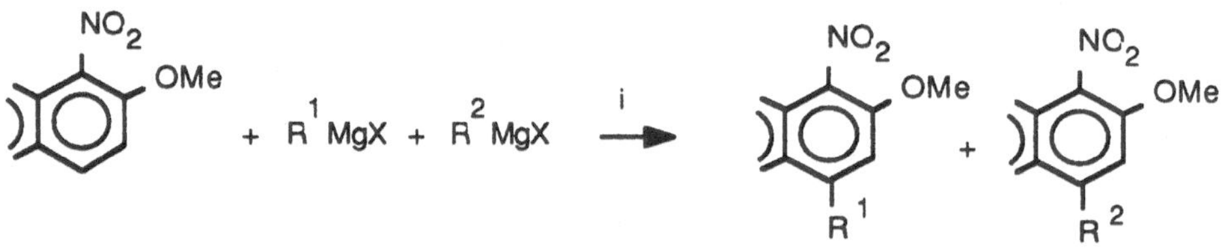

R^1MgX	R^2MgX				Overall yield(%)
i-PrMgBr	EtMgBr	70	:	30	78
EtMgBr	$PhCH_2MgBr$	51	:	49	76
EtMgBr	MeMgBr	97	:	3	73

[a] Data from ref. 3.
[i] in THF at 0°C for 10 min, followed by quenching with DDQ

There are further evidences supporting an SET mechanism. From competitive runs (see table II) it has been established[3] that the reactivity scale of Grignard reagents follows the order: i-Pr > $PhCH_2 \approx$ Et > Me. This trend is in good agreement with the oxidation potentials of the Grignard compounds detected by Holm (see table III).[16]

Furthermore, on the basis of the Marcus theory, Eberson[17] calculated the kinetic parameter for a possible SET process from 5-hexenyl magnesium bromide to 1-nitro-2-methoxy-naphthalene system. The second order rate constant value has been estimated to be 2×10^4 l. s.$^{-1}$ mol.$^{-1}$ at 0 °C. Thus, on the basis of Marcus theory, this reaction must be classified as feasible and occurring at a very fast rate.

Table III. Experimental values of reversible potential (E°) for R°/R- redox couples

Grignard reagent	E°/ V vs NHE[a]
i-PrMgBr	-0.95
$PhCH_2MgBr$	-0.73
EtMgBr	-0.66
MeMgBr	-0.25

[a]NHE = Normal Hydrogen Electrode

Table IV. Relative amounts of addition and reduction products from reaction of 4-chloronitrobenzene with lithium dithiane at various temperatures in THF[a].

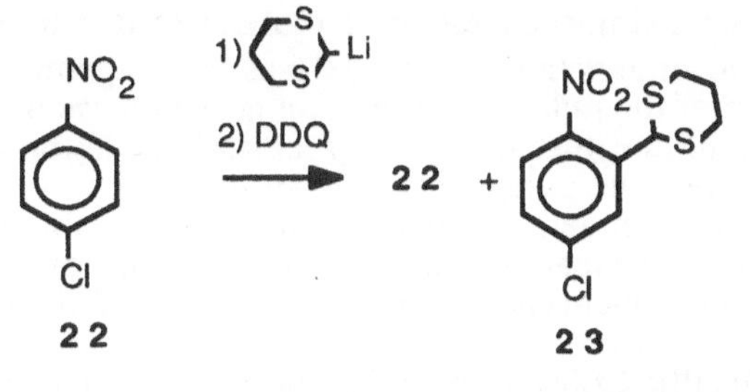

Reaction temperature (°C)	**23** yield (%)	**22** yield (%)
-5	13	67
-40	39	41
-70	48	21

[a] Data from ref. 4.

3.1 Reaction Temperature Effect on Product Distribution

The reaction temperature can affect reaction product distribution .At low temperatures, in fact, an increased proportion of uncyclized product is observed, while the overall yield remains substantially unchanged (see Table I). This finding indicates that a temperature decrease favours the in-cage reaction. At -70 °C, the amount of cyclized product accounts for only 8% of the addition products. This value is very close to the proportion of cyclopenthylmethyl isomer, present in the starting Grignard material (3-5%)[3]. Thus, in a fashion of an SET mechanism we can assume that in the present reaction in THF at -70 °C the addition product is almost exclusively formed from the in-cage reaction.

These results suggest that it is probably possible to affect the product distribution by acting on the temperature parameter.

In 1971, it has been reported[18] that lithium 1-phenyldithiane is quantitatively oxidized to bis-dithiane in THF at 0 °C by nitroarenes.

On the basis of 'probe' experiment results, the predominance in this system of redox products, 'exemplary' out-of-cage reaction, may be ascribed to the too high employed temperature. We recently[4] examined in detail the reaction of lithium dithiane with 4-chloronitrobenzene in THF at various temperatures followed by DDQ quenching. Our results are reported in Table IV.

As expected, a temperature decrease, favouring the in-cage reaction, favours addition product formation. Moreover these results show that it is now possible to adress the addition process to prevail over the redox one by carrying out the reaction at very low temperature (-70 °C).

The assumption that addition compounds must be considered as typical in-cage product needs a detailed discussion.

The probability of an efficient collision between two radical species when they return in the solvent cage is much less than that shown by the same radicals directly formed therein. In fact, in the latter situation the two radical species are generated close to the reaction centre with a favourable geometry and with an electronic arrangement (singlet state) fit for an immediate reaction. In the former situation, the radical pairs are formed with random geometries and it may be expected they undergo many collisions before finding a favourable one. In addition, only one-forth of the encounters between two free radicals have an electronic singlet character. In close related systems it has been estimated that the non-geminate recombination is about 1/16 as fast as geminate recombination.[19] As a consequence, if most of the radical pairs escape geminate recombination and diffuse in the solvent, it is very unlike that free nitroarene radical anion can scavenge all alkyl radicals, particularly when the independent decay processes of the latter radical species occur at a very fast rate. Conversely, when geminate recombination is a very efficient process, there is an higher probability that the little amount of nitroarene radical anions which had diffused in the solvent can scavenge a large part of alkyl radicals.

3.2 Influence of the Metallic Counterion on Reaction Product Distribution.

The different product distribution observed in reaction of lithium dithiane with 4-chloronitrobenzene with respect to reactions of magnesium reagents with 1-nitro-2-methoxynaphthalene suggests that the nature of the metallic counterion can affect the efficiency of the in-cage reaction.

In order to confirm this hypothesis on the basis of homogeneous data, we have compared the reactivity of lithium dithiane and butyl lithium with that of the corresponding Grignard reagents towards 4-chloronitrobenzene under the same experimental conditions.

As expected, in both cases when Grignard instead of lithium reagent is used, the extent of redox process becomes negligible, as demonstrated by the presence of only trace amounts of starting material among the reaction products after the DDQ quenching. Viceversa an appreciable increase in the addition product yield is observed (see Table V).

In the reaction of butyl reagents, we were able to confirm these results by direct ESR analysis of the reaction mixture before the DDQ quenching. A very intense signal is observed, in fact, in the case of butyl lithium. From quantitative measurements, the radical anion formation has been estimated to account for about the 19% of the reaction. On the contrary, a weak but persistent signal is observed in the case of the Grignard reagent. The spectra of the lithium and magnesium 4-chloronitrobenzene radical anion are reported in figure 1.

In a fashion of an SET mechanism, the greater tendency of Grignard with respect to the lithium reagent to give addition product can be explained in terms of a more neutral character showed by the radical anion , when the counterion is magnesium. In fact, it may be expected that nucleophilic radicals, such as alkyl ones, collapse with a radical anion the easier the less it has anionic character.

Table V. Influence of the metallic counterion on product distribution in the reaction of 4-chloronitrobenzene with RM in THF.

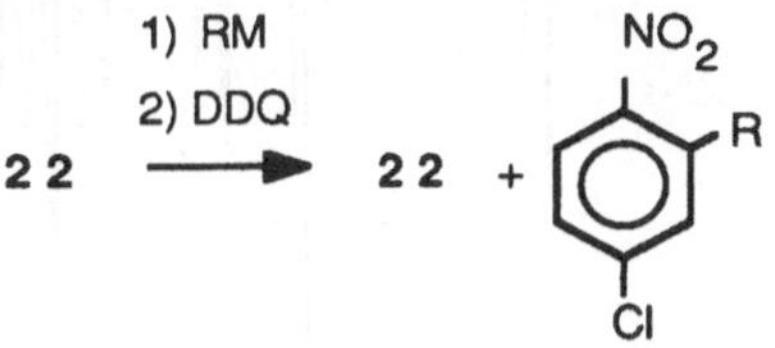

24

RM	Reaction temperature (°C)	**24** yield (%)	**22** yield (%)
BuMgBr [a]	0	71	traces
BuLi [a]	0	54	27
lithiumdithiane [b]	-40	39	41
dithianylmagnesium [a,c]	-40	59	traces

[a] Data from this work.
[b] Data from ref. 4.
[c] Prepared by adding one equivalent of $MgBr_2$ to a THF solution of lithium dithiane.[9]

3.3 Influence of Steric Factors on Reaction Product Distribution

Steric factors can affect product distribution,too. Unlike lithium dithiane (**25a**), the more sterically hindered 1-methylderivative (**25b**) leads almost exclusively to the formation of redox products (see Scheme 5), when it reacts with 4-chloronitrobenzene (**22**) in THF at -70 °C.

Figure 1. ESR spectra of magnesium and lithium 4-chloronitrobenzene radical anion

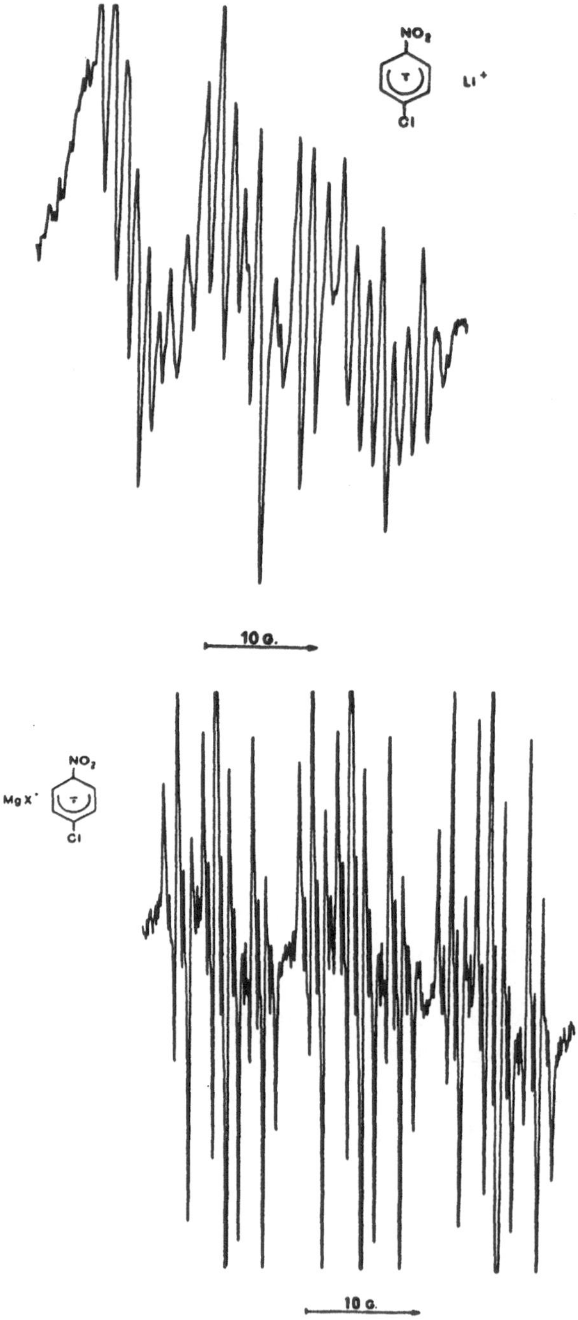

Furthermore, in the reaction with 1-nitro-2-methoxynaphthalene (**13**), which has a low sterically hindered reactive position, the methyl derivative **25b** gives the addition product in a yield comparable to that obtained with the less hindered unsubstituted **25a**.

Scheme 5. Reaction of lithium dithiane (**25a**) and lithium 1-methyldithiane (**25b**) with 4-chloronitrobenzene (**22**), 1-nitro-2-methoxynaphthalene (**13**) and nitrobenzene (**1**) in THF at -70 °C.[a]

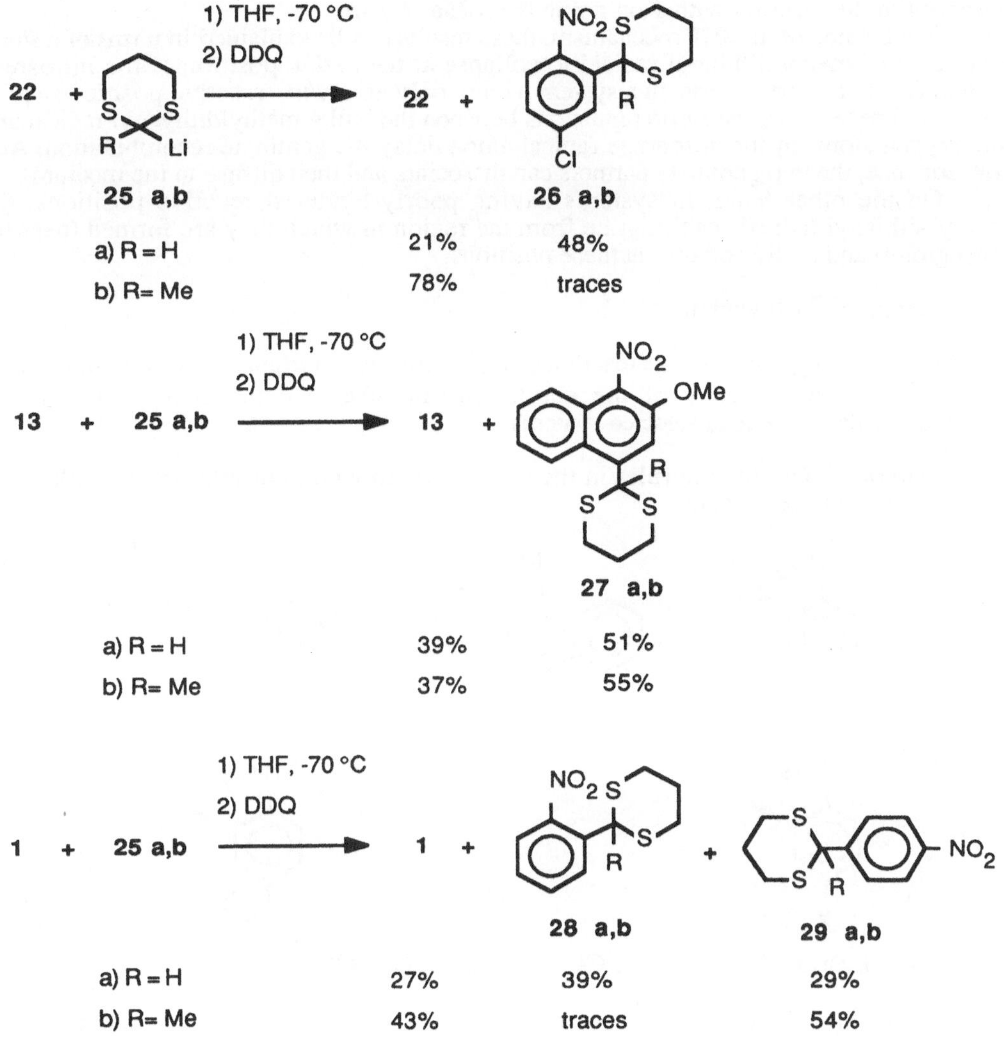

[a] data from ref. 4.

Finally, in the reaction with nitrobenzene (**1**), a substrate having two different reactive positions one of which sterically hindered, **25a** gives both 1,4 and 1,6 addition products in 39% and 29% respectively, while the bulky methyl derivative gives only the 1,6 addition product in 54% yield. The redox products account for a 27% of the reaction in the former case and for a 43% in the latter.

If we hypothesize a polar character for the addition and an SET for the redox process, the dramatic enhancement of the redox compounds observed in the reaction of **22** with **25b** must be ascribed to the greater ability of **25b** to promote the SET reaction than unsubstituted **25a**.[4,20] However, this hypothesis cannot account for both the comparable yields shown by **25a** and **25b** with **13** and the increase in the 1,6 addition product observed in the reaction with **1** on going from **25a** to **25b**.

In a fashion of an SET mechanism, these results can be explained in terms of a steric hindrance of methyldithianyl radical to collapse at the *ortho* position of the nitroarene radical anion. In other words, in a system having only an *ortho* reactive position such as 4-chloronitrobenzene, the steric repulsions between the bulky methyldithyanyl radical and the oxygen atoms of the nitroarene radical anion delay the geminate recombination. As a consequence, the two geminate partners can dissociate and then diffuse in the medium.

On the other hand, in systems having poorly-hindered reactive positions, the methyldithianyl radicals can migrate from the region in which they are formed (near the nitro group) and easily collapse at these positions.

3.4 Orientation of Alkylation

Another interesting aspect of the reaction of nitroarenes with Grignard and lithium reagents is the high regioselectivity of the alkylation.[1] In scheme 6 the arrows indicate the orientation rules in various selected systems.

Scheme 6. Orientation rules in the reaction of alkylation of nitroarenes with organometallic reagents.

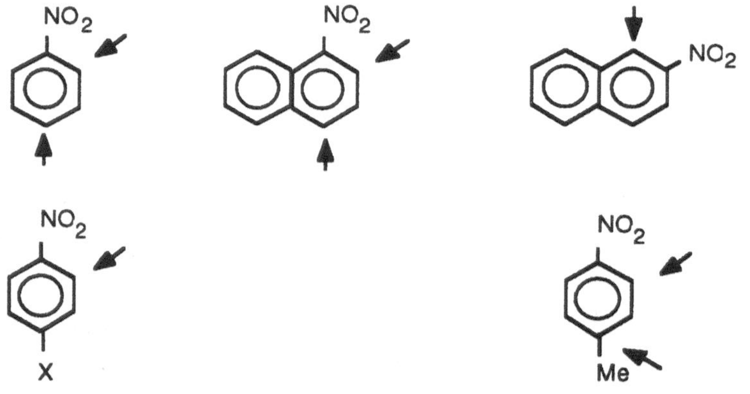

X = Cl, Br, I, OR, SR, NR_2, COOR, CHO, COR, CN

In a fashion of an SET mechanism, these rules can be explained on the basis of the π-spin density distribution of the corresponding radical anions.

In nitrobenzene[21] and 1-nitronaphthalene[22] systems, the unpaired electron is delocalized in the aromatic ring mainly in the *ortho* and *para* positions with similar spin density magnitude. In fact, an *ortho:para* alkylation ratio of 2:1 is generally observed[1] in nitrobenzene as predicted by statistical factors, while a 1:1 ratio is generally found in 1-

nitronaphthalene system. As above reported (see Section 3.3), these relative proportions can be drastically modified by steric factors. In an analogous manner, in 2-nitronaphthalene, the π-spin density is delocalized in the aromatic system mainly on the C-1 carbon[22] and thus it is not surprizing that alkylation occurs exclusively at this position. For sake of semplicity, we have discussed only three significative examples but the same interpretation applies to other systems such as indole, quinoline, benzothiazole, anthracene etc.

When a reactive position is hetero-substituted or substituted with cyano or carbonyl functions, the alkylation occurs almost exclusively at the unsubstituted reactive position.[1,23] According to an SET mechanism, the lack of reactivity at the substituted position can be justified by a drop in spin density in the corresponding radical anion, due to the great ability of the above substituents to delocalize the unpaired electron.

On the other hand, positions carrying a substituent with poor delocalizing effect, such as a methyl group, show a reactivity comparable with that of an unsubstituted one.[24]

4. CONCLUSIONS

We have interpreted our results in a fashion of an SET pathway. However, no conclusive experimental evidences exist to exclude that, at least in part, the addition products can arise from a polar pathway. In fact, it is very difficult to assign a definite kind of mechanism to the reaction part proceeding without clear SET evidences (the so-called cage-reactions).

Nevertheless, this question can be settled if the existence of a continuous mechanistic spectrum from 'full polar' to 'full SET' pathways is accepted on the basis of the electron shift theory.[25]

On the other hand, our mechanistic studies allow to find out the factors affecting the product distribution in the reaction of nitroarenes with carbanions which give an irreversible attack to the aromatic ring. In particular, low reaction temperatures and the use of Grignard instead of lithium reagents favour the conjugate addition product formation, while steric hindrance enhance the redox process.

Nevertheless, this reaction needs further mechanistic studies, since there is a relevant problem from a synthetic point of view. Grignard reagents warrant a high efficiency in addition products. However,they can be almost exclusively obtained through direct method (magnesium and alkyl halides), which suffers from several limitations. Thus, in the Grignard series, only a few functionalized reagents are available.

On the contrary, there is a wide spectrum of functionalized lithium derivatives, because they can be prepared with several efficient methods. Therefore, the use of lithium instead of Grignard reagents undoubtedly represents an important goal. As a consequence, further studies are necessary to find out experimental conditions to replace magnesium with lithium derivatives without enhancing the redox products and by-passing the preparation of RMgX from magnesium-lithium exchange between RLi and $MgBr_2$ which is a too complicate methodology for a large preparative scale.

Moreover we think to be interesting to direct further investigations towards other carbanionic systems. The recently reported mediated fluorine addition of silyl enol ether to nitroarenes[26] represents an intriguing approach to this problem.

REFERENCES

1. G. Bartoli, *Acc. Chem Res.*, 1984, **17**, 109, and literatures cited therein.

2. B.J. Wakefield 'Compounds of the alkali and alkaline earth metals in organic synthesis' in *Comprehensive organometallic chemistry*, G. Wilkinson, Ed.; Pergamon, New York, 1982; chapter 44.
3. G.Bartoli, M.Bosco, G.Cantagalli, R.Dalpozzo, F.Ciminale, *J. Chem. Soc. Perkin Trans. II*, 1985, 773.
4. G.Bartoli, R.Dalpozzo, L.Grossi, P.E.Todesco, *Tetrahedron,* 1986, **42**, 2563.
5. Y. Yost, H.R. Gutman, C.C. Muscoplat, *J Chem . Soc . C,* 1971, 2120.
6. V.I. Savin *Zh. Org. Chim.* ,1978, **14,** 2090. [*Chem. Abst.,* 1979, **90,** 71443r].
7. P. Buck, G. Kobrich, *Tetrahedron Lett .*, 1967,1563.
8. G.Bartoli, M.Bosco, R.Dalpozzo, unpublished results.
9. C. Agami, M. Chauvin, J. Levisalles, *Bull. Soc. Chim. Fr.,* 1970, 2712; H.M. Walborsky, A.E. Young, *J. Am. Chem. Soc.*, 1964, **86**, 3288.
10. E.C. Ashby, J.S. Bower jr., *J. Am. Chem. Soc.,* 1977, **99**, 6059.
11. T. Holm, *Acta Chem. Scand. Ser. B,* 1974, **B28**, 809.
12. J.F. Garst, C.D. Smith, *J. Am. Chem. Soc.,* 1977, **99**, 1520.
13. C. Walling, A. Cioffari, *J. Am. Chem. Soc.,* 1970, **92**, 6609; C. Walling, A. Cioffari, *J. Am. Chem. Soc.,* 1972, **94**, 6059.
14. D. Lal, D. Griller, S. Husband, K.U. Ingold, *J. Am. Chem. Soc.*, 1974, **96**, 6355.
15. R.H. Noyes, *J. Am. Chem. Soc.,* 1955, **77**, 2042.
16. T. Holm, *Acta Chem. Scand. Ser. B,* 1983, **B37**, 567.
17. L. Eberson, *Acta Chem. Scand. Ser. B,* 1984, **B38**, 439.
18. W. Baarschers, T.L. Loh, *Tetrahedron Lett .*, 1971, 3483.
19. J.F. Garst, C.D. Smith, *J. Am. Chem. Soc.,* 1977, **99**, 1526.
20. E. Juaristi, B. Gordillo, D.M. Aparicio, *Tetrahedron Lett .*, 1985, **26**, 1927.
21. W.M. Gulik, Jr., *J. Am. Chem. Soc.,* 1972, **94**, 29.
22. P. Furderer, F. Gerson, *Helv. Chim. Acta*, 1976, **59**, 2492.
23. G.Bartoli, M.Bosco, R.Dalpozzo, P.E.Todesco, *J. Org. Chem.,* 1986, **51**, 3694.
24. G.Bartoli, M.Bosco, G. Pezzi, *J. Org. Chem.,* 1978, **43**, 2932; N. Armillotta, G.Bartoli, M.Bosco, R.Dalpozzo, *Synthesis,* 1982, 836; G.Bartoli, M.Bosco, A.C. Boicelli, *Synthesis,* 1981, 570.
25. A. Pross, *Acc. Chem Res.,* 1985, **18**, 212.
26. T.V. RajanBabu, G.S. Reddy, T. Fukunaga, *J. Am. Chem. Soc.,* 1985, **107**, 5473.

REACTION OF ALKYL HALIDES WITH LITHIUM AND MAGNESIUM REAGENTS. NUCLEOPHILIC SUBSTITUTION VS. SINGLE ELECTRON TRANSFER

M. Nojima and S. Kusabayashi
Department of Applied Chemistry
Faculty of Engineering
Osaka University
Suita, Osaka 565, Japan

ABSTRACT. The reaction of ambident allylic halides with magnesium reagents was found to proceed by two alternative processes, i. e., nucleophilic substitution vs. single electron transfer, the contribution of each path being a marked function of various factors including steric and electronic effects of the substituent, solvents, and additives. Moreover, a remarkable difference in proportion of the two alkylation products was observed between these mechanistic alternatives. In the reaction of allylic lithiums with alkyl halides also the same trends were observed.

I. INTRODUCTION

With the realization that the allyl moiety is an integral feature of many natural products and biosynthetic intermediate, it has become important to develop methods that would lead to controlled C-C bond formation at either the alpha- and gamma-position of an allyl substrates [1]. For this purpose the reaction of allylic organometallics with alkyl halides and also the reaction of allylic halides with organometallic reagents seems to be promising [2]. But the regiochemistry of the products has been found to be a function of various factors including substituents, solvents, additive, etc [3]. To develop more sophisticated method, it must be, therefore, important to understand the detailed mechanism of the reaction. When stereochemistry, CIDNP, and cyclizable probes are used as the most definitive indicators of the mechanism, it has been confirmed that at least two reaction mechanisms, i. e., single electron transfer and direct nucleophilic substitution, are possible [4]. Evidence is available for both mechanisms in individual cases. However, a delicate balance must exist between these mechanistic alternatives, and therefore, a minor change of reaction conditions may alter the extent of each process. In this paper we present some clear-cut examples which demonstrate that the difference in reaction pathways is one of the origins for appearance of the effects of variables on the alkylation regiochemistry.

M. Chanon et al. (eds.), Paramagnetic Organometallic Species in Activation / Selectivity, Catalysis, 503–517.

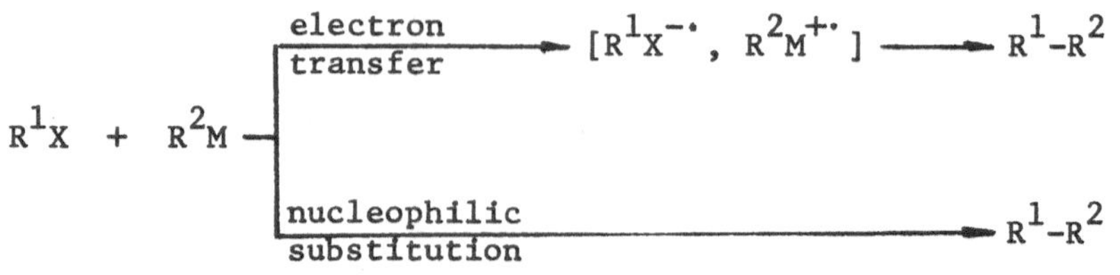

2. RESULTS AND DISCUSSION

2. 1. The Expected Substituent Electronic Effects on the Regiochemistry of Cross-Coupling Products

Stereochemistry, CIDNP, and cyclizable probe experiments are method of choice to differenciate the reaction processes. Alternatively, the dependence (or independence) of the alkylation regiochemistry upon the substituent electronic effects would be useful to identify the participating intermediate in the case of 1-arylallyl system.

In this respect, the spin- and charge-densities of 1-arylallyl radicals (1), 1-arylallyl cations (2), and 1-arylallyl anions (3), were calculated by an ab initio SCF MO method at the HF/STO-3G level (Table I) [5,6]. For a series of 1-arylallyl radicals (1a-c), the

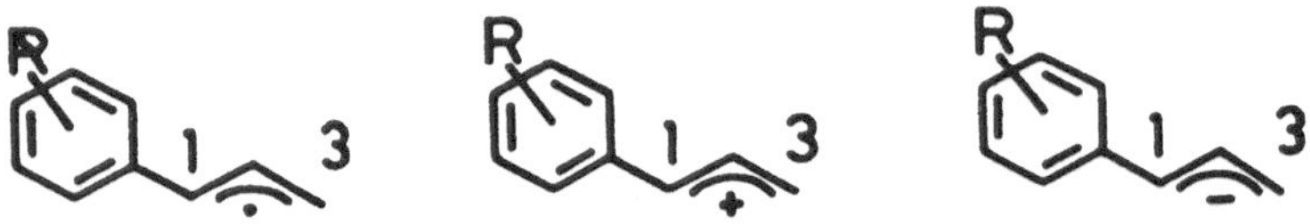

(1) (2) (3)

a; R = 4-OMe b; R = H c; R = 3-Cl

TABLE I. An ab Initio SCF MO Study for 1-Arylallyl Radical, Cation, and Anion

Substituent	(1) spin density C-1	(1) spin density C-3	(2) charge density C-1	(2) charge density C-3	(3) charge density C-1	(3) charge density C-3
4-OMe	0.904	1.130	0.086	0.058	-0.215	-0.298
H	0.905	1.130	0.092	0.069	-0.214	-0.296
3-Cl	0.904	1.130	0.095	0.076	-0.209	-0.286

substituents R exert no meaningful influence on the spin densities at C-1 and C-3. This would imply that if alkylation products are produced by cross-coupling of 1-arylallyl radical (1) and the alkyl radical, then the ratio of C-1 coupling vs. C-3 coupling would be substituent-independent. Perhaps in accordance with this, the reaction of halides, (4) and (5), with Ag in tetrahydrofuran, which proceeds via (9-anthryl)arylmethyl radical, afforded a mixture of two dimerization

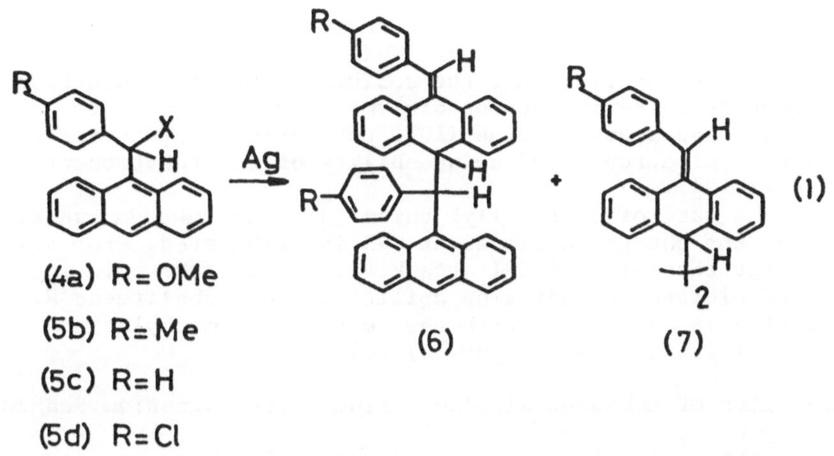

products, (6) and (7), the ratio 5:4 being substituent-independent (eq 1) [7].

For a series of 1-arylallyl cations (2a-c), the substituents R show a significant influence on the charge densities at both C-1 and C-3. The electron-donating 4-methoxy group decreases the charge densities at both C-1 and C-3 and in direct contrast, the electron-withdrawing 3-chloro substituent increases the charge densities at the same positions. More important is that the carbon far from the substituent (C-3) is more susceptible to the substituent electronic effects than C-1 attached directly to the aryl group. Thus, if reaction proceeds via a carbonium ion intermediate (2), then the product regiochemistry would be influenced by the substituent R such

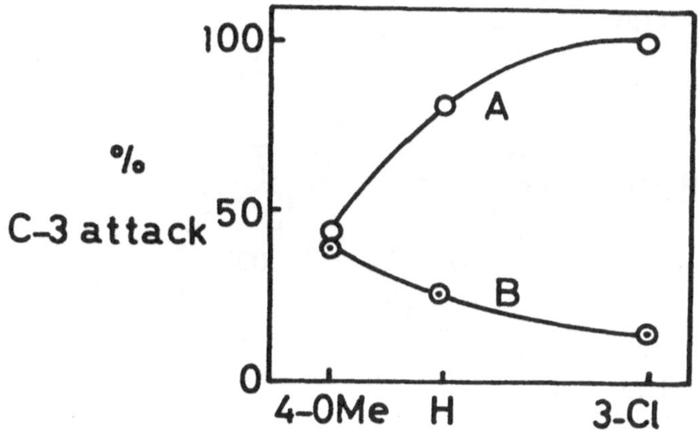

Figure 1. A: Sodium borohydride reduction of 1-aryl-3-chloropropene (8). B: Reaction of 1-arylallyllithium (11) with methyl bromide in ether in the presence of TMEDA.

that the increase in electron-withdrawing ability increases the proportion of C-3 attack product. This prediction is in good agreement with the fact that the sodium borohydride reduction of 1-aryl-3-chloropropene (8) in aqueous diglyme provided a mixture of 1-arylpropene (9) and 3-arylpropene (10), the ratio being increased with the increase in electron withdrawing ability of the substituent R (Figure 1) [5].

In the case of 1-arylallyl anion (3), the reverse substituent effect on the charge densities at C-3 is calculated. From this we deduce that the ratio of C-1 attack vs. C-3 attack would increase with increased electron-withdrawing ability of the substituent R. This expectation is consistent with the results for methylation of 1-arylallyllithium (11) (Figure 1) [6].

2.2 Reaction of Ambident Allylic Halides with Magnesium Reagents

2.2.1 Reaction of 1-Aryl-3-chloropropenes [5]. The reaction of ambident 1-aryl-3-chloropropenes (8) with a series of Grignard reagents, R'MgY (R'=Me, Pr, isopropyl, Ph, tert-butyl; Y=Br, Y) was carried out in diethyl ether (EE). The products were a mixture of two alkylation products (12; C-3 attack) and (13; C-1 attack), and three dimerization products (14)-(16) (eq 2). The alkylation:dimerization ratio and the composition of the two alkylation products were significantly influenced by the substituent electronic effects in the chloride (8), and R' or Y of R'MgY (TABLE II). The reaction of (E)-1-aryl-3-chloropropenes (8) with R'MgBr revealed the following. (a) The (12):(13) ratio increased with the increase of electron-withdrawing ability of the substituent R. The alkyl group of R'MgBr also affected the (12):(13) ratio, the proportion of C-3 attack product (12) being increased with the decrease of steric bulk of R': isopropyl < Ph < Pr < Me. (b) The yields of dimers (14)-(16) increased as the donicity of R'MgBr increased: Ph ≃ Me < Pr < isopropyl < tert-butyl [8]. The composition of dimers (14):(15):(16) (ca. 3:6:1) were almost insensitive to the substituent electronic effects and the structure of alkyl groups of R'MgBr. The addition of $FeCl_3$, a possible accelerator

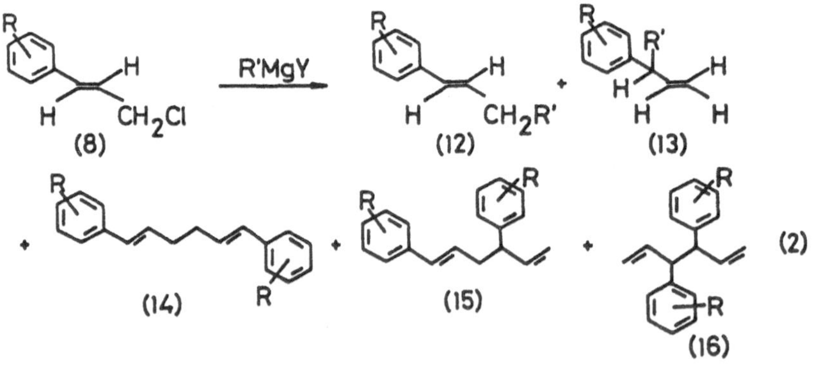

a; R=4-OMe   b; R=H   c; R=3-Cl

Table II. Reaction of (E)-1-Aryl-3-chloropropenes (8a-c) with Grignard Reagents in Diethyl Ether[a]

Chloride	R'MgY R'	Y	Products % Alkylation[b] [(12):(13)]	% Dimerization[b] [(14):(15):(16)]
(8a)	Me	Br	100 [65:35]	
(8b)	Me	Br	94 [96:4]	7[c]
(8c)	Me	Br	100 [100:0]	
(8b)	Me	I	41 [86:14]	59 [30:55:15]
(8c)	Me	I	36 [89:11]	64 [27:59:14]
(8a)	Pr	Br	100 [52:48]	
(8b)	Pr	Br	50 [83:17]	50 [26:62:12]
(8c)	Pr	Br	44 [94:6]	56 [30:60:10]
(8b)	Pr	I	21 [70:30]	79 [32:54:14]
(8c)	Pr	I	45 [70:30]	55 [30:58:12]
(8a)	i-Pr	Br	48 [53:47]	52 [25:61:14]
(8b)	i-Pr	Br	26 [64:36]	74 [25:70:5]
(8c)	i-Pr	Br	33 [67:33]	67 [28:69:3]
(8a)	Ph	Br	100 [63:37]	
(8b)	Ph	Br	89 [72:28]	11 [25:75:0]
(8c)	Ph	Br	76 [85:15]	24 [25:61:14]
(8b)	t-Bu	Br	4 [69:31]	86 [16:74:10][d]
(8c)	t-Bu	Br		60 [25:70:5][d]

[a] The reaction with 3 mol equiv of a Grignard reagent in diethyl ether at 30 °C for 1 h. [b] The total yield was ca. 70%, normalized; 100% = % alkylation + % dimerization. [c] The composition was not determined. [d] The formation of reduction products, (9) and (10), was accompanied.

of single electron transfer (SET) process [9], resulted in an increase in the yield of dimers (14)-(16). The ratio (14):(15):(16) was much the same as that from the reaction in the absence of $FeCl_3$.

In marked contrast to the reaction with R'MgBr, the reaction of chloride (8) with R'MgI (R'=Me, Pr) gave in each case a substituent-independent mixture of two alkylation products (12) and (13), the (12):(13) ratios being 9:1 for MeMgI and 7:3 for PrMgI. Another characteristic feature was the dimer yield. The proportion of dimers (14)-(16) in the reaction with MeMgI was as high as 60%, whereas the reaction with MeMgBr provided mainly alkylation products (12) and (13) (TABLE II).

In all the reactions of (E)-1-aryl-3-chloropropenes (8) with Grignard reagents, the corresponding 3-substituted 1-arylpropene (12) with E-configuration was obtained, together with 3-substituted 3-arylpropene (13) and dimers (14)-(16). In accordance with this, when the reaction of (Z)-1-phenyl-3-chloropropene (Z)-(8b) with PhMgBr was undertaken in EE, (Z)-1,3-diphenylpropene was obtained exclusively.

The reaction with isopropylmagnesium bromide also gave (Z)-1-phenyl--4-methyl-1-pentene together with 3-phenyl-4-methyl-1-pentene. In contrast, the reaction of (Z)-(8b) with propylmagnesium iodide resulted in the formation of an E-Z mixture of 1-phenyl-1-hexene (E/Z=3:2), together with 3-phenyl-1-hexene and dimers (14)-(16).

It is worth noting that (a) sodium borohydride reduction of (Z)-(8b) in aqueoud diglyme, which is known to proceed by 1-phenylallyl cation [10], gave a mixture of (E)- and (Z)-1-phenylpropene (9b) in a ratio of ca. 1:1, together with 3-phenylpropene (10b) and (b) tributyltin hydride reduction of (Z)-(8b) via 1-phenylallyl radical [11] also afforded an E-Z mixture of 1-phenylpropene (9b) in a ratio of ca. 1:1, suggesting that the reaction of (Z)-(8b), if it proceeds by prior carbon-halogen cleavage, provides an E-Z mixture of the corresponding 3-substituted 1-phenylpropene [12]. As might be expected, lithium aluminum hydride reduction of (Z)-(8b) in EE, in which an S_N2 pathway should predominate, gave exclusively (Z)-(9b) [12].

When the reaction of chloride (8) with 5-hexenylmagnesium bromide was performed in EE, a mixture of only non-cyclized products, 1-aryl-1,8-nonadiene (17) and 3-aryl-1,8-nonadiene (18), was obtained (eq 3). These results suggest that a radical-radical coupling process is not important for the formation of alkylation products (12) and (13) from the reaction of chloride (8) with primary Grignard reagents. Unfortunately, the reaction of chloride (8b) with 5-hexenylmagnesium iodide resulted in the formation of onlv dimers (14)-(16) and reduction products, (9) and (10).

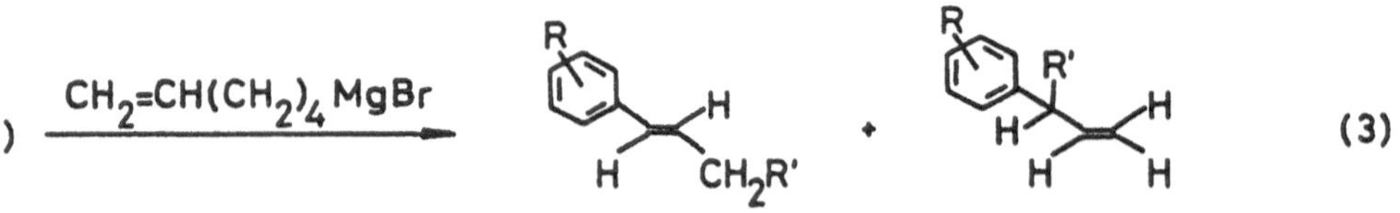

(17) (18)

a; R=4-OMe R' = -(CH_2)$_4$CH=CH_2

b; R=H R' = -(CH_2)$_4$CH=CH_2

c; R=3-Cl R' = -(CH_2)$_4$CH=CH_2

Dimers (14)-(16) are most likely to be produced by coupling of 1-arylallyl radicals (1) (Scheme I). As the MO study suggests, the substituent-independent composition of the dimers (14)-(16) is consistent with this hypothesis. The effect of the alkyl group of Grignard reagents on the product composition would be interpreted as that the Grignard reagents having a relatively higher donicity favors a SET process, with the dimers (14)-(16) being produced in a greater percentage [13].

For the reaction of chloride (8) with R'MgBr, the following facts require explanation. (a) The (12):(13) ratios are a marked function of

Scheme I

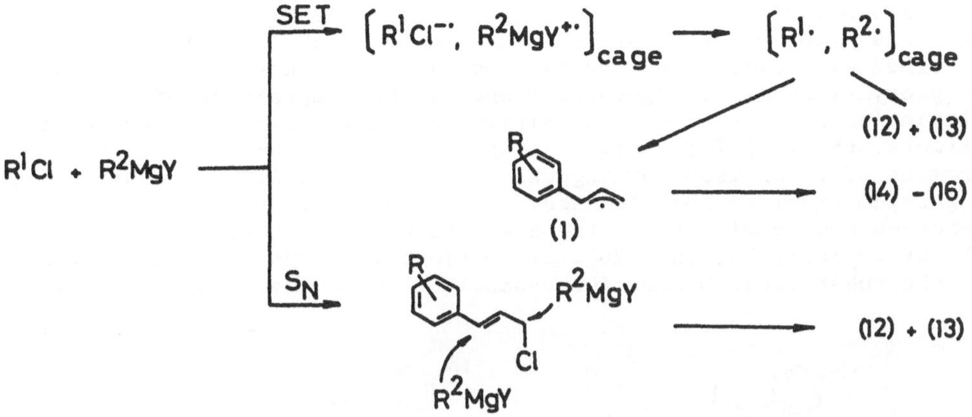

the substituent R, the ratio increasing with the increase in electron-withdrawing ability of the substituent R. (b) The reactions are stereospecific, suggesting that C-C bond formation occurs prior to C-Cl bond cleavage. In EE having a poor co-ordinating ability, the electropositive magnesium would co-ordinate to electronegative chlorine. Then the stereospecificity in alkylation and notable substituent electronic effects on the (12):(13) ratio leads us to deduce that the reaction of chloride (8) with Grignard reagents would proceed by a charge-developed species (19), in which the C-Cl bond is still retained; the Grignard reagent co-ordinated to chloride (8) would attack the chloride at C-1 to provide (13), whilst external attack of a Grignard reagent on chloride (8) at the 3-position would lead to the formation of (12).

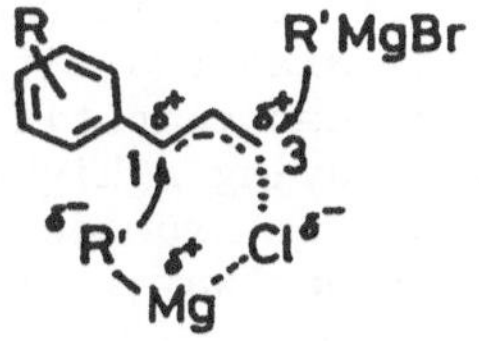

(19)

The reaction of chloride (8) with R'MgI seems to proceed by a different pathway. The observed non-stereospecificity in the alkylation suggests that C-Cl bond cleavage would occur prior to cross-coupling. Then, the substituent-independent composition of alkylation products would imply that a mechanism involving 1-aryl-allyl radical (1) is important. It is reasonable to expect that alkylmagnesium iodide, having probably a lower ionization potential than the corresponding alkylmagnesium bromide [4c], favors a SET pathway.

2.2.2 Reaction of (9-Anthryl)arylmethyl Chlorides with Grignard Reagents [7a]. The reaction of (9-anthryl)arylmethyl chloride (4) with a series of Grignard reagents gave a mixture of two alkylation products, (20; coupling at C-alpha) and (21; coupling at C-10), together with dimers (6) and (7) (eq 4). In the case of tert-butyl-, isopropyl-, and ethylmagnesium reagents the composition of the alkylation products were insensitive to the substituent electronic effects, the (20):(21) ratios being 3:7 for tert-butylmagnesium bromide, 3:7 for isopropylmagnesium bromide, and 1:1 for ethylmagnesium bromide. Phenylmagnesium bromide behaved differently. The reaction gave a mixture of the alkylation products (20) and (21) almost exclusively, the (20):(21) ratio being substantially increased as the substituent became increasingly electron-donating (TABLE III).

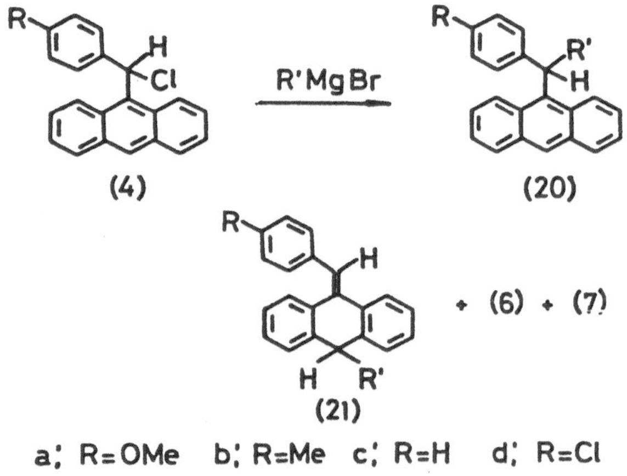

(4)

These results lead us to deduce that in the reaction of (9-anthryl)arylmethyl chloride (4) also two alternative pathways, i. e., SET and nucleophilic substitution, can competitively participate. In the case of the Grignard reagents having a higher donor power, tert-butyl-, isopropyl-, and ethylmagnesium bromides, a SET process predominates. In contrast, a polar pathway is important for phenylmagnesium bromide.

The behavior of MeMgBr, having an intermediate donor power, is also characteristic. The reaction with this reagent gave both the alkylation products, (20) and (21), and dimers, (6) and (7). The ratio (6):(7) was insensitive to the para substituent. In direct contrast, the ratio (20):(21) was a marked function of the electronic effects of the para substituent. A plot of the (20):(21) ratio vs. sigma plus gives, however, a curved line, the ratio (20):(21) being a minimum when the substituent is hydrogen (TABLE III). The alkylation vs. dimerization ratio was also affected by the para substituent, the ratio being increased with the increase of electron-donating ability of the substituent. Addition of p-dinitrobenzene (p-DNB), which is well known

TABLE III. Reaction of (9-Anthryl)arylmethyl Chloride with Alkyl-Magnesium Bromide in Ether[a]

R'MgBr R'	Chloride	Additive	Products[b] % Alkylation [(20):(21)]	% Dimerization [(6):(7)]
Ph	(4a)		96 [68:32]	4 [60:40]
Ph	(4b)		100 [42:58]	
Ph	(4c)		92 [23:77]	8 [56:44]
Ph	(4d)		96 [22:78]	4 [60:40]
Me	(4a)		100 [92:8]	
Me	(4b)		75 [62:38]	25 [59:41]
Me	(4c)		66 [41:59]	34 [50:50]
Me	(4d)		33 [46:54]	67 [50:50]
Me	(4b)	p-DNB[c]	90 [62:38]	10 [55:45]
Me	(4c)	p-DNB[c]	96 [43:57]	4 [50:50]
Me	(4d)	p-DNB[c]	89 [38:62]	11 [55:45]
Me	(4d)	$FeCl_3$[d]	29 [38:62]	71 [50:50]

[a] The reaction with the Grignard reagent from Grignard grade magnesium turning (99.5%) at 0 °C for 2 h. [b] Normalized: 100% = % alkylation + % dimerization. [c] As 4 mmol of MeMgBr is known to be consumed by each mole of p-dinitrobenzene (p-DNB), 9 equiv of R'MgBr was treated with a mixture of 1 equiv of chloride and 1 equiv of p-DNB. [d] $FeCl_3$ (1000 ppm) was doped.

to work as a "trapping agent" of organometallic reagents and radical anion [14], exerted a pronounced influence on product composition. The addition resulted in a remarkable increase of the alkylation products. Moreover, the ratio (20):(21) decreased by the sequence expected from a nucleophilic substitution mechanism; MeO > Me > H > Cl. These results would be interpreted as the electron-withdrawing para substituent facilitating the occurence of the SET pathway; by addition of p-DNB this process is significantly retarded and as a result, the alternative nucleophilic substitution process becomes to be important.

2.3 Reaction of Phenyl-Substituted Allyllithiums with tert-Alkyl Bromides [15]

The reaction of 1-arylallyllithiums (11a-c) with primary- and sec-alkyl halides gave in each case a corresponding mixture of two alkylation products in excellent yield, the ratio of C-1 attack vs. C-3 attack being increased as the substituent became increasingly electron-withdrawing (reference Figure 1) [5]. No evidence was obtained for the formation of dimers (14)-(16). These results clearly demonstrate that the reaction proceeds by a polar pathway.

A remarkably different trend was observed in the reaction with tert-alkyl bromides. The reaction with 6-bromo-6-methyl-1-heptene (22) is representative (TABLE IV). When the reaction of 1-phenylallyllithium

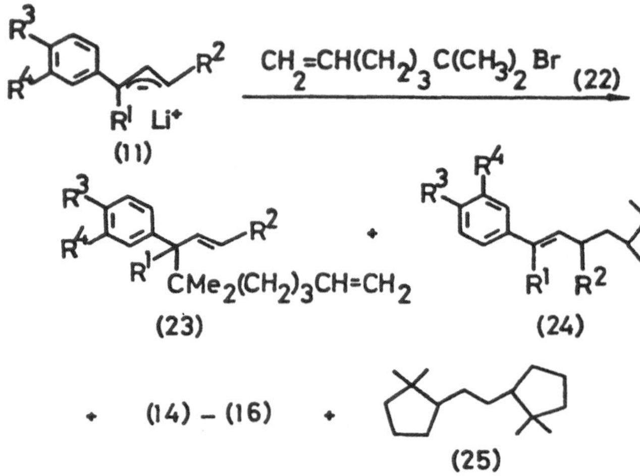

(5)

compound	R^1	R^2	R^3	R^4
a	H	H	OMe	H
b	H	H	H	H
c	H	H	H	Cl
d	Me	H	H	H
e	H	Me	H	H
f	H	Ph	H	H

(11b) with (22) was performed in ether in the presence or absence of tetramethylethylenediamine (TMEDA), a mixture of two alkylation products, (23; a straight chain product formed by coupling at the more hindered C-1) and (24; a cyclized product formed by coupling at C-3), was obtained in around 50% yield, the (23)/(24) ratio being 17:83. Dimers (14)-(16) and 1,2-bis(2,2-dimethylcyclopentyl)ethane (25) were also produced in yields of 10% and 8%, respectively (eq 5). The reaction of (11b) in ether in the presence of hexamethylphosphoramide (HMPA), however, gave mainly a straight chain product (23). Another characteristic feature of this reaction was the absence of (14)-(16) and (25). The electronic spectra demonstrated that under former conditions the lithium compound (11b) exists as contact-ion pairs, while under the latter conditions solvent-separated ion pairs are important. In the case of lithium compounds (11d) and (11e) also the same trends for the effects of solvent systems on the product composition were observed (TABLE IV).

It is surprising that of the two possible cyclized products, only one isomer (24) is isolated. This would imply that (a) a SET process is certainly involved in the reaction of lithium compound (11) with tert-alkyl bromides and (b) 2,2-dimethylcyclopentyl radical (28), formed from (22) by SET followed by cyclization, couples with

TABLE IV. Reaction of Phenyl-Substituted Allyllithiums with 6-Bromo-6-methyl-1-heptene[a]

Li Compound	Additive[b,c]	Products: Alkylation % yield [(23):(24)]	Dimerization % yield [(14):(15):(16)]	(25),% yield
(11b)		45 [16:84]	10 [22:59:19]	8
(11b)	TMEDA	68 [18:82]	10 [20:60:20]	7
(11b)	TMEDA, CHD	20 [51:49]	8 [31:62:7]	
(11b)	HMPA	35 [73:27]		
(11b)	HMPA, CHD	33 [84:16]		
(11c)	TMEDA	17 [77:23]		
(11c)	HMPA	30 [88:12]		
(11d)	TMEDA	53 [28:72]	17 [20:80:0]	12
(11d)	TMEDA, CHD	23 [48:52]	8 [31:69:0]	
(11d)	HMPA	46 [92:8]		
(11d)	HMPA, CHD	33 [94:6]		
(11e)	TMEDA	43 [16:84]	19 [7:73:20]	22
(11e)	TMEDA, CHD	12 [68:32]	29 [6:74:20]	
(11e)	HMPA	37 [80:20]		
(11e)	HMPA, CHD	29 [90:10]		
(11f)[d]		12 [100:0]		
(11f)[e]	TMEDA	40 [100:0]		
(11g)[f]		20 [100:0]		
(11g)[g]		14 [100:0]		

[a] The reaction with 2.5 equiv of 6-bromo-6-methyl-1-heptene (22) at 20 °C for 1 h. [b] TMEDA: The lithium compound was prepared by treating the hydrocarbon precursor with 1.2 equiv of BuLi in the presence of 1.2 equiv of TMEDA. HMPA: The lithium compound was prepared by treating the hydrocarbon precursor with 1.2 equiv of BuLi in the presence of 10 equiv of HMPA at -45 °C. [c] CHD (1,4-cyclohexadiene): The reaction in the presence of 3 equiv of CHD. [d] The reaction in THF/pentane. [e] In THF. [f] In pentane. [g] In ether.

phenyl-substituted allyl radical (26) at the less hindered C-3 exclusively (Scheme II).

Then, a question may arise whether the straight chain product (23) is also formed by a process involving SET or alternatively by a polar process. To differenciate these mechanistic alternatives, the reaction of lithium compound (11b) with a cyclizable probe (22) was undertaken in the presence of 1,4-cyclohexadiene (CHD), a radical scavanger [16] (TABLE IV). In the reaction in ether/TMEDA the additive CHD exerted a remarkable influence on the product composition; the yield of the cyclized product (24b) decreased significantly (46% → 10%), whereas no meaningful influence was observed in the case of the straight chain product (23b) (12% → 10%). In addition, the formation of 1,2-bis-(2,2-dimethylcylopentyl)ethane (25) was completely suppressed,

Scheme II

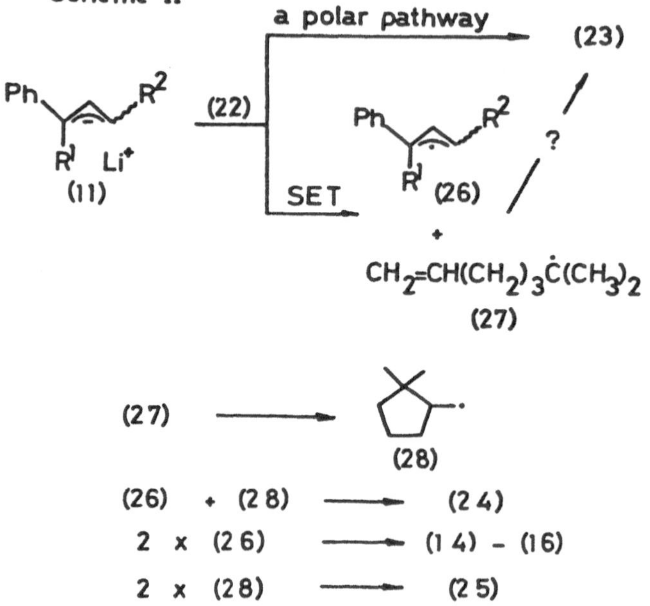

suggesting that CHD can effectively capture the radical (28). In the reaction in ether/HMPA, however, CHD exerted a relatively smaller effect on both the yield (35% → 33%) and the composition [(23b)/(24b) ratio; 73:27 → 84:16] of the alkylation products. Exactly the same trends were observed for (11d) and (11e) also.

The observed effect of the additive CHD leads us to deduce that a cyclized product (24) is formed by a SET pathway, CHD retarding the formation. In contrast, a polar pathway predominates for the formation of a straight chain product (23), and consequently, the additive CHD does not exert a meaningful effect on the production of (23).

The reaction of the relatively more stable 1,3-diphenylallyllithium (11f) with a tert-cyclizable probe (22) resulted in exclusive formation of a straight chain product (23f). In the case of 1-phenylindenyllithium (11g) also a straight chain product (23g), formed by coupling at the more hindered C-1, was the sole cross-coupling product under all the conditions (eq 6). In this framework the predominant formation of a straight chain product (23c) from the m-chloro derivative (11c) even in ether/TMEDA would be understood. Thus, the

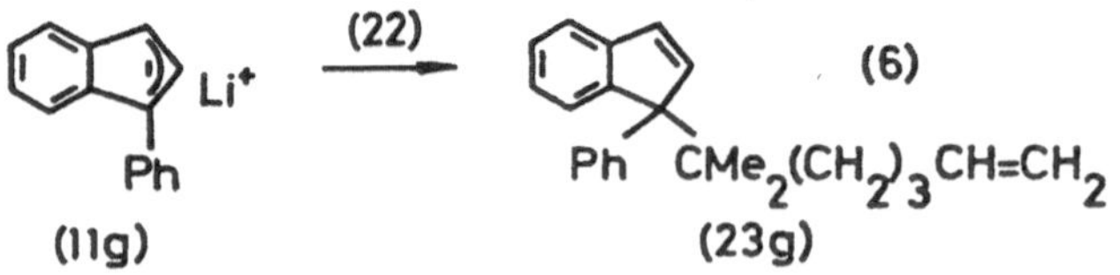

(6)

increase in stability of carbanions and the increase in donicity of the solvent systems seems to enhance the contribution of a polar pathway [17].

In the reaction of (9-anthryl)arylmethyllithium with benzhydryl halides also SET and nucleophilic substitution pathways were found to participate competitively, the significantly different product regiochemistires being observed between these mechanistic alternatives [18].

3. CONCLUSION

The systematic survey on the reaction of allylic chlorides with Grignard reagents and also the reaction of allyllithiums with alkyl halides, has revealed that the product regiochemistry observed in the reaction proceeding by SET would be remarkably different from that proceeding by a polar pathway (nucleophilic substitution). The concept itself is, however, not particularly novel. About 30 years ago Kornblum [19] and Russell [20] demonstrated that treatment of the salt of an aliphatic nitro compound with an alkyl halide may result in alkylation at carbon or oxygen, the $S_{RN}1$ process being important for carbon alkylation and in direct contrast, the oxygen alkylation product is obtained by a polar process. Our contribution would be the finding that even for the simple allylic chloride or allylic lithium having two reaction sites with apparent similar reactivities, different pathways would lead to notably different product regiochemistries. This leads us to deduce that as an approach to perform regioselective C-C bond formation, the control of reaction pathways would be attractive.

4. ACKNOWLEDGEMENT

I am indebted to the enthusastic efforts of my co-workers cited throughout the text. Calculations were carried out at the Computer Center of the Institute for Molecular Science by using IMSPAK (WF-10) program in the Computer Center library program package. Finally, special gratitude is due to Professor Shigeru Nagase of Yokohama National University whose help for the calculations and skilful discussion led to the body of work described herein.

5. REFERENCES

(1) Poulter, C. D.; Rilling, H. C. Acc. Chem. Res. 11, 307 (1978).

(2) (a) Seebach, D.; Geiss, K. H. J. Organomet. Chem. Libr. 1, 1 (1976). (b) Lever, O. W., Jr. Tetrahedron 32, 1943 (1976). (c) Gompper, R.; Wagner, H. U. Angew. Chem. Int. Ed. Engl. 15, 321 (1976). (d) Yamamoto, Y.; Yatagai, H.; Maruyama, K. J. Am. Chem. Soc. 103, 1969 (1981). (e) Tseng, C. C.; Yen, S.; Goering, H. L. J. Org. Chem. 51, 2892 (1986).

(3) (a) Negishi, E. Organometallics in Organic Synthesis; Wiley: New York, 1980; Vol. 1, Chapter 4. (b) Magid, R. M. Tetrahedron 36, 1901, (1980). (c) Bates, R. B.; Ogle, C. A. Carbanion Chemistry; Springer-Verlag: Berlin, 1983. (d) Biellman, J. F.; Ducep, J. Organic Reactions; Wiley: 1982; Vol. 27. (e) Bushby, R. J.; Ferber, G. J. J.

Chem. Soc. Perkin Trans. 2, 1688, 1695 (1976). (f) Evans, D. A.; Andrews, G. C. Acc. Chem. Res. 7, 147 (1974). (g) Hayashi, T.; Hori, I.; Oishi, T. J. Am. Chem. Soc. 105, 2909 (1983). (h) Hoppe, D.; Krämer, T. Angew. Chem. Int. Ed. Engl. 25, 160 (1986). (i) Hua, D. H.; Sinai-Zingde, G.; Venkataraman, S. J. J. Am. Chem. Soc. 107, 4088 (1985).

(4) (a) Guthrie, R. D. Comprehensive Carbanion Chemistry, Buncel, E.; Durst, T., Eds.; Elsevier: Amsterdam, 1980; Part A, Chapter 5. (b) Bank, S.; Bank, J. F. Organic Free Radicals, Pryor, W. A., Ed.; American Chemiscal Society: Washington DC, 1978; ACS Symp. Ser. No. 69. (c) Kochi, J. K. Organometallic Mechanisms and Catalysis; Academic Press: New York, 1978. (d) Chanon, M. Acc. Chem. Res. 20, 214 (1987). (e) Chanon, M. Bull. Soc. Chim. Fr. 197, 216 (1982). (f) Eberson, L.; Greci, L. J. Org. Chem. 49, 2135 (1984). (g) Bartoli, G. Acc. Chem. Res. 17, 109 (1984). (h) Maruyama, K.; Katagiri, T. J. Am. Chem. Soc. 108, 6263 (1986). (i) Yamataka, H.; Fujimura, N.; Kawafuji, F.; Hanafusa, T. Ibid. 109, 4305 (1987).

(5) Muraoka, K.; Nojima, M.; Kusabayashi, S.; Nagase, S. J. Chem. Soc. Perkin Trans. 2, 761 (1986).

(6) Tanaka, J.; Nojima, M.; Kusabayashi, S.; Nagase, S. J. Chem. Soc. Perkin Trans. 2, 673 (1987).

(7) (a) Takagi, M.; Nojima, M.; Kusabayashi, S. J. Am. Chem. Soc. 104, 1636 (1982). In accordance with this, it is known that the distribution of odd electrons in para substituted benzyl radicals are perturbed by the para substituent only to a small extent,[b] and addition of CCl_3Br to substituted stilbenes does not show significant para substituent effects on orientation.[c] (b)Gey, E. Z. Chem. 14, 279 (1974). (c) Gadogan, J. I. G.; Duell, E. G.; Inwald, P. W. J. Chem. Soc. 4164 (1962).

(8) Holm, T.; Crossland, I. Acta Chem. Scand. 25, 59 (1971).

(9) (a) Ashby, E. C.; Wiesemann, T. L. J. Am. Chem. Soc. 100, 189 (1978). (b) Ashby, E. C.; Bowers, J. R., Jr. Ibid. 103, 2242 (1981).

(10) (a) Bell, H. M.; Brown, H. C. J. Am. Chem. Soc. 88, 1473 (1966). (b) Ogata, F.; Takagi, M.; Nojima, M.; Kusabayashi, S. Ibid. 103, 1145 (1981).

(11) Kuivila, H. G.; Walsh, E. J., Jr. J. Am. Chem. Soc. 88, 571 (1966).

(12) Hirabe, T.; Takagi, M.; Muraoka, K.; Nojima, M.; Kusabayashi, S. J. Org. Chem. 50, 1797 (1985).

(13) (a) Nugent, A. W.; Bertini, F.; Kochi, J. K. J. Am. Chem. Soc. 96, 4945 (1974). (b) Chen, J. Y.; Gardner, H. C.; Kochi, J. K. Ibid. 98, 6150 (1976).

(14) Kornblum, N. Angew. Chem. Int. Ed. Engl. 14, 734 (1975).

(15) Tanaka, J.; Nojima, M.; Kusabayashi, S. J. Am. Chem. Soc. 109, 3391 (1987).

(16) Ashby, E. C.; Argyropoulos, J. N. J. Org. Chem. 50, 3274 (1985).

(17) (a) Alnajjar, M. S.; Kuivila, H. G. J. Am. Chem. Soc. 107, 416 (1985). (b) Troughton, E. B.; Molter, K. E.; Arnett, E. M. Ibid. 106, 6726 (1984). (c) Panek, E. J.; Rosgers, T. J. Ibid. 96, 6921 (1974). (d) Zieger, H. E.; Angres, I.; Mathisen, D. J. Ibid. 98, 2580

(1976). (e) Bordwell, F. G.; Clemens, A. H. J. Org. Chem. 46, 1035 (1981).

(18) Takagi, M.; Nojima, M.; Kusabayashi, S. J. Am. Chem. Soc. 105, 4676 (1983).

(19) Kornblum, N.; Michel, R. E.; Kerber, R. C. J. Am. Chem. Soc. 88, 5660, 5662 (1966).

(20) Russell, G. A.; Danen, W. C. J. Am. Chem. Soc. 88, 5663 (1966).

LIST OF PARTICIPANTS

C.AMATORE, ENS, Laboratoire de Chimie , 24 rue Lhomond, 75231 PARIS Cédex, FRANCE.

D. ASTRUC, Laboratoire de Chimie Moléculaire des Métaux de Transition, LA 35, Université de BORDEAUX I, 351 Cours de la Libération, 33405 TALENCE Cédex, FRANCE.

G. BALAVOINE, Chimie de Coordination, Université PARIS XI, Batiment 420, 91405 ORSAY, FRANCE.

D. BALLIVET-TKATCHENKO, CNRS, 205 Route de Narbonne , 31077 TOULOUSE Cédex, FRANCE.

G. BARTOLI, Department of Chemical Sciences, Via San Agostino, 1, CAMERINO (MACERATA) , 40 136 BOLOGNA, ITALIE.

D.H.R. BARTON, Texas A&M University, Department of Chemistry, COLLEGE STATION, TX 77843-3255, USA.

T.L. BROWN, Beckman Institute, University of Illinois, 1304 West Clark St., URBANA, IL 61801, USA.

M. CHANON, Université AIX-MARSEILLE III, Centre de Saint-Jérome, 13397 Marseille Cédex 13, FRANCE.

C.CHATGILIALOGLU, Istituto dei Composti del Carbonio, Contenenti et Eroatomi - CNR, 89 Via Tolara di Sotto, 40064 OZZANO EMILIA , BOLOGNA, ITALIE.

H. CHERMETTE, Institut de Pysique Nucléaire de Lyon, 43 bd du 11 Novembre, 69 622 VILLEURBANNE, FRANCE.

N.G. CONNELLY, Department Inorganic Chemistry, University of BRISTOL, School of Chemistry, BS 81TS BRISTOL, UK.

M. COSTANTINI, RHONE-POULENC ST-FONS, 85 Avenue des Frères Péret, BP 62, 69192 SAINT-FONS Cédex, FRANCE.

R.H. CRABTREE, Department of Chemistry, YALE University, PO Box 6666, NEW-HAVEN, CN 06511, USA.

P. DIXNEUF, Université de RENNES 1, Laboratoire de Chimie de Coordination Organique, Campus de BEAULIEU, 35042 RENNES Cédex, FRANCE.

M.H. GARCIA, Istituto Superior Technico, Av. Rovisco Pais, 1096 LISBOA Cédex, PORTUGAL

M. Chanon et al. (eds.), Paramagnetic Organometallic Species in Activation / Selectivity,Catalysis, 519–521.

C. GIANNOTTI, ICSN CNRS, 99190 GIF-SUR-YVETTE, FRANCE.

J. HALPERN, Department of Chemistry, University of CHICAGO, 5735 South Ellis Ave., CHICAGO, Il 60637, USA.

M. JULLIARD, Laboratoire de Chimie Inorg. Moléc., Centre de Saint-Jérome, Université AIX-MARSEILLE III, 13397 MARSEILLE Cédex 13, FRANCE.

H. KAGAN, Université de PARIS-SUD, Centre d'ORSAY, 91405 ORSAY Cédex, FRANCE.

W. KAIM, Institut für Anorganische Chemie, Universität, 6000 FRANKFURT AM MAIN 50, NIEDERUSELER HANG 1949, RFA.

J.K. KOCHI, Department of Chemistry, University of HOUSTON, HOUSTON, TX 77 004, USA.

J. KOTZ, Department of Chemistry, State University College, ONEONTA, N.Y. 13820, USA.

P.K. KRUSIC, Central Research and Development Dpt, E.I. du PONT de Nemours Cie, WILMINGTON, DELAWARE 19898, USA.

P. LEMOINE, Institut de Chimie, Université Louis Pasteur, 67008 STRASBOURG, FRANCE.

F. MINISCI, Politecnico di MILANO, Dipartimento di Chimica, 20133 MILANO, ITALIE.

J. OSBORN, Institut de Chimie, Université Louis Pasteur, 67000 STRASBOURG, FRANCE.

K.A. OSTOJA-STARZEWSKI, BAYER AG, Zentral Forschung, Wiss Hanptlabor, D-5090 LEVER KUSEN, RFA.

H. PATIN, ENSCR, Avenue du Général Leclerc, 35700 RENNES BEAULIEU, FRANCE.

G.F. PEDULLI, Università degli Studi di BOLOGNA, Dipartimento di Chimica Organica, 40127 BOLOGNA, ITALIE.

A.J. POE, J. TUZO WILSON Laboratories, Frindale College, University of TORONTO, MISSIAUGA ONTARIO, CANADA L5L IC6.

J.-C. POITE, Centre de Saint-Jérome, Université AIX-MARSEILLE III, 13397 MARSEILLE Cédex 13, FRANCE.

P.H. RIEGER, Department of Chemistry, BROWN University, PROVIDENCE, Rhode Island 02912, USA.

E. ROMAN, Facultad de Quimica, Pontificia Universidad Catolica de Chile, Casilia 6177, SANTIAGO DE CHILE, CHILI.

G.A. RUSSELL, Department Chemistry, IOWA State University, AMES, IOWA 50011, USA.

J. SAN FILIPPO Jr, Wright Rienam Chemistry Laboratories, Rutgers State University of New Jersey, NEW BRUNSWICK, NJ 08903, USA.

C.G. SCRETTAS, Institute of Organic Chemistry, National Hellenic Research Foundation , 48 vas Constantinou Av, ATHENS 116 35, GRECE.

D. SMITH, BP Research Center, SUNBURY-ON-THAMES, MIDDX, UK.

J.-M. SURZUR, Université d'AIX-MARSEILLE III, Centre de Saint-Jérome, 13397 MARSEILLE Cédex 13, FRANCE.

M.C.R. SYMONS, Department of Chemistry, University Road, LEICESTER, LE1 7RH, UK.

D. TYLER, Department of Chemistry, University of OREGON, EUGENE, OREGON 97403, USA.

G. VAN KOTEN, Vakgroep Organische Chemie, University of UTRECHT, PADUALAAN 8 Trans III, UTRECHT, NETHERLAND.

Subject Index, Formula Index, Author Index

A program working on IBM PC and written by **Dr. J.L. Larice** (same address as M. Chanon) has made possible the direct printing of these Indexes. Persons interested by this program may write to J.L. Larice.

Subject Index

Formula Index

Author Index